Thomas Bechtold, Tung Pham
Textile Chemistry

Also of interest

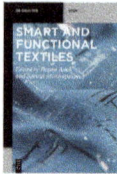

Smart and Functional Textiles
Bapan Adak, Samrat Mukhopadhyay, 2023
ISBN 978-3-11-075972-3, e-ISBN (PDF) 978-3-11-075974-7

Self-Reinforced Polymer Composites.
The Science, Engineering and Technology
Padmanabhan Krishnan and Sharan Chandran M, 2022
ISBN 9783110647297, e-ISBN (PDF) 9783110647334

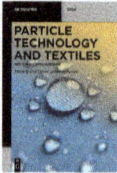

Particle Technology and Textiles.
Review of Applications
Editors: Jean Cornier and Franz Pursche, 2023
ISBN 9783110670769, e-ISBN (PDF) 9783110670776

Smart Polymers.
Principles and Applications
José Miguel García, Félix Clemente García, José Antonio Reglero Ruiz,
Saúl Vallejos and Miriam Trigo-López, 2022
ISBN 9781501522406, e-ISBN (PDF) 9781501522468

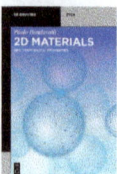

2D Materials.
And Their Exotic Properties
Paolo Bondavalli, 2022
ISBN 9783110656329, e-ISBN (PDF) 9783110656336

Thomas Bechtold, Tung Pham

Textile Chemistry

2nd Edition

DE GRUYTER

Authors
Prof. Dr. Thomas Bechtold
Research Institute for Textile Chemistry and Textile Physics
Universität Innsbruck
Hoechsterstrasse 73
6850 Dornbirn
Austria
Thomas.Bechtold@uibk.ac.at

Prof. Dr. Dipl.-Ing. Tung Pham
Research Institute for Textile Chemistry and Textile Physics
Universität Innsbruck
Hoechsterstrasse 73
6850 Dornbirn
Austria
Tung.Pham@uibk.ac.at

ISBN 978-3-11-079569-1
e-ISBN (PDF) 978-3-11-079573-8
e-ISBN (EPUB) 978-3-11-079577-6

Library of Congress Control Number: 2022950668

Bibliographic information published by the Deutsche Nationalbibliothek
The Deutsche Nationalbibliothek lists this publication in the Deutsche Nationalbibliografie;
detailed bibliographic data are available on the internet at http://dnb.dnb.de.

© 2023 Walter de Gruyter GmbH, Berlin/Boston
Cover image: brightstars/E+/Getty Images
Typesetting: Integra Software Services Pvt. Ltd.
Printing and binding: CPI books GmbH, Leck

www.degruyter.com

Contents

Preface of second edition

The chemistry around clothing and colouration belongs to the oldest "chemical" approaches undertaken by humans. The search for new fibres and colourants was one of the driving forces from which modern chemistry originated. For a long period, the research and development in textile chemistry addressed new processes and products. Improved sustainability was appreciated as an added value, however, with limited priority. In the period between the first edition and the second edition, clear requests for fibre-to-fibre recycling and textile reuse appeared. Design for recycling at the phase of textile product development requires substantial knowledge about the chemistry behind textile processes and products. In the new Chapter 16, aspects of circularity, recycling and disposal are brought into the context of textile and fibre chemistry, which is a highly interdisciplinary science, covering knowledge from physical and organic chemistry to polymer and macromolecular science. The content of the book has been selected in order to focus on concepts which are of high value for the understanding of textile chemical operations or represent technically relevant processes.

The book emphasises the chemistry behind materials and processes with the aim to provide understanding of the principles building a multidisciplinary field. This book has been designed as the first entry into the field of textile chemistry. Specialised monographs and scientific articles are then recommended as a source of further information.

Thomas Bechtold and Tung Pham

https://doi.org/10.1515/9783110795738-203

1 Textiles

1.1 Introduction

The architecture of a textile structure is a result of a combination of several production stages. The respective stage of production determines the principles of applicable textile chemical operations and also defines the final result. Application of some chemical treatment, for example, a colouration step, will lead to a completely different result when applied at the level of fibres or fabric.

This explains various properties and appearances of textile products, which also helps in continuous progress in the design of new products and exploring new properties. Examples of different levels of textile chemical modifications are summarised in Table 1.1.

In Table 1.1 various stages of textile processing and the materials are listed along the production chain. Each stage contributes to the development of characteristic properties, which are dependent on the choice of material, the combination of production steps and conditions.

Based on the stage where a textile chemical treatment is to be installed, the basic methodology and theoretical background as well as the technical equipment will differ.

With the rise of technical textiles, the scope of treatment and the field of applications for textile products have widened further. New techniques to obtain specialised properties, for example, electrical conductivity, surface modification by plasma or specific barrier properties through coating were introduced in textile chemical operations.

An almost endless number of variations and combinations of material and processing steps make a straight ordering and arrangement of textile chemical steps impossible; thus in this chapter, definitions of important textile technical terms and techniques are given to formulate a basic set of fundamental terms that will be useful for the other chapters.

1.2 Spinning

With the exception of silk, all other natural fibres that are of technical relevance are obtained in the form of staple fibres. Staple fibres exhibit a characteristic average length. In the spinning process, fibres are entangled, for example, by rotation, to increase the friction between the individual fibres and thus build a staple fibre yarn.

Different techniques are used to introduce the required number of turns per length into a spun yarn: ring spinning, open-end spinning, friction spinning and air-jet spinning. The characteristics of a yarn will depend on the spinning technology used. This will also influence the outcome of textile chemical operations as accessibility for chemicals will

https://doi.org/10.1515/9783110795738-001

Table 1.1: Relevant stages in textile processing, materials and processes, relevant characteristic properties and examples for products.

Stage	Material/process	Characteristic properties	Example
Spinning			
Fibre formation	Native, synthetic polymers	Chemical reactivity, physical properties	Cellulose, polyester
Thread formation	Filament production, staple fibre spinning	Longitudinal structure	Yarn, ply, wrapped yarn
Fabric formation	Weaving, knitting, nonwovens	Physical properties, accessibility	Plain weave, twill, satin, single jersey, fleece, web
Embroidery	Pattern formation	Localised modifications	Tailored fibre placement, conductive lines
Textile chemical modification			
Dyeing, printing, finishing	Possible at every stage of production	Chemical, physical properties, surface modification	Coloration, pattern formation, antimicrobial properties, coating
Garment production	Assembly of material to a final product, sewing, taping	Final form and functionality	Clothing, technical products, tents, sails

depend on the physical structure of a yarn (e.g., fineness of fibres, yarn count and number of turns per metre).

Synthetic fibres are fabricated as continuous filament fibres, namely, cables. The primary product of a fibre-spinning process is the endless fibre, the filament that, for example, has been obtained through melt extrusion. In rare cases, a single filament, the so-called monofilament, is the final product, which, for example, could be a string for a guitar or a fishing line. When several filaments are produced at the same spinneret, a multifilament is formed. A multifilament yarn, for example, is used to knit tights. For reasons of productivity, a very high number of holes are built in the same spinneret, which then result in a high number of filament yarns, forming a tow. The tow is then cut to staple fibres or is torn in a converter to deliver staple fibres.

Staple fibres are processed to yarns using similar procedure as that for natural fibres. At the stage of fibre spinning, fibre blends can also be formed; these blends are called intimate blends, for example, cotton/polyester (50/50) and cotton/polyamide (40/60).

When two or more yarns are wound together, a ply is formed. When different materials are wound together, another level of material blending is realised.

Core–spun yarns consist of two material layers. The core forms the inner layer, for example, highly elastic polyurethane filament yarn, which then is wrapped with another material, for example, polyamide fibres, to cover the core material. In the given case, the core material is responsible for the high elasticity of the yarn, while the outer layer is required to achieve dyeability of the yarn.

1.3 Linear density – yarn count

A characteristic measure to classify fibres and yarns is given by the linear density, which defines the mass of polymer per length of fibre or yarn [1]. Different systems for linear densities are in use; their use depends on the material and region.

Tex system (Tt): In the *tex* system, the mass of fibre, yarn in grams per 1,000 m length of fibre, yarn is considered (eq. (1.1)). A yarn count of 1 tex corresponds to a mass of 1 g of material for a yarn length of 1,000 m. Thus in the tex system, the reference length l = 1,000 m. One kilometre of a yarn with 20 tex yarn count will exhibit a mass of 20 g.

$$Tt = \frac{m}{l} \quad \text{(units: tex, g/1,000 m)} \tag{1.1}$$

For fibres the subunit dtex is often used, which then represents the mass of fibre in grams, which is obtained by a length of 10,000 m of fibres.

Metrical number (Nm): An older but still widely used unit is the metrical number (Nm), which considers the length of a fibre, yarn that is obtained by 1 g of fibres, namely, yarn (eq. (1.2)). The unit of the yarn count in Nm thus is metre per gram. Thus, one gam of a yarn with Nm 20 will exhibit a length of 20 m:

$$Nm = \frac{l}{m} \quad \text{(units: m/g)} \tag{1.2}$$

For filament yarns, still an older system is quite often used, the *French denier* (D). Similar to the Tt system, this count considers the mass of material per reference length; however, the sampling length in the case of the denier system is 9,000 m (eq. (1.3)). Thus, a fibre with a count of 1.3 denier will exhibit a mass of 1.3 g for a length of 9,000 m.

$$D = \frac{m}{l} \quad \text{(units: g/9,000 m)} \tag{1.3}$$

There are still a number of other systems in use to describe the linear density, for example, the Number English (Ne), which relates the mass to a certain defined length.

1.4 Fabric formation

Different techniques are in use to produce plain textile structures. Main techniques are summarised in Table 1.2.

Table 1.2: Main techniques in textile fabric formation.

Technique	Characteristics of formation	Examples
Weaving	Rectangular crossing of warp and weft yarn	Plain weave Twill Satin/atlas
Knitting	Fabric structure formed by multiple loops/stitches	Flat knitting Round knitting Warp knitting
Fibre entangling	Web structure built through entanglement of fibres	Non-wovens: melt blown, hydroentangled, needle punched

Weaving: A woven fabric consists of warp (longitudinal threads) and weft (fill yarn, transversal yarn system). The w*arp beam* is the longitudinal set of yarns that is processed to a fabric by the weaving loom by inserting the weft threads. Based on the construction of the weave, the physical properties of the fabric will change: the plain fabric being the most rigid structure, satin offering the highest drapeability and twill being in between (Figure 1.1). Variations in fabric construction are possible by combining different fabric constructions and warp beams, for example, velvet and terry cloth.

Knitting techniques build the textile fabric through loops (stitches), which result in high flexibility and drapeability. In flat and round knitting, the direction of the threads in the fabric is perpendicular to the direction of production, while in warp knitting the direction of the threads is parallel to the direction of production. Combination of different threads leads to system-blended fabric, where, for example, polyamide filament yarn is combined with polyurethane filaments yarn.

The production of *nonwovens* directly begins with a stage where fibrous material at first is laid in the form of a web, which is then processed into a fabric structure through fibre entanglement. Different techniques are used to achieve durable fibre entanglement, for example, mechanically by needles, hydraulic by means of high-pressure water injection or thermally by using meltable fibres or melt glue.

In spunbond production, the filament cable from the fibre spinning is cut and opened to obtain a layer of non-oriented fibres, which is laid down on a moving belt. Consolidation of the fibres is obtained mechanically by needle punching, calendaring or thermal bonding. In the case of thermal bonding (thermobonding), meltable binding fibres with lower melting temperature T_m are blended into the fibre web. Typical data for the mass per area of nonwovens are between 20 and 500 g/m^2. Needle-punched webs are usually thicker as more intense fibre entanglement is required to

Woven fabric plain weave		Knitted fabric	
Twill		Warp Knitted fabric	
Satin		Nonwoven layer	

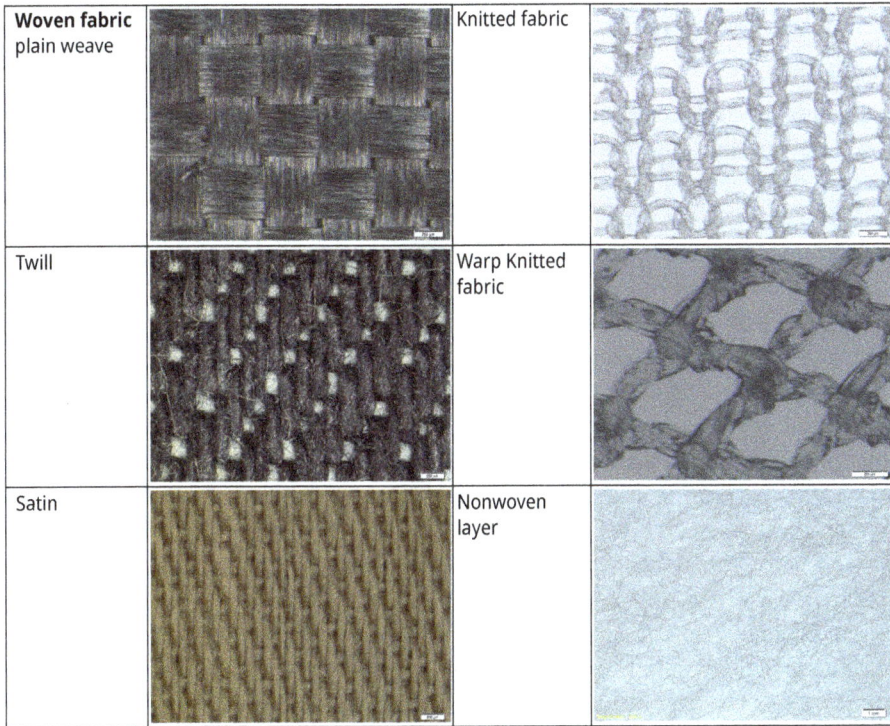

Figure 1.1: Microphotographs of basic structures of woven and knitted fabrics.

achieve sufficient fabric bonding. Thermal bonding allows the production of thinner webs as fabric bonding is achieved through adhesion of the meltable fibres.

1.5 Textile chemical aspects in spinning and fabric formation

A number of chemicals are used for preparation in the course of textile fibre, yarn and fabric production, with the only reason to develop material properties for better processability. Spin finishes and yarn preparations are used to control friction and static charge during yarn spinning and knitting, while sizing is used to strengthen the warp yarns to better resist the abrasive stress during weaving.

On the contrary to other textile chemical treatments, these auxiliaries have to be removed completely before further textile chemical processing, for example, bleaching and dyeing. The majority of these chemicals is hydrophobic; thus, such additives would prevent uniform wetting by aqueous processing baths, leading to negative consequences in uniformity of dyeing and finishing results.

1.5.1 Spin finish

During mechanical processing of fibres and fabric production, a lot of tension and friction is applied to the material. As a result of the high friction and abrasive stress, a number of negative effects appear, which are as follows:
-   undesired temperature increase of fibres and yarn
-   yarn breakage and machine stops, leading to lower product quality
-   uncontrolled yarn tension during fabric formation, leading to crease formation and shrinkage
-   electrostatic charge that leads to ballooning of multifilament yarns and adhesion to machine parts
-   development of fibre dust leading to machine failure.

Thus, it is important to prevent such effects and to control fibre friction and yarn tension during processing and to maintain well-controlled yarn tension during fabric production. Besides accurate control of climate (temperature and relative humidity) to stabilise the respective fibre properties, specific auxiliaries are also added on the fibres and yarn.

During production of synthetic fibres, for example polyester, polyamide, polyolefin or elastomer fibres, spin finishes are added as the last stage of the fibre production. Such spin finishes may include lubricants, emulsifiers, antistatic agents, bactericides and antioxidants. The effect of these chemicals is mainly based on the formation of a thin but uniform layer, which covers the fibre surface and thus reduces surface friction and flexural rigidity of the textile material.

Chemically spin finishes may contain the following:
-   surface active agents (non-ionic or anionic surfactants)
-   various mineral oils and fatty esters of low viscosity as lubricants, for example, for polypropylene fibres, alkoxylated long chain alcohols, alkoxylated trigycerides and fatty acid polyglycol esters can be used [2].

For elastomer fibres, high-performance lubricants based on polysiloxanes are in use to permit controlled tension and processing during stitch formation in knitting. Representative examples for chemicals used as spin finishes and lubricants are listed in Table 1.3.

The spin finishes have to be removed completely as the presence of oils and waxes and, in particular, silicones causes serious problems in later processing baths; for example, in dyebaths dye reservation effects due to hindered access of dyestuff will occur, and the formation of oily droplets causes spots and irregular dyeings.

Fewer difficulties arise from the use of ethoxylated surfactants, which are more easy to remove in the pre-washing steps. The hydrotropic properties of this group of lubricants also impart reduced static charge, which is a result of an increased water sorption on the surface of hydrophobic fibres (e.g., PA and PES).

Table 1.3: Representative examples for spin finishes and lubricants used to control friction and static charge. From Baldwinson [3].

Dimethylsiloxane	
Silicon polyether	$-(EO)_m-(PO)_n-[SiO(CH_3)_2]-(PO)_n-(EO)_m-$
Alkoxylated long chain alcohol	
Fatty acids	
Fatty acid esters	
Polyglycol esters	Ethoxylates Propoxylates Random and block copolymers

(continued)

Table 1.3 (continued)

Alkoxylated trigycerides	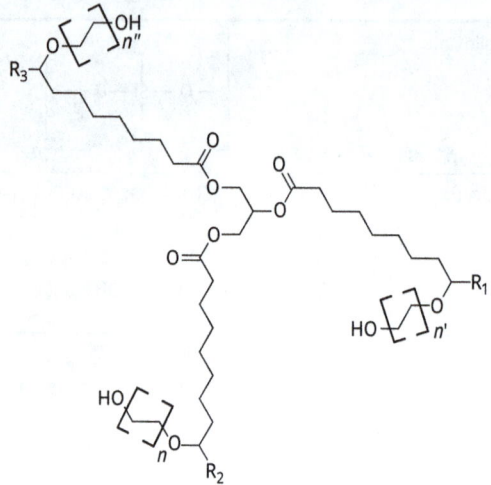
Polyalkylene glycols	$RO-(EO)_m-(PO)_n-(EO)_m-H$ $RO-(PO)_{m'}-(EO)_{n'}-(PO)_{m'}-H$

EO, ethyleneoxide; PO, propyleneoxide; R, long alkyl chain; R_2, methyl.

All these lubricants have to be removed before further textile chemical processes can be begun, which thus leads to a secondary problem in the effluents of the textile plant. The presence of hydrocarbons and of silicones in the waste water released from a textile dyehouse is monitored regularly as part of the effluent control and strict limitations with regard to the content of these substances have to be considered.

1.5.2 Sizing

Sizing represents another not permanent textile chemical yarn preparation. The warp yarn has to stand intensive abrasion during weaving as a result of the movement of the weaving reed and the heald frame forming the shed. Weaving a typical fabric, for example, for a shirt, bedlinen or a jeans, easily involves processing of 2,000–3,000 warp yarns.

Breakage of a single warp yarn will stop the process of weaving until the broken yarn has been repaired. Thus, warp yarn breakage directly influences the productivity and efficiency of the weaving step. As a representative value, breakage of a warp yarn should not occur more often than every 10,000 turns of the loom, which corresponds to an intake 10,000 of weft yarns without any machine stop.

Thus, the quality of sizing is highly relevant for the efficiency of a weaving process. In sizing, the warp yarn is coated with film-forming polymers that protect the threads against abrasive damage during weaving.

Typical polymers for sizing are as follows:
- natural polymers: starch and modified starch, for example, carboxymethyl starch (CMS)
- carboxymethyl cellulose (CMC)
- polyacrylate-based sizes
- polyvinylaclohol (PVA)

For sizing the warp beam is dipped into the hot polymer solution (representative concentrations of the sizing bath are at 10–15 wt% of size), squeezed off through padding and dried on Teflon® coated drying cylinders.

A careful preparation of the sizing bath through dissolution at the boil in combination with high shear rate is necessary to obtain a homogenous solution of the polymer, which is a prerequisite for a uniform coating of the individual threads of the warp beam. After weaving the size has to be removed before further textile chemical processes, for example, bleaching and dyeing are begun. The conditions for desizing depend on the type of size used. Starch sizes would need a longer time of hot water treatment to dissolve; thus, the commonly used processes are based on the enzymatic hydrolysis of the starch by using amylases. The enzyme preparation is padded on the fabric and after several hours of enzymatic reactions the degradation products are removed by rinsing with hot and cold water. The released carbohydrate products are easily biodegradable; however, they represent a substantial share of the total chemical oxygen demand and biological oxygen demand in the waste water (compare Chapter 15) [4]. The organic load released from desizing can cause difficulties because of its high capacity demand for biodegradation in the responsible communal waste water treatment plant.

Water-soluble sizes, for example, PVA, polyacrylate sizes and CMC sizes, are simply washed off. Following pre-swelling in water, the size is washed off with hot water. No polymer degradation is required to remove synthetic sizes; thus regeneration and reuse of the size is possible after purification and re-concentration with ultrafiltration plants. Such processes, however, require several conditions to be economical: the sizing, weaving and desizing must be located in short distance. The fabric production should be rather uniform to allow the use of a main recipe for sizing. Quality control of recovered sizes is necessary. The re-concentration step alters the average molecular weight of the polymer through molecular filtration effects, which then also changes the viscosity and film-forming properties of the regenerated size, and thus influences the efficiency in weaving. In addition, care has to be taken with regard to time of intermediate storage of regenerates as rapid microbial growth in the regenerates can cause substantial odour problems in storage tanks before reuse.

1.5.3 Embroidery

In embroidery techniques, a pattern or structure is formed on a flat basic fabric or non-woven layer. The embroidery is formed by crossing two threads (front yarn and back yarn). The front yarn is transported to the back side of the basic layer by means of a needle. The stitch is formed by crossing with the front yarn with the back yarn. In contrary to weaving or knitting techniques, the pattern formation can be executed in any direction. Thus, in embroidery a 2D pattern can be constructed in any direction of the flat basic material (Figure 1.2). Embroidery originally was developed as a technique to produce decorative textile elements; today the flexibility of embroidery techniques makes them highly interesting for technical textiles as tailored fibre placement and near net-shaped structures can be produced rather easily.

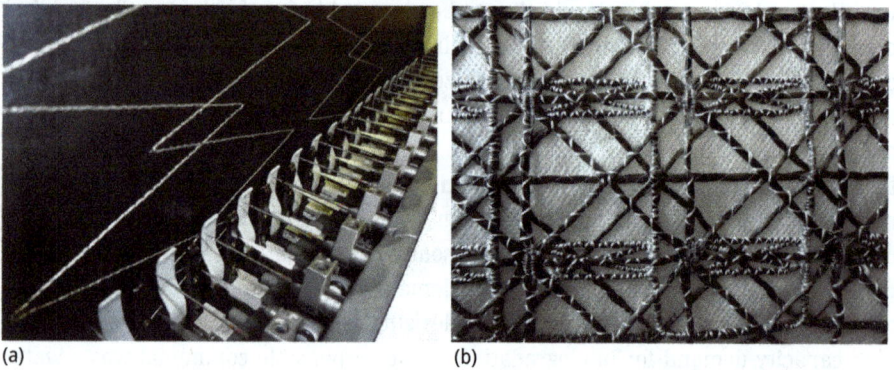

(a)   (b)

Figure 1.2: Embroidery: (a) front view of an embroidery machine and (b) technical embroidery (carbon roving positioned on water-soluble PVA nonwoven).

1.5.4 Printing/coating

A localised modification of a textile product can be achieved by means of printing techniques, if a decorative or functional component is placed on the surface of a textile structure. The range of functionalities that can be placed through printing techniques is extremely wide:
- coloration and pigmentation
- antimicrobial function and deposition of active pharmaceutical ingredients
- coatings, polymer foams and hot-melt adhesives.

Usually printing is performed on one side of the textile. Different techniques can be applied to transfer the dyestuff and chemicals to the surface of the fabric: screen printing, ink-jet printing and thermal sublimation (thermotransfer printing).

1.5.5 Garment production/assembly/joining

In the last stage of textile and garment production, different materials are combined to form the final product. Based on the construction of the product, the appropriate manufacturing steps to combine the textile components have to be chosen. Representative examples are as follows: *Laminating* of textile fabric and membranes to form barrier textiles. In such processes, different layers are bound together through combination of reactive binding systems or hot-melt glue, with the aim to build a water-tight barrier. Such textiles exhibit a textile surface at their outer side and show excellent water tightness through membranes inside the layered structure.

Sewing represents the standard process to join flexible textile materials by stitching. The stitches can be formed by regular crossing of two or more threads. Based on the expected mechanical properties, a seam can be constructed as very rigid or highly elastic joint. Even in a simple pair of jeans, more than six different types of sewing are combined [5].

Bonding/taping In sewing every stitch perforates an in initially water-tight structure; thus for applications where water-tight sewing is required, a conventional seam is covered by a tape that then seals the sewing tightly. In addition, thermal bonding, for example, based on ultra-sonic sewing machinery is in use.

1.5.6 Technical textiles

There is a huge difference between the economic impact and relevance of research activities in traditional textile industry when compared to technical textiles (Figure 1.3). In traditional textile industry, the vertical production from fibre through fabric to garment ends up with a consumer product, for example, underwear, clothing and household textiles. Commercial impact finally is achieved in the consumer market for textiles and garment.

In the emerging field of textile-based technical products, the classical textile production contributes with specialised expertise to almost every producing industry. Examples are construction and architecture, automotive and lightweight applications, health and medical application, sensors and electrical devices. The relevance of fashion and design is reduced; however, an enormous number of additional aspects have to be considered for the development of technical/functional aspects required for proper functioning of the material.

Compared to a classical textile production, the field of technical textiles exhibits five characteristic differences:

- Textile production is understood as a general manufacturing technique, valuable to deliver functional components for a wide range of applications.
- Application of the components often is intended for "non-textile" products, for example, lightweight structures for mobility applications.

Traditional vertical textile production

Advanced textiles/lightweight composites

Fibre production

Yarn

Fabric/nonwoven

Dyeing/finishing

Garment/textiles

Consumer

Fibres/polymers/matrix
Structure formation
Integration into product

Application as functional element

Intelligent textiles/sensors
Electrical engineering

Safety and health
Protection/medical appl.

High-performance materials
Construction
Lightweight application
Automotive

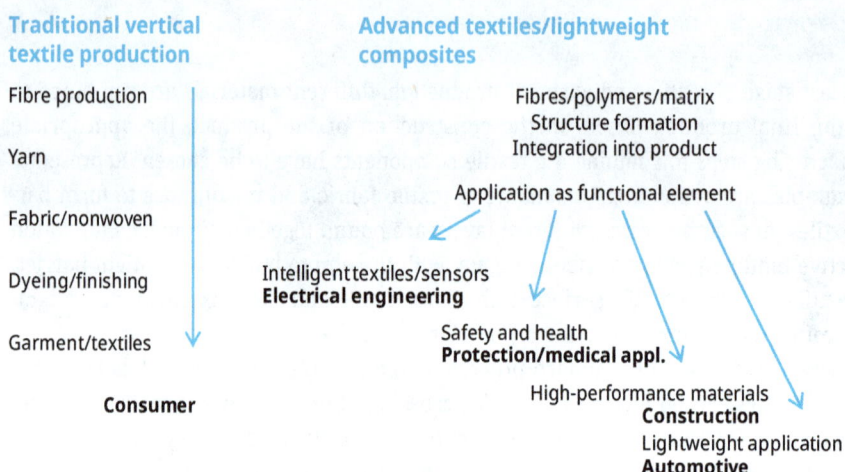

Figure 1.3: Technical textiles versus conventional textile production.

- Multidisciplinary approaches are required, for example, between textile chemical processing, polymer chemistry and textile machinery and material sciences.
- Development is organised by consortia, where the presence of textile and non-textile partners is a characteristic sign.
- There is no restriction with regard to choice or combination of materials; any useful material will be considered, for example, carbon, copper wires, aluminium profiles, epoxy resins, textile fibres, elastomeric foam and concrete.

The general technological and textile chemical questions will be very similar, while the applications of the textile material and the intended use of the final product will be completely different to the existing applications, for example, garment. Thus, concepts of textile chemistry will remain valid as the general principles of fibre accessibility, sorption, surface modification and chemical modification. Major changes will happen with regard to the type of polymer material processed, the chemical environment of the manufacturing process and the conditions of the application.

References

[1]   ASTM D1907/D1907M-12, Standard Test Method for Linear Density of Yarn (Yarn Number) by the Skein Method, ASTM International, West Conshohocken, PA, 2012.

[2]   Li, Y., Hinestroza, JP. Chapter 12. Boundary lubrication phenomena in coated textile surfaces, Gupta, BS., ed., Friction in textile materials, Woodhead Publishing Ltd., Cambridge, England, 2008, ISBN 978-1-85573-920-8

[3]   Baldwinson, TM. Chapter 10, Classification of dyeing and printing auxiliaries by function, Agents for fibre lubrication, softening, antistatic effects, soil release and soil repellency, Colorants and

Auxiliaries, Vol. 2, Shore, J., ed. Society of Dyers and Colourists, BTTG-Shirley, Manchester, England, 1990.

[4]   Bechtold, T., Burtscher, E., Hung, Y. Chapter 8 Treatment of Textile Wastes, in Handbook of Industrial and Hazardous Wastes Treatment, Second Edition, Wang, LK., Hung, Y-T., Lo, HH., Yapijakis, C., ed. Marcel Dekker, Inc. New York, 2004, ISBN: 0-8247-411-5

[5]   Hayes, S., McLoughlin, J. Joining techniques for denim jeans, Chapter 8 in Denim. Manufacture, Finishing and Application, Paul R, ed. Woodhead Publishing, Cambridge, UK, 2015.

Take home messages

A textile product is a result of a hierarchic series of processing stages, beginning with fibre polymers and ending with the final product.

An enormous variety of products can be designed by combination of fibres, yarn and fabric construction, textile chemical processing and product assembly.

During the mechanical processing of fibres and yarns, auxiliaries are added, which have to be removed during wet-chemical treatment. The function of these auxiliaries mainly addresses aspects of technical manufacturing such as control of frictional properties, electrostatic charge, adhesion and resistance to abrasive forces (sizing).

Thus in the major part of textile chemical processing, such auxiliaries have to be removed.

Quiz

Question 1. Explain the terms: satin, twill and plain weave.
Question 2. Why a core-shell yarn is of interest?
Question 3. Explain the mechanical differences between a woven fabric and a knitted product on the basis of the construction.
Question 4. Which auxiliaries are used during fabric formation and what are relevant functions of these products?

Exercises

1.   Try to define conversion factors among units of *Tt, Nm* and *D*.
2.   A fabric with mass per area of m_a = 150 g/m^2 and yarn density of 31 threads per cm in warp and 29 threads per cm in weft direction has been formed by the same yarns. The crimp of the yarn in the textile reduces the effective length in the fabric, 1 m of yarn corresponds to 0.96 m fabric. Calculate the yarn count in the *Tt* and *Nm* system.

3. A yarn with metrical number *Nm* 50 has been produced by polyester fibres that had a linear density of *Tt* = 1.3 dtex. Calculate the average number of fibres in the cross-section of the yarn.
4. How long would be 1 kg of a single filament fibre with 1 dtex?

Solutions

1. (Question 1): Satin, twill and plain weave are different types of woven fabric. The differences are in the style of threads cross. In plain weave, the yarns bind at every crossing, in twill a regular diagonal pattern is formed through floating of yarns, which is shifted in the neighbouring yarn sequentially by one binding point and in satin the floats are at least over four yarns and the shift of the neigh-bouring yarn is at least for two binding points.
2. (Question 2): A core-shell yarn allows to cover the core yarn with another mate-rial. Thus, fibres that are difficult to dye (e.g., polyurethane filaments) can be cov-ered with the same fibres as the other yarns in a fabric.
3. (Question 3): In knitted fabric the yarn structure is formed by loops, which thus permit a substantially higher elongation than woven fabrics, where yarns are combined in rather straight form. Of course, it is possible to reduce elongation and elasticity of knitted material by technical construction. The straight structure of a fabric, however, makes the production of thin and light material more easy.
4. (Question 4): Fibres and threads are covered with spin finishes and oils to control friction and static charge. Warp yarn is covered with sizes to improve resistance against wear during weaving.
5. (Exercise 1): $Nm = 1,000/Tt$, $Nm = 9,000/D$, $D = 9\ Tt$.
6. (Exercise 2): At first we have to calculate the overall length of yarn in a certain area of the material, for example, a 10 × 10 cm piece: Number of yarns: warp n_w = 10 cm × 31 y/cm = 310 yarns, fill: n_f = 10 cm × 29 y/cm = 290 yarns. Thus, a total length of yarn in the fabric will be l = (290 + 310) × 10/0.96 = 6,250 cm = 62.5 m = 0.0625 km. The mass of such a piece (100 cm^2) will be 1.5 g. Thus, the yarn count in tex will be Tt = 1.5/0. 625 g/km = 24 tex. Nm = 1000/24 = 41.7 m/g.
7. (Exercise 3): The yarn count in Tt = 1,000/Nm = 20 tex. Thus, the number of fibres in the yarn cross-section results as n = 20/0.13 = 154 fibres.
8. (Exercise 4): 1 dtex corresponds to a mass of 1 g per 10 km. Thus, 1 kg of a 1 dtex fibre will have a length of 10,000 km. (Comment: This is an important issue when the productivity of a fibre spinning unit is assessed, as 1 spin bore will produce 1 kg of fibres when having spun a length of 10,000 km).

List of abbreviations/symbols

T_m	melting temperature
Tt	yarn count in tex, dtex
m	mass
l	length
Nm	yarn count as metrical number
D	yarn/fibre count in den
CMS	carboxymethyl starch
CMC	carboxymethyl cellulose
PVA	polyvinylaclohol
EO	ethyleneoxide
PO	propyleneoxide

2 Textile fibres

2.1 Fibre production

In 2021, the worldwide fibre production reached 113 million metric tonnes. Approximately, a share of 54% of the total production volume was contributed by polyester (PES) fibres. The annual production of cotton (Co) varies between 27 and 29 million tonnes, which corresponds to a share of 24% of the total fibre production. The third position with regard to production volume is taken by cellulosics, which reached 7.2 million tonnes (6.4%). Other fibres such as polyamide (PA) and polyacrylic fibre (PAC), and polypropylene fibres hold a share in the range of 1–5% each, which still corresponds to an amount of 1–6 million tonnes for every type of fibre. Protein-based fibres such as wool and silk take minor shares with a volume of 1 million tonnes (1%) and 0.1–0.2 million tonnes (0.1%), respectively (Figure 2.1).

The average consumption of textile fibres per capita is in the magnitude of 10 kg textiles per year, but can reach more than double this value in highly developed counties, for example, Europe and the United States.

Based on their origin, textile fibres can be classified into natural fibres and man-made fibres (also called chemical fibres).

Cellulose is the most important polymer in natural fibres. Cellulose-based fibres include cotton, flax, hemp and jute. Protein-based fibres are wool and other hair, silk and spider silk. While the term macromolecule is appropriate for these proteins, the term "polymer" usually would not be understood as a correct classification because of the irregular order of amino acids in the protein chain.

Man-made fibres are based on natural polymers (e.g. cellulose, chitosan, alginate and casein) or synthetic polymers (e.g. PES, PA and PP).

A rather wide diversity in textile fibres exists with regard to fibre polymer, fineness, cross section, surface characteristics, additives and chemical modification.

The variability of properties and specialisation to a given purpose explains the available numerous types of fibres, which is growing due to additional aspects of sustainability, renewable resources and recycling.

Tables 2.1 and 2.2 summarised the selected properties of important textile fibres to give an impression about the wide range of variability of properties and to demonstrate the individuality of the different materials. This also explains why a single type of fibre will not be able to take over the full market and all possible applications.

The physical characteristics of the different fibres are the result of a number of polymer chemical and mechanical aspects which also form the backbone which determines the final properties of the fibres and the behaviour in textile chemical operations.

https://doi.org/10.1515/9783110795738-002

Textile Fibres
113 Mio. t/a (2021)

Natural fibres   **Man-made fibres**

Cellulose   **Synthetic polymers**
Cotton (Co)
(27 Mio. t/a)
                            Polyester (PES)
other plant based   (61 Mio. t/a)
(6.5 Mio. t/a)   **Natural polymers**
                            Polyamide (PA)
Protein   (5.7 Mio. t/a)   Cellulosics
Wool (Wo)   (7.2 Mio. t/a)
(1.0 Mio. t/a)   Viscose (CV)
                            Polyacryl (PAC)   Lyocell (CLY)
                            (1.7 Mio. t/a)   Acetate fibre (CA)
Silk (Ms)
(0.11 Mio t/a)
                            Polypropylene (PP)
                            (2.9 Mio. t/a)

                            Elasthane (PUE)
                            (1.1 Mio. t/a)

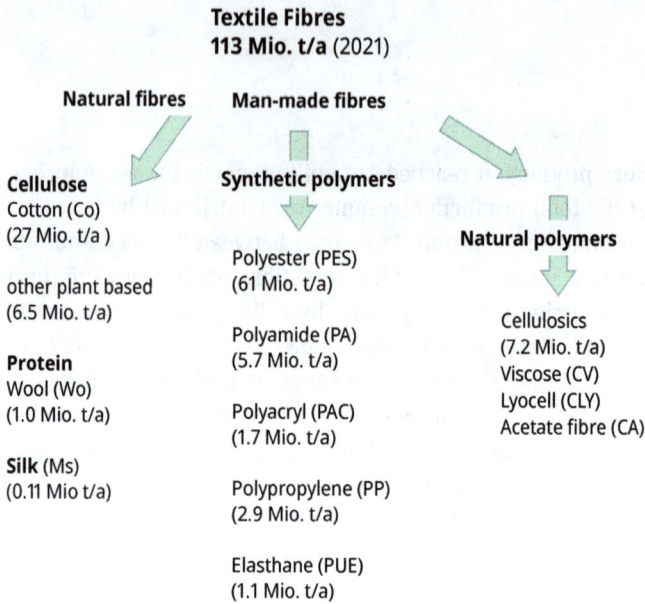

Figure 2.1: Important textile fibres, classification and production volume (2021).

Table 2.1: Selected physical characteristics of relevant textile fibres.

Fibre	Density (g/cm^3)	Tenacity (cN/dtex)	Ultimate tensile strength (daN/mm^2)	Wet strength % of dry	Elongation %
PES	1.36–1.41	3–9.5	35–130	95–100	8–55
PA 6,6	1.14	3.5–9	40–100	80–90	15–60
Arom. PA	1.38–1.45	4.4–27	60–400	75–80	2–30
PAC	1.17–1.19	1.8–4.5	20–53	75–100	15–60
PP	0.9–0.92	2.5–6	22–55	100	15–50
CV	1.52	1.6–3	25–45	40–70	7–40
CLY	1.5	3.5–7.2	50	75–85	9–17
CA	1.29–1.33	1.0–1.5	13–20	50–80	20–45
Co	1.50–1.54	2.5–5	35–70	100–110	6–15
Wo	1.32	1–2	15–25	70–90	25–50
Steel	7.8–8.0	1.1–4.8	100–240	100	1–35
Carbon	1.8–2.0	10–17	200–300	100	<1
Glass	2.5–2.6	4–14	100–350	90	2–5

Some remarks to the values given in Tables 2.1 and 2.2 are as follows:

Density: Density properties will depend on polymer interactions and density of packaging in the fibre structure, as well as on degree of polymer order and crystallinity.

Table 2.2: Selected physical characteristics of relevant textile fibres.

Fibre	Specific electrical resistance (Ω cm)	Specific heat (J/g K)	Thermal conduct. (W/m K)	Moisture content*(%)	WRV (%)	Thermal decomposition (°C)
PES	10^{11}–10^{14}	1.1–1.4	0.25	0.2–0.5	3–5	283–306
PA 6.6	10^{9}–10^{12}	1.5	0.21–0.4	3–4	9–15	310–380
Arom.PA		1.2	0.13	1.5–5	12–17	370–550
PAC	10^{8}–10^{14}	1.2–1.5	0.2	1–2	4–6	300
PP	$>10^{13}$	1.6	0.22–0.3	0	0	328–410
CV	10^{6}–10^{9}	1.4–1.5	0.3–0.6	11–14	85–120	175–205
CLY	–	–	–	12–14	50–70	–
CA	10^{9}–10^{14}	1.3–1.5	0.3–0.6	2–7**	10–28	255–260
Co	10^{6}–10^{8}	1.3	0.45	7–11	45–50	175–200
Wo	10^{8}–10^{11}	1.3–1.6	0.38	15–17	40–45	130–140
Steel	0.7×10^{-4}	0.46	15	0	0	1,400
Carbon	10^{-3}	0.7	15–100	–	–	3,650
Glass	10^{12}–10^{15}	0.7–0.8	0.8	0.1	0	1,100–1,500

*Normal climate (20 °C, 65% relative humidity).
**Dependent on substitution of CA-CT.

Weak interactions in PP are due to the absence of polar groups, which lead to comparable low fibre density (<1 g/cm^3).

Ultimate tensile strength: High tenacity is a result of highly ordered and rigid polymer structure with strong interactions between the polymer chains. Such properties, for example, are observed at aromatic PAs and carbon fibres. However, ultimate tensile strength compares only the strength for a certain cross-sectional area, thus lightweight material apparently lose ground when compared to steel, because specific weight does not influence this value.

Wet strength: A number of fibres loose strength considerably in wet state, which is due to the effects of water in the molecular structure of the fibre. Thus, in uptake of water through swelling, hydrogen bond interaction can loosen the structure, for example, in viscose fibres, while hydrophobic fibres (e.g. PP) do not exhibit changes in their tensile properties in wet state. A relative increase in strength in wet state is observed, for example, with cotton fibres and indicates a better ability for distribution of stress to all load-bearing parts of a fibre.

Electrical resistance: When the specific electrical resistance of the major part of textile fibres is very high, above a value of 10^9 Ω cm, static electricity becomes a problem, leading to unwanted static attraction of dust and formation of sparks. Thus, many synthetic fibres tend to develop more static electricity than natural fibres. The wide range of variation in values is due to the influence of climate, which is more distinct for fibres that exhibit higher absorption of moisture.

Specific heat and thermal conductivity: Interestingly, the majority of fibres show rather similar specific heat and thermal conductivity, which indicates that the development of thermal insulating properties and a warm or cold handle will be more due to later stages of product design, for example, texturising and formation of highly porous nonwoven structures with a lot of air trapped inside.

Moisture uptake: Fibres with high moisture uptake will offer better functionality with regard to wear comfort; however, in many cases their physical and chemical properties will be more dependent on the actual climate conditions.

Water retention: The water retention value (WRV) characterises the ability of a fibre to bind water through swelling and inside pores. While capillary water is present in fibre pores and cavities, swelling of amorphous areas will change fibre dimensions, alter the textile chemical properties of a fibre and also lead to dimensional changes of textile fabrics.

Chemical stability: The chemical stability of a polymer is reflected by the thermal stability, which in general is rather disappointing, as the vast majority of fibres will withstand temperatures above 350 °C for short time only. Long-time stability of majority of fibres is even much lower, and ranges between 90 °C(!) and 150 °C.

2.2 Fibre formation

For successful production of fibres from a type of macromolecule, a number of conditions and prerequisites need to be considered. For simplification in the following discussion, the term polymer will include synthetic polymers as well as bioresources-based material, for example, cellulose, chitosan, alginate and protein-based fibres like wool, hair and silk.

2.2.1 Polymer-dependent fibre properties

The final application of a fibre-based material defines the requested portfolio of physical mechanical, chemical, thermo- and photochemical stabilities.

Based on the functional groups present in a respective polymer and the chemical environment of application, a certain pattern of chemical reactions will be possible. These reactions will alter the polymer chemically, lead to chain scission and initiate degradation and ageing of the material.

Natural fibres are already obtained in a fibrous form, which sometimes requires extraction steps to isolate them from a stem (e.g. flax and hemp), and by further chemical processing; however, usually no structural reorganisation at the molecular level of the polymer chain is required during the step of fibre production.

Synthetic fibres and man-made fibres from natural polymers require an appropriate shaping process to form a fibrous structure. Thus, the basic polymers of production of synthetic fibres (including regenerated cellulosics) have to be
- either thermally stable enough to be melted without thermal degradation (melt extrusion spinning), which means that the temperature of decomposition $T_{decomp.}$ is substantially higher than the melting temperature T_m;
- or the formation of highly concentrated solutions (spin dopes) is possible, which means that an appropriate solvent for polymer dissolution is available. From the spin dope, the fibre is formed during the step of polymer regeneration and solidification. Regeneration, for example, is achieved through solvent evaporation (dry spinning) or solvent extraction (wet spinning), which thus initiate polymer precipitation.

A completely insoluble and inert polymer like polytetrafluoroethylene (PTFE, Teflon®) requires use of alternate processes, for example, coagulation and sintering of particles to obtain fibrous structures.

For successful formation of a fibrous shape some general requirements have to be fulfilled by the polymer:

– *Linearity*: Optimum fibre spinnability is obtained with linear and unbranched polymers. Side chains or a branched polymer structure prevents formation of fibres with useful properties for textile applications. Even if a longitudinal shape could be achieved, the particular properties of a textile fibre with regard to longitudinal order and high specific strength will not be achieved.

Cellulose and starch are built by the same monomeric unit. Cellulose is a linear homopolymer of anhydroglucose, while starch consists of the linear amylase molecule (20–25 wt%) and to a major part of the branched amylopectin (75–80 wt%). In the presence of low-molecular-weight plasticisers, for example, glycerol or sorbitol starch can be thermally shaped into films for packaging, disposable eating utensils, however, not into fibres.

– *Intermolecular attraction*: Substantial molecular interactions between adjacent polymer chains are a basic requirement to achieve physical strength and load transfer between polymer chains inside the fibre structure. Based on the fibre type, different principles of interaction between functional groups can be distinguished (Table 2.3).

During fibre spinning, shear forces are applied to the polymer chains which thus become oriented in the direction of the fibre axis. This leads to more intensive interactions between neighbouring polymer chains and as a result an increase in ultimate tensile strength is observed. As demonstrated in Table 2.3, different attractive forces between polymer chains appear. The energy of attraction depends on the functional groups present in the polymer. In wool besides hydrogen bonding and salt bridges even covalent bonding exists, which have to be opened before the protein could be dissolved.

Table 2.3: Interaction between fibre polymers: Fibres, type of polymer, functional groups, attraction, binding energy/energy of attraction, degree of polymerisation (DP) and average molecular mass of polymers.

Fibre	Polymer	Binding	Interaction	Energy (kJ/mol)	Degree of polymerisation n	Molecular mass (g/mol)
Wool	Protein	$-S-S-$ Disulphide bonds $-CO_2H_3N^+-$ Salt bridges $-CO_2^- \cdots {}^{..}C=O-$	Covalent Ionic attraction Ion–dipole	250 30 15	–	10–60,000 [1]
Cotton	Cellulose	$-OH \cdots {}^{.}OH-$	Hydrogen bond	20	1,500–3,000	240,000
PA6	$-[(CH_2)_5CONH]-$	$-NH \cdots {}^{..}O=C-$	Hydrogen bond	20	380–770	15–30,000
PES	$-[COC_6H_4CO-OCH_2CH_2O]-$	$-C=O$ dipole interaction Aromatic rings	Van der Waals forces π-Interaction	2	220–260	22–26,000
PAC	$-[CH_2-CHCN]_n-$	$-C\equiv N$ dipole interaction	Van der Waals forces	2	>1,100–2,500	30–70,000
PP	$-[CH_2-CHCH_3]_n-$	Hydrophobic interaction	The London dispersion forces	2	4,000–7,500	170,000–300,000

A similar situation is observed in cellulose where numerous hydrogen bond interactions exist along the long molecular chains. For dissolution, these hydrogen bonds have to be opened which requires specific measures. In addition, substantial reduction in the DP of the cellulose molecule has to be achieved before successful fibre formation through dissolution, and regeneration processes are possible. The attractive forces between cellulose chains increase the melting point of cellulose above its temperature for decomposition; thus, thermal shaping will be difficult as long as the high number of hydrogen bridges is not chemically blocked.

In the representative examples for synthetic fibres, a general relation between the magnitude of molecular interactions and the DP required for fibre formation can be observed. Polymers such as PA and PES exhibit stronger interactions (hydrogen bonds, dipole–dipole and π-interactions) than hydrocarbon-based fibres. Thus, a rather short chain length (e.g. DP ≈ 500) is already sufficient to form textile fibres with high strength. As long as the DP is kept within the appropriate limits, the melting point of the polymer and the viscosity of the polymer melt remain below the temperature of decomposition and the melt extrusion process can be applied for fibre spinning.

The weaker the specific interactions between the adjacent polymer chains become, the longer an individual polymer chain has to become for fibre production. In PP, only the London dispersion forces are generated; thus, a very long chain length is required to accumulate small attractive forces to the minimum required for fibre formation.

If we imagine a PP chain in the form of a tube, with a single propylene unit as an element with a length of 0.5 m and a diameter of 0.5 m, the PP in Table 2.3 will exhibit a total length of 3,500 m, or 3.5 km.

The development of intermolecular forces is also supported by appropriate orientation of side groups present in the polymer. In case of PP the methyl group favourably is oriented in its isotactic conformation to permit maximum packing density and higher interaction. As a result, the physical properties of catalytically polymerised PP are substantially better compared to atactic PP obtained by the radical polymerisation process.

In dissolved or molten state, a flexible linear molecule will be present in random coil conformation. An important function of the shear stress applied on the polymer melt during fibre formation is to uncoil the polymer chains and to increase orientation in the direction of flow.

Some linear polymers form liquid crystalline phases in solution or molten state, for example, PP, which then also will determine the state of order achieved during solidification of the polymer melt.

– *Orientation and crystallisation*: During fibre spinning shear forces will appear in the spinneret and also later during fibre drawing. This leads to orientation of the macromolecule chains along the fibre axis. The presence of intermolecular attraction forces and certain chain flexibility allows formation of highly ordered crystalline domains, which then are surrounded by less ordered amorphous areas. The formation of highly ordered crystalline parts is of substantial relevance to achieve high tensile strength as polymer chains will participate in several crystalline domains, thus forming a network of interlinkages. From the point of textile chemical considerations, the amorphous zones of a fibre represent the more interesting regions due to their higher accessibility, absorbancy and chemical reactivity.

– *Supramolecular order*: Through oriented arrangement of crystallites a supramolecular structure is formed leading to formation of fibrils and fibril bundles. Supramolecular structures are observed both in natural fibres (e.g. cotton, flax, wool and silk) as well as in man-made fibres (e.g. viscose, lyocell and PES) independent of their completely different origin.

2.3 Molecular mass/degree of polymerisation

The DP defines the number of molecular subunits (monomers) that form the macro-molecule. Due to the biochemical or chemical synthesis, fibre polymers will not exist as a uniform chemical substance containing identical polymer molecules with the same DP. The distribution of polymer chains with different molecular weight influen-ces the fibre properties through formation of amorphous and crystalline regions and thus also determines the physical and also chemical properties of a material.

In a real polymer, the DP represents an average value that is calculated from the distribution of the different chain lengths. The non-uniformity and the distribution of DP are described in the distribution function.

Based on the chemical basis on which the average molecule size is calculated, and on the methodology which has been applied to determine the molecular mass, different average degrees of polymerisation result [2].

In a polymer, the monomer is the repetitive subunit that combines to form the polymer chain. When the molar mass of the monomer unit M (g/mol) is multiplied with the number of monomers in the polymer P_i, the mass of a polymer chain M_{Pi} (g/mol) is calculated as follows (eq. (2.1)):

$$M_{P_i} = MP_i \tag{2.1}$$

Based on the method used to determine average DP, respectively, average molecular weight, the reference basis differs.

An average DP can be calculated as number-average DP (M_n), which considers the number of polymers of a certain length (eq. (2.2)). Using this calculation every poly-mer will contribute with the same impact to the result independent of its respective mass:

$$M_n = \frac{\sum_i n_i M_{P_i}}{\sum_i n_i} \tag{2.2}$$

The number average DP is obtained from absolute analytical methods, which deter-mine molecular masses. No assumptions with regard to the polymer structure are re-quired. Typical methods are membrane osmometry, cryoscopy and chemical titration of end groups.

The mass average DP (M_w) considers the presence of polymers based on their con-tribution to the total mass of a material (eq. (2.3)). Thus, while low-molecular-weight polymers and oligomers have a substantial impact on the number average DP, their contribution to the mass average DP M_w is due to their lower mass:

$$M_w = \frac{\sum_i m_i M_{P_i}}{\sum_i m_i} = \frac{\sum_i n_i M^2{}_{P_i}}{\sum_i n_i M_{P_i}} \tag{2.3}$$

Typical methods that deliver an average molecular mass as M_w are inverse size-exclusion chromatography, light scattering methods and X-ray small-angle scattering. In these methods, already assumptions about the structure of the polymer have to be introduced, for example, arrangement of the dissolved polymer chain in uncoiled linear form or as spherical polymer coil.

Average molecular masses of higher order are obtained from the measurement of sedimentation rates in special ultracentrifuges at very high gravitational fields (e.g. 900,000g). The coefficient of sedimentation is a combined value that for example includes assumptions about the hydrodynamically effective mass and the volume of the polymer, which then permit calculation of the sedimentation rate. From sedimentation velocities in an ultracentrifuge the z-average (M_z) can be extracted, which then is weighted towards larger polymer molecules (eq. (2.4)):

$$M_z = \frac{\sum_i z_i M_{P_i}}{\sum_i z_i} = \frac{\sum_i n_i M_{P_i}^3}{\sum_i n_i M_{P_i}^2} \qquad (2.4)$$

A widely used method to determine the average molecular weights is based on the viscosity measurement of the polymer in diluted solution or on the determination of the melt viscosity. In both cases, viscosity measurements deliver another average DP, which requires calibration with absolute methods.

M_z, M_w and M_n follow a general order $M_z \geq M_w \geq M_n$. By comparison of these values, a measure for the polymolecularity, the non-uniformity of a polymer, can be obtained, which usually is expressed as polymolecularity index D (eq. (2.5)):

$$D = \frac{M_w}{M_n} \qquad (2.5)$$

The absolute value of the polymolecularity index will always be larger than 1, with values of 1–2 for rather uniform polymers. Technically used polymers may exhibit a polymolecularity index D of up to 20–40.

The distribution of the DP or molecular masses of a polymer is represented in distribution functions, which are often represented as curves, but are discrete functions, as the molecular weight changes in the discrete intervals of a single monomer (Figure 2.2).

Examples for textile chemical applications of distribution functions are as follows:
– Distribution of cellulose molecular weight in different plant sources used as substrates for regenerated cellulose fibre production
– Analysis of fibre degradation during ageing and identification of chain scissoring
– Analysis of molecular weight distribution during polymer synthesis
– Identification of oligomeric components in synthetic fibres, for example, PES and PA fibres.

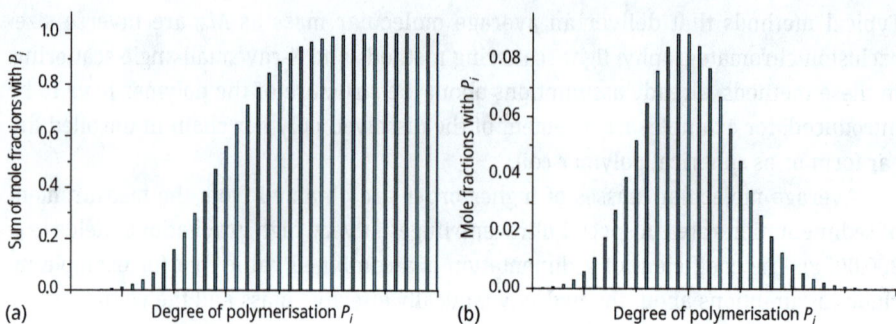

Figure 2.2: Molecular mass distribution of polymers. Differential representation (a) and sum function (b) of the discontinuous function.

2.4 Important fibres and their chemistry

2.4.1 Cellulose fibres

Cellulose is the most abundant natural polymer. The annual yield of cellulosic matter is in the dimension of 1.3×10^9 metric tonnes.

The major part of cellulose is used for manufacturing of paper, boards and nonwovens (approximately 150×10^6 tonnes/year). For textile fibres, a substantially lower amount of 30×10^6 tonnes is processed per year. Only a minor part is processed as high-quality cellulose, for example, from cotton linters or wood pulp to produce regenerated cellulose fibres such as viscose, lyocell, cellulose acetate fibres, viscose sponges and films.

The longitudinal character of the cellulose polymer can be visualised by the amount of cellulose produced in average by a single tree, that is, 13.7 g cellulose per day. If this amount of polymer is lined up to a single chain this would result in a length of 2.62×10^{10} km, which corresponds to 175 times the distance between the Earth and the Sun (Figure 2.3).

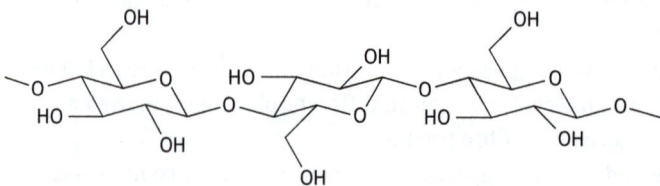

Figure 2.3: Molecular structure of cellulose.

Only few sources for cellulose contain the polymer in rather pure form, and the most important representative being cotton. The composition of raw cotton fibres is shown in Table 2.4. The share of other components present in the raw fibre depends on the quality, maturity of the cotton fibre as well as on the technology of harvesting, thus different values have been reported in the literature.

Table 2.4: Composition of raw cotton. From Rath [3], Schultze-Gebhart and Herlinger [4], Haudek and Viti [5].

Component	Content (%)	Content (%)	Content (%)
Cellulose	83.7	87	92.7
Bound water	6.7	7	–
Hemicelluloses and Pectins	5.8	1.2	5.7
Proteins	1.5	1.3	–
Waxes	0.6	0.6	0.6
Water-soluble parts	–	1.7	1.0
Ash	1.65	1.2	–
Other	ad 100	ad 100	–

Other natural cellulose fibres such as flax, hemp, jute, sisal contain higher contents in hemicelluloses and lignin (Table 2.4).

2.4.2 Structure of cotton/flax/hemp fibres

The structure of cotton fibres can be used as a representative case to discuss the general structure of native cellulose fibres as many similarities to the other natural cellulose fibres exist (Figure 2.4).

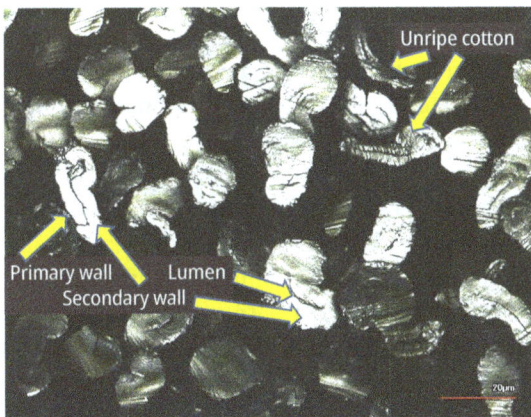

Figure 2.4: The structure of a cotton fibre (cotton/modal fibre blend).

In dried state the fibre exhibits a characteristic oval shape, which results from the collapse of the fibre during drying of the liquid cell plasma. The inner part of the fibre is filled with cell plasma which dries during the last stage of seed ripening and leaves an empty hollow channel, the so-called lumen. Thus, the fibre is not round, but has a diameter between 12 and 22 μm. The cell wall has a thickness of 2–5 μm, the values being dependent on the quality of the fibre and the status of fibre ripeness.

The outer layer of the fibre is called cuticula and mainly consists of waxes and pectin. The next layer is the primary wall, followed by the secondary wall. Both layers consist of cellulose fibrils and form the structural body of the fibre.

Fibres that have been harvested too early exhibit a thinner cell wall and are called unripe fibres. Fibres that exhibit a very thin cell wall have been interrupted in their growth at a very early stage, for example, due to insects and are called "dead" cotton.

The content of immature and dead cotton in the harvested cotton is a relevant quality determining factor, as the dyeing properties of these fibres are substantially different. The lower cellulose content in the cell wall of these fibres leads to lower dyestuff fixation, the fibres appear light and thus reduce the levelness of the dyeing and appear as lighter fibres in a dark fabric.

The introduction of machines for cotton harvesting led to an economically favourable high productivity in cotton farming. However, the three main aspects that have to be considered based on fibre quality and ecology are as follows:
- The automated harvesting increased the content of immature cotton fibres as no selection with regard to status of cotton ripeness is achieved.
- Through intensive use of pesticides and insecticides cotton fibre production became an important consumer of such products. A considerable part of these substances is transported with the raw cotton fibre to the dyehouse which undertakes the first wet processing, for example, scouring and bleaching. Thus, pesticides and insecticides are released into the wastewater at the site where cotton pre-treatment is undertaken.
- The water consumption of cotton fibre farming is dependent on the climatic region in which the fibres are grown. Up to 10,000 L of water is consumed for purposes of irrigation during the period of growth to produce 1 kg of cotton fibres.

Bast fibres (e.g. flax, hemp or jute) have to be extracted from the stem of the plant. The fibres grow in form of long fibre bundles with the function to mechanically strengthen the stem. After harvesting the non-fibrous parts of the stem have to be degraded, retted by microbial attack, treated chemically or with enzymes to facilitate mechanical isolation and purification of fibre bundles. Wooden parts and non-fibrous components have to be removed. The fibre bundles built from elementary fibres are held together by woody gum. As the elementary fibres are shorter than cotton, the separation of the fibre bundles into elementary fibres (cottonised bast fibres) is usually not desired, as only a low-quality "cotton"-type fibre would be the result.

The presence of fibre bundles explains the characteristic irregularly structured appearance of flax textiles.

The presence of larger fibre bundles of short elementary fibres explains the coarse and hard character of jute and sisal. The high lignin content of these fibres requires more intensive bleaching, which leads to reduction of fibre strength and durability. This still limits applicability of these fibres in garment.

The extraction of fibres from stem reduces the effectively useable amount of material substantially. Typical values for fibre yield in case of ground retted flax are in the magnitude of 11 wt% of the total crop. As an example 18 tonnes of green jute deliver approximately 1 tonne of retted fibre, which corresponds to 6 wt% of the total crop.

The representative natural fibres shown in Table 2.5 take over different functions in nature.

Table 2.5: Composition of selected natural fibres in wt%. From Schultze-Gebhart and Herlinger [4].

Component	Cotton	Flax	Hemp	Jute	Sisal
Cellulose	92.7	62.1	67	64.4	65.8
Hemicelluloses	5.7	16.7	16.1	12.0	12.0
Pectin	0	1.8	0.8	0.2	0.8
Lignin	0	2.0	3.3	11.8	9.9
Waxes	0.6	1.5	0.7	0.5	0.3
Water solubles	1.0	3.9	2.1	1.1	1.2
Length of elementary fibre (mm)	10–56	25–40	10–30	1–5	1–5

Cotton fibres are seed fibres. The unicellular fibre grows firmly attached to the cotton seed and has to be removed after harvesting by ginning. Residual short fibres are called cotton linters. The cotton linters cannot be used for spinning of yarn but represent a source of high-quality cellulose.

Flax, hemp and jute are stem fibres that take the function of strengthening the stem mechanically. These fibres exhibit higher lignin content and the wooden characteristics increase in the order flax < hemp < jute.

Sisal is obtained from the leaves of agaves and also serves as strengthening component.

Coconut fibres are relatively hard and coarse fibres that are removed from the fruit.

A completely different source for cellulose is *bacterial cellulose*, which for example is produced by *Acetobacter xylinum* [6]. At present, bacterial cellulose is of interest for medical applications; the costs for fibre production and the limited yield, however, prevent a wider use of bacterial cellulose. At present, an important non-textile application is in the field of wound care.

Regenerated cellulose fibres are produced from cellulose which has been isolated from wood by pulping and bleaching. Wood can be understood as a composite material consisting of the polyphenolic polymer lignin, hemicelluloses and cellulose. A representative formula of lignin is given in Figure 2.5.

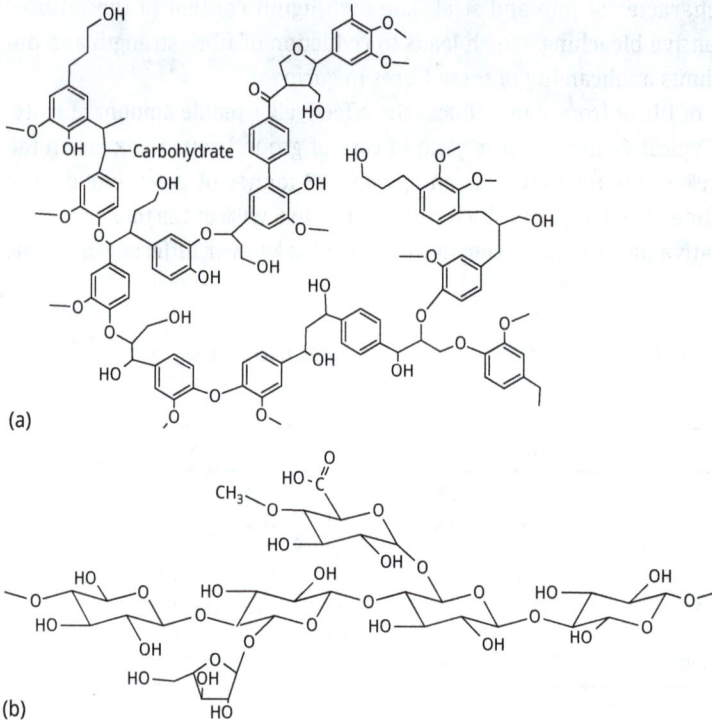

Figure 2.5: Representative structures of lignin (a) and hemicelluloses (b).

Representative values for the cellulose content of wood are 40 wt% for birch and 53 wt % for poplar. In pulping, lignin and hemicelluloses are removed, which means that in total 50% or more of the mass of wood processed becomes removed. Two major processes for pulp production that are in use today are the sulphate (Kraft) process and the sulphite process.

In the sulphate process, the wood chips are pre-hydrolysed by steaming/cooking, for example, at 140–170 °C in water, or 110–120 °C in mineral acid, which depolymerises the lignin and dissolves the hemicelluloses. The exact conditions depend on the type of wood processed (e.g. beech and eucalyptus). Then the extraction is completed in a solution of NaOH and Na_2S at 130–180 °C. The S^{2-} and HS^- ions lead to a cleavage of lignin ether bonds (eq. (2.6)). When lignin and hemicelluloses dissolve, the cellulose remains as solid residue:

$$R-O-R' + HS^- \rightleftharpoons R-SH + R'OH \qquad (2.6)$$

The spent extraction liquors are concentrated to solids content of 60–80 wt% by evaporation and then burned to produce energy. In the reducing conditions during combustion, sulphide is regenerated and also part of the alkali is obtained from the ashes. The volume of collected spent liquor and the concentration of organic matter therein are sufficient in height to deliver a considerable share of the total energy required for the pulping process.

In the sulphite process, alkali sulphites, earth alkali sulphites and bisulphites are used for degradation and extraction of lignin and hemicelluloses. Representative conditions for this process are: pH value between 1.5 and 5 and cooking temperature between 130 and 160 °C. The removal of the lignin occurs through formation of soluble derivatives, the ligno-sulphonates. An example for the acid-catalysed cleavage of the ether bridge in lignin and the formation of a sulphonate is given as follows (eqs. (2.7), (2.8)):

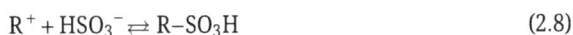

$$R-O-R' + H^+ \rightleftarrows R^+ + R'OH \tag{2.7}$$

$$R^+ + HSO_3^- \rightleftarrows R-SO_3H \tag{2.8}$$

Similar to the sulphate process, re-concentration and incineration of the spent liquors permit generation of thermal energy (steam) and partial recovery of the process chemicals.

In the case of cellulose fibre production, the pulping process has to be modified to deliver "dissolving pulp", a cellulose type with high purity and appropriate molecular weight distribution.

In all variations of production, the pulp has to be bleached to achieve a sufficient degree of whiteness for use, for example, in paper making or regenerate cellulose fibre production. The major part of bleaching chemicals are oxidants, which can be oxygen-based (e.g. H_2O_2, O_3, O_2, organic and inorganic peroxo-acids) or halogen-based chemicals (e.g. Cl_2, HOCl and OCl$^-$, ClO_2^-). When halogen-based oxidants are used, the formation of halogenated compounds can lead to critical concentrations of adsorbable halogenated compounds (AOX) in the wastewater. Thus, the use of halogen-based bleaching agents has decreased substantially, leading to development of elemental chlorine free and totally chlorine-free pulp.

Besides the oxidative treatment also washing with alkali, complexing agents, enzymes (e.g. laccases and oxidoreductases) or reductive treatment ($Na_2S_2O_4$, sodium dithionite) are in use.

Important parameters in quality testing of pulp are:
- alkali solubility in 5–18 wt% NaOH solution;
- hydrolysis and sugar analysis;
- DP and molecular mass distribution;
- lignin content;
- in organic solvents extractable materials;
- inorganic ingredients that form ashes (ash content);
- carboxylic group content.

For the majority of these parameters, standard methods have been defined (e.g. Tappi methods) [7].

2.5 Relevant aspects of cellulose chemistry (chemical reactions, derivatisation and chemical stability)

Cellulose represents the basic polymer for almost 50% of the total textile fibres produced. While substantial differences exist between the different fibres with regard to fibre composition, fibrillar structure, porosity and surface, as well as physical properties, the basic chemistry of the main polymer is the same (Figure 2.6).

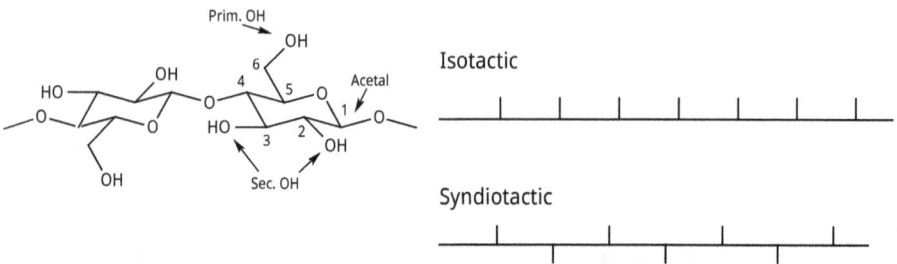

Figure 2.6: Basic chemical structure of cellulose (numbering of carbon atoms and functional groups).

Different approaches can be applied to describe the molecular structure of cellulose under chemical aspects:

The basic chemistry is defined by the presence of functional groups:

– Hydroxyl groups: Every anhydroglucose unit contains three hydroxyl groups.
– Acetals: The linkage between two anhydroglucose units is formed by an acetal group.
– Aldehydes = the reducing end: One side of the cellulose chain exhibits a half acetal group, which easily opens to the corresponding aldehyde. This is the "reducing end" of a cellulose chain, while the other end is formed by a free hydroxyl group at C4.
– As result of oxidative processes the reducing end is transferred into a carboxylic group. In wood pulp usually one carboxyl group is determined analytically per 100–1,000 anhydroglucose units, and in cotton one carboxyl group is determined per 100–500 anhydroglucose units. Widely used analytical methods can be acidimetric titration methods or based on the sorption of a cationic dye (e.g. methylene blue).

Cellulose is a polyacetal of glucose. Every second glucopyranosyl unit is rotated along the polymer axis by an angle of nearly 180°. Thus, as shown in Figure 2.6, the repetitive unit in cellulose is formed by two anhydroglucose units, which is also called cellobiose.

As the conformation of the rings and side groups then is repetitive, the polymer is isotactic with regard to cellobiose and syndiotactic with regard to the glucopyranosyl unit.

In general, the chemical description of cellulose given up to this point is ambiguous and thus further details have to be specified in the correct chemical name: The bridge between two glucopyranosyl units is formed from the aldehyde group (C1) to the C4 of the next glucopyranosyl unit. The acetal is present in β conformation and the glucose is stereochemically correct D-glucose.

Thus in the simple case of cellobiose the correct name is 4-*O*-β-D-glucopyranosyl-D-glucose.

Cellulose is the isotactic β-1,4-polyacetal of cellobiose or the syndiotactic polyacetal of glucose.

When the end groups are neglected, the chemical formula of cellulose is $C_{6n}H_{10n+2}O_{5n+1}$ ≈ $(C_6H_{10}O_5)_n$ with n represents the number of anhydroglucose units in the chain, respectively, the DP.

This then results in an elemental composition:

44.4% C, 6.2% H, 49.4% O and a mass of a glucopyranosyl unit $M(C_6H_{10}O_5)$ = 162 g/mol.

The chemistry of cellulose in solution corresponds to the usual behaviour of alcohols, aldehydes and acetals. In solid state, the fibre chemistry exhibits accessibility-controlled reactivity and follows the rules of a topochemical reaction.

Reactions will occur at sites only where chemicals will have access. Amorphous parts of the fibre will be accessible through pores and these parts of the polymer thus exhibits 1,000–5,000 times faster reactivity than crystalline parts.

In a simplified approach, a cellulose fibre can be understood as consisting of two phases, the amorphous and the crystalline parts, both consisting of cellulose, however with different reactivity.

As a result of the different accessibility, the amorphous parts of the cellulose will be the relevant sites for textile chemical operations. The crystalline parts are less re-active. Reactions in the crystalline parts can occur after swelling of the cellulose in swelling agents, solvents or following derivatisation reactions.

Textile chemical operations that modify the structure of the fibre (e.g. alkalisa-tion) will not only modify the physical properties of the fibres but also change the chemical behaviour substantially.

An increase in cellulose reactivity can be achieved through a so-called activation treatment. Such processes intend to increase the accessible internal surface (e.g. through fibre swelling) in Li-salt solution (e.g. LiCl and LiSCN); inorganic acids (e.g. H_3PO_4); inorganic and organic bases (e.g. NaOH and LiOH); tetramethyl ammonium hydroxide, ethyl-amine, solvent exchange, cellulose structure-weakening additives, degradation and mechanical grinding.

An important technical process that leads to cellulose activation is called mercerisa-tion. In this process, cellulose textiles (mainly cotton) are treated in concentrated NaOH

(10–24 wt%). In many cases, the process is executed at room temperature. During the process, a reorganisation of the crystalline parts of cellulose from the allomorph cellulose I to cellulose II occurs. In parallel, activation of the amorphous parts is achieved, which results in higher dyestuff uptake and fixation.

Hornification is the reverse process, which results in reduced accessibility and reactivity, for example, as a result of prolonged treatment at elevated temperatures. The removal of tightly bound water and parallel molecular reorganisation, for example, at temperatures of 130–150 °C leads to partial collapse of the pore structure of the fibre, and an overall reduction in reactivity is observed.

The average DP of cellulose differs based on the respective plant source (Table 2.6):

Table 2.6: Representative average values for the degree of polymerisation P_n. From Fras Zemljic et al. [8].

Source	P_n	Crystallinity
Cotton	3,000–6,000	
Cotton scoured, bleached	1,500–300	0.74
Flax	8,000	
Beech	3,050	
Bacterial cellulose	2,700	
Viscose	200–250	0.39
Modal	400–600	0.39
Lyocell	600–800	0.62

Besides the average DP also the polymolecularity or polydispersity of the cellulose differs based on the material source. This explains the need for careful selection of the appropriate cellulose raw material for production of dissolving pulp for manufacturing of regenerated cellulose fibres.

Under normal climate conditions cellulose is not a pure carbohydrate material, but will contain substantial amounts of water in sorbed state. The sorbed water can compete with chemical reactions and can cause side reactions. When higher amounts of water are absorbed in the amorphous parts of the fibres with a dimensional change, swelling is observed.

In aqueous caustic soda solutions or in other concentrated hydroxyl ions containing solutions, for example, tetramethyl-ammoniumhydroxide $((CH_3)_4N^+OH^-)$, more extensive swelling can occur and at lower degrees of polymerisation even dissolution of the cellulose polymer is observed.

2.6 Chemical reactions

The chemical properties of cellulose are determined by the presence of
– primary hydroxyl (–OH) group at C6;
– two secondary hydroxyl (–OH) groups at C2 and C3; and
– an acetal of an aldehyde at C1 and an acetal of a secondary alcohol at carbon C4.

2.6.1 Derivatisation

Sterical reactions are favoured to occur at C6. In derivatisation reactions, the degree of modification can be characterised by the degree of substitution (DS). With regard to the hydroxyl groups available per monomer unit, a theoretical maximum DS of 3 can be achieved.

When a partially substituted cellulose molecule has been synthesised using fibres as insoluble substrate, the substitution (symbol B) will be concentrated on the more accessible amorphous parts, while the less accessible crystalline parts will remain in the unsubstituted state (symbol A).

As a result, the modified cellulose will appear as a block polymer, with the general structure AAAAA-BBBBBB-AAA-BBB-AAA.

Uniform derivatisation requires homogenous solution conditions to achieve the same probability for reaction of each unit along the polymer chain. Thus, the polymer has to be present in dissolved form.

2.6.2 Hydrolysis

The glycosidic bond is sensitive to hydrolysis. Acetals are mainly sensitive to acid-catalysed hydrolysis; thus, acid-catalysed chain hydrolysis and reduction in DP mainly occur in the presence of acids. The reaction rate increases with concentration; thus, in textile operations volatile acids (e.g. acetic acid and formic acid) are used for neutralisation of residual alkali to prevent formation of higher acid concentrations. When a slight excess of a volatile acid is applied on a fabric, the surplus will evaporate during drying. Non-volatile acids (e.g. sulphuric acid) will concentrate on the textile and initiate serious damage through hydrolysis reactions.

Acid hydrolysis of cellulose in homogenous state leads to statistical molecular mass distribution and follows first-order reaction rate (Figure 2.7).

In the heterogeneous hydrolysis of cellulose fibres, a change in hydrolysis rate is observed at a DP of 25–100, which corresponds to the length of the elementary crystallites. The morphology and crystallinity of the fibre thus determine the hydrolysis reaction as highly ordered parts will hydrolyse much slower. The selective acidic hydrolysis of cellulose material has found application in production of microcrystalline cellulose

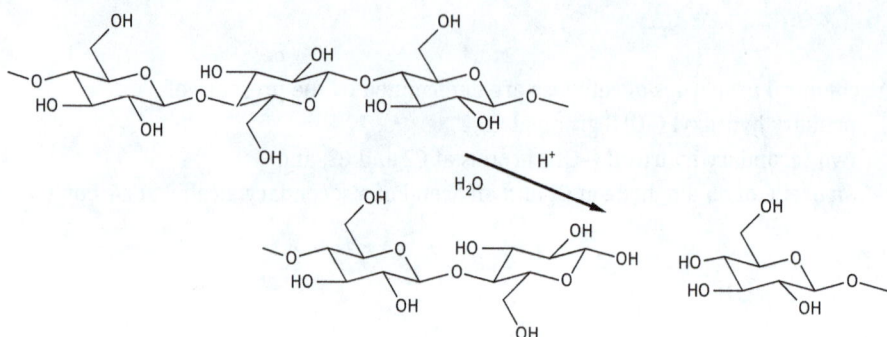

Figure 2.7: Acid-catalysed hydrolysis of the cellulose chain.

through hydrolytic removal of the amorphous parts of the cellulose. Microcrystalline cellulose is used in many applications, for example, as disintegration agents for laundry detergents, tablets, as anticaking aids, filters and tabletting aids and as food additives to adjust for a desired texture.

Degradation in alkaline solution occurs much slower and is observed with concentrated alkali solutions at elevated temperatures. The process chemistry is rather complex and involves hydrolytic reactions at the glycosidic bond, oxidative effects due to the presence of dissolved air (oxygen), which is a quite strong oxidant. In alkaline solution a series of rearrangements occurs, which is summarised under the term peeling reactions (Figure 2.8).

2.6.3 Oxidation

Chemical oxidation can occur at any of the six different carbon atoms in the glucopyranose unit.

Oxidation reactions of a hydroxyl group can occur at C2, C3 and C6 groups. In addition, oxidation of the C4 hydroxyl group forming the non-reducing end group and of the aldehyde group forming the reducing end of the cellulose chain can occur.

The site of attack as well as the functional groups formed will both depend on the oxidation chemicals used.

Extensively oxidised cellulose, for example, through intensive oxidation in bleach processes or repetitive laundry with heavy-duty washing agents is called oxycellulose.

Specific oxidants for the C1 aldehyde are hypoiodite OI^- and chlorite ClO_2^- leading to the corresponding carboxylic group (eq. (2.9)):

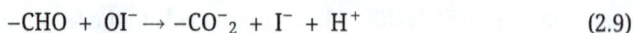

$$-CHO + OI^- \rightarrow -CO_2^- + I^- + H^+ \tag{2.9}$$

Periodate specifically attacks C2 and C3 with ring splitting and formation of a dialdehyde (Malaprade reaction) (eq. (2.10)):

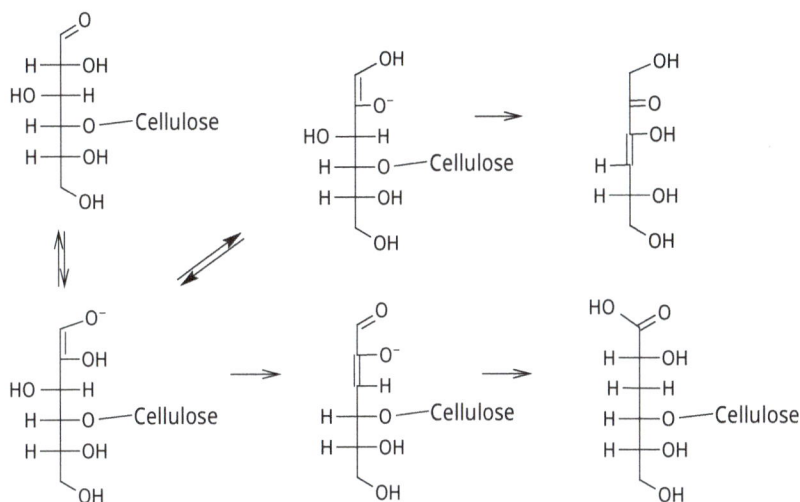

Figure 2.8: Peeling reactions – rearrangements of carbohydrates in alkaline solution. From Mozdyniewicz et al. [9].

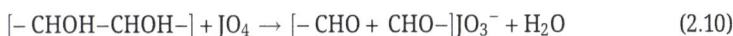

$$[- CHOH–CHOH–] + JO_4 \rightarrow [- CHO + CHO–]JO_3^- + H_2O \tag{2.10}$$

Regioselective oxidation of the C6–OH to the corresponding carboxylic group can be achieved by the use of a combination between TEMPO (2,2,6,6-tetramethylpiperidine-1-oxyradical) NaBr and NaOCl. Due to sterical effects, the oxidation of the C6 hydroxyl group is a regioselective reaction. The presence of carboxylic groups can be visualised by staining with a cationic dye (e.g. methylene blue). The sorption of methylene blue proceeds as a stoichiometric process. A dye molecule binds to a carboxylic group; thus, this method is also useful to quantify the carboxylic group content in cellulose pulp and cellulose fibres (Figure 2.9).

2.6.4 Esterification

Esterification is an important reaction between alcohols and acids, thus leading to cellulose esters. Organic or inorganic acids can be used to form esters.

With inorganic acids as examples, cellulose nitrate, sulphate or phosphate esters are obtained. With organic acids, cellulose acetate, propionate and so on are formed (Figure 2.10). The synthesis follows the usual chemical procedures for alcohols; however, it has to be considered that if the reaction is carried out as heterogeneous process, the reaction conditions applied and the accessibility of the cellulose structure will determine the overall DS and the uniformity of esterification along the cellulose chain.

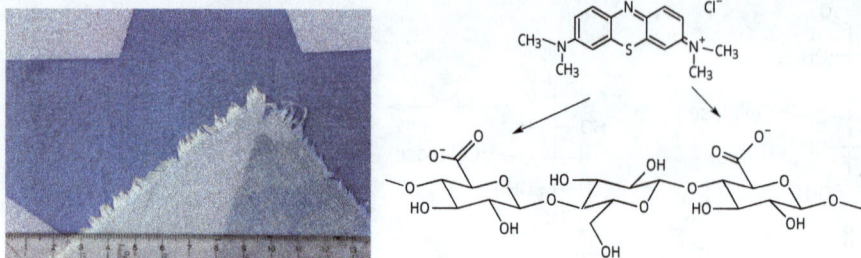

Figure 2.9: Localised oxidation of cellulose fabric with TEMPO/NaBr/NaOCl and localisation of carboxylic groups formed by staining with methylene blue (dark blue areas = cellulose with high carboxylic group content at C6, stained by methylene blue).

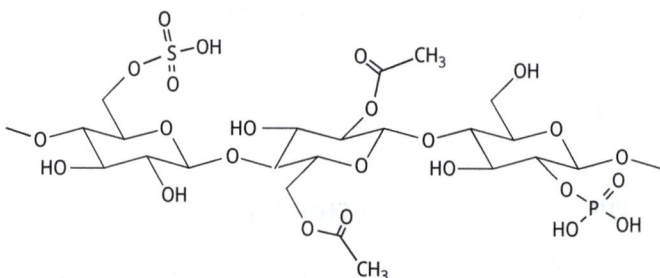

Figure 2.10: Cellulose esters with inorganic acids: C6-monosulphate, C2- and C6-acetate, C2-phosphate ester.

The esterification of hydroxyl groups with sulphate or phosphate groups alters the solubility of the corresponding esters in water. Dependent on the DS swelling, gel formation and dissolution can occur.

Acetylation increases the solubility in organic solvents; thus, cellulose diacetate (CA, DS = 2) and cellulose triacetate (CT, DS = 3) are soluble in acetone or chloroform.

2.6.5 Etherification

Typical strategies for etherification can be based on substitution of halogenides in alkyl halides, addition on epoxides or Michael addition to activated double bonds (eqs. (2.16)–(2.18)) (Figure 2.11).

While in organic chemistry usually the reactivity of alcohols follows the order primary alcohol > secondary alcohol > tertiary alcohol, the reaction rates in cellulose etherification depend on the reagents used. During etherification a distribution pattern of products is obtained.

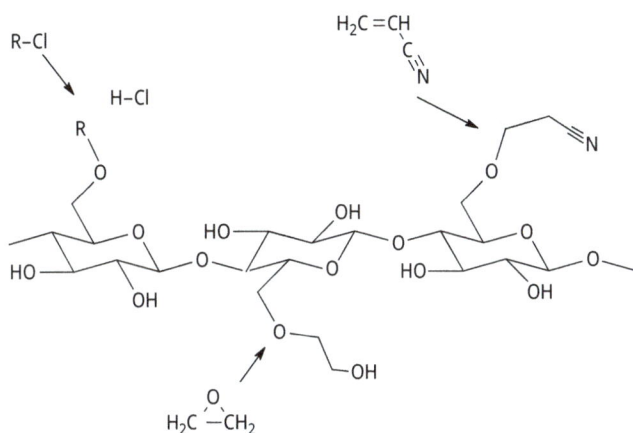

Figure 2.11: Reactions of cellulose with an alkyl chloride, epoxide and acrylonitrile.

In many cases, etherification at C2 is the preferred reaction (e.g. in the synthesis of methylcellulose). The distribution during formation of carboxymethylcellulose follows the ratio 2:1:2.5 for C2, C3 and C6 substituted products, respectively.

Similar to the esterification reaction, heterogeneous etherification will lead to a block substitution, delivering a mixture of fully substituted, irregularly and unsubstituted glucose pyranose rings.

In many cellulose finishing reactions as well as in reactive dyeing both etherification and esterification reactions are used as cross-linking principles. In many cases, also the reaction of anchor groups for covalent binding of a reactive dye chromophore to the cellulose structure follows these principles.

2.6.6 Complex formation

Cellulose can act as a multi-dentate ligand, thus particularly in alkaline conditions stable metal complexes can be formed. In many cases, the solubility of the cellulose increases substantially with metal complex formation and dissolution of the polymer can be achieved. The formation of Cu–ammonia complexes and iron–tartaric complexes with cellulose can be used to regenerate cellulose fibres from their concentrated solution (e.g. cuoxam solution). Cellulose dissolution through complex formation is also of interest to prepare solutions for determination of the DP by viscosimetry. Figure 2.12 shows a proposed structure for the Fe^{3+} complex between cellulose and tartaric acid present in the alkaline solution.

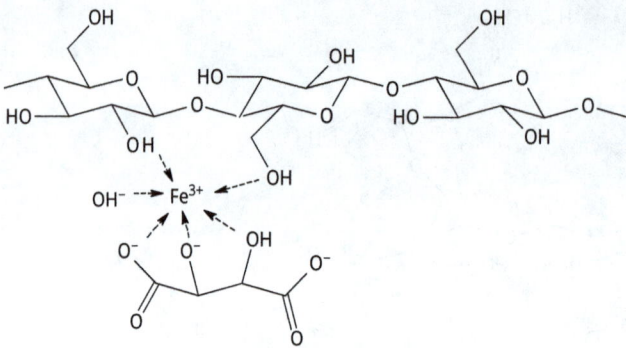

Figure 2.12: Scheme of an iron complex formed with cellulose and tartaric acid in aqueous alkaline solution. From Bechtold et al. [10].

2.6.7 The secondary structure of cellulose

In solid state, a 3D network of hydrogen bonds is established between neighbouring cellulose chains. An ordered crystalline lattice is formed and crystalline areas develop over certain length (Figure 2.13).

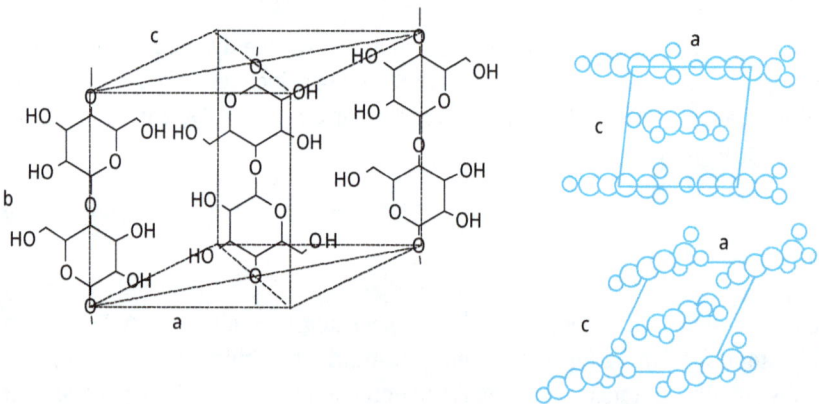

Figure 2.13: Crystal lattice of cellulose, unit cell cross section of cellulose I (top right) and cellulose II (bottom right) [according 26].

Cellulose exhibits the phenomenon of polymorphism, which means that cellulose crystals with different allomorph crystal lattices can be observed.

Native cellulose grows in the polymorph Cellulose I, in which the pyranose rings are assumed to form a crystal lattice with antiparallel arrangement of the cellobiose units. Regenerated cellulose as well as native cellulose which has been treated in concentrated alkali solution (c(NaOH) > 3–4 M) is present in the cellulose II form. Here the pyranose rings of a unit cell are arranged in twisted orientation. Cellulose that has been treated in liquid ammonia at −29 °C is present in cellulose III form. Representative values for the size of elemental crystallites in cellulose have been determined with 12–20 nm for the length and 2.5–4.0 nm for the width of a crystallite. The actual dimensions depend on the type of fibre and the textile chemical treatments applied.

The crystallites and crystallite strands (elementary fibrils) form the basic elements to build the supermolecular structure. Resulting microfibrils and macrofibrils are aggregates of elementary fibrils.

The fibrillar structure of a fibre depends on the type of natural fibre and is also determined by its growth. In regenerated cellulose fibres, the principle of production and the conditions of cellulose regeneration determine the build up of a certain fibrillar structure.

During treatment of cellulose fibres in swelling solutions, for example, tetramethyl ammonium hydroxide, mechanical effects can split the fibre and the fibrillar structure becomes visible (Figure 2.14).

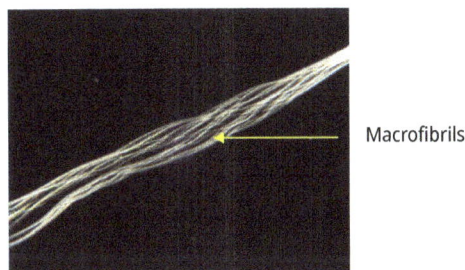

Macrofibrils

Figure 2.14: Fibrillar structure of a lyocell fibres after treatment in concentrated tetramethyl-ammonium hydroxide solution.

The overall supermolecular structure of a cellulose fibre thus is built by a series of structural elements:
– Crystalline/amorphous parts
– Elementary fibrils
– Micro- and macrofibrils

This architecture finally determines the overall physical properties and chemical reactions. Despite being made of the same cellulose molecule, substantial textile chemically relevant differences will result from differences in the individual structure of a cellulose fibre.

2.6.8 Methods of structural characterisation

There are a number of methods available which permit characterisation of the structure of a cellulose fibre.

Crystallite dimensions and degree of order can be determined by small-angle X-ray scattering and wide-angle X-ray scattering. Besides X-ray scattering, degree of crystallinity can be determined by infrared spectroscopy, from water sorption experiments or from deuterium exchange experiments. It is important to consider that the method applied also influences the outcome of the results. Thus, interpretation is always based on reference experiments and compare results obtained with the same methodology.

2.7 Regenerated cellulose fibres

2.7.1 Dissolution of cellulose

The production of regenerated cellulose fibres follows a general scheme:
1. Production of pulped cellulose or use of another highly pure cellulose source (e.g. cotton linters)
2. Dissolution of cellulose
3. Spinning of the cellulose solution and regeneration of the cellulose as fibre
4. Recovery of the solvent system

To dissolve cellulose, the intensive attractive forces between neighbouring cellulose chains have to be opened by the solvent system. Main forces between cellulose chains result from intramolecular and intermolecular hydrogen bonds that have to be replaced by interactions with the solvent. As the position of the hydroxyl groups at C2, C3 and of the C6 hydroxy methyl group all are equatorial, the pyranose ring exhibits polar sites in equatorial direction. Above the centre of the glucopyranose ring the pyranose ring exhibits only low polarity and appears as rather hydrophobic molecule. This phenomenon, for example, is demonstrated by the ability of cyclodextrines (cyclic low-molecular-weight oligosaccharides), to host hydrophobic substances inside their ring.

Thus, cellulose solvents also will have to exhibit both sites of high polarity and hydrophobic properties at the same time [11].

Different strategies to dissolve cellulose have been utilised to shape regenerated cellulose in form of fibres, films and sponges.

Derivatisation: The hydroxyl groups are masked by formation of a chemical derivate, which is decomposed after the step of fibre formation. For viscose and modal fibre production, solubility of the cellulose is achieved by derivatisation through formation of a cellulose xanthogenate.

Formation of cellulose carbamate allows dissolution of the derivatised polymer in diluted alkaline solution.

Through acetylation the hydroxyl groups become esterified and solubility in organic solvents (e.g. acetone) can be achieved.

In any case, a certain degree of derivatisation is required to alter the properties of the cellulose chain sufficiently, for example, in case of cellulose acetate, a DS of 2–3 is required.

Complex formation: Through formation of Cu–amino complexes with cellulose the hydroxyl groups of the cellulose participate in a metal complex system and thus solubility in aqueous alkali solution is achieved.

Direct dissolution: Use of highly polar solvents [e.g. *N*-methyl-morpholine-*N*-oxide (NMMO)] allows direct dissolution of cellulose and regeneration through dilution with a non-solvent. Direct dissolution also can be achieved in ionic liquids (e.g. 1-butyl-3-methylimidazolium chloride) [12] (Figure 2.15).

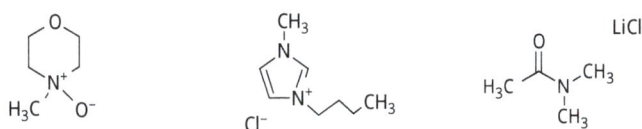

Figure 2.15: Cellulose solvents: NMMO (*N*-methyl-morpholine-*N*-oxide), BuMIMCl (1-butyl-3-methylimidazolium chloride), LiCl/DMAc (lithium chloride/dimethylacetamide).

Also other solvent systems such as urea/NaOH and LiCl/dimethylacetamide can dissolve cellulose without degradation [13, 14].

2.7.2 The viscose process

In the first step, high-quality cellulose (α-cellulose content >90 wt%), for example, from wood pulp or cotton linters is impregnated in 20 wt% aqueous NaOH solution (Figure 2.16). This so-called alkalisation step is an exothermal step. Sodium cellulose I is formed with a stoichiometric composition. Frequently, the alkali cellulose is written as Cell-O⁻Na⁺ which for an organic chemist would correspond to formation of an alcoholate in aqueous solution. From the pK_a values of aliphatic alcohols, such a reaction however is not expected to occur in aqueous systems. In contrary to the behaviour of aliphatic alcohols in aqueous solution, the formation of alkali-cellulose is a heterogeneous reaction between solid cellulose and NaOH, and the regular stoichiometry and structure in solid state has been proven by X-ray diffraction. This indicates that reaction chemistry of solid cellulose can differ substantially from the usual behaviour of similar compounds in solution.

A representative pseudostoichiometric composition of alkali cellulose is according to:

$$(C_6H_{10}O_5).(NaOH).(H_2O)_{3-5}.$$

After alkalisation the excess NaOH solution is pressed off and the alkali cellulose is shredded to flakes. Residual hemicelluloses present in the pulp are dissolved in the excess NaOH solution and removed therein. The presence of oxygen leads to oxidative chain degradation and degree of polymerisation reduces substantially. This process is called pre-ripening (Figure 2.16).

The viscose process

Pulp from wood, α-Cellulose content >89–93%
       Cotton linters (in cotton producing countries)
Alkalisation in cellulose (20 wt% NaOH) exothermal 12 kJ/mol AGU
       Sodium cellulose I is formed
       pseudostoichiometric compound: $(C_6H_{10}O_5).(NaOH).(H_2O)_{3-5}$
Pressing to 30 wt% alkali-cellulose content
Ageing/shredding: DP reduces to 300 for CV, 450 for modal
       hemicelluloses are washed out
Pre-ripening: Reduction of DP to adjust viscosity
Xanthation: Treatment with CS_2 to DS 0.5–0.6
       150–400 kg CS_2 per 1,000 kg fibres
Viscose solution: Xanthate dissolution in NaOH, honey like solution
       7–12% cellulose
       5–8% caustic
Maturing/Filtering: 18 °C, 50–80 h
       Sulphidation becomes more uniform
       Ability to coagulate increases
Fibre spinning: Injection into diluted sulfuric acid
       Cellulose coagulates viscose
Washing
Drying

Figure 2.16: General scheme of viscose fibre production.

This first reduction in DP is necessary to achieve solubility of the xanthogenate and to reduce the viscosity of the spinning dope to the level required for filtration and fibre spinning. The shredded alkali cellulose then is treated with carbon disulphide (CS_2) to form the xanthogenate. The product resulting from the xanthation exhibits an average DS between 0.5 and 0.6, which means that in average every second glucopyranose ring carries a xanthogenate group. The cellulose xanthogenate is soluble in diluted aqueous alkali solution and thus can be dissolved in 5–8 wt% aqueous NaOH solution, thereby viscous solutions with a cellulose content of 7–12 wt% are formed. The solution is filtered and deaerated to remove any bubbles. During this maturing step, ripening of the spin solution is achieved. During xanthogenation, the position of substitution in the pyranose ring is determined kinetically, and during the phase of ripening, the reallocation of substitution occurs and the sulphidation becomes more uniform. The overall DS decreases as a result of reactions and the tendency of the dissolved cellulose to coagulate increases.

Regeneration of the cellulose is achieved by injection of the viscose dope into an aqueous acid bath containing sulphuric acid and sodium sulphate (wet spin process).

Due to the corrosive environment, the spinneret has to be formed from precious metals such as Au, Pt/Ir and Pt/Rh and from selected materials (e.g. glass and PTFE). Due to shear effects in the spinneret and drawing in the spin bath, the cellulose polymer molecules become oriented along the fibre axis.

Two steps occur in the precipitation bath. At first, the alkali present in the extruded dope is neutralised, which leads to coagulation of the cellulose xanthogenate. In the second step, the xanthogenate decomposes and CS_2 is released (eqs. (2.11), (2.12)). The CS_2 has to be recovered because of safety reasons as the solvent is highly flammable and toxic, and also because of environmental concerns:

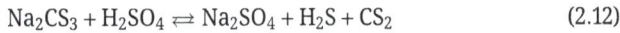

$$2\text{Cell}-\text{O}-\text{CS}-\text{S}^-\text{Na}^+ + H_2SO_4 \rightleftharpoons 2\text{Cell}-\text{OH} + 2CS_2 + Na_2SO_4 \qquad (2.11)$$

$$Na_2CS_3 + H_2SO_4 \rightleftharpoons Na_2SO_4 + H_2S + CS_2 \qquad (2.12)$$

Na_2SO_4 is formed as a result of the neutralisation step and accumulates in the precipitation bath. A typical composition of a neutralisation bath is 7–12 wt% H_2SO_4 and 12–24 wt% Na_2SO_4. A reduction of the Na_2SO_4 concentration could be achieved by an increased exchange of the filling of the bath, which however would increase consumption of sulphuric acid released with the bath overflow (Figure 2.17).

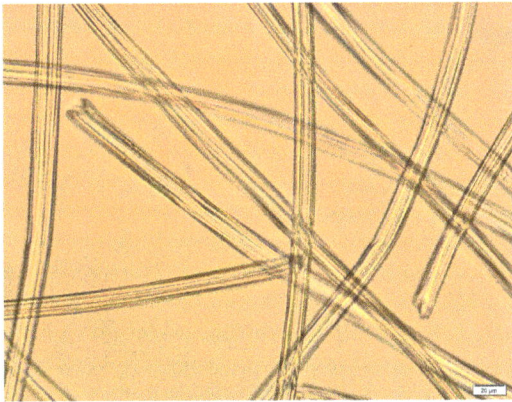

Figure 2.17: Photomicrograph of viscose fibres.

In the case of viscose fibre, the rapid coagulation at the fibre surface leads to a diffusion barrier between precipitation baths and fibre core. As a result, a core shell structure is formed, which is a characteristic of viscose fibres. The mantle region is higher ordered, more compact and contributes more to the fibre strength than the less ordered, more porous fibre core, which contributes more to water absorption.

The fibre formation begins from cellulose solutions with solids content of 7–12 wt%; thus, a considerable shrinkage occurs during fibre formation. The regeneration of the fibre core proceeds later than the initial formation of the compact fibre skin. As the dimensions of the skin are already stabilised when the fibre core is still in process of coagulation, the skin form folds and the characteristic multilobal cross section of viscose fibres results.

The regenerated cellulose exhibits cellulose II structure.

When modifiers, for example, Zn salts or aliphatic and aromatic amines or EO-PO-condensation products are added to the coagulation bath, fibre formation and xanthogenate decomposition become synchronous. As a result, more uniform and highly oriented, full mantle fibres are obtained (modal fibres and high wet modulus fibres). The characteristic sign of these fibres is their round cross section. Due to the absence of a highly absorbent fibre core, the water sorption and water uptake of the fibres are lower.

Modal-type fibres are in use for garment production, high-wet-modulus fibres are representatives for important technical fibres, for example, for tyre cord.

After regeneration, the cellulose fibres are washed, neutralised, desulphurised and dried. The fibres are used either as endless filament yarn or cable, or the cable is cut into staple fibres and delivered as bale.

Fibre modifications can be obtained by addition of functional components into the spin dope, for example, white pigments (TiO_2) to reduce gloss, $BaSO_4$ to obtain X-ray contrast fibres, dyes and pigments for spun-dyed fibres, flame-retardant substances to obtain reduced flammability (Figure 2.18) and carbon to achieve electrical conductivity.

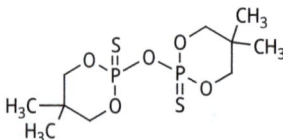

Figure 2.18: (2,2′-Oxybis)-5,5-dimethyl-1,2,3–dioxadiphosporinane–2,2′–disulphide (Sandoflam® 5060).

Physical modification of the fibre shape is possible through use of appropriately shaped spin nozzles, which for example leads to a trilobal fibre with high water absorbency (Galaxy®). Hollow fibres, for example, can be manufactured by injection of a bore fluid inside the fibre core, which then is removed after the fibre has been formed.

2.7.3 Lyocell fibres – the NMMO process

NMMO (Figure 2.15) is a highly polar solvent that dissolves cellulose in a direct dissolution process. In the first step, pulp is mixed with a mixture of NMMO and water. A typical composition of the slurry is 13 wt% cellulose, 20 wt% water and 67 wt% NMMO. The water content of the mixture is then removed by means of thin film evaporation. During this concentration step at elevated temperature the cellulose dissolves and the spin dope is obtained (14 wt% cellulose, 10 wt% water and 76 wt% NMMO). At higher temperatures, uncontrolled exothermal redox reactions between NMMO (oxidant) and cellulose (reductant) could start, which are prohibited by the addition of stabilisers (e.g. propyl gallate) and careful monitoring of the process temperatures.

The extrusion of the cellulose solution through the spinneret into the coagulation bath (NMMO water) is performed at temperatures above 100 °C; thus, an air gap between the spinneret and the coagulation bath is required (dry jet-wet spinning). Orientation of the cellulose molecules is the result of high shear rates in the spinneret and consecutive drawing in the air gap. In the coagulation bath, the fibre regeneration occurs as a complex process: The decrease in solution temperature increases viscosity substantially and solidifies the solution. The NMMO is washed out into the non-solvent water, which also accesses to the fibre structure. Coagulation and shrinkage of the porous structure leads to development of a distinct fibrillar fibre structure, which also explains the tendency of the fibre to fibrillate when mechanical action is applied in the swollen state. Once the solvent NMMO has been washed off, a highly ordered soft filament is obtained (Figure 2.19).

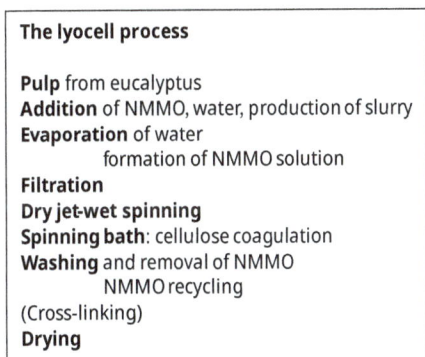

The lyocell process

Pulp from eucalyptus
Addition of NMMO, water, production of slurry
Evaporation of water
          formation of NMMO solution
Filtration
Dry jet-wet spinning
Spinning bath: cellulose coagulation
Washing and removal of NMMO
          NMMO recycling
(Cross-linking)
Drying

Figure 2.19: Production scheme of lyocell-type fibres.

Before drying, the fibre is present in a highly hydrated expanded state, the so-called never dried state. Upon drying the removal of the hydrate water leads to a collapsing of the open fibre structure, and the final dimensions of the lyocell fibre are obtained.

In the never dried state, cross-linking of the lyocell fibres with bifunctional cross-linkers (N-hydroxy-dichloro-s-triazine or Tris-acryloyl-hexahydro-s-triazine; Figure 2.20) is performed to establish chemical bridges between adjacent cellulose chains in the fibre. Through formation of chemical linkages between the cellulose, the fibrillation of the fibres in wet state is reduced.

A decisive step for the successful commercialisation of the lyocell process was the development of a technically feasible solvent recovery and purification system. The expensive NMMO-based solvent system requires recovery rates in the dimension of 99% and higher.

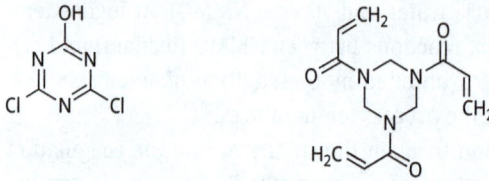

Figure 2.20: Formula of *N*-hydroxy-dichloro-*s*-triazine, NHDT; Tris-acryloyl-hexahydro-*s*-triazine, THAT.

The current regeneration of the NMMO utilises the fact that amine oxides can be protonated to form the corresponding organic hydroxyl ammonium ion, which then can be collected and purified by means of a cation exchange process (eq. (2.13)):

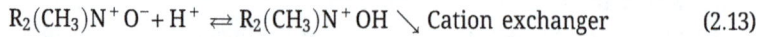

$$R_2(CH_3)N^+O^- + H^+ \rightleftharpoons R_2(CH_3)N^+OH \searrow \text{Cation exchanger} \qquad (2.13)$$

Table 2.7 summarises the relevant characteristics of native and regenerated cellulose fibres.

Table 2.7: Structural characteristics of native and regenerated cellulose fibres. From Schuster et al. [15].

Parameter	Unit	State	Viscose	Modal	Lyocell	Cotton
Degree of polymerisation, P_w	–		200–250	400–600	600–800	2,000
Crystallinity	1	–	0.39	0.39	0.62	0.74
Water retention value	%	Wet	70–80	50–53	50–70	38–45
Void volume	mL/g	Wet	0.68	0.49	0.60	0.30
Inner surface	m^2/g	Wet	444	407	418	207
BET surface	m^2/g	Dry	276	277	246	132
Mean pore diameter	Å	Wet	31	23	28–30	28

As shown in Table 2.7, the lowest mass average DP is found in viscose fibres. The higher tensile strength of modal fibres in wet state can be explained with the higher DP of the cellulose used and with a more compact structure due to controlled xanthogenate coagulation – decomposition. This also results in the lower void volume, lower inner surface and the smaller diameter of pores of modal fibres when compared to viscose fibres.

The WRV characterises the uptake of water of the swollen fibre state. The amount of water held in the fibre structure is determined gravimetrically after an excess of water had been removed by centrifugation. Modal fibres show substantially less expansion in the wet state; thus, the water retention reduces from 70 to 80 wt% for viscose to 50 to 53 wt% for modal.

For comparison of such experimental results, the same state of material (e.g. fibres, yarn and fabric) has to be used always, as in fabric structure restrictions with regard to maximum swelling exist due to the more compact structure.

Lyocell-type fibres exhibit higher crystallinity which is due to its distinct longitudinal order, while the other parameters are similar to viscose and modal.

Cotton is the most compact fibre among the four cellulose fibres compared in Table 2.7, showing the highest crystallinity and lowest values for pore volume and inner surface.

Thus, in terms of chemical reactivity cotton fibres will exhibit lowest reactivity and will also show the least changes in properties between dry and wet states.

Viscose fibres exhibit highest expansion in wet state, which also leads to the so-called "wet stiffness" of viscose fabrics. As a result of the high water uptake, a substantial decrease in wet tensile strength is observed with viscose fibres. The high water uptake of viscose fibres is the reason for their extensive use in hygiene products where high absorbency is expected, while maximum tensile strength in wet state is less important.

2.7.4 Ionic liquids and other cellulose solvents

During the last decade, a high number of alternative solvents for cellulose have been investigated and proposed for cellulose shaping into fibres and films [12].

Ionic liquids are salts that exhibit a melting point at temperatures near to ambient temperature and as another characteristic consists of a large organic ion and a smaller counterion. A representative example is 1-butyl-3-methylimidazolium chloride.

While successful formation of fibres has been achieved even in pilot scale, difficulties remain unsolved in economic aspects as regeneration and purification of the rather expensive solvents still are a challenge.

Also molten salt hydrates (e.g. $ZnCl_2 \times 3\text{--}4\ H_2O$, $LiClO_4 \times 3\ H_2O$), mixtures of alkali with urea [13] or Li salt containing solvents [e.g. LiCl/N,N-dimethylacetamide (LiCl/DMAc)] [14] have been studied as cellulose solvents. LiCl/DMAc is widely used as a solvent for homogenous derivatisation of cellulose and for analytical purposes, for example, determination of molecular weight distribution by gel permeation chromatography. The application of such alternative solvents in fibre production, however, is still a challenge as these solvents are expensive, corrosive or difficult to regenerate.

2.7.5 Cupro fibres – the cuprammonium process

The cuprammonium process belongs to the older techniques for production of regenerated cellulose fibres and is based on the solubility of cellulose in an alkaline solution of Cu^{2+}, NH_3 and NaOH. The solvent system is the copper–diamino complex Cu $(NH_3)_4(OH)_2$ which dissolves cellulose with complex formation at the C2 and C3 OH groups of the polymer. An alkaline solution containing 8–12 wt% of cellulose, 6–8 wt% NH_3 and 3–5 wt% Cu^{2+} serves as a spinning dope. The viscose solution is injected into streaming water, which serves as a non-solvent and applies stretch during precipitation of the cellulose (stretch spin process). While the use for textile applications has almost disappeared, the cupro process still is in use to produce porous membranes and hollow fibres, for example, for filtration application and extracorporeal dialysis.

The biocompatibility of such fibres can be improved through grafting, for example, with polyethylene (PE) glycols or heparin.

A major problem in the cupro-fibre technology lies in the need for complete recovery of the copper-based solvent. The rather high copper concentrations in the process solvent require overall recovery rates for Cu^{2+} above 99%, otherwise the copper load released into the wastewater stream would be inacceptably high. The recovery is achieved by electrochemical methods, precipitation and ion-exchange processes. The recovery of the ammonia is performed by distillation.

2.7.6 Cellulose diacetate and triacetate

Another strategy to shape celluloses is based on esterification of cellulose with acetic acid (acetylation) (eq. (2.14)). The derivatisation is performed with the use of acetic acid anhydride. The reduction in hydrogen bond interaction and the introduction of less polar ester groups increase the solubility in organic solvents (e.g. acetone):

$$Cell–OH + CH_3–CO–O–CO–CH_3 \rightleftharpoons Cell–O–CO–CH_3 + CH_3–CO–OH \qquad (2.14)$$

Through derivatisation of two to three hydroxyl groups of each glucopyranose ring (DS = 2–3), the hydrophobic character of the cellulose decreases and the modified polymer becomes soluble in organic solvents and thermoplastic.

To shape fibres, the spinning dope is prepared by dissolution of the cellulose–acetate in the solvent. The polymer solution is pressed through the spinneret and the solvent is evaporated (dry spin process).

With increasing degree of esterification, the swelling in the presence of liquid water reduces. This effect also increases with the number of carbons in the aliphatic residues of the organic acid used (Table 2.8) [4].

The reduction in water uptake and the increased hydrophobicity can be understood, when the chemical composition of the fibres is written as given in Table 2.8. While the polar character of the glucopyranose ring with six carbon atoms remains constant, the apolar character of the hydrophobic substituents increases with chain length of the ester groups. In cellulose triacetate, the number of "apolar" carbon atoms is already as high as the number of carbon atoms of the glucopyranose rings. In cellulose trivalerate, 15 carbon atoms form the hydrophobic side chains; in cellulose tristearate, 54 (!) carbon atoms determine a strongly hydrophobic behaviour, which predominates over the six carbons in the glucopyranose ring.

Table 2.8: Swelling of cellulose fibres and cellulose esters in water.

Fibre type	Formula	Swelling%wt
Cotton	$[C_6H_{10}O_5]_n$	18
Viscose	$[C_6H_{10}O_5]_n$	74
Cuoxam fibre	$[C_6H_{10}O_5]_n$	86
Cellulose triacetate	$[C_6H_{10}O_5(COCH_3)_3]_n$	10
Cellulose tripropionate	$[C_6H_{10}O_5(COCH_2CH_3)_3]_n$	2.5
Cellulose tributyrate	$[C_6H_{10}O_5(CO(CH_2)_2CH_3)_3]_n$	1.8
Cellulose trivalerate	$[C_6H_{10}O_5(CO(CH_2)_3CH_3)_3]_n$	1.6
Cellulose tristearate	$[C_6H_{10}O_5(CO(CH_2)_{16}CH_3)_3]_n$	1.0

Thus, the water-related properties of cellulose tristearate are more near to PP or PE than to cellulose.

The modification of cellulose by esterification seems to be a straightforward strategy to obtain cellulose polymer-based fibres from renewable resources. However, it has to be considered that in cellulose triacetate six of 12 carbons are already derived from the acetyl groups and with longer chain length the non-cellulosic parts increase substantially. As a result, a higher mass of acid components will be required for production than of cellulosic raw material.

2.8 Protein fibres

2.8.1 General aspects of protein fibres

A large number of protein-based fibres are used for textile products, technical and medical applications (e.g. wool, cashmere goat hair, angora rabbit hair and silk) as well as regenerated protein fibres (e.g. from soya protein or casein).

In this chapter the properties of wool and silk will be discussed as most important representatives for the protein-based fibres.

In general, the textile chemical behaviour of these fibres is defined by the chemical reactions of the respective protein chains. The fibre proteins are built from amino acids, which are linked to the macromolecule through peptide bonding. The natural protein fibres differ with regard to three major characteristics, which explains their distinct differences:

1. Physical structure of the natural protein fibres: diameter, surface characteristics and fibre structure [e.g. shed-like cuticula, presence of a fibre core (cortex) in wool, combination of different major proteins: fibroin and sericin in silk].
2. Protein conformation, secondary and tertiary structure: the secondary structure of the protein can be in random coil, sheet, twisted and helical conformation.

3. Primary structure of the protein. The major amino acids present in the protein differ with regard to composition and distribution between fibres, but also between different parts of a fibre (e.g. cuticula and cortex of wool, sericin and fibroin of raw silk). Differences are also observed between different types of hair and between breeds of the same fibre group of hair, for example, wool (Merino, Cheviot wool).

2.8.2 Protein structure – basic properties

In the protein macromolecule, different amino acids are linked, which forms the amino acid sequence of the protein. The average distribution of amino acids analysed in different protein sources is given in Table 2.9.

Table 2.9: Average distribution of relevant amino acids in protein fibres and selected proteins. From Rath [3], Schultze-Gebhart and Herlinger[4], McGrath and Kaplan [16].

Amino acid	Wool keratin (mol%)	Fibroin (mol%)	Sericin (mol%)	Casein (milk) (mol%)
Glycine	6.5	42.9	13.5	1.9
Alanine	4.1	30.0	5.8	1.5
Valine	4.8	2.5	2.9	7.2
Leucine and isomers	11.3	0.6	0.7	9.4
Aspartic acid/asparagine	6.6	1.9	14.6	6.7
Glutamic acid/glutamine	14.1	1.4	6.2	15.6
Hydroxyglutaminic acid	–	0.2	–	3.8
Arginine	10.3	0.5	3.1	2.5
Lysine	2.7	0.4	3.5	1.5
Histidine	0.7	0.2	1.4	5.9
Proline	6.8	0.5	0.6	10.5
Tryptophan	1.8	–	–	0.7
Phenylalanine	3.8	0.7	0.4	3.2
Tyrosine	4.7	4.8	3.6	4.5
Serine	10.3	12.2	34.0	0.5
Threonine	6.4	0.9	8.8	4.5
Cysteine	11.9	–	–	0.2
Others	ad 100	ad 100	ad 100	ad 100

The basic chemistry of protein fibres: In terms of polymer chemistry, we could understand wool as a co-polymer of approximately 20 different α-amino acids. The chemical behaviour is thus determined by the reactivity of the peptide backbone, the functional groups present in the side chains and the end groups of the protein (Figure 2.21).

Similar to cellulose, the end groups of the proteins exhibit a characteristic chemical behaviour. At the amino end of the protein, the free amino group of the last α-amino acid exhibits basic properties and the pH-dependent reactivity of an amino or ammonium

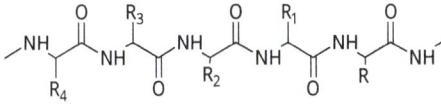

Figure 2.21: General formula of a peptide chain.

group. The carboxylic end of the protein exhibits acidic properties and the chemical reactivity characteristic for carboxylic groups.

Major functional groups present in the side chains are carboxylic groups (aspartic and glutamic acid), amino groups (arginine, lysine), hydroxyl groups (serine), phenolic groups (tyrosine), disulphide groups and mercapto (thiol) groups (cysteine and cysteine). Hydrophobicity of the protein is dependent on the presence of long-chain aliphatic or aromatic groups (e.g. leucine and phenylalanine).

Due to the presence of both basic and acidic groups, a betaine structure is present in the neutral state. In this state, the acidic groups dissociate and form a salt with neighbouring amino groups, which become protonated to the corresponding ammonium form (eq. (2.15)). In the following scheme, the neutralisation and zwitterion formation are formulated for the amino end group and the carboxylic end group; however, the reaction occurs throughout the full protein chain with involvement of the basic and acidic side chains:

$$H_2N-R-CO-NH-R'-COOH \rightleftharpoons H_3N^+-R-CO-NH-R'-COO^- \tag{2.15}$$

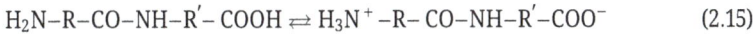

Acidic pH:

$$H_3N^+-R-CO-NH-R'-COO^- + H^+ \rightleftharpoons H_3N^+-R-CO-NH-R'-COOH \tag{2.16}$$

Alkaline pH:

$$H_3N^+-R-CO-NH-R'-COO^- + OH^- \rightleftharpoons H_2N-R-CO-NH-R'-COO^- + H_2O \tag{2.17}$$

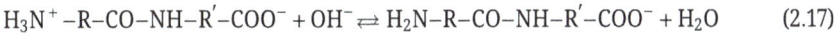

When pH is lowered by the presence of stronger acids, the relatively weak carboxylic group becomes protonated, and the ammonium group will remain as a cationic group (eq. (2.16)). Thus, the overall charge of the protein fibre becomes positive.

When the pH is increased by addition of a stronger base, the ammonium group dissociates and the protein fibre charge becomes negative (eq. (2.17)).

The pH value at which the positive charges and negative charges in a protein structure compensate each other to neutrality is called isoelectric pH of a protein and in case of protein fibres indicates the pH range of highest chemical stability and reduced fibre swelling. At such pH, strong interactions in the fibre structure due to ionic attraction between positively and negatively charged groups (e.g. glutamic acid and lysine, eq. (2.18)) lead to a compact protein structure:

$$R-CO_2^- \ H_3N^+R' \tag{2.18}$$

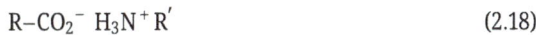

The ionic bonds in the protein fibres can be opened and rearranged through the action of heat and steam, which is utilised in thermal setting of wool fabric.

Representative isoelectric pH value for wool is 4.9, for silk 5.0 and for human skin pH 5–6.

Due to the presence of a high number of charged sites wool and silk both exhibit properties of a weak ion exchanger. These exchange properties are pH dependent (anion exchanger in acidic conditions and cation exchanger in alkaline conditions).

The cationic groups in wool can serve as positive anchor groups for anionic dyes, which is utilised in dyeing of wool with acid dyes.

As shown in Table 2.9 silk does not contain significant amounts of cysteine; thus, the absence of disulphide bridges is a characteristic difference to wool. The disulphide bridges form intramolecular and intermolecular linkages between protein chains. Disulphide bridges can be opened by reduction, for example, with thioglycolic acid to yield the corresponding thiol (=cysteine):

$$R-S-S-R' + 2\,HS-CH_2-CO_2H \rightleftharpoons R-SH + HS-R + HO_2C-CH_2-S-S-CH_2-CO_2H \qquad (2.19)$$

The reductive cleavage of the cysteine bonds in wool can be reversed and new disulphide bonds are formed (eq. (2.19)). Such processes are used to achieve permanent fixation of fabric dimensions and creases. A similar process is used in curing of hair protein for setting of permanent waves.

Oxidative damage of cysteine bonds, for example, through action of chlorine or persulphuric acid, leads to formation of the corresponding sulphoxides and sulphones. A representative reaction scheme is given as follows (eqs. (2.20), (2.21), (2.22)):

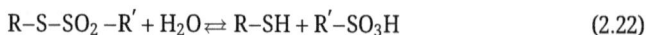

$$R-S-S-R' + Cl_2 + H_2O \rightleftharpoons R-S-SO-R' + 2\,HCl \qquad (2.20)$$

$$R-S-S-R' + 2\,Cl_2 + 2H_2O \rightleftharpoons R-S-SO_2-R' + 4\,HCl \qquad (2.21)$$

$$R-S-SO_2-R' + H_2O \rightleftharpoons R-SH + R'-SO_3H \qquad (2.22)$$

Following the oxidative attack of the disulphide groups, chemical rearrangements and hydrolytic cleavage of the thiosulphuric ester bond lead to irreversible cleavage of the former disulphide bridge, thus leading to damage of the fibre structure and losses in mechanical performance of the fibres.

In the side chains of the protein, the presence of hydroxyl and amino groups is of relevance to understand the reactivity of wool towards anchor groups present in reactive dyes. Also other reagents for chemical functionalisation of wool can react with these functional groups (also compare chemical reactions of reactive dyes in Chapter 10).

The peptide chain is sensitive towards hydrolytic attack, which leads to degradation and dissolution of protein fragments. Hydrolysis can be catalysed by acid and alkali; however, the reaction rate in alkaline conditions is substantially higher, thus leading to sensitivity of wool against treatments in alkaline conditions. There is an additional influence of fibre and protein structure on sensitivity of protein against hydrolysis. For example, silk fibres are less sensitive to hydrolytic attack of alkaline aqueous solutions than wool, but less resistant against action of concentrated acid solutions compared to wool. This is utilised in the basic operations during pre-treatment of wool and silk.

As an example, carbonisation of wool is performed with relatively concentrated sulphuric acid, while degumming of silk is performed with aqueous sodium carbonate solutions at the boil.

Due to the high density of complex forming ligands in the protein chain (carboxylic, amino, hydroxyl and phenolic groups), protein fibres can act as solid complexing agents for heavy metal ions (e.g. Sn salts).

This property is utilised in weighing of silk, where an increase in fibre weight up to 200% is obtained through immersion in concentrated salt solutions. In a similar approach, wool and silk are loaded with metal ions during pre-mordanting in dyeing with natural colorants [17].

2.8.3 Wool

Wool hair is obtained by shearing of sheep. The raw wool is called sweat wool. The purity of the sweat wool depends on the region of sheep farming, the type of wool and the respective part of the fleece. The quality of the wool also depends on the age of the sheep, for example, the first cut obtained from lambs (lambswool) is substantially finer than wool from sheep with an age above 1 year. The fineness of wool hair is usually described in terms of average fibre diameter. Typical average diameters for fine wool are in the range of 16–18 µm, and the standard quality wool for clothing is in the range of 20–28 µm. Coarse wool with an average diameter above 30 µm is used for carpets and technical products.

Representative values for the composition of raw wool are summarised in Table 2.10.

Table 2.10: Representative composition of sweat wool.

Component	Share (wt%)
Raw wool	33
Soil/dirt	26
Sweat	28
Fat (lanolin)	12
Minerals	1

In the first textile chemical treatment, the sweat wool is washed to remove the major part of impurities. This process is usually performed in the country of origin. The wasted water carries a high load of biodegradable organic components, however, is a highly valuable source to recover lanolin, which then after purification is used for cosmetics and other skin care products.

In wool the protein chains are present in α-helical conformation. Three peptide chains then form a protofibril, surrounded by a protein matrix. Protofibils arrange further to form micro- and macrofibrils. These structural elements then build the cortical cells that form the cortex, the inner part of the wool hair. Two different parts of cortex exist in wool. The more ordered ortho cortex is more reactive and can be stained with anionic dyes, and the less structured para cortex is stained more with cationic dyes.

The cortex is covered by a roof-like arrangement of cuticular cells. The structure of these epithelial cells causes the direction-dependent frictional behaviour of wool hair, which is the basis for the felting processes as well as the undesired shrinkage of woollen fabric in wash operations (Figures 2.22 and 2.23).

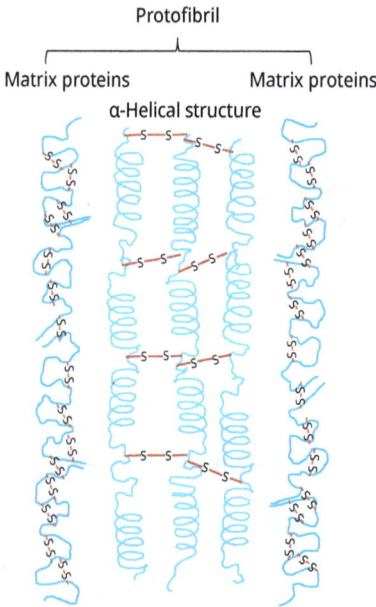

Figure 2.22: Structure of a protofibril – the secondary and tertiary structure of wool proteins.

Processes that intend to improve the dimensional stability of woollen goods in wash processes and to prevent undesired felting and shrinkage thus have to attack the roof-like tile structure of the cuticula and reduce directional friction effects. Technical strategies include localised oxidative attack and weakening to the wool surface without chemically damaging the cortex. In the second approach, the surface structure of the wool hair is covered by deposition of polymers.

Carbonisation: While soluble impurities, fat and soil can be removed through appropriate washing processes, cellulosic impurities from leaves, straw and burrs are

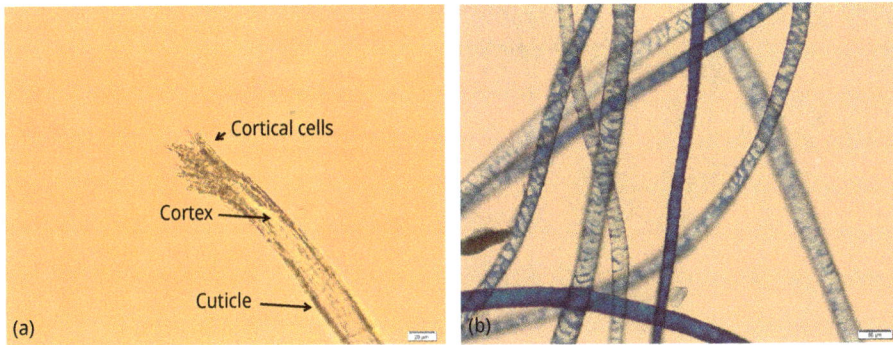

Figure 2.23: Wool structure: (a) cortical cells at the end of a broken wool fibre and (b) cuticle cells stained with methylene blue (cationic dye).

removed through hydrolytic degradation of the cellulose by treatment with sulphuric acid. The degraded parts are then removed mechanically and through washing.

2.8.4 Silk

In general, a number of protein-based fibres are collected under the term silk. The threads can be spun by caterpillars, spiders and mussels. Among the most relevant silk producers are the larvae of the silk moth. In this chapter, the silk fibres formed by the mulberry silk moth (*Bombyx mori* L.) are used as a representative model to discuss the structure and chemistry of silk. Besides mulberry silk which is reported to be domesticated since several thousands of years, also tussah silk (wild tussah) is a relevant class of silk.

The silkworm (larva) produces the cocoon as a protective housing for the pupa from which the silk moths finally emerge. The growth of the pupa has to be interrupted by heat and steam as the cocoon would be destroyed by the silk moth hatching out.

The silk fibre is produced as filament in spinning glands. Raw silk mulberry actually consists of two major protein components: Fibroin (approximately 75–83 wt%) and sericin (25–17 wt%). Two filaments of fibroin are embedded into a layer of silk gum (sericin) (Figure 2.24).

The fibre is highly oriented and the protein chains are arranged in pleated-sheet conformation (β-keratin), which then forms microfibrils and macrofibrils.

Three different proteins form the major components of fibroin from *Bombyx mori*: a high molecular mass H-chain, which is bound to the lower molecular mass L-chain through disulphide bonds and the smaller P25 protein.

The interlinkage between the proteins is mainly based on hydrogen bond attraction and salt bridges, and no covalent bridging through disulphide bonds exist. Thus,

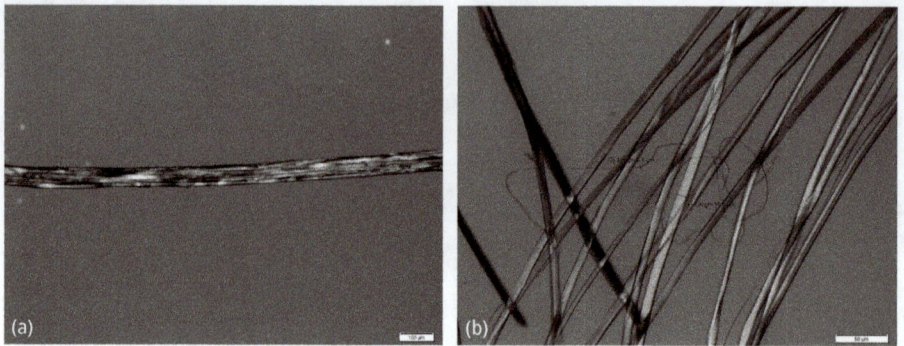

Figure 2.24: Raw silk (a, scale bar 100 μm) and degummed silk (b, scale bar 50 μm).

on the contrary to wool keratin, fibroin can be dissolved in concentrated salt solutions (e.g. $CaCl_2$/EtOH/water, $CaNO_3$/MeOH/water, LiBr and NaSCN). Also ionic liquids (e.g. 1-allyl-3-methylimidazolium chloride) can dissolve fibroin. Dissolution occurs without the formation of defined complexes between the metal salt solution and the fibroin through formation of an ion-rich hydration layer and interaction of ions with charged and highly polar groups of fibroin. Similar to the present understanding of silk formation in the spinning glands, the regeneration of fibroin can be initiated after reduction of ion concentration present in solution and application of high shear rates [18].

Dissolution of silk keratin and regeneration has been studied extensively for the production of composite materials for bioresorbable implants and scaffolds for medical applications. Also regeneration of fibres and membranes can be achieved from concentrated fibroin solutions.

In a similar approach, also other protein sources can be shaped into fibrous structures, for example, casein or soya protein.

2.8.5 Casein fibres

Both keratin and fibroin are proteins designed as structural proteins to provide hair with form and strength. This is a general difference to proteins collected from milk (casein) or soya (soya protein) which are sphero-proteins and thus present in coiled form. Enrolling to a more stretched linear form is achieved through shear forces applied during extraction and coagulation in fibre spinning. The usage of proteins from sources that serve as food must be assessed very critically; however, in Germany every year approximately 2 million tonnes of milk is not consistent with the regulations for delivery into the food market and thus have to be disposed [19]. In case of milk the content of 2.8–3.2 wt% of protein is of value for production of casein fibres. As a first step, casein has to be isolated from the milk by coagulation and filtration. Coagulation, for example, can be achieved by pH reduction through addition of acid or by addition of proteases (rennin, chymosin). The filtered casein then is dried and a powdery product is formed. The casein is then dissolved in caustic soda solution and the spinning dope becomes extruded into an acidic precipitation bath (e.g. sulphuric acid) [20]. To stabilise the coagulated protein structure, tanning agents and cross-linkers (e.g. formaldehyde and glutaraldehyde) are added.

Five amino acids represent the major components of casein (Table 2.9): Glutamic acid/glutamine (15.6%), proline (10.5%), leucine and isomers (9.4%), valine (7.2%) and aspartic acid/asparagines (6.7%). Chemical reaction with formaldehyde occurs mainly with amino groups, amide groups and hydroxyl groups present in side chains as well as with imino groups of the peptide bond, thus leading to cross-linking with methylene bridges. The hydrocarbon side chains of valine, leucine and proline will not contribute to chemical cross-linking operations.

There is a considerable similarity in the amino acid composition between casein and wool; however, the presence of disulphide linkages between the protein chains in wool explains the higher wet strength of wool and the need of cross-linkers to strengthen the regenerated casein fibre. Also metal salt mordanting with the use of aluminium, titanium or zircon salts can be used to increase the wet strength of the fibre.

2.9 Synthetic fibres

2.9.1 General

Traditionally, synthetic fibres were understood as fibres from polymers that were synthesised from monomers that had been obtained from petrochemical sources. Major concerns with regard to lack in sustainability led to introduction of alternative fibre polymers, which use monomers that were obtained from natural resources. Bio-based plastics derive from man-made or man-processed organic macromolecules collected from biological resources with application in plastics, films and fibre production [21]. The field of bio-based textile fibres developed substantially during the last decade and an enormous variety of new polymer concepts has appeared.

Synthetic fibres can be tailored in many properties and offer impressive mechanical properties and rather high durability. The rather compact fibre structure and the limited polarity of building blocks also define restrictions with regard to the dyeability of such fibres.

In general, water sorption of the synthetic polymers is low, which is an advantage with regard to stability of form, wet strength, speed of drying and high water repellency. However, due to low water sorption, wear comfort and capacity for regulation of wear climate are limited. The major synthetic fibres are electrical insulators, which also lead to the undesired appearance of electrostatic charging.

2.9.2 Polyamide

The characteristic sign of PAs is the presence of amide groups along the polymer chain. Two major groups of PAs exist: aliphatic and aromatic PAs.

In the class of aliphatic PAs, most frequently, a defined number of methylene groups are present between the amide linkages. As the correct IUPAC nomenclature of these polymers can become complicated, shorter names are in use. For aliphatic PAs, the "Nylons", usually the number(s) of C-atoms of the starting monomer unit(s) is mentioned in the name. For example (Figure 2.25):

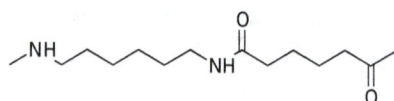

-[CO-(CH$_2$)$_4$-CO-NH-(CH$_2$)$_6$-NH]$_n$-
Polyamide 6,6 (Nylon)
Poly (hexamethylene adipamide)
IUPAC: Poly[imino(1,6-dioxohexamethylene) iminohexamethylene]

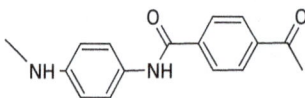

Poly-para-phenyleneterephthalamide (Kevlar®)
IUPAC: poly(imino-p-phenyleneimino= terephthaloyl)

Figure 2.25: Structure units and names of representative polyamides.

The major part of aliphatic PAs is processed into fibres with use of the melt-extrusion spinning. Such a technique requires that the melting point of the polymer lies substantially below the temperature for polymer decomposition.

When aromatic groups are present in the chain, the polymer chain becomes more rigid and an increase in the melting point of the fibre occurs. Thus spinning of aromatic PAs requires solution-based spinning techniques instead of spinning by melt extrusion.

Side chains disturb interactions between the polymer chains and thus lead to a decrease in the melting point.

The physical strength of the polymer fibre increases with DP; however, limitations exist with regard to maximum viscosity of the polymer melt to be processed in the extrusion process.

The melting point of aliphatic PAs decreases with the number of methylene groups between the amide groups. At the same time, the water sorption decreases. Similar to the behaviour of cellulose-tri-esters (Table 2.8), the properties of an aliphatic polymer (e.g. PE) become more important with increasing number of –CH$_2$– groups in the chain, while the influence of the PA groups on the fibre properties decreases (Table 2.11).

Table 2.11: Physical properties of polyamides as a function of methylene groups in polymer chain (water sorption at 65% r.h.). Schultze-Gebhart and Herlinger [4].

Polyamide	–CH$_2$– groups	Melting point (°C)	Water sorption (wt%)
4	3	265	9.1
6	5	223	4.3
8	7	200	1.7
10	9	188	1.4
12	11	179	1.3

The amide groups lead to strong hydrogen bond interactions between adjacent chains. These interactions are intensified further through orientation of the polymer chains along the fibre axis as a result of the drawing during fibre spinning.

2.9.3 Aminocarboxylic acid-based PAs

The general synthesis of PAs can follow two chemical principles:
1. Polycondensation of an ω-aminocarboxylic acids with more than five carbon atoms
2. Hydrolytic or anionic polymerisation of lactams

An important aspect for successful formation of longer PA chains is that the formation of cyclic products with low molecular weight is chemically not favoured in comparison to the polymer formation.

Examples for PAs

Nylon 1   -CO-NH-CO-NH-CO-   "Polyuret"

Formally, this substance would belong to the PAs; however, the properties differ from the other PAs, being more a polyurea, than a PA.

Nylon 2   H-[CH$_2$-CO-NH]$_n$-H   "Polyglycine"

Nylon 2 would be of interest as a polymer that had been constructed from a single amino acid (glycine) and thus directly shows the chemical relations between PAs and protein fibres. While being of interest as fibre polymer, the tendency of glycine to favour formation of cyclic products makes the synthesis of high-molecular-weight polymers expensive.

Nylon 6 (polycaprolactam, poly(hexano-6-lactam), Perlon) IUPAC poly(hexano-6-lactam.

The major route for synthesis of Nylon 6 is shown in Figure 2.26, starting from cyclohexanone.

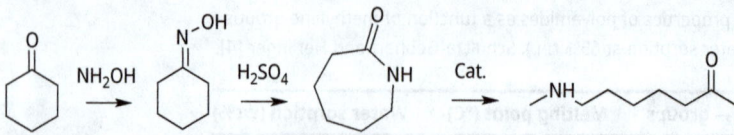

Figure 2.26: Synthesis of polyamide 6 (nylon 6).

The synthesis of PA 6 is based on the polymerisation of ε-amino-caprolactam in the presence of small amounts of water or a catalyst. The polymerisation of the caprolactam at 250 °C ends up in an equilibrium mixture, which consists approximately of 89% of the desired linear PA, 8.5% of un-reacted caprolactam and 2.5% larger ring amides. The concentration of these two minor components has to be lowered by washing of the extruded polymer granules in boiling water, followed by drying in the vacuum.

A particular advantage of PA 6 lies in the fact that the polymer chain has been built from a single building block. This makes recycling through depolymerisation and regeneration of the caprolactam more easy than in cases where two chemicals were combined in the polymer, e.g. PA 6,6, which is a product of hexamethylene di-amine and adipic acid.

When the polymerisation is initiated by anionic catalysts, the lower reaction temperature permits in situ polymerisation of PA 6, which is of particular interest for use as matrix for fibre-reinforced composites in the resin transfer moulding technique. The low viscosity of the ε-amino-caprolactam allows deep penetration of the textile pre-form, for example, made from carbon fibres. Polymerisation to PA 6 then can be initiated at temperatures near 150 °C.

Nylon 11 (Rilsan, poly-11-aminoundecanoic acid): The production of nylon 11 is based on castor oil as a natural raw material (Figure 2.27).

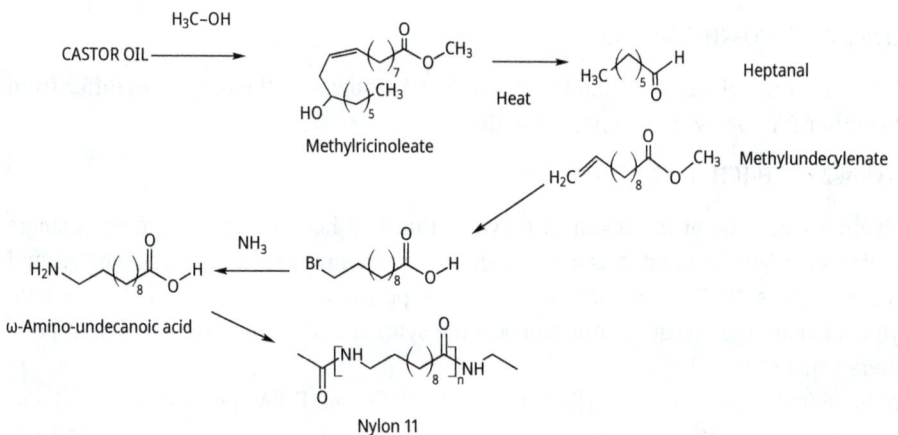

Figure 2.27: Production scheme for nylon 11.

After transesterification of the castor oil into methyl-ricinoleate, the fatty acid is split at the double bond to deliver *n*-heptaldehyde and undecylenic acid methyl ester. The C11 component is then processed further to the ω-aminoundecanoic acid, which is the raw material for polycondensation to PA 11.

PA 11 is more expensive compared to the standard PAs (PA6 and PA6,6), however, is a representative for the emerging group of biobased polymers.

2.9.4 Diamine–dicarboxylic acid-based PAs

These PAs mainly are produced through polycondensation of a diamino component with a dicarbonic acid. The numbers behind the generic name indicate the number of carbon atoms of the diamino component and of the dicarboxylic acid, respectively. The major representative is PA 6,6 (nylon).

Due to the straightforward synthesis and the good material properties, PA 6,6 is one of the favourable PAs with regard to raw material costs and performance (Figure 2.28).

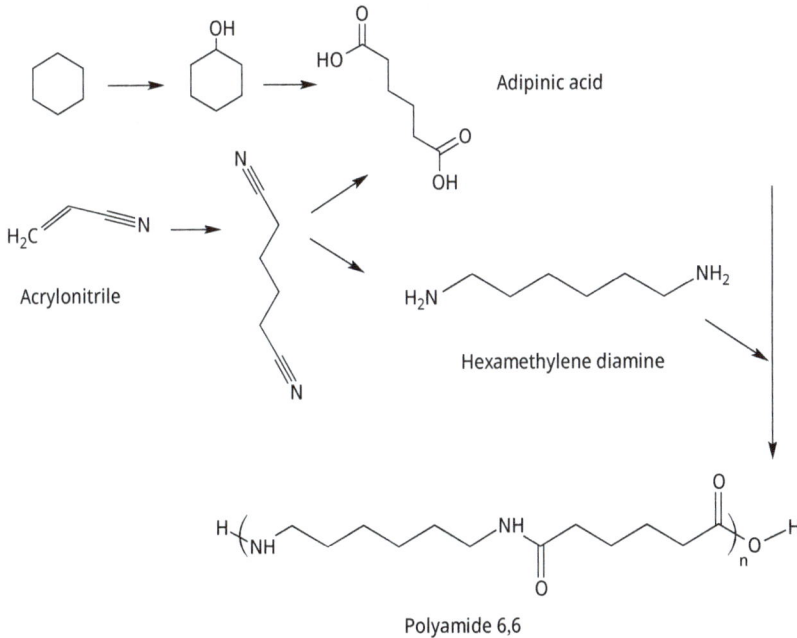

Figure 2.28: Synthesis of polyamide 6,6.

Polycondensation is initiated from aqueous solution of AS-salt (adipinic acid) and HMDA (hexamethylenetetramine) in a pressurised autoclave, then followed by reduction of pressure and polycondensation under vacuum.

The purity of the starting materials is of high relevance to avoid uncontrolled chain growth or increased tendency to form coloured products (yellowing). During polycondensation, monofunctional and bifunctional components (e.g. primary and secondary aliphatic amines, diamines and carboxylic acids) are added to control chain growth and thus influence the melt viscosity. As the polymer is sensitive to oxygen, the processing steps at elevated temperature have to be carried out under protective atmosphere (<0.001% oxygen). The polymer also has to be dried carefully as presence of traces of water already initiate chain hydrolysis reactions and reduction in DP.

Fibres are shaped with the use of the melt extrusion process (melt spinning). Representative conditions are as follows: The polymer granulate is molten in an extruder at temperatures near 300 °C, pressed through a spinneret with bores of 200–400 µm and drawn away, thus reducing the fibre diameter by a factor of 5–50.

Textile fibres are cooled on air, while technical fibres with a diameter above 30 dtex are cooled in liquid (e.g. water and solvent).

During the step of fibre extrusion, additives can be introduced into the polymer melt. Examples are:

- Delustering agents such as TiO_2 (anatase, 0.03–2 wt%) and $BaSO_4$ are added to increase light scattering and reduce gloss of fibres. The addition of inorganic components contributes to undesirable abrasive effects observed during yarn processing, for example, on thread-guiding elements and breaks of textile machinery.
- Additives for light protection and to improve ageing, for example, manganese(II) hypophosphite $Mn(H_2PO_2)_2$.
- Additives are also added to improve heat resistance, for example, 0.1–1 wt% of antioxidants and radical scavengers such as sterically hindered phenols, 3,5-dialkyl-4-hydroxyphenylcarboxylic acid, Cu(I)iodide (5×10^{-4} wt%) can be incorporated.
- Spun dyeing/mass colouration of fibres and optical brightening: Optical brightening and colour also can be introduced in the stage of fibre formation. Dyestuff dispersions as well as carbon black are added in a concentrated form to a polymer melt to form the so-called colour master batches, which then are added to the colourless polymer melt to obtain the desired final colour depth.

Properties in brief (main representatives: PA 6 and PA 6,6):

Compared to other textile fibres, the mechanical properties of PA fibres used for textiles are outstanding in tenacity, elongation and abrasion resistance. The fibres can be heat set and through the high polarity dyeing is comparably easy (compare Chapter 10).

Compared to other synthetic fibres, the moisture uptake is relatively high with 3–4 wt% at standard conditions (20 °C, 65% r.h.), which explains the better wear comfort and the wider use of PA in baselayer applications.

A major disadvantage of PAs used today is its limited stability against light and weather, which reduces lifetime in outdoor applications. This is of relevance for ropes and slings which, for example, are in use for climbing and are highly safety-relevant items.

The biological stability against degrading microorganisms (bacteria and fungi) and insects (e.g. moths) is good; however, this also includes a very limited biodegradability.

Recycling of PA fibres is possible; however, polymer ageing and presence of incorporated additives has to be considered.

2.9.5 PES fibres

PES fibres contain an ester group as linkage between the hydrocarbon or aromatic chain segments of the polymer.

The most important representative is polyethylene terephthalate (PET) poly(oxy-1,2-ethanediyloxycarbonyl-1,4,-phenylenecarbonyl).

The PET fibres hold the biggest share in production volume among all textile fibres. Besides a straightforward and thus cost-effective synthesis from petrochemical raw materials (ethylene and toluene), the fibre can be obtained by highly efficient melt-extrusion spinning (melt spinning) and also offers highly valuable textile fibre characteristics (Figure 2.29).

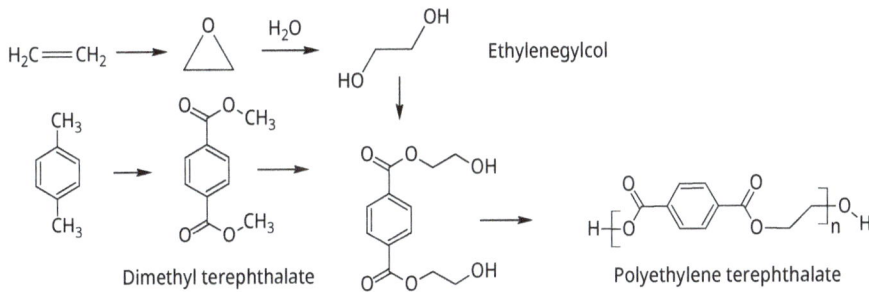

Figure 2.29: Synthesis of polyethylene terephthalate.

The primary intermediate products are dimethyl terephthalate (DMT) and ethylene glycol (1,2-dihydroxyethane). In the first transesterification step, methanol in DMT is replaced by glycol molecules. Through removal of the liberated methanol by distillation the equilibrium is shifted towards the formation of bis-glycolterephthalate. As the next step, catalytic polycondensation is initiated at elevated temperature and the ethylene glycol which results as product is removed by vacuum distillation. Examples for catalysts used in polycondensation are antimony(II)oxide Sb_2O_3 or other oxides of amphoteric materials.

The polymer then is transferred either directly to the melt spinning unit or extruded into highly amorphous chips and dried.

The DP in PET is in the dimension of 80–150 formula units.

The polycondensation reaction follows statistical rules; thus, a considerable amount of low-molecular-weight oligomers still is present in the polymer mass. Representative values are up to 2 wt% of cyclic trimers and linear oligomers. During textile operations using hot water, for example, high-temperature dyeing, these low-molecular-weight substances are released into the process bath. The presence of PES oligomers requires attention as deposition at sites of lower temperature (valves) and on the dyed material (deposits) can cause technical problems as well as negative influence on appearance of the dyed material (uneven dyeing, lowered fastness).

Molecular interactions between the chains are based on dipole–dipole interactions between carbonyl groups of the ester bond and on π-bond interactions between the aromatic rings. In the monoclinic unit cell of the lattice of PET, the aromatic rings are oriented parallel, which permits strong π-electron interaction. This explains the rather short-chain length of the polymer which is already sufficient to yield high fibre strength.

For comparison of the DP with other polymers (e.g. cellulose), it has to be considered that one formula unit in ethylene-terephthalate is more than twice as long as an anhydroglucose unit. Thus, a simple comparison of DP values between two different polymers is not an appropriate way to compare polymers.

The end groups of a PET molecule can be either hydroxyl or carboxylic groups. Average values for the carboxylic group content of PET are in the dimension of 10–40 mmol/kg of polymer, which is similar to the carboxylic group content of cellulose. The presence of acidic groups explains the tendency of an acid-catalysed ester hydrolysis at elevated temperatures (e.g. 270 °C). Such temperatures, for example, are applied during fibre extrusion. Thus moisture content of the raw PES polymer has to be reduced to less than 0.01 wt% to avoid acid-catalysed reduction in DP (Figure 2.30).

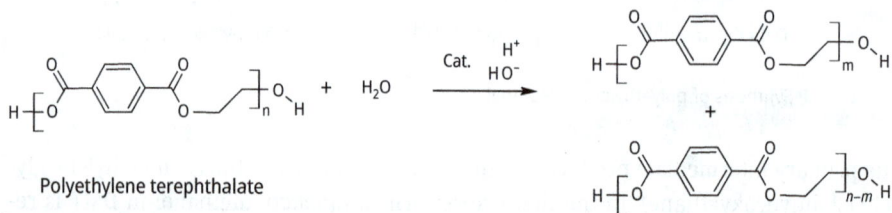

Polyethylene terephthalate

Figure 2.30: Hydrolysis of polyethylene terephthalate.

Fibre spinning of PET fibres is performed at temperatures above the melting point which is below the decomposition temperature. A typical value for melting range of PET is between 255 and 265 °C. The melting range depends on crystallinity and crystal size of the polymer.

After fibre extrusion through the spinneret, the fibres are drawn in length, for example, by a factor of 2. During this operation fibre properties change substantially:

- Longitudinal orientation of the polymer in direction of the fibre increases
- Crystallinity increases with applied stress
- Tenacity increases and elongation decreases

At high spinning speed, a crystallinity of fibres of up to 50% can be achieved. Based on the degree of orientation, one can distinguish between highly oriented yarns, partially oriented yarns and low-oriented yarns.

Highly oriented fibres exhibit a fibrillar structure and a high degree of crystallinity.

The relaxation, orientation and crystallisation processes during fibre spinning are determining factors that influence later dyeability of a fibre; thus, every change in process conditions (e.g. temperature profile during fibre drawing, draw rate, rate of cooling and fibre fineness) will influence both mechanical properties and textile chemical characteristics.

Fibres also can be drawn later in a separate process at a temperature above T_g (e.g. 80 °C), which increases orientation and order in the molecular network. Highly oriented fibres, which had been spun at high production speed (e.g. 5,000 m/min), even start to lose orientation at temperatures above T_g.

With regard to molecular order, the drawn fibres are present in a frozen state, where amorphous and more flexible parts of the polymer chain remain immobilised. When T_g is reached, fibre reorganisation occurs in these regions, orientation lowers and crystallinity in ordered regions increases. At the same time the fibre starts to shrink.

To avoid later shrinkage of textiles, for example, during washing procedures, heat setting of the PES fibres thus has to be achieved at temperatures between 110 and 180 °C. Maximum stabilisation is obtained during shrinkage without dimensional restraints; however, in practice tension is applied to a fabric, to avoid completely uncontrolled dimensional changes and to achieve stabilisation of a certain fabric width.

With increasing crystallinity, the dye uptake in a fibre decreases, while the reduction in orientation of the amorphous parts contributes to an increased dyestuff uptake.

Maximum dyeability is achieved with setting temperatures between 150 and 180 °C.

Heat setting also reduces tendency to develop creases in a fabric, for example, during washing and hot ironing.

Textile properties: The PES fibre is robust and exhibits high tenacity which results in good mechanical performance. The high strength of the fibres and the rather low frictional coefficient in yarn, however, leads to undesired pilling effects. As a result of surface friction, single fibre ends are pulled out of the fabric structure and become entangled to form small fibre spheres at the surface of the material. The pills remain attached to the surface through the bound fibre end. The presence of fibre pills at the fabric surface leads to an undesirable appearance.

Two strategies were applied to control pilling: During abrasion, higher tension is applied to the entangled fibre pills; thus, controlled reduction of fibre strength was used to facilitate release of pills during abrasive forces. This leads to a certain "cleaning" of the surface and formation of a dynamic equilibrium between pill growth and release. Use of finishing chemicals with the intention to increase fibre–fibre friction also helped to reduce the tendency of fibre ends to become pulled out of the fabric surface.

Due to the absence of charged functional groups or a significant number of hydroxyl or amino groups in the polymer, the dyeing of PES is performed with disperse dyes. For disperse dyeing, the glass transition temperature T_g of the fibre polymer has to be exceeded. In case of PET, T_g is near to 70 °C. T_g depends on the level of molecular order. While isotropic amorphous PET exhibits a T_g of 68 °C, highly oriented fibres exhibit a T_g of up to 100 °C.

The PES fibres offers high fibre volume; thus, besides usual textile applications, low-density PES fleece also serves as thermal insulation layer (e.g. in outdoor jackets and sleeping bags).

Particularly high fibre volume can be achieved by texturing and crimping. In this process, highly curled fibres are obtained from straight fibres through a combination of thermal and mechanical treatment.

Through its rather non-polar character the moisture uptake of PES fibres is comparably low, for example, 0.4% water (20 °C, 65% r.h.). Thus, capacity for sweat absorption is low, which also influences the capability for climate regulation negatively.

Water retention is low as the fibre swelling in water is minimal, which helps to explain positive effects like high crease recovery after washing (non-iron properties), rapid drying and application in high-performance sports where high sweat volumes have to be handled and transported.

Similar to PA the low electrical conductivity leads to formation of static charge, which influences wear properties as well as soil adsorption behaviour negatively.

Fibre modification through addition of TiO_2 (anatase sometimes the harder rutile) to dull the fibres, integration of pigments, dyes and carbon is similar to the techniques for PA fibres.

PET fibres exhibit good resistance against many chemicals and solvents, and high stability against ageing under UV light.

Texturing: For many applications, the straight spun fibres do not develop sufficient volume, and artificial crimping has to be done to provide the material with a higher volume. The general principle of all methods for crimping utilises the thermoplasticity of the fibres. By thermal fixation of highly twisted or bent fibres a crimped state can be stabilised. The application of the turns can be achieved by a number of different processes, for example, using rotating elements (false twist texturing), friction (friction false texturing), air-jet, gears, in a stuffer box or by knit-de-knit techniques (Figure 2.31).

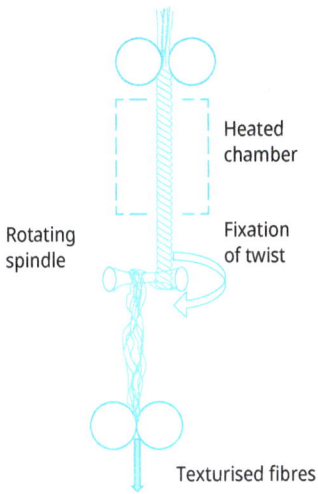

Figure 2.31: Texturising of thermoplastic fibres.

2.9.6 Polylactic acid and polyhydroxybutyrate

The need to introduce a higher amount of bio-based and biodegradable polymers into commercial applications led to a growing interest in renewable sources for production of PES fibres. A considerable number of different natural raw materials and polymers have been proposed as a potential material to substitute classical synthetic fibres by renewable and non-petrol-based resources.

Two representative examples polylactic acid and poly(3-hydroxybutyrate) P(3HB) will be discussed in more detail (Figure 2.32).

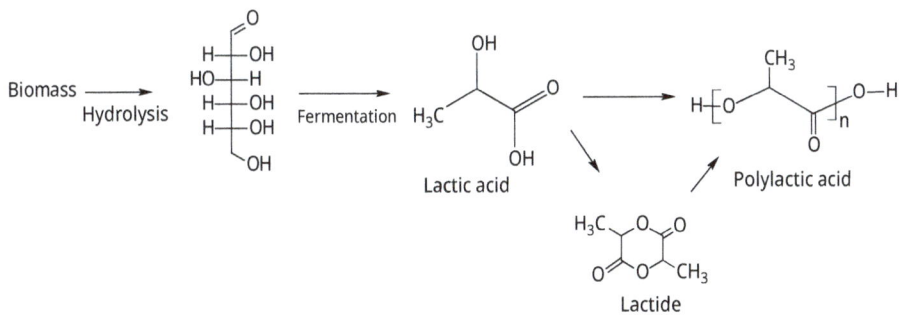

Figure 2.32: Synthesis of polylactic acid (PLA).

The production of polylactic acid starts with glucose which is then converted into lactic acid by a microbial process. Lactic acid can be present in two stereoisomeric forms D- and L-lactic acid. The polymer formed only with pure L-form is poly(L-lactic acid), denoted as PLLA, while the polymer formed by the corresponding D- enantiomer will be PDLA. Based on the combination of D- and L-forms, the polymer properties vary substantially as molecular order in the solid material is dependent on the optimum arrangement of the different conformational isomers. In Table 2.12, different PLA-based polymers and their respective melting points are shown.

Table 2.12: Monomer order and melting point of different forms of PLA. From Shen et al. [21].

Name	Abbreviation	Constitution	$T_m(°C)$
Stereocomplex PLA	scPLA	-L-L-L-L-L-L-L-L-	230
		-D-D-D-D-D-D-D-	
Stereo-block PLA	PLLA	-L-L-L-L-D-D-D-D-	200
L-Isomer (pure)	PLA	-L-L-L-L-L-L-L-L-	180
		-L-L-L-D-L-L-L-L-	130
Amorphous PLA	A-PLA	-L-L-D-L-L-D-D-D-	No melting point
		-L-D-D-D-L-L-D-L-	
		D-L-L-L-D-D-D-L-	

The stereocomplex PLA (scPLA) shows highest crystallinity and melting point, which can be explained with an energetically favourable arrangement of polymer chains in the solid polymer. The other polymer blends and copolymers with different composition and constitution of L- and D-forms permit only less ordered solid states and result in a lower melting point.

Raw materials are obtained from fermentation processes of various sources such as carbohydrates, alcohols, alkanes, organic acids (lactic acid from dairy industry), fats and oils (plant and animal wastes).

P3HB (R = methyl, Figure 2.33) exhibits poor properties for direct production of textile fibres (T_m = 180 °C, T_g = 4 °C, tensile strength 40 MPa); thus, copolymers with other hydroxyalkanoates are used to optimise fibre properties.

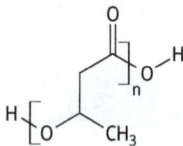

Figure 2.33: Polyhydroxybutyrate (P3HB).

2.9.7 Polyurethanes/elastomer fibres

Polyurethane fibres contain a urethane group R–NH-CO-O-R in their polymer backbone. The synthesis of polyurethanes usually follows the addition of an alcohol component to an organic isocyanate group (eq. (2.23)):

$$R-OH + R'-N = C = O \rightleftharpoons R'-NH-CO-O-R \qquad (2.23)$$

For the synthesis of a polymer, bifunctional reagents are used, usually one component is a diisocyanate and the other a diol. Polyurethane fibres were originally developed to compete with PAs; however, this application is no more relevant. The major application of polyurethane-based fibres is for production of elastomers (e.g. Lycra®, Spandex® and Dorlastan®). Representative technical data are given in Table 2.13. The major characteristic of these fibres is their high elongation with very high elasticity. Thus, nowadays considerable amounts of elastomeric fibres are used in textile fabrics where high elasticity is requested.

Table 2.13: Technical properties of polyurethane elastomeric fibres. From Schultze-Gebhart and Herlinger [4].

Property		Unit
Elongation	400–800	%
Tenacity	0.45–1.2	cN/dtex

Polyurethane-based elastomeric fibres are block-co-polymerisates. A diisocyanate component, often 4,4'-methylene-bis(phenyl isocyanate) MDI, reacts with a pre-polymer, which contains hydroxyl groups as end groups (polyol). As a pre-polymer either a PES- or a polyether-based polyol is used (Figure 2.34).

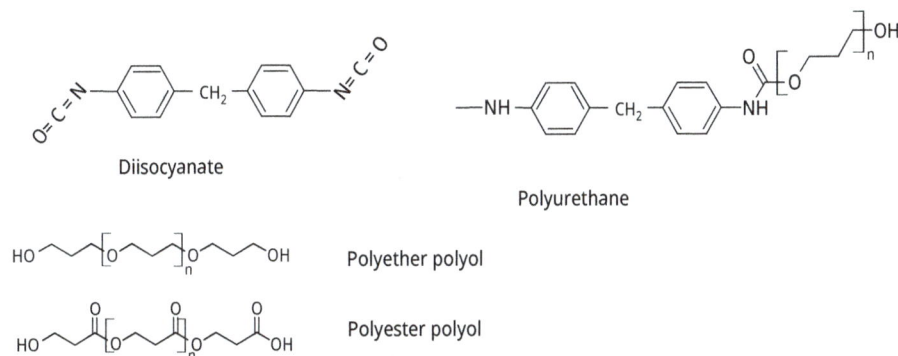

Figure 2.34: Representative polyurethane forming components: a diisocyanate and polyols.

The hydroxyl end groups of the polyol react with the isocyanate groups and thus form the polyurethane. Modification of polymer chain length can be achieved through addition of chain extenders such as short aliphatic diamines (e.g. 1,2 diaminoethane).

Different interactions between the urethane and the polyether/PES polymer blocks lead to a particular structure. The stronger hydrogen bond interaction between urethane elements initiates the formation and organisation of ordered "hard" segments. The attractive forces between the polyether/PES segments are smaller; thus, organisation of these segments is determined by the structure of the "hard" segments. As a result, a two-phase structure forms in the polymer, which is a combination of "hard" and "soft" segments. In the "soft" segments, the polyether/PES segments are present in the partially coiled state. Under load, the coiled segments uncoil with the linear extension of the fibre. The load elongation curve shows a distinct increase in stress resistance at the point of strain, where the "soft" segments have reached their maximum extensibility, and irreversible fibre deformation begins.

When the load is lowered before this point has been reached, the flexibility of the polyether/PES segments permits complete recovery of the original dimensions.

The principle of an elastomeric fibre is shown in Figure 2.35.

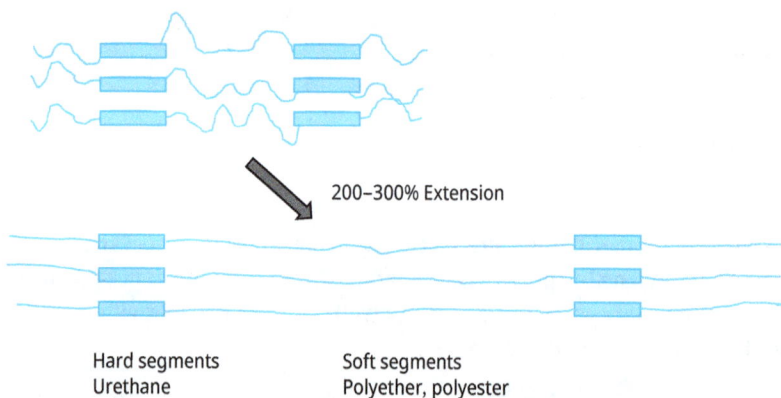

200–300% Extension

Hard segments Soft segments
Urethane Polyether, polyester

Figure 2.35: Soft and hard segments of polyurethane fibres, shown in relaxed form and under tension. The soft segments become expanded.

Representative examples for the PES-based soft segments are PESs of adipic acid with poly(tetramethylene ether glycol) as the diol component. From the point of polymer chemistry, these block polymer fibres can be understood as a polymer blend with two separated phases: a polyether/PES phase and a urethane phase, which are however linked together by chemical bonds.

Each of these polymer phases exhibits individual thermal characteristics in terms of T_g and T_m. The T_g of the elastic soft segment is in the range of 210–250 K (−73 to −23 °C). Below this range of temperature, the flexibility of the soft segments disappears and the elasticity of the material disappears. The upper temperature for thermal processing in textile operations is in the range of 150–170 °C; above this temperature, substantial thermal decomposition of the polyurethane becomes relevant.

For application in textile materials for garment, this range of temperature is acceptable. For technical applications, the temperature range of a chosen elastomeric material has to be considered with care, as for example an elastomeric sealing will lose its elasticity when temperatures fall below T_g of the elastic soft segments.

Polymer production: As a first step, a pre-polymer containing isocyanate end groups is synthesised by the reaction of an α,ω-OH polymer, the so-called macrodiol with an aromatic diisocyanate (e.g. MDI), to obtain the α,ω-NCO-functional pre-polymer. The reaction is carried out in a polar solvent at 50–80 °C and leads to formation of a pre-polymer with a typical molecular mass of M = 4,000–8,000 g/mol. In the next step, the chain extension (polymerisation) is started by the addition of a diamine to the macro-diisocyanate.

When trifunctional components are introduced to the reaction, cross-linked polyurethanes are obtained.

The polyurethane cannot be melted; thus, fibre formation has to be achieved by using solvent spinning from DMAc (*N,N*-dimethylacetamide) or DMF (*N,N*-dimethylformamide) solution. The spinning process can be performed as a dry spinning process, where the solvent is removed by evaporation at elevated temperature and is then recycled.

In wet spinning, the spinning dope is pressed into coagulation bath, which contains a non-solvent. The initially formed single filaments are rather sticky and cement to a multifilament fibre. Thus, many polyurethane elastomeric fibres exhibit a characteristic shape and cross section as these fibres consist of a bundle of filaments that stick together to form the final fibre (Figure 2.36).

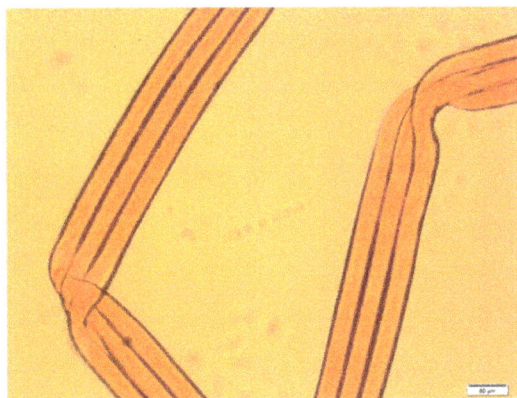

Figure 2.36: Characteristic shape of fibre bundles in a polyurethane fibre.

Reaction spinning is a variation of the wet spinning process. After extrusion of the spinning dope into the precipitation bath, further chain extension and cross-linking is initiated. This technique allows the production of cross-linked fibres, while the cross-linked polymer cannot be processed anymore into fibres.

The freshly spun polyurethane fibres are slightly tacky; thus, preparations (e.g. silicone oil and metal soaps) are added as lubricants and separating agents. The addition of such components improves processability of the highly elastic material remarkably; however, in case of incomplete removal during pre-wash operations substantial problems in aqueous dyeing operations occur.

General properties: Compared to rubber-based elastomeric fibres, PUR elastomers exhibit higher chemical stability. In case of rubber the disulphide/polysulphide bridges between the polyisoprene chains and residual double bonds are sensitive to oxidative attack. The polyether, PES and urethane groups exhibit lower reactivity; thus, degradation occurs only in rather concentrated acidic media (e.g. 10 wt% HCl). Block polymers that contain PES-based soft segments are more sensitive to hydrolytic attack than elastomer fibres that contain on polyether segments. This is an indication that hydrolytic attack does not start at the urethane segments, which are packed tightly in ordered segments, but occurs faster at the low-ordered flexible elements, which also exhibit higher accessibility for chemicals.

The absence of accessible polar functional groups makes dyestuff sorption and binding difficult; thus, polyurethane fibres are difficult to dye in textile chemical processes. In particular for dark-coloured fabrics with uniform surface structure, the appearance of undyed white elastomeric fibres is unwanted. Then staining of the polyurethane fibres by the colourants used to dye the major fibre component in the textile (e.g. PA 6,6) is attempted, or core-spun yarns are used to cover the elastomeric fibres with better dyeable fibres.

Oxidative chemicals (e.g. chlorine and NO_x) and intensive exposure to light induce fibre damage, which lead to strength loss and colour change (yellowing). Polyether-based elastomeric fibres exhibit higher strength loss in the presence of diluted aqueous solutions of chlorine (e.g. 20 mg/L Cl_2, 20 °C, 24 h). Such conditions are representative for tests that simulate chlorine content in indoor swimming pools [22].

The sensitivity for photo oxidative damage is increased by the presence of unsaturated fatty acids. Light stabilisation can be achieved through addition of benzotriazoles and phosphites to the spin dope.

2.9.8 Polyolefin fibres, PE and PP

Polyolefins can be synthesised through polymerisation of the respective unsaturated monomer (eqs. (2.24), (2.25)). The polymerisation reaction can be started through radical forming initiators or by means of catalysators (Ziegler–Natta system, $TiCl_4/AlR_3$):

$$nCH_2CH_2 \rightleftharpoons -(-CH_2-CH_2-)_n-LDPE, HDPE \tag{2.24}$$

$$n\mathrm{CH_2CHCH_3} \rightleftharpoons -(\mathrm{CH_2-CHCH_3-})_n - \text{atactic, isotacticPP} \tag{2.25}$$

The mechanism of polymerisation determines the properties of the polymer material.

PE fibres: In case of PE, the catalytic mechanism leads to more uniform DP and thus mechanical properties are on higher level, compared to the product obtained by radical polymerisation. The catalytic polymerisation yields HDPE (high-density PE) while through radical polymerisation low-density PE is obtained. In case of PE, the low melting point, the waxy character and the limited textile character restrict the wider use of this material for textile applications. Through its low price and the efficient processing by melt extrusion, fibres and split fibre bands have found wide application in more technical uses, for example, for packaging, woven bags, films and composites. The share of HDPE in textile applications of polyolefins is in the dimension of 10%, while PP fibres are predominant with a share of 90%.

PP fibres: The production of PP through catalytic polymerisation delivers the isotactic polymer. The structure of PP contains a chiral centre at the CH– group. The configuration of this optically active centre is of decisive relevance for the physical properties of the material. The catalytic synthesis defines the conformational structure of the propylene formed during catalytic chain propagation and the isotactic configuration results. In the product less than 5% of the polymer is present in atactic configuration. In the isotactic polymer the methyl side groups all are oriented on the same side of the polymer chain, which permits more densely packaging of the polymers and increases the relatively weak interactions between the apolar hydrocarbon chains. A high molecular weight of the polymer is required to develop sufficient interactive forces (Van der Waals forces) between adjacent molecules, which is a condition to obtain satisfactory fibre strength for textile applications.

Representative values for the average molecular mass are between 170,000 and 300,000 g/mol, which corresponds to a DP P_w of 4,000–7,500.

Fibres are produced by melt extrusion spinning at spinning temperatures of 220–300 °C (Figure 2.37). The molten polymer exhibits non-Newtonian behaviour and the melt viscosity decreases with shear stress. Strong pseudoplastic behaviour is observed, which means that an elastic component leads to considerable expansion of the polymer at the end of the bore in the spinneret (die swelling). The mechanical properties of the fibres depend on the spinning conditions, for example, spinning speed, draw ratio (1:3–1:5) and cooling rate.

Crystallinity of fibres is usually between 40% and 50%. Processing of spun fibres includes crimping and thermal setting (100–130 °C), which also defines the maximum temperature applicable in later stages before thermal shrinkage, and relaxation will begin to appear.

General properties: The absence of polar groups in the polymer chain makes both PE and PP resistant to most chemical agents (acids and bases). The PP fibre is highly hydrophobic with a surface energy (surface tension) of 28–30 mN/m and the water absorption of PP is extremely low with less than 0.5 wt%.

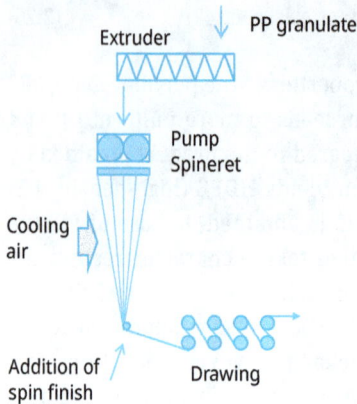

Figure 2.37: Melt extrusion process for PP-fibre production.

The surface energy of PP, human skin, sweat water and oil is summarised in Table 2.14. This explains the wider use of PP fibres as a baselayer in sports textiles, which is designed to hold only a minimum of sweat in the fabric structure and thus supports rapid drying of the baselayer. The rapid release of excess sweat into other layers of fabric is intended to minimise undesirable sensations of wetness and chill during a phase of lower physical activity following a period of intensive exercise.

Table 2.14: Surface energy of PP and relevant liquids. From Agache and Humber [23].

Substance	Surface energy(mN/m)
Polypropylene	28–30
Water	73
Skin (front arm)	>28
Sweat	32
Olive oil	32
Polyester	43
Polyamide 6	45
Polyacrylonitrile	46

The results given in Table 2.14 demonstrate the hydrophobic character of PP. Remarkably, the surface energy of sweat and also of skin is comparable to values of plant oil as well as to PP. This is a prerequisite for the controlled transportation of sweat in PP-fibres. The hydrophobicity of PP fibres also explains their use as surface layer in nonwovens for hygiene applications e.g. disposable diapers.

As a disadvantage the polymers show limited resistance to thermo-oxidative and photo-oxidative degradation.

Stabilisation is required for uses where light exposition at elevated temperature has to be considered e.g. requirements of hot light fastness for car interior. As a general strategy additives are incorporated to stop radical reactions and absorb UV irradiation. Representative examples for additives are:

- free radical scavengers and chain terminators to form non-reactive radicals (phenolic antioxidants), hindered amine light stabilisers;
- hydroperoxide deactivators, which decompose peroxide without radical formation (disulphides and trialkyl phosphites);
- UV absorbers (hydroxybenzophenone and benzotriazole);
- quenchers that quench excited states of ketone groups (Ni-chelate-based stabilisers); and
- antiacids to neutralise any formed acidic products (e.g. Ca-stearate).

The absence of polar groups leads to severe restrictions in dyeability from aqueous solution; thus, mass dyeing through addition of pigments and dyes during melt extrusion and fibre formation is the major technique to produce coloured PP or PE fibres. In mass colouration, colour master batches containing high concentration of colorant are mixed to the colourless PP granules. Examples for pigments are inorganic substances such as TiO_2 (rutile and anatase form), iron oxides, carbon black, and examples for organic pigments are insoluble azo pigments, copper phthalocyanines, vat pigments and carbazole dioxyazine pigments (Figure 2.38).

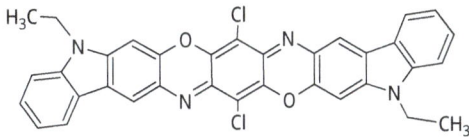

Figure 2.38: Pigment Violet 23.

Improvement of dyeability can be achieved through formation of copolymers with maleic acid, through incorporation of dye-adsorbing compounds and through surface modification, for example, by deposition of adsorbent layers or plasma modification.

The low density of 0.91 g/cm^3 makes the fibres of interest for maritime applications as the products will remain swimming on the water surface. Monofilaments with a diameter of 0.1–3 mm are used to produce fishing nets, packaging nets, sieves, filter fabrics and bristles.

PP fibres exhibit favourable mechanical properties (e.g. abrasion resistance, tensile strength in dry and wet state and high dimensional stability). Due to the low water uptake, wet fabrics dry rapidly. The hydrocarbon-based chemical structure explains the good resistance to microorganisms and insects, which, on the other hand, leads to a low biodegradability in the environment.

The thermoplastic behaviour of the fibres permits thermal bonding in garment/textile processing, however, causes difficulties when ironing of the fabric is attempted.

A distinct disadvantage arises from the flammability of the material, which also releases substantial amounts of heat due to a purely hydrocarbon character.

The high frictional coefficient causes difficulties in staple fibre spinning, thus spinning lubricants have to be added to reduce friction between fibres.

The similar surface tension between fatty substances makes the removal of oil-based soil difficult.

2.9.9 Polyacrylonitrile fibres

Fibres that contain at least 85 wt% acrylonitrile monomer in the polymer chain are called polyacrylonitrile fibres, and modacryl fibres exhibit an acrylonitrile content of 50–85 wt%. The major part of fibres is synthesised as ter-polymer; thus, the material is a copolymer that contains three different monomers (e.g. 89–95 wt% of acrylonitrile, 4–10 wt% of non-ionogenic co-monomer and 0.5–1% of ionogenic co-monomer). Often the ionogenic monomer bears a sulpho ($-SO_3H$) or sulphonate ($-SO_3Na$) groups, which impart cation-exchange properties to the fibre. This modification is a prerequisite to make the PAC fibres dyeable with cationic dyes.

The pure acrylonitrile polymer cannot be dyed in conventional textile dyeing operations; thus, modification of the polymer is necessary. Fibre types present in the market represent a compromise between improved dyeing properties through synthesis of copolymerisates and loss of valuable characteristics of the polyacrylate fibre through the presence of other monomers.

A ter-polymer is formed by the combination of

$$-[-CH_2-CHCN-CH_2-CHCOOCH_3-CH_2-C(CH_3)(CH_2SO_3Na)-]-$$

1. Monomer: Acrylonitrile

$$CH_2 = CH-CN$$

2. Monomer: Methyl esters of acrylic or methacrylic acid

$$CH_2 = CH-COOCH_3$$

$$CH_2 = C(CH_3)-COOCH_3$$

$$CH_2 = CH-OCOCH_3$$

$$CH_2 = CH-CONH_2$$

3. Monomer methallylic sulphonic acid $CH_2 = C(CH_3) - CH_2SO_3Na$

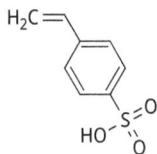

Vinylstyrene sulphonic acid

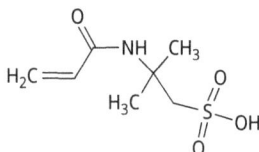

2-Methyl-2-sulphopropyl acrylamide

The polymerisation/co-polymerisation reaction is an anion- or radical-initiated chain reaction. Equal amounts of isotactic and syndiotactic sequences are obtained.

Due to the high melting point, decomposition of the polymer begins already at a temperature below the melting temperature T_m. Thus, polar solvents (e.g. DMF, DMAc, dimethylsulphoxide or ethylene carbonate) are used to dissolve the polymer for dry or wet spinning processes. Polyacrylonitrile also dissolves in concentrated aqueous salt solutions (e.g. NaSCN).

In dry spinning, the solvent is evaporated in the spinning shaft; in wet spinning coagulation is initiated through extrusion of the solution into a spin-bath which is filled with a non-solvent (e.g. water).

Typical values for polymer concentrations in the spinning dope are 25–30 wt% for DMF in dry spinning and 20–25 wt% in wet spinning.

The high boiling point of DMF (153 °C) requires rather high evaporation temperatures of 200–240 °C. The evaporation has to be performed in an inert atmosphere as polymer degradation can occur through oxidative effects at this temperature. In addition, precautions against ignition of the flammable solvent vapour have to be considered. A considerable part of solvent (7–30%) remains in the fibres and be removed during the after-treatment.

In wet spinning a solvent exchange occurs in the precipitation bath, where the solvent is replaced by water. The coagulation of the polymer occurs from a diluted spin dope; thus, the fibres are initially present in a highly swollen gel-like state, which collapses later during washing and drying. This gel state is of particular interest for incorporation of chemicals, for example, dyes into the fibre structure. PAC fibres can be dyed in this stage (gel-dyeing). Cationic dyes diffuse into the gel structure of the highly swollen yarn. This dyeing process leads to good uniformity of the colour because of the stable process conditions. However, the limited flexibility of the technique prevented a wider application of the process.

After formation of a fibrous structure the orientation of the polymer chains is increased by drawing, followed by relaxation at elevated temperature.

Typical finishing agents added to improve processability of the fibres in spinning, knitting and so on are ethoxylated fatty acids, ethoxylated sulphonic acids and phosphoric acid esters of long-chain aliphatic alcohols with an amount of 0.2–0.4 wt%.

Solvent spun fibres from DMF and DMAc solution exhibit wool-like hand; however, moisture sorption is low with 1–2 wt% water present in the fibre at standard climate conditions (20 °C, 65% r.h.). The fibres exhibit high tensile strength and abrasion resistance and are extremely resistant against exposure to light and weather.

Usually basic dyes (cationic dyes) are used for dyeing; light shades can also be obtained with disperse dyes.

Modacrylic fibres that contain vinyl chloride as co-monomer exhibit a reduced flammability (eq. (2.26)):

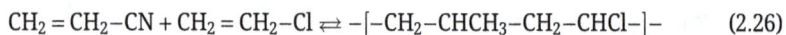

$$CH_2 = CH_2-CN + CH_2 = CH_2-Cl \rightleftharpoons -[-CH_2-CHCH_3-CH_2-CHCl-]- \qquad (2.26)$$

Acrylic fibres are an important source for production of carbon fibres for lightweight composite. The polyacrylonitrile cable at first is partially oxidised and then carbonised under inert atmosphere.

2.9.10 High-performance fibres

High-performance fibres are mainly in use for technical textiles, such as ropes and belts, for reinforcement of composites in protective garment applications [24]. These fibres deliver a particular property, for example, high thermal stability for protective cloths, high strength for climbing ropes and bullet proof vests.

The focus on a certain outstanding property that has to be delivered with the product explains why numerous limitations with regard to cost, dyeability and ecology in production have been accepted.

Typical textile fibres exhibit tenacities from 0.1 to 0.4 N/tex and a modulus between 2 and 5 N/tex. Technical PA fibres and PES fibres reach tenacities in the range of 0.8 N/tex and moduli of 9 N/tex. Para-aramid fibres already have reached tenacities over 2 N/tex and carbon fibres even over 3 N/tex with moduli of 80 and 400 N/tex, respectively.

Important groups of high-performance fibres are:
-   aramids;
-   gel-spun PE fibres;
-   glass fibres; and
-   carbon fibres.

Aramids: While polymerisation of PA and PES can be achieved through direct polycondensation, the formation of aramids requires a substantially more complex reaction chemistry.

The synthesis of *p*-phenylene terephthalamide (PPTA, Kevlar®) requires activation of the terephthalic acid either in the form of acid chloride (terephthaloyl dichloride), or by special activation reagents (e.g. triaryl-phosphites $(RO)_3PO$), which then become consumed during the reaction in a stoichiometric amount (Figure 2.39).

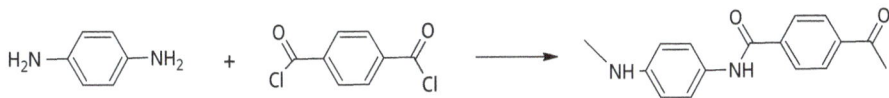

Figure 2.39: Synthesis of *p*-phenylene terephthalamide (PPTA, Kevlar®).

The solvents used for the aramid synthesis are remarkable, as for example *N*-methyl-pyrrolidone with the addition of pyridine, LiCl and $CaCl_2$ are used. Such solvents are related to the concentrated salt-based solvents known for fibroin and PA dissolution and for swelling of cellulose fibres.

Spinning of these fibres is performed from a 10–20 wt% solution of the polymer in concentrated sulphuric acid at temperatures near to 80 °C. At this condition, the polymer solution forms a nematic liquid crystalline phase. The fibre spinning is performed as dry jet-wet spinning process, which is basically similar to the lyocell fibre spinning process. During spinning, coagulation with water and drawing of the fibre, an orientation of the original liquid crystalline phases occurs. The fibre obtains its highly ordered structure, which is the key property for extraordinary mechanical properties.

The high tenacity (e.g. 2.1 N/tex), the high thermal stability (decomposition temperature of 550 °C) and the high modulus of 50–70 N/tex make these fibres of interest for safety applications (protective clothes), technical applications (ropes, cables) and high-performance composites.

From the view of textile chemistry the highly ordered state restricts rapid and easy access of dyes into amorphous zones and thus also limits the dyeability of such fibres. Also mass dyeing during fibre spinning is difficult as dyes and pigments have to be used, which exhibit stability against decomposition in the concentrated sulphuric acid spin dope.

The limited stability against ageing under UV/visible light leads to rapid degradation under outdoor conditions, which thus requires special precautions to guarantee longer lifetime of the high-performance products.

Gel-spun PE fibres: Ultra high molecular weight PE (UHMWPE) is used to produce these fibres (molecular mass 2×10^6). While aramid fibres are already present in an ordered liquid crystalline phase, the orientation and structure formation of the flexible and non-polar PE molecules is achieved by physical methods during fibre shaping. The high molecular weight of the PE leads to high melt viscosity, which thus prevents the use of melt extrusion processes. The highly entangled molecules cannot be ordered to the requested orientation by drawing of the viscose melt.

Thus, a solvent system is used which then initially forms a PE gel. The PE gel is extruded and drawn to yield a fibre with highly oriented polymer molecules. The orientation is required to develop the mechanical properties available from use of a high-molecular-weight polymer.

Hydrocarbons (decaline, paraffin oil and paraffin wax) at elevated temperature are used to form the polymer gel, which is then extruded through the spinneret and drawn. The fibre formation, gelation and crystallisation are supported by solvent evaporation, cooling and solvent extraction. Maximum order is then achieved by further drawing (super drawing, draw ratio 1:50–1:100).

In a high-performance polyethylene (HPPE) fibre orientation of more than 95% and crystallinity of up to 85% can be achieved. In normal melt-extruded PE, the orientation is low and crystallinity is below 60%.

Commercial representatives are the Dyneema® and the Spectra® fibre. Representative values are 2.5 N/tex for the tenacity and 100 N/tex for the modulus of such fibres. The density is 0.97 g/cm^3, which is an important property for marine applications. Among the standard textile fibres only PE- and PP-based fibres exhibit a density lower than water.

Chemical resistance and resistance to UV and visible light is substantially higher compared to aramid fibres. The use of cheap, less corrosive and hazardous solvents and the more simple recycling makes this fibres favourable for many applications. A major drawback arises from the limited thermal stability and shrinkage at temperatures above 120–140 °C. Another disadvantage is the flammability of the fibres. Thus curing temperatures in textile operations should not exceed 140 °C.

The high crystallinity and the absence of any polar functional groups limit dyeing processes for UHMWPE to spun dyeing.

However, in many applications both aramid fibres and HPPE fibres are used as core material for core-spun yarns. The outer layer then consists of standard fibres (e.g. PA and PES), which can then be dyed with usual techniques.

An increase in heat resistance is possible through radiation-induced cross-linking.

The UHMWPE fibres are used for high-strength ropes and cables, as fabric for sails, as industrial filters, as bullet proof vests and as reinforcing layers in composite material.

Melt spun aromatic PESs: Vectran® is a representative for fully aromatic PES fibres. The polymer is synthesised through condensation of 4-hydroxybenzoic acid and 6-hydroxy-2-napthoic acid to form poly(4-hydroxybenzoyl-6-oxynaphtoic acid). The polymer can be processed by use of the melt extrusion process. The polymer exhibits properties of a thermotropic liquid crystalline polymer (TLCP), which means that liquid crystalline phases are present in molten state. The formation of these liquid crystalline phases reduces the viscosity of the melt to a level which permits the use of conventional shaping procedures (Figure 2.40).

As a result of the highly ordered liquid state, the fibre is well organised and exhibits high tenacity (0.9 N/tex), which can be increased further to values of 2–2.5 N/tex by thermal treatment under inert atmosphere conditions. The modulus of TLCP fibres

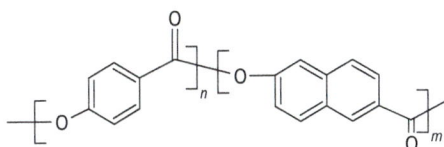

Figure 2.40: Chemical structure of Vectran®.

is in the range of 80–90 N/tex. Similar to other highly oriented fibres, flex fatigue can be a problem, however, at higher level compared to aramids. A particular strength of these fibres is their excellent cut resistance, which makes them of interest for protective applications (e.g. working gloves).

Also these fibres are difficult to dye in conventional textile chemical approaches; thus, coatings have to be used to fix dyes on the fibre surface.

Poly(p-phenylene-benzobisoxazole) fibres (PBO): The synthesis of the PBO fibre is based on the condensation polymerisation of 4,6-diamino-1,3-benzenediol dihydrochloride with terephthalic acid in polyphosphoric acid. The PBO polymer exhibits a rigid rod-like molecular structure. Fibres are spun using this solution in the dry-jet wet-spinning process with water as non-solvent in the coagulation bath. The rigid structure of the polymer lifts the melting temperature T_m above the temperature for decomposition, thus conventional processes (e.g. melt extraction) cannot be applied. Again a highly ordered polymer structure is achieved in the fibre, as the dissolved polymer already forms a liquid crystalline phase in polyphosphoric acid.

Besides a high mechanical strength (tenacity 3.7 N/tex and modulus 150 N/tex), PBO fibres exhibit high thermal stability, which makes them of interest for protective cloths (Figure 2.41) [25].

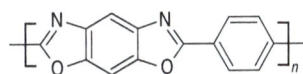

Figure 2.41: Formula of a poly(p-phenylene-benzobisoxazole) fibre (PBO).

References

[1] Wang, K., Li, R., Ma, JH., Jian, YK., Che, JN. Extracting keratin from wool by using L-cysteine. Green. Chem. 2016, 18, 476–481, doi I: 10.1039/C5GC01254F.
[2] Elias, H-G. An Introduction to Polymer Science, VCH Wiley, Weinheim, Germany, 1997, ISBN 3-527-28790-6.
[3] Rath H. Lehrbuch der Textilchemie, Springer Verlag, Heidelberg, Germany, 1972, ISBN 3-540-05587-8.
[4] Schultze-Gebhart, F., Herlinger, K-H. Chapter 1 Survey, ULLMANN's fibres, Volume 1, Fiber Classes, Production and Characterization, Wiley-VCH, Weinheim, Germany, 2008 ISBN: 978-3-527-31772-1.
[5] Haudek, W., Viti, E. Textilfasern – Herkunft, Herstellung, Aufbau, Eigenschaften, Verwendung, Johann L. Bondi & Sohn, Melliand Textilberichte, Heidelberg, Germany, 1978, ISBN 3-87529-018-6

[6]   Esa, F., Tasirin, SM., Abd Rahman, N. Overview of bacterial cellulose production and application. Agric. Agricu. Sci. Procedia. 2014, 2, 113–119.

[7]   https://www.tappi.org/standards-and-methods/test-methods/ accessed 22.05.2018.

[8]   Fras Zemljic, L., Hribernik, S., Manian, AP., Öztürk, HB., Persin, Z., Sfiligoj Smole, M., Stana Kleinscheck, K., Bechtold, T., Siroka, B., Siroky, J. Chapter 7: Polysaccharide fibres in textiles, in The European Polysaccharide Network of Excellence (EPNOE) by. P. Navard (ed.), Springer-Verlag, Wien, 2013, 401, ISBN: 978-3-7091-0420-0

[9]   Mozdyniewicz, DJ., Nieminen, K., Sixta,, H. Alkaline steeping of dissolving pulp. Part I: Cellul. degradation kinet. Cellulose 2013, 20/3, 1437–1451, doi:10.1007/s10570-013-9926-2.

[10]   Bechtold, T., Manian, A., Oeztuerk, H., Paul, U., Široka, B., Jan, Š., Soliman, H., Vo, TTL., Vu-Manh, H. Ion-interactions as drying force in polysaccharide assembly. Carbohydr. Polym. 2013, 93, 316–323, http://dx.doi.org/10.1016/j.carbpol.2012.01.064.

[11]   Lindman, B., Karlström, G., Stigsson, L. On the mechanism of dissolution of cellulose. J. Mol. Liq. 2010, 156/1, 76–81.

[12]   Isik, M., Sardon, H., Mecerreyes, D. Ionic liquids and cellulose: dissolution, chemical modification and preparation of new cellulosic materials. Int. J. Mol. Sci. 2014, 15(7), 11922–11940, doi:10.3390/ijms150711922.

[13]   Zhou, J., Zhang, L. Solubility of cellulose in NaOH/Urea aqueous solution. Polym. J. 2000, 32, 866–870.

[14]   Potthast, A., Rosenau, T., Buchner, R., Röder, T., Ebner, G., Bruglachner, H., Sixta, H., Kosma, P. The cellulose solvent system N,N-dimethylacetamide/lithium chloride revisited: the effect of water on physicochemical properties and chemical stability. Cellulose. 2002, 9/1, 41–53.

[15]   Schuster, KC., Rohrer, C., Eichinger, D., Schmidtbauer, J., Aldred, P., Firgo, H. Environmentally friendly lyocell and rayon fibers – in: Natural Fibers, Polymers and Composites – Recent Advances – Wallenberger, F. T. and Weston, N. E. ed. – Kluwer Academic Publishers, Boston/Dordrecht/London, 2004.

[16]   McGrath, K., Kaplan, D. Protein-Based Materials, Birkhäuser Boston, USA, 1997.

[17]   Bechtold, T, Mussak, R. Handbook of Natural colorant, John Wiley & Sons, Chichester, UK, 2009.

[18]   Hardy, JG., Römer, LM., Scheibel, TR. Polymeric materials based on silk proteins. Polymer (Guildf) 2008, 49, 4309–4327.

[19]   Bier, MC., Kohn, S., Stierand, A., Grimmelsmann, N., Homburg, SV., Ehrmann, A. Investigation of the casein fibre production in an eco-friendly way. Aachen, Dresden, Denkendorf International Textile Conference, Dresden, November 24–25, 2016.

[20]   Mishra, S.P. A Text Book of Fibre Science and Technology, New Age International, New Delhi, India, 2000.

[21]   Shen, L., Haufe, J., Patel, MK. Product Overview and Market Projection of Emerging Bio-Based Plastics. Pro-Bip 2009, Utrecht University, Utrecht, The Netherlands, 2009.

[22]   ISO 105-E03:1987, Textiles – Tests for colour fastness – Part E03: Colour fastness to chlorinated water (swimming-bath water)

[23]   Agache, P., Humber, P. Measuring the Skin, Springer Verlag, Berlin, Germany, 2004.

[24]   Hearle, JWS. High-Performance Fibres, Woodhead Publishing Ltd. Cambridge, England. 2004.

[25]   http://www.toyobo-global.com/seihin/kc/pbo/zylon-p/bussei-p/technical.pdf, accessed 04.06.2018.

[26]   Klemm D., Philipp B., Heinze T., Heinze U., Wagenknecht W. Comprehensive Cellulose Chemistry, Volume I and II. Wiley-VCH, Weinheim, Germany, 1998, ISBN 3-527-29413-9.

Take home messages

Among the different classes and types of fibres, every individual fibre exhibits its unique characteristics. This explains the high number of fibre variations.

Reactivity of a fibre is determined by its chemical nature and by the accessibility of the fibre structure. Non-accessible parts are non-reactive parts. Changes in accessibility thus modify fibre reactivity.

A textile fibre consists of highly ordered crystalline parts and the amorphous parts, which are actually the textile chemically more active sites.

Interactive forces between polymer chains determine the minimum length of a polymer required to form fibres.

Fibre modification can be achieved by the use of different polymers/copolymers, by polymer blends, through grafting operations. Also the conditions of fibre formation, the thermal treatment and drawing of fibres during production determine the final properties of a product.

Polymer orientation in fibrous structures is a key factor to obtain high tenacity.

Quiz

Question 1. Name two natural fibres and two synthetic fibres and sort with regard to strength, moisture uptake and electrical resistance (Tables 2.1 and 2.2).

Question 2. Which fibres are outstanding with regard to the following characteristics? Highest/lowest water sorption at standard climate, low flammability, high tenacity, decrease of strength in wet state.

Question 3. Name four fibres that are easily biodegradable.

Question 4. Which fibres are difficult to dye and why? What options we have then to produce coloured fibres?

Question 5. What happens during alkalisation/mercerisation of cellulose? Why this process is not used for wool, PES or PP? Explain on a chemical basis.

Exercises

1. Try to visualise the length of a single polymer in dimensions of a tube, which we can imagine better than a molecule. Take the data from Table 2.2 for DP of cotton and PES and estimate the length of a polymer molecule for the dimensions of a single formula unit given below:
   Co: Anhydroglucose unit: length (l) = 0.6 m, width (w) = 0.6.
   PES: PET: l = 1 m, w = 0.5 m. PA 6: l = 1 m, w = 0.3 m; PP: l = 0.5 m, w = 0.3 m.
2. Calculate the molecular mass of a cellulose chain. The formula of the anhydroglucose unit is $C_6H_{10}O_5$ and the DP P is 1,200 units per molecule. If the cellulose

contains 8.5 wt% moisture sorbed, what is the molar ratio between water and an anhydroglucose unit?

3. Prove the statement that 13.7 g cellulose molecules in one line would have a length of 2.62×10^{10} km, which corresponds 170 times the distance to the sun. The length of a cellobiose unit can be assumed with 10 Å (1 Å = 10^{-10} m = 0.1 nm).

4. Estimate the chemical oxygen demand (mg O_2 per 1 L of effluent) for the substances released during cotton scouring. Assume 3% of the cotton is released into the wastewater in form of (a) polysaccharides and (b) in form of a saturated C18 fatty acid. The treatment is performed at a liquor ratio of 1:10, and three rinsing baths are used in addition.

5. When a viscose fibre consists of 100% cellulose and all reducing end groups have been oxidised to carboxylic groups during the fibre manufacturing, what will be the expected carboxylic group number in mmol COOH/kg fibre material?

6. Please try to explain, why under steady state conditions the Na_2SO_4 concentration in the viscose fibre precipitation bath is rather high (12–24% Na_2SO_4), when only 7–12% H_2SO_4 is present? How could the concentration be reduced and what would happen to the overall volume balances?

7. Assume the degree of PES (PET) is in the dimension of 80–100 formula units and the DP of cellulose in viscose is 200–250. Could you try to estimate the relative chain length of the polymers by comparison of the carbon/oxygen numbers arranged in a row (PES: 2C, O, C, 4C, C, O = 10 atoms, cellulose: O 4C = 5 atoms).

8. Assume a DP_w of 150 for PET and presence of 0.1 wt% of water as residual moisture. Estimate the reduction in average DP during melt extrusion, when the complete amount of water is used to hydrolyse ester groups.

9. A PES fibre spinning line produces 8 dtex fibres, using a 1,125 holes spinneret and 16 producing spinnerets, each working with a speed of 1,500 m/min. Estimate the linear density of the cables produced per spinneret, daily (20 h/day) production in km tow and metric tonnes.

10. Assume every methylene/methine/methyl unit in PP is a sphere with a diameter of 1 m. Estimate the length of an isotactic PP polymer with molecular mass of 250,000 g/mol. Compare the chain length with UHMWPE (molecular mass 2×10^6).

Solutions

1. (Question 1) Co, Wo, PES, PP.
   Tenacity: PES > PP > Co > Wo; moisture uptake: Wo > Co > PES > PP; electrical resistance: PP ≥ PES > Wo > Co.

2. (Question 2) Highest water sorption: Wo; lowest water sorption: PP, glass fibres, carbon; low flammability: glass, carbon, aromatic PAs; high tenacity: aromatic PAs; UHMWPE, carbon, glass; decrease of strength in wet state: Wo, CV, CA, aromatic PAs.

3. (Question 3) Wool, silk, cotton, flax, hemp, jute, viscose, lyocell.
4. (Question 4) PP: hydrophobic fibre, absence of polar groups; aromatic PAs and aromatic PESs: very compact structure with low accessibility for dyes. Polyurethanes: the amorphous elastic segments exhibit only few polar groups that do not permit sufficient dyestuff absorption. Dyeing: mass coloration, pigment dyeing with binders, production of core spun yarns with dyeable fibres as outer layer.
5. (Question 5) During alkalisation, intensive swelling of the amorphous parts of the fibres occurs and alkalicellulose is formed (hydroxyl ions containing stoichiometric cellulose compound), the crystal lattice of cellulose I is transformed into cellulose II. In case of wool or PES fibres, a hydrolytic degradation would occur and dissolution of lower molecular weight fragments leads to dissolution of the full fibre structure. In case of PP, no chemical reactions will happen.
6. (Exercise 1) Cotton, DP = 1,500–3,000; total length = $l \times$ DP = 900–1,800 m.
   PES: PET, DP = 220–260; total length = 220–260 m.
   PA 6, PA 6, DP = 380–770; total length = 380–770 m.
   PP; DP = 45,000; total length = 22,500 m (!).
7. (Exercise 2) Molecular mass of $C_6H_{10}O_5$ = 162 g/mol.
   Mass of the polymer = DP × $M(C_6H_{10}O_5)$ = 162 × 1,200 = 194,400 g/mol.
   Molar ratio of anhydroglucose:water: 1 kg of cellulose contains 8.5% moisture.
   M(cellulose) = 915 g; this corresponds to $n = m$(cellulose)$/M$(anhydroglucose) = 5.65 mol anhydroglucose.
   A mass of 85 g water corresponds to 4.7 mol water. Thus, one water molecule is absorbed per every 1.2 anhydroglucose units.
8. (Exercise 3) A mass of 13.7 g cellulose corresponds to n = 13.7/162 = 0.0846 mol anhydroglucose and 0.0423 moles of cellobiose. This amount of moles corresponds to 0.0423 × 6.022 × 10^{23} = 0.254 × 10^{23} cellobiose units with a length of 1 nm each. Thus, the total length of units will be 1 × 10^{-9} × 0.254 × 10^{23} m = 0.254 × 10^{14} m = 2.54 × 10^{10} km. Distance of the Earth from the Sun: 1.49 × 10^8 km. Thus, a factor of 170 is correct.
9. (Exercise 4) Treatment with a liquor ratio of 1:10 means 1 kg goods is processed in 10 L of liquid. Release of 3 wt% of cotton impurities, thus, will lead to a concentration of 3 g/L of impurities in the wastewater. (a) In case of polysaccharides (($(C_6H_{10}O_5)_n$)) with $M(C_6H_{10}O_5)$ = 162 g/mol total oxidation to $6CO_2$ and $5H_2O$ will require 6 moles of O_2 per formula unit. $m(O_2)$ = (3/162) × 6 × 32 = 3.555 g/L oxygen or 3,555 mg/L oxygen. Including the 3 rinsing baths the total dilution factor will be 4, thus the final result will be 3.555/4 = 0.89 g/l O_2 and 2.197 g/l O_2 respectively. (b) In case of a $C_{17}H_{35}CO_2H$, the required amount of oxygen will be $n(O_2)$ = (3/284) × (17 + 9) = 0.27 mol. The COD value thus will be $m(O_2)$ = 0.27 × 32 = 8.78 g/L oxygen or 8,788 mg/L.
10. (Exercise 5) DP = 200–250, $M((C_6H_{10}O_5)_{n = 200–250})$ = 162 × (200 or 250) = 32,400 or 40,500 g/mol. About 1 kg of fibre material thus represents
    N = 1 × 10^3/(32,400 or 40,500) = 30.86–24.7 mmol/kg.

11. (Exercise 6) The concentration of 12 g/L Na_2SO_4 in the precipitation bath is a result of a dynamic equilibrium between the intake of NaOH with the viscose dope and feed of fresh H_2SO_4. At a given production rate the concentration of NaOH can be reduced only by an increase in the dosage volume of fresh sulphuric acid. As a result, the overflow of spent precipitation bath would increase substantially, thus leading to higher losses in non-used sulphuric acid.

12. (Exercise 7) Along a PET unit, the following atoms form a row: 2C, O, 6C, O = 10 atoms. A chain with DP 80–100 will be approximately 800–1000 atoms long. In CV along the chain we find: O4 C = 5 atoms. A chain with DP 200–250 will be 1,000–1,250 atoms in a row.

13. (Exercise 8) The molecular mass of a PET unit is 192 g/mol. The molecular mass of a polymer is thus $M(PES)$ = 28,800 g/mol. About 1 kg of polymer thus contains n = 1,000/28,800 = 0.0347 mol PES and $n(H_2O)$ = 1/18 = 0.055 mol water, which means that the DP will be reduced to less than half.

14. (Exercise 9) At first the count of the total cable production is calculated: T_t = 0.8 tex × 1,125 fibres × 16 = 14,400 tex (g/km). Daily production: 14,400 g/km × 1.5 km/min × 60 min/h × 20 h/day = 25,920 kg/day. Length of cable per spinneret per day = 1,500 m/min × 60 min/h × 20 h/day = 1,800 km.

15. (Exercise 10) DP = 250,000/42 = 5,950 length = 1 m × 5,950 × 2 = 11.9 km. UHMWPE: DP = $1 × 10^6/28$ = 35,714 length = 1 m × 35,714 × 2 = 71.4 km.

List of abbreviations/symbols

PES	polyester
PA 6.6	polyamide 6.6, polyhexamethylene adipate
Arom.PA	aromatic polyamide
PAC	polyacrylnitrile
PE	polyethylene
HDPE	high-density polyethylene
LDPE	low-density polyethylene
PP	polypropylene
CV	viscose
CLY	lyocell
CA	cellulose diacetate
CT	cellulose triacetate
Co	cotton
Wo	wool
EO	ethylene oxide
PO	propylene oxide
NMMO	N-methyl-morpholine-N-oxide
P_w	mass average degree of polymerisation
T_g	glass transition temperature
PET	polyethylene terephthalate
P3HB	poly(3-hydroxybutyrate)

PPTA	*p*-phenylenediamine terephthalamide
DMF	*N,N*-dimethylformamide
DMAc	*N,N*-dimethylacetamide
Et	ethyl group
Me	methyl group
DMT	dimethyl terephthalate
DP	degree of polymerisation
UHMWPE	ultra high molecular weight polyethylene
TLCP	thermotropic liquid crystal polymer
PBO	poly(*p*-phenylene-benzobisoxazole)
M	molecular mass

3 Structure of textile fibres

3.1 General aspects

The reactions during the formation of a fibre have determining influence on the final structure of the material. Naturally grown fibres such as cotton, flax, wool and silk exhibit a complex structure with outer layers (primary wall, secondary wall, lumen; cuticula and cortex; fibroin and sericine), which is due to natural growth, the complex reactions during fibre formation in spinning glands, and so on.

At first, the formation of a man-made fibre appears to lead to a more simple fibre structure; however, considerable structural complexity arises from the spinning process, which yields differences between fibre shell and core, for example, with regard to orientation, crystallisation as well as porosity and accessibility.

A huge difference in fibre structure also comes from the spinning technique applied, which can be melt extrusion spinning (PA and PES), wet spinning (CV and CC), dry jet-wet spinning (Lyocell) and dry spinning (CA CT).

3.2 Crystallinity versus amorphous regions

A major part of fibre properties is governed by the state of molecular order in the fibre; thus, it is important to understand the different factors that determine order and as a consequence properties of the resulting fibre. The state of order is reflected in the crystallinity of the fibre, which is a measure for the approximate share of crystalline (highly ordered) regions in a material.

From the point of fibre mechanics, the crystalline parts are of importance as tenacity and modulus directly are related to these factors.

For a textile chemist the reactive amorphous parts of a fibre are of higher interest, as accessibility of chemicals, sorption and chemical reactivity primarily will be observed in the less ordered regions of a fibre. To understand the reactivity/stability of a fibre polymer, it is thus of relevance to consider all factors that lead to a redistribution in the share of ordered/crystalline/less reactive parts and less ordered/amorphous/more reactive parts.

The overall structure of a textile fibre is determined by a series of factors that contribute. These are as follows:
- The constitution of the fibre polymer
- Molecular mass and distribution of the polymer
- Uniformity in composition (protein fibres, native cellulose fibres)
- Configuration and conformation of the polymer in solid state
- Amorphous and crystalline regions
- Supermolecular order, fibril formation

https://doi.org/10.1515/9783110795738-003

- Porosity and fibre anisotropy
- Fibre dimensions: diameter, cross-sectional shape

3.3 Constitution of a polymer

The constitution of an organic molecule describes the identity and connectivity of atoms in a molecule. For a polymer the description of the constitution could become complicated; thus, usually only the constitution of a monomer is described. Therefore, a monomer is the building block that forms the polymer.

A polymer that consists only of one monomer unit is called a *homopolymer*, and when two or more monomers form the polymer it is called a *copolymer*.

In an *alternating bipolymer*, two monomers alternate in the chain. In a *statistical bipolymer*, two monomers have been linked together to form a copolymer following the statistics of the polymerisation reaction. In a *gradient copolymer*, a compositional gradient appears along the polymer chain. In a *block copolymer*, blocks of homopolymer sequences are combined to form a polymer. In a *homochain polymer* all chain atoms are the same, for example, carbon or sulphur; in a *heterochain polymer*, different atoms are part of the polymer-forming chain, for example, C and O in polyester, polyether, Si and O in siloxanes. Important definitions and examples are listed in Table 3.1.

Table 3.1: Polymer constitution – definition and examples.

Constitution	Principle	Example
Homopolymer	–A–A–A–A–	$- [- (CH_2-CH_2) -]_n-$ PE
Copolymer		
Alternating bipolymer	–A–B–A–B–A–B–	$- [-CO(CH_2)_4CONH(CH_2)_6NH-]_n-$ PA6,6
Statistical bipolymer	–A–B–B–A–B–A–A–A–B–A–B–B–	$- [CH_2CH_2-CH_2CHCH_3-]_n-$ PP–PE
Gradient bipolymer	–A–A–A–B–A–B–B–A–B–B–B–B–	
Block copolymer	–A–A–A–A–B–B–B–B–	$[- (OCH_2CH_2)_{m1}O-CONH$ $(CH_2)_{m2}NHCO-]_n$ elastomer
Homochain polymer	$-[- (CH_2-CH_2) -]_n-$	PE
Heterochain polymer	$- [-O-CH_2-]_n-$ $-O-Si(CH_3)_2-$	Polyoxymethylene silicone

Star polymers exhibit three or more branches that develop from a centre molecule. A particular example of a star molecule is a dendrimer, in which all polymer arms exhibit the same length. This can be achieved by using an appropriate synthesis strategy, which permits a step-by-step growth of the polymer and results in a very high degree of molecular uniformity (Figure 3.1).

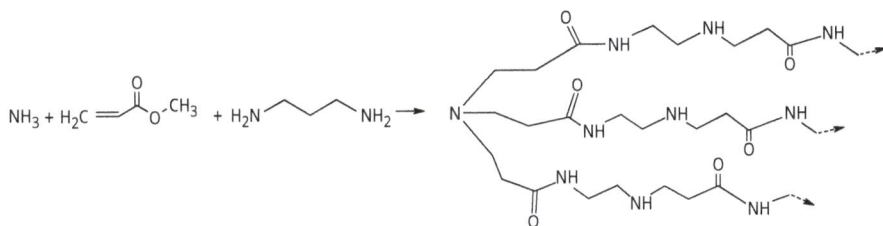

Figure 3.1: Synthesis of a dendrimer from ammonia, methacylate and ethylenediamine.

A homopolymer offers best conditions to form ordered, crystalline parts, as all monomer units have the same structural components. In copolymers or block copolymers, the formation of ordered structures is more complicated. For example in the case of block polymers between polyether and polyurethanes, only the polyurethane segments form ordered parts and fibres with elastomeric properties result.

The presence of minor amounts of a second monomer with other steric requirements can be used to steer crystallisation. For example in a copolymer of ethene and norbornene, an optically transparent polymer (cyclic olefin copolymer, COC) is obtained. The formation of crystalline parts in dimensions that cause light scattering is prevented by the copolymerisation with norbornene (Figure 3.2).

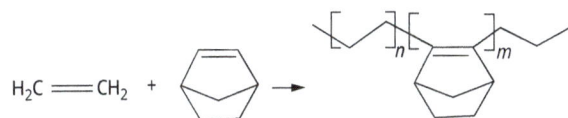

Figure 3.2: Polymerisation of ethane and norbornene to form COC.

A similar situation arises from the presence of constitutional failures during polymerisation. In the polymerisation of polypropylene (PP), besides the head-to-tail polymerisation, irregularities (regioisomerism) can also arise from accidental head-to-head/tail-to-tail combinations (Figure 3.3).

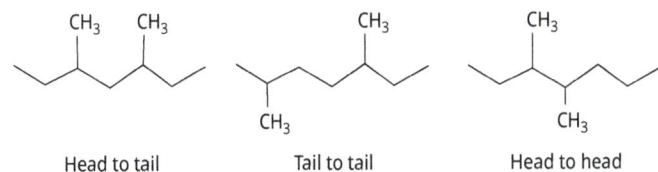

Figure 3.3: Different constitutional irregularities in a PP.

The presence of such constitutional irregularities will influence the formation of crystalline regions and vice versa, thus supporting the development of amorphous regions.

Disorder in a regular constitution also can arise from branching of the polymer chain, which then contains the main polymer chain and side chains (Figure 3.4).

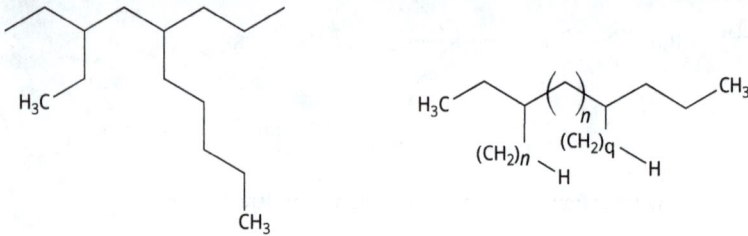

Figure 3.4: Irregular branching of polymer (left: short chain branching; right: long chain branching).

Substituents at the side of the polymer are not regarded as branched polymers, for example, polyacrylates or copolymerisates of ethene with iso-octene (Figure 3.5).

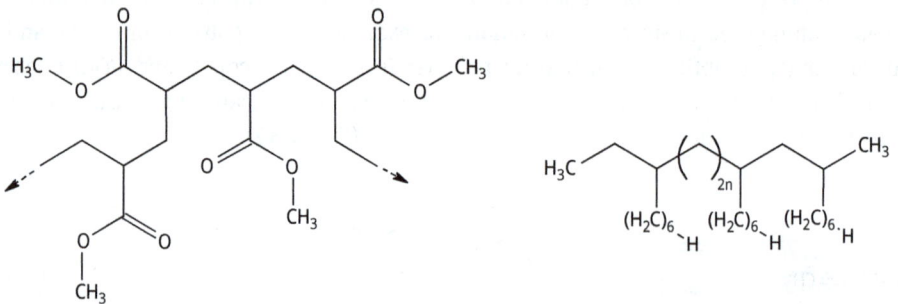

Figure 3.5: Structure of poly(acrylic ester) (left) and a copolymerisate between polyethylene and iso-octene (right).

3.4 High-performance fibres

The particular mechanical properties (high tenacity and high modulus) of high-performance fibres are due to their molecular structure.

– The polymer chain is highly oriented and order extends over a long distance.
– Strong axial bonding exists.
– Only minor parts of the fibre are present in lower ordered amorphous segments.

Linear molecules such as polyethylene must be fully oriented in the fibre structure and chain folding must be avoided. Long polymer molecules that have rigid rod-like structure tend to form liquid crystals in molten state or in solution. These pre-ordered structures contribute to the development of highly organised molecular structures during the fibre spinning process. Examples are Kevlar® (para-amide, polyphenylene terephthalamide), Vectran® (poly-*p*-hydroxy benzoyl-6-hydroxy-2-napthoic acid) or Dyneema® (high molecular weight polyethylene).

Another strategy to achieve a highly ordered molecular structure in fibres utilises gel-spinning. In this process extrusion of a solvent polymer gel is followed by drawing of the formed fibres.

Sheet structures: Large two-dimensional structures are already present as building elements in carbon fibres. During the high-temperature processing of acrylic fibres, graphite sheets are formed that build a rigid structure. Graphite-type building blocks then are kept together by disordered parts, which contribute to the cohesion and avoid slipping off of graphite layers.

Three-dimensional (3D) networks can result from polymerisation of monomers with more than two reactive sites. Different material form such 3D networks, though not all belong to the high-performance materials, for example, thermoset resins, rubber; however, typical representatives for fibre materials are glass, silicon oxides or carbides.

The highly ordered structure of high-performance polymers leads to a different resistance during bending. The high tenacity of the fibre prohibits extension and thus leads to compression of the inner part of a bent fibre. As the compressive yield strength of the fibre is lower than the tensile strength, the fibre becomes compressed and deformed. When the fibre is bent back to straight form, the initial state is achieved only in case the compressed part was pulled back to the initial length. Upon multiple bendings, this may not be the case and the outer site of the fibre compensates the shorter length of the inner part by forming a buckle, which is also the beginning of fatigue failure. In another case, the multiple bending of the fibres leads to axial splitting and fibrillation at the bent part of the fibre.

Crosslinked polymers: Crosslinking of polymer chains leads to the formation of a 3D molecular network. Attractive forces in a molecular network can be as follows:
- intramolecular covalent bonding, which leads to an insoluble network, for example, of a duroplast;
- intermolecular physical interactions, for example, ion-clusters, crystallites, microphases, mechanical polymer chain entanglement, which can open through appropriate solvents, thus leading to swelling or dissolution of the polymer.

3.5 Molecular weight distribution

The synthesis of polymers both in nature and in technical synthesis follows the rules of statistics. As a result, the degree of polymerisation and the molecular weight of the different polymer molecules are not uniform both in a natural grown and in a technically synthesised product. This non-uniformity is described by the term polymolecularity or polydispersity.

The characteristic number for the polymolecularity is the polydispersity index D (eq. (3.1))

$$D = \frac{M_w}{M_n} \tag{3.1}$$

with M_w being the mass average molecular mass and M_n being the number average molecular mass of the polymer.

When all polymer molecules have the same degree of polymerisation, the polydispersity D will be 1, which is a theoretical minimum. Very narrow distributions will exhibit a D of 1.04; linear polymers from polycondensation reactions can exhibit D of 2.0, while industrial polymers can have values of D between 20 and 40.

The distribution of different molecular masses is presented as distribution functions. Figure 3.6 shows differential and integral distribution functions.

In fact, the distribution function of a polymer is a discrete presentation of the probability that a certain degree of polymerisation appears in the material. Because of the high degree of polymerisation usually present, the graphs are drawn as continuous lines.

The experimental determination of a distribution function requires separation of the polymer material into fractions of different molecular weight. An important method for the determination of molecular mass distribution is gel permeation chromatography) that is a separation method, which then is coupled to a molecular size-dependent detector using light scattering or viscometry.

3.6 Consequences of polydispersity

From the point of textile fibre technology, polydispersity is an important factor that determines the formation of ordered crystalline regions. The lower the polydispersity factor D, the more compact the polymer chains will be arranged through the shear forces applied during fibre formation and drawing. A high polydispersity will support the formation of amorphous region with lower state of order. Compared to a more uniform polymer, such a material will exhibit lower tenacity and higher molecular accessibility and reactivity.

Thus even for the same type of polymers, the actual distribution of molecular mass in the raw material will influence the properties of a fibre.

Minor components with high molecular weight will disturb complete dissolution of the polymer in wet spinning and form aggregates with high viscosity in melt extrusion spinning.

Low molecular weight oligomers will leave the polymer matrix and dissolve when appropriate conditions are applied, for example, dissolution of oligomers in polyester dyeing and dissolution of lower molecular weight cellulose fragments during alkalisation.

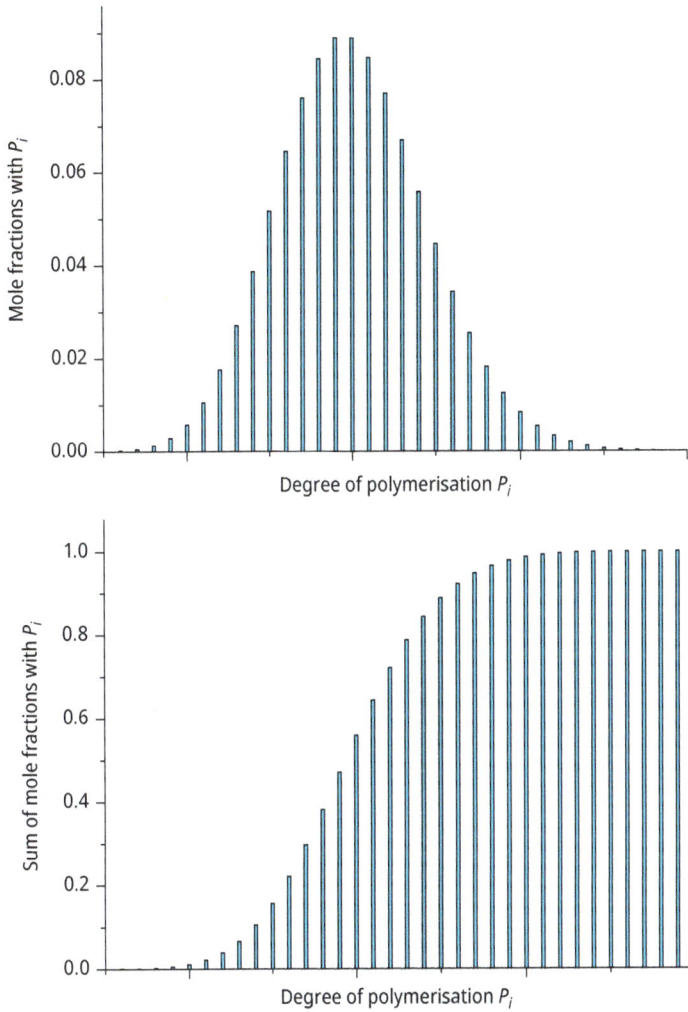

Figure 3.6: Differential and integral molecular mass distribution functions.

3.7 Configuration

While constitution considers only the connections of atoms in a molecule, the configuration includes the description of the spatial arrangement of atoms/groups. Configurational isomers cannot be converted to each other without breaking a chemical bond. Thus, the energy barrier between the isomers is high enough to permit the separation and isolation.

Configuration isomers define another level in fibre structure with distinct impact on the disorder in a polymer material (compare Chapter 2).

Different types of stereoisomers can be distinguished as follows: Enantiomers exist in two configurations that can be transformed to identity through the symmetry operation of mirroring. In a simple case, one chiral centre is present in the molecule as it is, for example, in D-lactic acid, which is transferred to L-lactic acid through mirroring. In Fisher projection, these molecules can be drawn as shown in Figure 3.7.

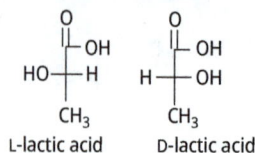

L-lactic acid   D-lactic acid   **Figure 3.7:** Fischer projection of L- and D-lactic acid.

In diastereomers at least two or more chiral centres are present and one of these has identical configuration, while the second one is present in mirrored configuration. While enantiomers, for example, D-glucose and L-glucose, exhibit identical chemical reactivity towards non-chiral reagents, a pair of diastereomers, for example, D-mannose and D-galactose, exhibits different physical and chemical properties.

Another system for the definition of chiral centres based on the R- and S- nomenclature is according to the Cahn–Ingold–Prelog convention. In this system, the substituents are ordered according priority rules that then define the configuration. A similar approach could be chosen to identify and name chiral centres in polymer chains.

However, the number of chiral sites in a polymer makes the situation more complex. PP can serve as an example to analyse these structural aspects (Figure 3.8).

The repeating units Ia and Ib can be transferred into identify by rotation along single bonds; thus, these are only conformational isomers and not configurational isomers. IIa and IIb as well as IIIa und IIIb are enantiomers. II and III are constitutional isomers.

In a polymer such as PP, a high number of stereorepeating units are present along the chain. Every propylene unit forms a chiral centre of the polymer chain.

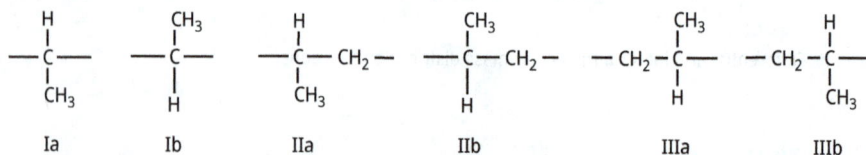

Ia   Ib   IIa   IIb   IIIa   IIIb

Figure 3.8: Repeating units in PP.

In a stereoregular polymer, only one stereorepeating unit forms the polymer. When the configuration of the repeating unit is defined in a repetitive manner, a tactic polymer is obtained. This can be an isotactic polymer (IT), a syndiotactic polymer (ST) or a heterotactic polymer (HT). If no regular order is present, the polymer has an atactic configuration (Figure 3.9).

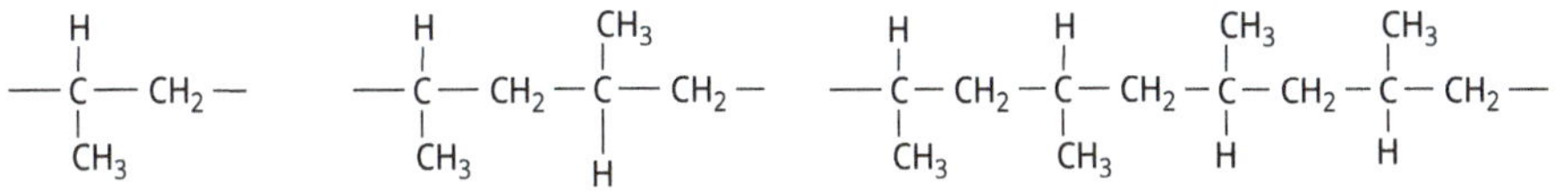

Figure 3.9: Stereorepeating units in PP.

It is important to consider that these classifications are based on a relative configuration. If a classification according Fischer projection or R- and S-system is attempted, the length of the polymer chain and the number of chiral sites (the same as the degree of polymerisation, e.g., 3,000) will make such an approach almost impossible.

However, the stereochemical aspects of a polymer have significant relevance for the properties of the material. As a result of the presence of optically active sites, polymers with chiral centres can be optically active or form a racemate. In the case of an isotactic polymer, a mirror plane in the middle of the polymer transfers the two optically active branches into each other. This means that one half is R-configuration and the other half is S-configuration. Thus, the polymer exhibits a meso structure, such as meso-tartaric acid, and it will not be optically active. In the case of an syndiotactic polymer, R- and S-configurations alternate; thus, the polymer is not optically active [1].

Besides the different optical properties of polymers that contain stereorepeating units, the configuration and the uniformity of configuration will contribute to the formation of ordered and crystalline phases; thus, tenacity, elongation, accessibility and thermal properties, for example, glass transition temperature and melting temperature can be determined.

A representative example for determining the influence of configuration on physical properties of a polymeric material is given by polylactic acid (PLA). In case equal amounts of D-form and L-form are present in the polymer, the meso form of PLA is obtained, which is not optically active. The probability to form larger crystalline sectors is very low; thus, the polymer is amorphous and transparent, as no light scattering occurs at the crystallites. PLA that consists only of one stereoisomer, for example, L-form, is the optically active PLLA (Poly-L-Lactic acid), which already exhibits a melting point of 180 °C. Interestingly a combination of PLLA and PDLA (Poly-D-Lactic acid) forms a highly organised stereocomplex scPLA, which exhibits the highest melting point of all PLA variations (230 °C).

In real polymers, tacticity always exhibits defects in the configurational order. Thus, the overall level of uniformity in configurational order will be a characteristic property of a polymeric material.

Important methods to determine the uniformity of tacticity are infrared spectroscopy, NMR methods, analysis of optical activity and determination of crystallinity and thermal properties (T_g and T_m).

3.8 Conformation

In organic chemistry, conformation describes the spatial arrangement of all groups (atoms) in a compound. Through rotation along single bonds, different conformations can be achieved, where all are separated only through a low-energy barrier (Figure 3.10).

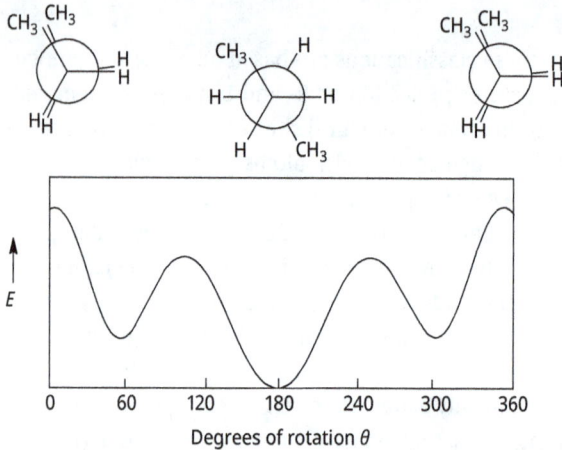

Figure 3.10: Energy level of micro-conformations and rotational barriers for the centre chain bond in butane [1].

In theory an endless number of micro-conformations exist; however, because of the differences in energy, some conformations exhibit higher probability of formation.

The energy difference ΔE_{TG} between a trans (T) and gauche (G) conformation in PP is in the magnitude of 4 kJ/mol. The energy of activation to pass the rotational barrier between the states $\Delta E^{\neq}_{TG}$ is around 13 kJ/mol. Thus, thermal energy at 25 °C of 2.48 kJ/mol is sufficient to enable molecules to pass the energy barrier through rotation. As a result at room temperature, a statistical distribution of conformations exists in a dynamic equilibrium as long as the molecule is in liquid state, which can be a polymer solution or a melt.

Several factors will influence the distribution of configurations:

– the presence of intermolecular attractive forces, for example, oppositely charged sites and hydrogen bonding;
– steric effects, for example, voluminous side groups such as in polystyrene (PS) and
– solvent effects (weakly bound hydrated layers and complex formation).

3.8.1 Conformational statistics

1. As a result of these considerations, we can understand that in dissolved or molten state the shape of a linear polymer chain will follow the rules of statistics.

    Few configurations will lead to a configuration as linearly arranged molecule, but a high number of arrangements will be possible for coiled structures. Thus as a result of statistics, the probability to observe a flexible polymer chain in solution in its coiled state is high.

    Shear effects support uncoiling and disentanglement of polymers and are of distinct relevance in nature (silk) and technical fibre production.

    In the presence of strong attraction between polymer chains, other arrangements will be favoured, for example, a right-handed helical structure in poly (L-alanine)–[–NH–CH(CH$_3$) –CO–]$_n$–.

2. Upon cooling of a melt or precipitation of a polymer from the solution, the formation of the structure in a solid state will be influenced by the conformation present in the molten/dissolved state.

A disentangled linear molecule will form a solid state with higher degree of crystallinity compared to a polymer that solidifies without application of any shear forces. Upon rapid cooling, the structure in the amorphous parts will be similar to a frozen liquid phase. Thus, we can envisage the structure of an amorphous region to represent all stages between a fully ordered crystalline phase and the order of a frozen melt. By an increase in temperature above T_g, the flexibility of the chains in the amorphous zones increases, while the crystalline parts still remain stable. In a simplified model above T_g, the polymer consists of a solid crystalline phase and of a more "liquid" amorphous phase. Many textile chemical processes in particular with synthetic fibres require the flexibility in the amorphous phases as a prerequisite to permit chemicals access into the bulk of the fibre.

3.9 Polymer assemblies

In solid state, a low molecular weight substance or a metal exhibits reversibility upon minor deformations. Liquid substances are completely deformable and exhibit viscose behaviour.

Polymer materials exhibit elastic and viscose behaviour at the same time; the respective level being dependent on temperature, pressure and time. This behaviour is a consequence of the presence of two different phases in the polymer (crystalline domains = solid phase and amorphous domains = viscous liquid phase).

In an idealised state, the polymeric material consists of two completely different states: the ordered, hard crystalline segments, in which different polymer chains join to build the crystal, and the amorphous, non-ordered state, which contains an irregular arrangement of polymer chains, polymer ends and all constitutional, configurational and conformal irregularities that do not fit into the formation of a regular crystalline order.

The amorphous part exhibits no regular order. The polymer chains achieve flexibility at the glass transition temperature T_g, above which a rubber elastic behaviour is observed.

The crystalline long range order disappears first at the melting point T_m, above which the polymer chains achieve their full mobility. The polymer melt in an ideal state exhibits no order and is a fluid with high viscosity. Non-long range order remains in the molten state, which, for example, can be proven by X-ray diffraction characteristics of the molten state.

The degree of crystallinity usually describes the part of a fibre that is assumed to be present in ordered/crystalline state. Different methods can be applied to determine the degree of crystallinity:
– Analysis of the relative intensity of selected absorption bands in FITR spectra.
– From X-ray diffraction analysis crystallinity as well as dimensions of crystallites can be determined.
– Density analysis permits the calculation of crystallinity.

It is important to consider that results from different methods cannot be compared directly. The different physical principles that are applied to assess the order in a fibre emphasise different structural characteristics; thus, comparison of results obtained with different methods have to be made with caution.

Because of its enormous length, a single polymer chain can participate in several crystalline and amorphous domains of a fibre at the same time.

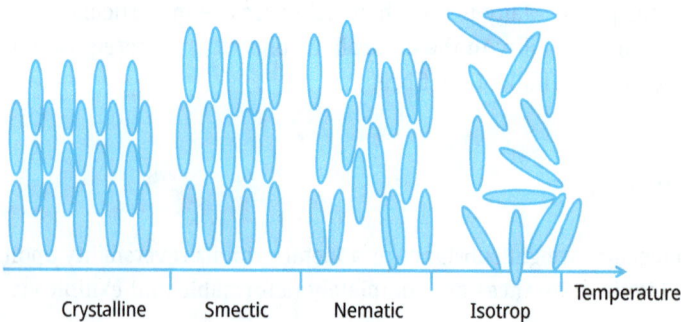

Figure 3.11: Liquid crystalline phases during melting of a polymer.

Besides these two idealised states, ordered states also can result from the formation of liquid crystalline phases already present in molten state (Figure 3.11). Liquid crystalline phases can be understood as conformationally unordered crystallites or plastic crystals. The presence of liquid crystalline phases can be observed by polarisation microscopy.

In a thermotrope liquid crystalline material, the formation of liquid crystalline phases is observed in a certain temperature range above the melting temperature. Above the clearing temperature, the thermal energy is high enough to destroy these ordered aggregates and the melt becomes isotrop.

Rapid cooling of a melt (quenching) and absence of crystallisation nuclei lead to the formation of a fully amorphous glass phase.

3.10 The thermodynamic non-equilibrium state of polymers

If a polymer melt could theoretically be crystallised to form a 100% crystalline material, then all monomer units participate to the ordered state in crystallographic equivalent positions.

In reality, the monomer units in the polymer chain are not independent, as they are linked through chemical bonds with others monomers of the chain. In addition, the high viscosity of the polymer melt makes chain disentanglement and uncoiling slow.

As a result crystallisation can start at the same time at different sites of one polymer. During the growth of the crystalline regions, a certain part of the polymer chain will have to remain in non-organised state and will thus form a second phase. This results in a partially crystalline mass with two different phases, which however is not in a state of thermodynamic equilibrium.

For a one-component material according the phase rule (eq. (3.2)), only one phase should be present at certain temperature and pressure.

Phase rule for a single component material:

$$\text{Phase rule}: P + F = K + 2 \tag{3.2}$$

The number of phases P should be unity for a single component because the number of components $K = 1$ for a single component, and the degree of freedom $F = 2$ (pressure, temperature).

As discussed, the approach that a polymer consists only of one type of chemical compound is not strictly valid (e.g., with consideration of the molecular mass distribution). The important conclusion, however, remains that a polymeric material will be only partly present in its thermodynamic equilibrium state. While the energy of the crystalline parts can be assumed to define a minimum state of energy, the energy of the amorphous regions is at a higher level, thus leading to

- a locally higher chemical reactivity
- the tendency for reorganisation above T_g
- higher chain mobility in this domains at elevated temperature.

Dependent on cooling conditions (temperature and drawing) and the concentration of nuclei for crystallisation, a polymeric material thus can be present in energetically different states, which then will also influence the overall textile chemical behaviour.

Because of the presence of a statistically most probable conformation, the structure of the molten polymer will be coiled and entangled, which then is the starting point for crystallisation. As a result chain folding will occurs during crystal formation instead of solidification in the straight conformation (Figure 3.12a).

In case a regular crystallisation occurs without preferential directions, spherulites are formed, which are highly ordered spherical structures (Figure 3.12b).

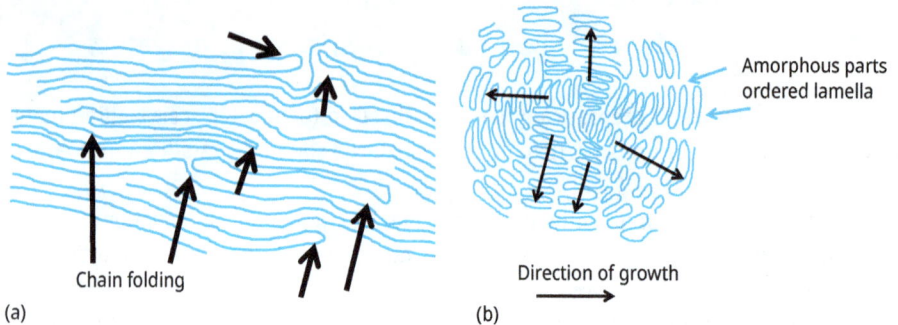

(a) (b)

Figure 3.12: Chain folding in polymer assemblies (a) and sperulites (b).

Crystallisation under shear stress increases uncoiling and orientation of the polymers in the direction of drawing, as a result a fibrillary structure is obtained.

Elastomers: The principle of crystalline and amorphous regions can be utilised to produce synthetic elastomers. As an example a triblock from PS (St_n) with polybutadiene (Bu_m) contains spherical PS domains, which exhibit a glass transition temperature of $T_g = 80\,°C$ in a continuous matrix of poly(butadien) with $T_g = -10\,°C$. In the temperature range between $-10\,°C$ and $80\,°C$, the material behaves like an elastomer, as the butadiene matrix offers flexibility and the "hard" PS segments maintain structural integrity. Above the T_g of the PS segments, the polymer can be processed like an thermoplast. Thus, such a material can be used as a thermoplastic elastomer for the production of sealings.

Below $-10\,°C$ both polymer blocks loose chain flexibility, the material is not longer elastic and the tightness of the sealing is insufficient.

3.11 Models of fibre structure

The characteristics of natural fibres are complex and dependent on their natural functionality; thus, application of fibre models is more straightforward for synthetic fibres.

In a general approach, the fibre structure can be visualised through a fringed fibrillar structure. A schematic presentation is shown in Figure 3.13. In this model, the fibre is built from a more or less ordered arrangement of crystalline/ordered domains that are linked together through polymer chains building amorphous less ordered areas.

The crystalline parts can be understood as highly ordered structures, which contain defined crystallographic unit cells. The situation of the amorphous parts is highly complex, offering a gradual change with regard to molecular order and environment. The chemical environment of polymer chains leaving the crystalline domain will be dependent on the distance to the crystallite and only in the fully amorphous part more or less constant conditions will be present.

The monomeric units in the amorphous part thus will be present at a different level of energy. Thus, in the amorphous part of the polymer, the reactivity as well as the sorption behaviour of monomer units will be dependent on the position of the

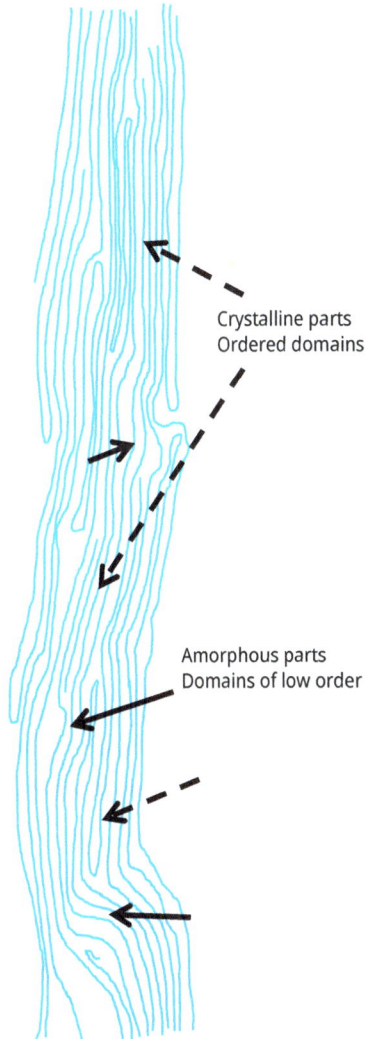

Crystalline parts
Ordered domains

Amorphous parts
Domains of low order

Figure 3.13: Schematic presentation of a fringed fibrillarystructure.

respective unit. A gradual change in properties will also determine the interaction with sorbates, for example, moisture, chemicals and the chemical reactivity.

For example, in an ideal crystalline substance, the sorption behaviour for moisture will be characterised by well-defined steps, for example, the formation of a molecular mono-layer and multi-layer adsorption.

The gradual change in chemical energy of monomeric units in amorphous parts thus will lead to different energy of interaction with sorptives, thus sorption isotherms can become more complex and the definition of a mono-layer adsorption can be difficult.

From the point of textile chemical processes, two different models, the pore model and the free volume model, can be formulated on the basis of the fringed fibrillar structure (Figure 3.14).

In the *pore model*, the polymer forms the body of the fibre, which also contains pores and voids (Figure 3.14a). Accessibility and transport of chemicals can then proceed from liquid or gaseous phase through diffusion into the pores along concentration gradients. The pores thus serve as transport channels for chemicals and dyes. Dependent on the porosity of the fibre, a very high internal surface can be observed, which, for example, in case of regenerated cellulose fibres reaches up to 200–400 m^2/g.

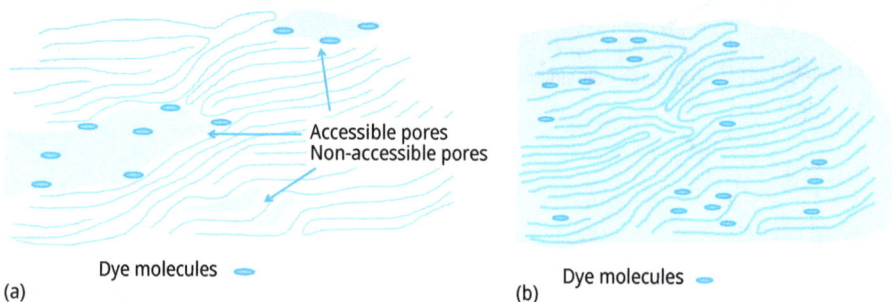

Figure 3.14: Models for fibre structure: (a) pore model and (b) free volume model.

The formation of pores and caves is a result of the fibre production process. Usually such a model is applicable for fibres that were obtained by coagulation and regeneration of a polymer from a diluted solution. When a spinning dope contains 20 wt% polymer, during coagulation and regeneration 80% of the extruded volume disappear during fibre formation. The removal of solvent and coagulation thus leads to a substantial shrinkage in the diameter of the regenerated fibre and also leads to the formation of caves and voids. These voids also serve as transport channels for solvent diffusion into the regeneration bath and for access of chemicals, for example, sulphuric acid in the case of the viscose fibre production or non-solvents in the case of lyocell or polyacrylonitril fibre production. The release of solvent and products (e.g., Na_2SO_4) from the inner part of the fibre occurs along these transport channels.

The *free volume model* describes a rather dense polymer structure (Figure 3.14b). Usually representative types of fibres for which this model applies are obtained by melt-extrusion processes. As there is no removal of chemicals or solvents, the polymer mass is shaped because of its thermoplastic behaviour. During cooling and drawing, the ordered crystalline parts form and become embedded by the amorphous parts. The structural behaviour of such fibres is characterised by two temperatures: above the melting temperature T_m the physical form of the fibre disappears and chain segments in crystalline zones become mobile. Above T_g only the chain segments in the amorphous parts are flexible and thus permit access of chemicals with appropriate size and polarity into the amorphous domains. The low density of the amorphous zones permits the formation of a free volume that can be accessed by dyes, optical brighteners and so on. Above T_g the amorphous parts can be understood as a "liquid" phase embedded in a solid fibre structure. Uptake of a disperse dye, for example, then is more like an extraction process from an aqueous dyebath into the "liquid" part of the fibre.

The free volume model thus is appropriate for the explanation of the behaviour of melt spun fibres, for example, polyester, polyamide and PP.

3.12 Consequences of polymer order in a fibre

The accessibility and the amorphous parts of a cellulose fibre will determine the capacity for *dyestuff sorption*. Thus, colour uptake and final shade also will be dependent on the molecular weight distribution of the polymer and the ratio between crystalline and amorphous parts of a fibre.

The fine structure of cotton as a natural fibre will always be dependent on the variety of plant, the growth conditions and the actual climate in the year of growth. To avoid drifting properties of spun cotton yarns, a lot of effort is placed at the beginning of the spinning process to mix bales and level out minor differences between lots.

However, even minor structural differences within the same batch of cotton yarn will lead to a slightly varying uptake of dyes. Apparently a spinning plant for cotton yarn produces a constant type of product; in fact, the overall quality of production is slightly drifting in its properties, with a very low frequency of weeks and months. Usually such slight differences in yarn quality are not visible in case two batches are compared. A critical situation can appear when different production batches of the same yarn, for example, produced in a time distance of few months are combined to a uniform round knitted fabric. In this case, stripes can appear after the dyeing process, which result form a very slight difference in dyestuff fixation, which is due to very minor structural differences in the cotton fibre.

Similarly the order of a polyamide fibre will also be dependent on polydispersity, spinning conditions, temperature control during fibre spinning and draw rate. The molecular order in the fibre will thus be directly dependent on the conditions of the fibre spinning process. Slight differences in production conditions will cause formation of visible colour differences in uniformly dyed fabric in case different production lots had been combined to one fabric.

3.12.1 Moisture/adsorption/swelling

The different models of fibre structure can be used to explain the general behaviour of fibres with regard to sorption or uptake of low molecular weight substances, for example, moisture, crosslinkers and catalysts for finishing of cellulose fibres, the function of carriers in polyester dyeing and softeners for PVC coatings.

Using of the pore model for cellulose fibres, the access of chemicals into the fibre can be understood to proceed by wicking and diffusional transport into the pores.

During wetting with water an expansion of the cellulose fibre structure occurs. Water attempts to dissolve carbohydrate segments in the non-crystalline parts, which however will not be successful, as the cellulose polymer remains bound in crystalline parts and also partly in the amorphous zones. As a result the accessible parts and the inner surface of the fibre begins to develop a gel-like structure with high water content, and the fibre expands and swells. Swelling can be understood as a first stage on the way to full dissolution of a polymer, which however only will be successful if the crystalline domains also can be dissolved.

As a result for textile chemical, considerations we can summarise the following general aspects for fibres, which are represented by the pore model:
- Accessibility for chemicals will be possible only through voids and pores.
- Reactivity will be concentrated to the amorphous parts.
- The presence of swelling agents will control the depth of access of chemicals.
- Fibre expansion will be important to increase access and thus reactivity.

An important example for a technically used process is the "alkali activation" for cellulose fibres. A pre-treatment of cellulose material in concentrated alkali is used to transfer the fibres into a highly swollen state, which exhibits higher reactivity and accessibility [2, 3].

The reactivity of a fibre in dried solvent-free state can thus be substantially different compared to a fibre in wet, swollen condition. Besides overall reactivity, the sites of reaction will also differ.

Swelling operations will modify the accessibility and reactivity of a cellulose fibre (Figure 3.15). As a consequence, the accessible volume of a swollen structure is dependent on the molecular size of the respective reagent. Thus, the accessible volume of a swollen fibre for finishing reagents will be different and decreases in the following order: water > magnesium chloride > crosslinker (dimethylol-dihydroxyethylene urea).

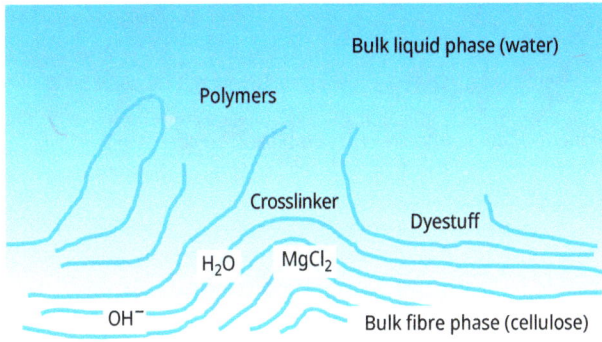

Figure 3.15: Accessibility of molecules into swollen "surface" of a cellulose fibre in transition between fibre structure and polymer dissolution.

Operations that change the accessibility of a cellulose fibre will thus influence the outcome of a crosslinking operation in a later stage of finishing.

The gradual difference in attractive forces of monomer units in the amorphous zones also explains the climate-dependent moisture sorption behaviour of cellulose and the difficulty to define a water-free state of cellulose. Dry cotton stored at 20 °C, 65% r.h. contains approximately 8.5 wt% of sorbed water. Technically dry cellulose is obtained through drying at 105 °C, however, still contains strongly bound water molecules, which will desorb at higher temperature.

The presence of water also increases the flexibility of cellulose chains, an effect that is utilised to achieve plasticisation of cellulose during steam ironing. This effect allows steam forming of cellulose fabrics at temperatures of 100–120 °C. The glass transition temperature of cellulose itself is about 225 °C, where degradation of the carbohydrate occurs [1].

Carrier: The effect of plasticisation is also observed in fibres where the free volume model is applicable. Plasticising chemicals enter into the amorphous domains and thereby lower the glass transition temperature. As a result the flexibility of the polymers chain increase, which, for example, is used to soften polyvinylchloride-based material. In polyester (polyethylenterephthalate) dyeing, high temperature and increased pressure (HT-dyeing, thermosol dyeing) are required to achieve sufficient rate of dyestuff uptake, or the carriers are added to the dyebath to lower the dyeing temperature below 100 °C. The carrier molecules reduce T_g of the polyester to such a extent that an acceptable dyeing rate is observed at the boil and permits use atmospheric dyeing apparatus.

References

[1]   Elias, HG. An Introduction to Polymer Science, VCH Wiley, Weinheim, Germany, 1997, ISBN 3-527-28790-6

[2]   Jaturapiree, A., Ehrhardt, A., Groner, S., Öztürk, HB., Siroka, B., Bechtold T. Treatment in Swelling Solutions Modifying Cellulose Fibre Reactivity – Part 1: Accessibility and Sorption. Macromolecular Symposia: Zellcheming. 2007 conference proceedings 2008, 262, 39–49.

[3]   Bui, HM., Lenninger, M., Manian, AP., Abu-Rous, M., Schimper, CB., Schuster, KC., Bechtold, T. Treatment in Swelling Solutions Modifying Cellulose Fibre Reactivity – Part 2: Accessibility and Reactivity. Macromolecular Symposia: Zellcheming. 2007 conference proceedings 2008, 262, 50–64.

Take home messages

Textile fibres are built from polymer molecules. The structure of the long polymer chains influences the order in a polymer material by different factors: polymolecularity, constitution, configuration and conformation, molecular interactions, orientation, crystalline and amorphous parts and liquid crystalline phases.

The final physical properties of a fibre as well as sorption behaviour and chemical reactivity are determined by the internal order.

Two major models to visualise the internal structure of a polymer can be used: the pore volume model and the free volume model. The choice depends on the type of polymer and the technique of fibre formation.

Quiz

1. Explain the terms constitution, configuration and conformation.
2. Name four types copolymers and draw general examples.
3. What is the influence of liquid crystalline phases present in the polymer melt on the properties of a fibre?
4. At temperatures above T_g of a thermoplastic fibre reorganisation can take place. Why this process is an continuously ongoing process?

Exercises

1. Apply the definitions for polymer constitutions on the following polymers: cellulose, polyethylene terephthalate, wool, silk, polyamide 6, polysiloxane (fibre) and carbon.
2. Polymerisation of acrolein (CH_2=CH–CH=O) can proceed through different pathways, thus leading to polymers with different functionality and constitution. Try to draw at least five polymers with different constitution.

3. A molecular 3D network represents a single molecule. Taking a block of 1 kg mass that is completely crosslinked, the mass of a single molecule then is 1,000 g. Could we produce 1 mol of this substance on earth? How big this block would be (mass of earth = 5.9×10^{24} kg)? What mass corresponds to 1 mol of water and 1 mol of PP ($P_w = 7,000$)?

Solutions

1. (Question 1): Constitution: sequence of combination of atoms in a molecule
Configuration: considers the stereochemistry
Conformation: considers rotations around single bonds
2. (Question 2): Alternating bipolymer   A–B–A–B–A–B–
Statistical bipolymer   A–B–B–A–B–A–A–A–B–A–B–B–
Gradient bipolymer   A–A–A–B–A–B–B–A–B–B–B–B–
Block copolymer   A–A–A–A–B–B–B–B–
3. (Question 3): The formation of LCP in a polymer melt influences the order of polymer chains in the solid fibre.
4. (Question 4): Upon solidification of the polymer melt the fibre polymers cannot reach the perfect equilibrium state, part of the polymer chain remains in less ordered amorphous domains, which tend to stabilise further above T_g as chain mobility is increased. This reorganisation is a continuously ongoing process; of course the extent of stabilisation decreases, and thus a fibre can be stabilised and relaxed by a controlled thermal treatment.
5. (Exercise 1): Explain the covalent bindings in each polymer using its structure formula: e.g. cellulose = homopolymer, heterochain polymer; PET = alternating bipolymer from terephthalic acid and glycol, or homopolymer of terephthalic acid monoglycolester, heterochain polymer; wool = heterochain polymer; PA6 = homopolymer, heterochain polymer; polysiloxane = heterochain polymer; carbon: homochain polymer (complex structure!).
6. (Exercise 2):

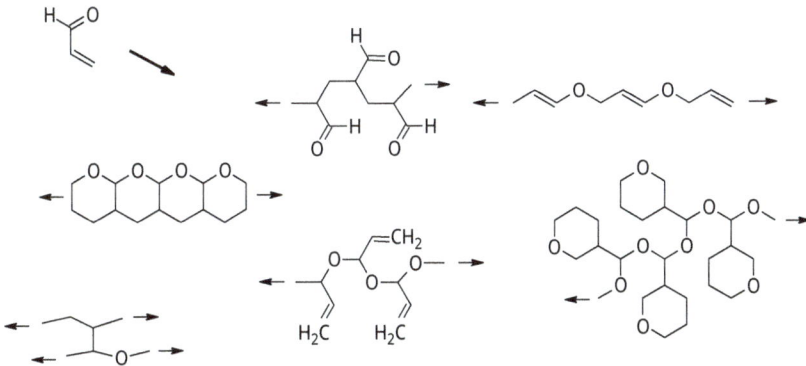

7. (Exercise 3): The mass of 1 mol water is 18 g, the mass of 1 mol PP = 7,000 × 42 = 294,000 g = 294 kg. The mass of the block = mass of one molecule (1 kg) x Avogadro number (6.022 × 10²³) = 6.022 × 10²³ kg! The mass will be 10% of the mass of the earth; thus, this will be an impossible project.

List of abbreviations/symbols

PP	polypropylene
PA	polyamide
PES	polyester
PVC	polyvinylchloride
CV	viscose
CC	cupro fibre, cuprammonium fibre
CA	cellulose diacetate
CT	cellulose triacetate
COC	cyclic olefin copolymer
T_g	glass transition temperature
T_m	melting temperature

4 Basic interactions between fibre polymers and sorptives

4.1 General

In Chapters 2 and 3, the basics of the structure of a textile fibre and interactions leading to attractive forces between molecule chains have been discussed.

In a simplified approach, we can assume that attractive intermolecular and intramolecular forces within crystallites are saturated through the formation of crystal lattice.

In the amorphous parts of a fibre structure, the pattern of interactions with neighbouring monomer units is incomplete and functional groups remain available for interaction with other molecules; for example, sorptives or reactants from vapour phase or liquid phase. The uptake of small molecules into the fibre structure forms the basis for a wide range of phenomena observed in textile chemical processes. Examples are:

–   Vapour sorption from gas phase
–   Swelling through uptake of water, liquid
–   Dyestuff sorption in textile colouration

In this chapter the fundamental physical–chemical interactions leading to interactions of functional groups present in a polymer with low molecular weight substances will be summarised to form a general basis for the understanding of physical–chemical principles in textile chemical operations.

4.2 Dipoles

The majority of polymers and macromolecules – for example, polyester, polyamide, cellulose, wool – contain hetero elements such as oxygen, nitrogen or sulphur as part of their chemical structure. The difference in electronegativity between carbon, hydrogen and the heteroelements leads to an imbalance in the distribution of electron density along the chemical bond (Table 4.1). This causes the appearance of partial charges $+q$ and $-q$ at the two elements bound through the polar covalent bond.

A dipole is formed. In Figure 4.1 relevant dipoles are indicated in the formulas of important polymers.

The dipole μ moment is calculated according to eq. (4.1):

$$\mu = q \times l \qquad\qquad (4.1)$$

with μ being the dipole moment in Coulomb meter (C m), q being the charge in Coulomb and l(m) the distance between the two partial charges. Often still the unit Debye (D) is used instead of Coulomb with $1\,D = 3.33 \times 10^{-30}$ C m. As can be recognised by the

https://doi.org/10.1515/9783110795738-004

Table 4.1: Electronegativity of important elements present in polymers.

Element	EN
Carbon	2.5
Hydrogen	2.1
Oxygen	3.5
Nitrogen	3.0
Sulphur	2.5
Chlorine	3.0
Fluorine	4.0

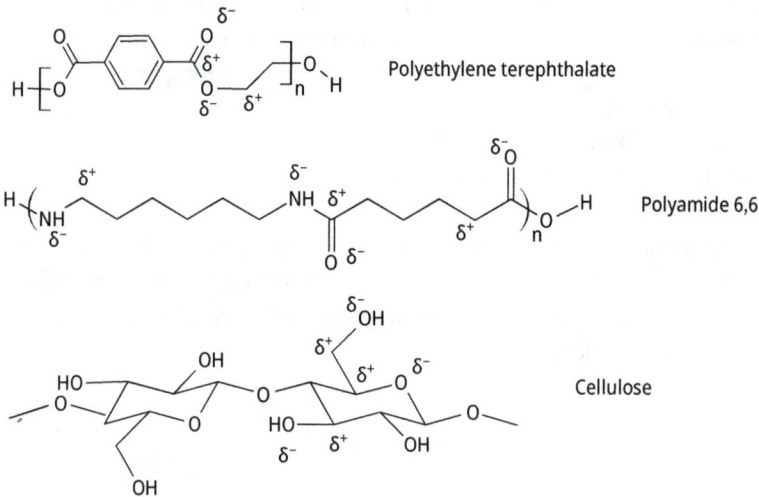

Figure 4.1: General scheme of a dipole, dipoles in polyester, polyamide, cellulose (δ^+ positive partial charge, δ^- negative partial charge).

transformation, the SI unit is a rather big one, compared to Debye, which is more appropriate for description of dipole moments on molecular scale.

When a positive and a negative charge is separated by a distance of 0.1 nm, then a dipole moment of 1.6×10^{-29} C m is achieved, which corresponds to 4.8 D. In polar molecules, the dipole moment thus will be smaller and usually less than two (Table 4.2).

A dipole can also arise from differences in atomic radii, as this causes an imbalance in electron distribution. This so-called homopolar contribution explains formation of dipoles in case of carbon–hydrogen bonds. However, an exception exists for molecules which fulfil defined symmetry criteria belonging to a D-point group. While small molecules, for example, methane or benzene, would fulfil such criteria, the symmetry of a polymer is more complex.

Table 4.2: Dipole moments of models for small molecules in gas phase and polymers with related structures.

Material	μ/D	Analogous to
Ethylene glycol	2.28	Cellulose
Glycineethylester	2.11	Wool
Valineethylester	2.11	Polyester
Acrylonitrile	3.87	Polyacrylonitrile
Caproamide	3.90	Polyamide
Urea	4.56	Polyurethane
Ammonia	1.47	–
2-Methyl butane	0.13	Polypropylene
Water	1.85	–
Methanol	1.71	–
Ethanol	1.69	–
Benzene	0	–

In a simplified approximation, the pattern of dipole moments in a polymer can be understood as result of a combination of more or less independent functional groups.

In polymer chains, the presence of functional groups containing heteroelements or other bonding characteristics thus lead to the development of a high number of polar sites which bear localised partial charges.

The majority of fibre polymers are polar molecules which exhibit a repetitive pattern of dipole moments in every repeating monomer unit.

4.3 Polarisability

A dipole moment can result from distortion of a molecule in an electric field. The induced dipole moment μ^* is dependent on the magnitude of the electrical field E (V/m) (eq. (4.2)) [1].

$$\mu^* = aE \tag{4.2}$$

The proportionality constant α is called polarisability (unit C^2m^2/J). The higher the polarisability of a molecule, the higher the induced dipole will be. Instead of α, often the polarisability volume α' is used (eq. (4.3)).

$$\alpha' = \alpha/4\,\pi\,\varepsilon_0 \tag{4.3}$$

with ε_0 being the vacuum permittivity (unit C^2/Jm). Thus the polarisability volume α' has the dimension of volume. The polarisability volume correlates with the separation of the HOMO (highest occupied molecular orbital) and LUMO (lowest unoccupied molecular orbital). Molecules with large distance between HOMO and LUMO show low polarisability. A distortion of charge distribution is possible, if the distance between

HOMO and LUMO is low, which is the case for large molecules with a high number of electrons.

As the orientation of permanent dipoles requires time, polarisation also is dependent on the frequency of the electromagnetic field leading.

Orientation polarisation: In case rotation of a dipole is permitted by the structure of the macromolecule, the dipoles can change their direction according to the applied electrical field. Above a frequency of approximately 10^{11} Hz (microwave irradiation), this part of polarisation disappears.

Distortion polarisation: This polarisation bases on a distortion of nuclei in the applied electrical field. As this frequency must lie in the region of molecular vibration, the distortion polarisation is lost at frequencies above the infrared region.

Electronic polarisation: This polarisation bases only on distortion of electron distribution. As an example in the visible range of electromagnetic irradiation, only electrons can redistribute rapid enough to achieve distortion.

Through the definition of the molar polarisation P_m (cm^3/mol) (eq. (4.4)) and the Debye equation (eq. (4.5)) the dipole moment is related to the relative permittivity ε_r.

$$P_m = N_A / 3\varepsilon_0 (\alpha + \mu^2 / 3kT) \tag{4.4}$$

$$(\varepsilon_r - 1)/(\varepsilon_r + 2) = \rho P_m / M \tag{4.5}$$

ρ is mass density of sample (g/cm^3), M is molar mass (g/mol), k is Boltzmann constant (J/K) and T is temperature (K).

The relative permittivity of important polymer materials is summarised in Table 4.3.

Table 4.3: Relative permittivity of relevant polymer materials.

Material	ε_r	Frequency kHz	n_r
Air	1.00		–
Water	81.1		1.33
Cotton	18*	1	1.56
	3.2**		
Viscose	8.4*	1	1.54
Polyamid 6,6	3.7*	1	1.55
Polyacryl	4.2*	1	1.52
Polyethylene terephthalate	2.3*	1	1.63
Polypropylene	2.4	–	1.49
Polyvinychloride	3.5	1	1.53
Wool	5.5*	1	1.55

*65% r.h. **0% r.h.

The refractive index n_r for a given frequency is related to the relative permittivity ε_r of the material (eq. (4.6)).

$$n_r = \varepsilon_r{}^{1/2} \qquad\qquad (4.6)$$

Thus, an interlinkage between structural properties, which also determines dipole moment, polarisability and relative permittivity, and the optical properties exists.

As a result, the polymer order in a fibre also determines the refractive index, which depends on the direction of measurement, related to the fibre axis. The refractive index of polyamide 6.6 parallel to the fibre axis $n_\parallel$ is in the dimension of 1.580, while the refractive perpendicular to the fibre axis $n_\perp$ is 1.519.

Measurements of refractive index and microscopy of fibres with polarised light thus belong to the standard methods to assess order and internal structure in synthetic fibres; for example, as function of spinning conditions, draw rate, level of thermofixation.

Another important property of the permittivity is its strong dependency on sorbed water, which can be seen in case of cotton, where a permittivity $\varepsilon_r = 8$ is observed under conditions of higher humidity (65% r.h.) while a decrease to $\varepsilon_r = 3.2$ occurs in a dry atmosphere (0% r.h.).

4.4 Molecular interactions

Molecular interactions form the physical-chemical basis for attractive forces between polymer chains as well as between the polymer molecule and smaller molecules; for example, solvents, moisture, dyes and chemicals.

Depending on the origin and involvement of functional groups, different attractive forces can be distinguished as follows:
- Ion–ion interactions
- Hydrogen bonds
- Van der Waals forces
- π-electron interactions
- Hydrophobic interactions

4.4.1 Ion–ion interactions

The Coulombic attraction/repulsion between charged sites in polymers or ions in solutions belong to the group of strong forces which are effective over long distance.

In Table 4.4, relevant interactive forces and examples for textile chemical processes are given.

Table 4.4: Dependence of potential energy on distance r for different types of interaction (from Atkins and de Paula) [1].

Interactions	Decrease of potential energy with distance	Magnitude of energy/kJ/mol	Example
Ion–ion	$1/r$	250	Acid dye on wool [2]
Hydrogen bond		20	Cellulose
Ion–dipole	$1/r^2$	15	Wool
Dipole–dipole Stationary	$1/r^3$	2	Polyacrylics
Dipole–dipole Rotating	$1/r^6$	0.6	NO_x sorption on polyester
London forces, dispersion forces	$1/r^6$	2	All polymers
π-interactions			Polyester
Hydrophobic interactions			Polypropylene

The energy level of an ion–ion interaction is in the magnitude of a chemical bond. As an example, the dissociation energy of a covalent bond between two hydrogen atoms is 432 kJ/mol, and for a C–C single bond 368 kJ/mol. With increasing bond length the decrease in potential energy is linearly related to the distance, thus the range of this type of interaction is comparably long.

A considerable number of textile chemical effects base on the interaction of two ions of opposite charge:

– The ion–ion interaction of charged groups in wool is an important element in the structure formation and stability of the fibre. Shift of pH away from the isoelectric point leads to protonation of carboxylic groups in the acid region or dissociation of ammonium groups in the alkaline region. In both cases, the physical strength of the fibre is decreased as stabilising ionic linkages in the fibre disappear.

– Ionic bonds in the protein structure can be opened and closed easily, and also through thermal effects; for example, steaming, which is utilised to form permanent creases in woollen clothes.

– The binding of acid dyes in wool and silk bases on the formation of a ion bonding between positively charged ammonium groups in the protein structure and the negatively charged acid dye. Shift of pH into the alkaline region will lead to dissociation of the ammonium groups and thus weakens the sorption of the acid dye which becomes mobile and begins to bleed out.

– Cellulose can be modified chemically to bear cationic groups (quaternary ammonium ions), which then serve as binding sites for sorption of anionic dyes.

– Cationic polyelectrolytes can serve as agents to improve fastness of anionic dyes – for example, acid dyes, reactive dyes – through formation of insoluble aggregates on the fibre surface.

Also, a considerable number of processes utilise the repulsion of molecules/particles with same charge:

– Washing agents in laundry often utilise the sorption of negatively charged polyelectrolytes or the sorption of negative ions to increase electrostatic repulsion between adsorbed soil particles and support release into the wash liquor.
– Stabilisation of a vat dye dispersion, including indigo, can be achieved through
 the use of negatively charged lignosulphonates as dispersing agents (Setamol
 WS®, BASF), which prevent agglomeration of finely dispersed dyestuff pigments
 by adsorption at the particle's surface and mutual repulsion through their negative charge.

4.4.2 Van der Waals forces

Van der Waals forces include attractive forces between closed shell molecules. This
includes attraction between dipoles as well as repulsive forces between molecules,
which increase with short distance.

The attractive forces decrease by $1/r^6$ while the repulsive forces increase by $1/r^{12}$.
Both the interactions are summarised in the Lennard–Jones potential (Figure 4.2).

$$V(r) = 4\, E_{max} \left[(r_0/r)^{12} - (r_0/r)^6 \right] \tag{4.7}$$

r distance between dipoles (m)
V potential energy (J)
r_0 distance at potential energy $V = 0$
E_{max} maximum energy of attraction (J)

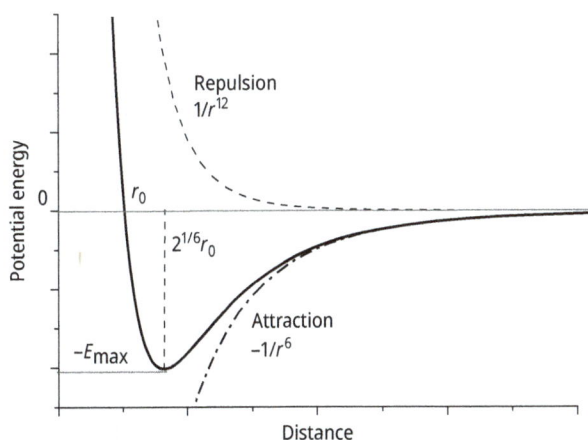

Figure 4.2: The Lennard–Jones potential (eq. (4.7)).

The attractive forces include a number of principles:
- Attraction between two dipoles, as an example, between two nitrile groups of neighbouring chains in poly acrylonitrile.
- Attraction between a dipole and an induced dipole which results from the polarisation of an atom adjacent to the dipole.
- Attraction between a transient dipole present in a non-polar molecule as result of fluctuation in electron distribution and an induced dipole (London interaction, dispersion forces). Such effects even include attractive forces between noble gas atoms.

Van der Waals forces are present in every polymer material; however, the contribution to the total attractive forces varies.

Compared to the attractive forces of salt bridges (250 kJ/mol) and hydrogen bonds (20 kJ/mol), the contribution of Van der Waals forces is low (2 kJ/mol). These differences in energy are of relevance for fibre polymers as well as for interaction of a polymer chain with smaller molecules.

As a result of the low attraction between monomer units of a non-polar polymer which does not bear polar groups, such as polypropylene, the average chain length of such a polymer must be substantially longer than in case of polyacrylonitrile or polyester, where stronger attractive forces appear.

In comparison to non-polar molecules, polar substances such as formaldehyde, sulphur dioxide and nitrous oxides adsorb more polymers which contain polar groups.

4.4.3 Hydrogen bonds

Hydrogen bonds represent a special case of Van der Waals forces. In case a hydrogen atom is bound to a heteroelement with higher electronegativity, for example, O, N, S, the hydrogen atom bears a positive partial charge. When another atom with negative partial charge is near to this hydrogen atom, attractive forces lead to overlap of the hydrogen electron orbitals with both orbitals of the two heteroelements. As a result, a hydrogen bridge is formed, which links both heteroelements together.

The strength of this special case is remarkably higher compared to other Van der Waals forces as typical values are near to 20 kJ/mol.

Hydrogen bonds are of enormous relevance in many textile chemical applications:
- The hydroxyl groups in cellulose build intra- and intermolecular interactions which contribute to the formation of highly ordered crystalline assemblies and also prevent sufficient thermal mobility which would be required to achieve thermoplastic behaviour.
- Hydrogen bonds are relevant structures determining forces in protein-based polymers, polyamides and polyurethanes.

-   The sorption of moisture as well as many other hydroxyl and amino group containing molecules on cellulose, wool and polyamide is supported by formation of hydrogen bonds between sorbent and sorbate.
-   Many textile dyes, which were not bound through covalent bond formation (reactive dyes) or are present in pigment form (vat dyes, sulphur dyes and azoic dyes, naphtholes), remain strongly bound to the fibre molecules due to the presence of multiple hydrogen bonds.
-   At standard conditions like 20 °C and 65% r.h., water will be adsorbed at the surface of cellulose and will be strongly bound by hydrogen bonds. This interaction modifies the structure of the bound water and as a result three types of water can be distinguished in cellulose structures: capillary bound water, which is kept in voids and capillaries and freezes at 0 °C, freezing bound water, which exhibits a freezing point below 0 °C. The third type of water is the non-freezing bound water, which does not exhibit any thermal event in DSC, which could be attributed to the event of freezing.

The hydrogen bonds exhibit thermal dependency in their length and organisation, while a tetrahedral geometry with 90% intact hydrogen bonds is observed near to the freezing point of water, the length of the hydrogen bonding as well as the statistical probability for a hydrogen bond reduces to 20% at the boiling point [1].

4.4.4 Hydrophobic interactions

Attractive forces are also observed between sites of low polarity; for example, hydrophobic non-polar long chain residues of surfactants that form aggregates in polar solvents. Another example is the cave of a cyclodextrin. A cyclodextrin is a cyclic oligomer of a carbohydrate and is used as carrier and sorbent for apolar low molecular weight substances; for example, fragrances and pharmaceutically active agents.

In case of cellulose, the polar hydroxyl groups are arranged in equatorial position, and the sites above and below the pyranose ring are considered to be non-polar, thus causing hydrophobic interactions [3, 4].

An analysis of the thermodynamic principles behind these attractive forces shows that the hydrophobic attraction is mainly due to entropic effects (Figure 4.3).

For a spontaneous reaction the overall free energy $\Delta G = \Delta H - T\Delta S$ for the reaction must be negative. This can be achieved either through a change in enthalpy $-\Delta H$, in this case the reaction is exothermal or through an increase in entropy $+\Delta S$. To estimate the overall change in entropy, the individual changes in entropy of both reaction partners has to be considered. On molecular level, the entropy change of a surfactant which forms aggregates will be negative as the level of order increases. The situation of the solvent is different. A polar solvent such as water is forced to form a highly ordered state in the neighbourhood of the long hydrophobic tail of a surfactant as no

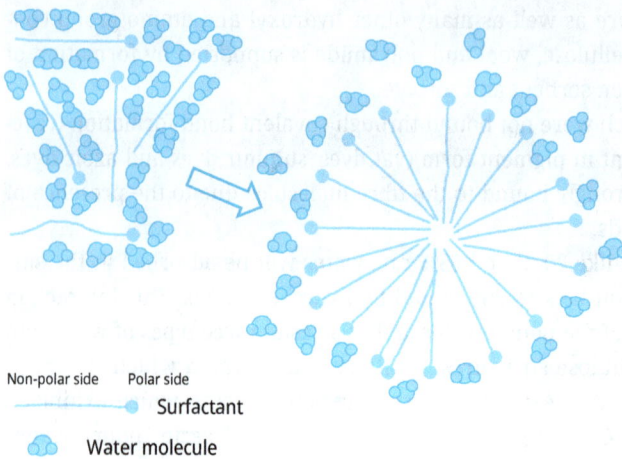

Non-polar side   Polar side
—————————● Surfactant

🔵 Water molecule

Figure 4.3: Surfactant aggregation as result of hydrophobic interactions – lower structuring of water molecules leads to increase in entropy.

hydrogen bond or dipole interaction to the non-polar organic tail can be established. Thus, in these domains the state of order of a high number of water molecules is at elevated level compared to the bulk solvent. The formation of surfactant aggregates allows the water molecules to reduce their order nearer to the level of bulk water. As the aggregation of a long chain surfactant releases a lot of smaller water molecules, the overall contribution of the water molecules leads to an increase in entropy which exceeds the contribution of the ordering of the surfactant.

Thus, what we observe as attractive force actually is an entropic effect.

A typical example of the effects of hydrophobic interaction is demonstrated in the literature where model calculations for the state of water molecules surrounding cellulose crystallites have been published. As a result, it could be shown that above the apolar pyranose rings the water molecules are forced to establish highly ordered structures with the aim to saturate all hydrogen bonds within the liquid water phase [5].

4.4.5 π-electron interactions

The interactions between larger conjugated systems of aromatic components also lead to attractive forces. The aromatic electron ring system is a site of high electron density. π-electron interactions can be divided into three different groups:

– Interactions between cations and the face of the aromatic ring, such effects are observed in proteins where cations can induce structure forming effects through interaction with π-electron systems of aromatic amino acids [6].

- Interactions between the two aromatic ring systems π–π interaction, as for example, are observed as structuring elements in crystalline parts of polyethylene terephthalate.
- Interactions between positively polarised protons and the π-electron systems: X–H . . . π-electron system. As an example, an interaction between water and the benzene ring system is energetically favourable (approx. −8 kJ/mol).

π-electron interactions can contribute substantially to the stability of material as for example is observed in aromatic polyamides, polymers with high number of condensed aromatic systems. Important examples are the BBB fibre (Poly-bis-benzimidazol-benzo-phenanthrolin; Figure 4.4), or in carbon fibres, which partially already exhibit a graphite-like structure.

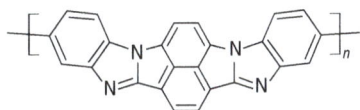

Figure 4.4: The chemical structure of the BBB (Poly-bis-benzimidazol-benzophenanthroline) fibre.

Summarising the number and principles of possible interactions between molecules, the surface of a textile fibre will exhibit a complex pattern of attractive forces, which will determine the interactions with a gaseous or liquid environment.

Interactions with the surrounding atmosphere mainly concentrate on the sorption of polar gaseous components, the most important among them being water (Chapter 5).

Substantially lower sorption is observed for the other atmospheric components such as nitrogen, oxygen, carbon dioxide or noble gases which only adsorb through Van der Waals interactions.

In the interactions with liquid phases, a boundary layer at the interface between the fibre polymer and the liquid phase is formed.

As the vast majority of textile chemical processes are performed from an aqueous environment, the boundary layer at a fibre–water interface will be discussed in more detail as a representative case for the phenomena which also will occur in other fibre–solvent systems in a similar way.

4.5 The polymer–solvent boundary layer

In the bulk phase of a polymer, the intra- and intermolecular attractions lead to the formation of solid polymer network. The same attractive forces also determine the interactions between solvent and polymer surface. As a consequence of the interactions between the solvent and the polymer surface, different effects will result:

- Depending on the respective surface energies, both of the solid and the liquid phases being in direct contact wetting of the polymer surface can occur.

- A structured solvent layer on the surface of the polymer and a boundary layer which contains oriented molecules form. Dipole–dipole interactions, hydrogen bonds, for example, account for a structuring of the boundary layer between water and cellulose. The boundary layer can be imagined as a zone next to the fibre surface in which a gradient in the concentration of two phases – 1. solid cellulose and 2. liquid bulk water – exists. The gradient begins at the surface of the cellulose crystallite with the relative concentration of cellulose = 1 and of water = 0, which then gradually changes to the bulk liquid phase where the concentration of cellulose = 0 and the concentration of water = 1. Thus, the surface of a cellulose fibre in water resembles more to a gel structure, than to a solid material surface.
- Diffusion and interaction of solvent molecules into the polymer matrix lead to expansion of the fibre volume, and thus swelling occurs. Many solvents lead only to a swelling of the polymer, thus access occurs only into the amorphous parts of the structure, but no dissolution occurs [7].
- In case the overall interactions between solvent and polymer permit formation of a preferential state in terms of Gibbs free energy ΔG, the solid polymer state is less favourable and the polymer begins to dissolve (Figure 4.5).

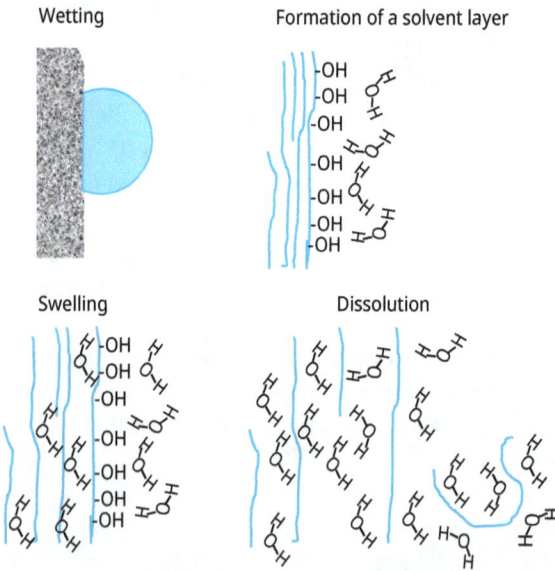

Figure 4.5: The different phenomena at the fibre–liquid interface (a) wetting, (b) formation of a sorbed solvent layer, (c) swelling and formation of concentration gradients, (d) initial swelling proceeds with dissolution (blue lines = polymer chains).

On the surface of a fibre polymer, the polar groups are present in partially unsaturated state, which causes the development of a certain surface energy. The surface energy of a textile material is of considerable impact on a number of textile chemical reactions and processes.

The wettability of a polymer material for process solutions such as water-based pretreatment, dyeing and finishing solutions will directly depend on the surface energy of the material. Rapid and uniform access of process chemicals to the fabric structure and into the fibre structure is only possible in case sufficient wettability is granted.

Besides a careful pretreatment of material through a prewash, for example, to remove oily substances which exhibit low surface energy, an increase in surface polarity is possible through plasma treatment. Also, the use of surfactants is another strategy to achieve wetting, in this case by reduction of the surface energy of water to a level which permits stable process conditions in terms of wetting. In Table 4.5, relevant surface energies of textile fibres are summarised.

Table 4.5: Surface energies of textile fibres, skin and selected liquids.

Material	Surface energy mN m^{-1}
Water	72.8
Viscose (32% r.h.)	62.2
Polyamide 6	43.2
Polyester (PET)	37.0
Polyacryl	37.4
Polypropylene	28–30
Oil	32
Sweat	32
Skin (forearm)	28 (and higher)

From the data in Table 4.5 it becomes evident that wetting of a viscose fibre, and similarly of cotton, with water should not be a problem as long as greasy substances (spin preparations, knitting oils), and in case of cotton the waxes, at the surface of the fibre have been removed.

On the contrary, the wetting of a polypropylene fibre with water requires use of appropriate surfactants. Remarkably, the surface energy of sweat is much lower compared to water, thus the uptake of sweat in a PP-baselayer of sports garment is not that problem.

In a similar way, the adhesion of textile chemicals such as pigment binders, coatings, melt glues will depend on the respective surface energy of the fibres. Plasma treatments permit introduction of polar groups at the surface of a polymer. In case of a polymer with low surface energy, this will lead to an increase in surface energy. The time stability of such functional groups is rather limited, and ageing is observed during short time [8]. In case of plasma activation thus the following printing processes, coating with adhesives and so on have to be executed within a short period of time after plasma activation has been performed.

A frequently used method to characterise the surface energy of a plane material is the determination of the contact angle.

It is important to consider that the measurement of a contact angle in principle is possible on any material; however, the influence of the structure of a textile material must not be neglected. While the theoretical background of the concept is valid for plane materials, the complex fibrous structure of a textile fabric will make a 'real' determination of a contact angle difficult.

The determination of the contact angle utilises the Young equation (Figure 4.6; eq. (4.8))

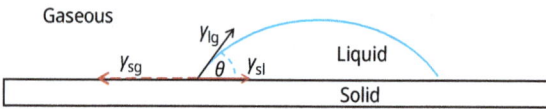

Figure 4.6: Equilibrium state of wetting: compensation of the different surface tensions between solid/liquid/and gas phase.

Young equation:

$$\gamma_{sg} = \gamma_{sl} + \gamma_{lg} \cos \theta \tag{4.8}$$

γ_{sg} interfacial energy solid–gas phase (mN/m)
γ_{sl} interfacial energy solid–liquid phase (mN/m)
γ_{lg} interfacial energy liquid–gas phase (mN/m)
θ contact angle (degree °)

Spreading stops at the moment when equilibrium between the three vectors has been established. When liquids with known surface energy are used, the respective surface energy of a material can be determined according to the Young equation from a measurement of the contact angle, for example, by microscopy.

4.6 Zeta potential

Besides dipoles and polar sites which are available establishing hydrogen bondings, charged species are also present in the boundary layer of a polymer.

These charged species can have different origin:

– Charged groups can be formed through dissociation of carboxylic groups, sulphonic acids, phosphate esters or protonation of basic groups such as amino groups of the polymer (eqs. (4.9–4.11)).

$$-\text{COOH} \rightleftharpoons -\text{COO}^- + \text{H}^+ \tag{4.9}$$

$$-\text{SO}_3\text{Na} \rightleftharpoons -\text{SO}_3^- + \text{Na}^+ \tag{4.10}$$

$$-\text{N}(\text{CH}_3)_2 + \text{H}^+ \rightleftharpoons -\text{NH}(\text{CH}_3)_2{}^+ \tag{4.11}$$

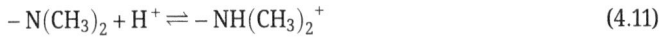

Charged molecules and ions also adsorb at the surface of the polymer, thus contributing to the surface charge. Examples being negatively charged dyes, cationic softeners and finishing chemicals, adsorption of metal ions such as Ca^{2+}, Mg^{2+} or heavy metal ions [9, 10].

– Ions like cations can form complexes with the polymer structure. Such phenomena, for example, are observed during adsorption of heavy metal ions on cellulose or protein fibres (wool, silk).

– Polyelectrolytes adsorb at the surface of the fibre. The use of carboxymethyl cellulose as an anti-soil-redeposition agent and for improved soil release, for example, utilises this principle. The adsorption of polycations on dyed fibres as auxiliaries to improve dyestuff fastness of the anionic dyes leads to formation of positive fibre surface charge.

For reasons of electroneutrality the charged groups bound to the surface of the fibre also requires the presence of oppositely charged counter ions. Through thermal mobility and possible effects of external fields, these counter ions can diffuse into the bulk liquid phase. As a result, charge separation between surface bound ions and mobile counter ions occurs, which leads to development of an electrical field. The electrical field represents a compensating force which limits the diffusion of the mobile counter ions into the bulk liquid. A stable electrical double layer is established, in which a dynamic equilibrium between diffusion of mobile ions into the liquid phase and attraction through the resulting electrical field exists.

The electrical double layer is formed at the surface of a solid material in contact to a liquid material. The electrical double layer exhibits a number of characteristic elements (Figure 4.7). In the example given in Figure 4.7, the negative ions sorb on the fibre surface and the mobile cations diffuse into the bulk solution.

Within the hydrodynamically stagnant layer, we can distinguish between the inner and outer Helmholtz plane. The inner Helmholtz plane represents the distance of minimum potential in the potential curve (most negative potential in Figure 4.7) and also represents the distance between the surface and the centres of the solvated molecules which adsorb on the surface (in this case the negatively charged ions). The outer Helmholtz plane represents the distance of closest approach of the respective counter ions (in this case the positively charged ions). The potential curve thus passes through a minimum at the inner Helmholtz plane and then rises towards zero at a distance of $d = \infty$.

The potential at the distance of the shear plane is called ζ-potential (zeta potential).

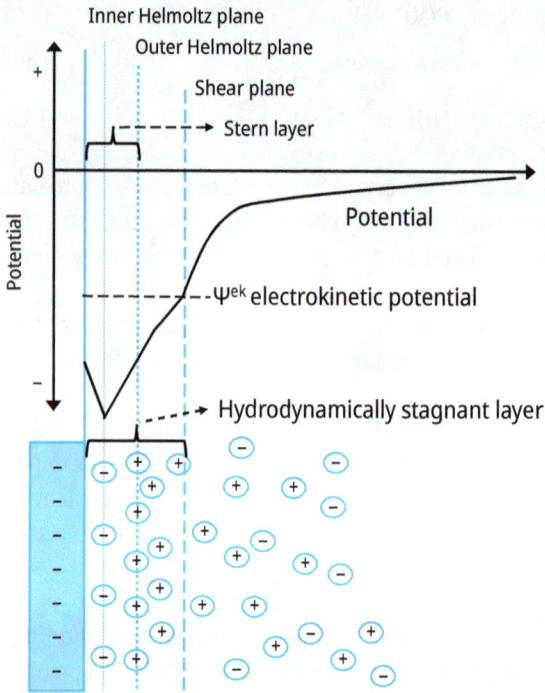

Figure 4.7: The electrical double layer of a fibre in contact to water.

As a result, the fibre exhibits a highly negative potential at the surface and the potential curve approximates to zero first at infinite distance to the surface. Also, the potential at the distance of the shear plane is negative, thus the fibre exhibits a negative ζ-potential.

Often the ζ-potential is also called streaming potential. The name is related to one of the important methods to determine the ζ-potential.

Relevant techniques are:

- Determination of the streaming potential. In a typical application of this method, a solution of known pH and conductivity is pressed through a fibre plug and the potential difference between the beginning and end of the fibre plug is measured. From the pressure difference, the measured potential difference and further solution characteristics and geometric parameters, the ζ-potential can be calculated according to eq. (4.12).

$$\zeta = \frac{dU}{dp} \cdot \frac{\eta}{\varepsilon \cdot \varepsilon_0} \cdot \frac{L}{Q \cdot R} \tag{4.12}$$

ζ	zeta potential (mV)
$\mathrm{d}U/\mathrm{d}p$	slope of streaming potential vs. pressure (mV/bar)
η	electrolyte viscosity (mPas)
ε_0	permittivity (C^2/J m)
ε	dielectric constant of electrolyte
L	length of capillary system (m)
Q	cross-sectional area of capillary system (m^2)
R	AC resistance of cell using electrolyte solution (Ohm)

- Electrophoretic mobility of charged particles (dispersions, emulsions) allows the determination of the ζ-potential form an analysis of the velocity of particles under defined external electrical field.
- In a similar approach, a potential measurement during settling of particles allows the determination of the ζ-potential.
- In a porous structure which bears charges on its pore surface, a plug flow appears in presence of an external electrical field. The movement of ions in the pores leads to an electroosmotic flow. The height of the electroosmotic flow can also be used to determine the ζ-potential of a porous material.

The formation of an electrical double layer is a determining factor in all solid–liquid interfaces. Water is the main solvent in textile chemical operations. The high polarity and the presence of dissolved ionic components in water both contribute to the development of a certain ζ-potential.

The properties of the electrical double layer are of high relevance for a number of processes and phenomena observed at the boundary layer between a textile fibre and a liquid phase:

- The presence of a certain ζ-potential leads to adsorption of ions with opposite charge. As an example, wool fibres are charged positively in acidic pH, which is utilised to support sorption of anionic dyes (acid dyes, metal complex dyes, reactive dyes).
- The uptake of charged molecules (e.g. cationic softeners) or dispersed substances in form of charged particles is directly determined by the ζ-potential of the fibre material. The adsorption of cationic softeners on the negatively charged cotton fibre is used to exhaust softeners during the last rinse in household laundry.
- Through dissociation of carboxylic groups and protonation of basic amino acids in wool, the overall charge of the fibre changes based on the pH value of the solution. In acidic pH, minimal dissociation of the carboxylic groups occurs and the amino groups are protonated, thus positive charges are present on the fibre surface, as well as inside the accessible parts of the fibre. Thus, a positive ζ-potential of wool is observed. With increasing pH value, the dissociation of the carboxylic groups begins. The pH range for dissociation is dependent on the respective pK_a value of the amino acids. With increasing pH, the ζ-potential reduces to zero. When a ζ-potential of zero is measured, the positive and negative charges at the

surface compensate each other and the overall surface potential of the fibre is zero. This point then is called the isoelectric point. The isoionic point of a protein is defined as the pH value at which the total amount of negatively charged groups in the protein structure is the same as the amount of positively charged groups, thus the overall charge of the protein is neutral. The isoionic point is not necessarily the same as the isoelectric point as a different distribution of amino acids between core and surface of a protein fibre will lead to different pH-dependent charging.

– When the pH increases above this range, the deprotonation of the ammonium groups begins and the ζ-potential of the fibre becomes negative. A negative ζ-potential weakens the attraction of the negatively charged dyes used for dyeing of wool, as a result the mobility of the dye is increased and the fastness properties of wool dyeing in alkaline wash baths are lowered compared to neutral washing conditions.

– Particularly at alkaline conditions fibres can exhibit a negative ζ-potential (Figure 4.8). This potential can result from the dissociation of functional groups, as well as from sorption of negative ions, such as hydroxyl ions, polyelectrolytes (carboxymethyl cellulose, CMC). The principle of increase of ζ-potential through sorption of polyelectrolytes and high pH values is applied in household laundry, where repulsion between the negatively charged fibre surface and deposited soil is maximised through alkaline washing conditions and addition of anionic polyelectrolytes.

– The selective sorption of charged surfactants such as alkyl-sulphates on the surface of dispersed particles or hydrophobic surfaces of oil droplets increases the charge density on the particles which thus can form stable dispersions in water. Repulsion of surface charges then prevents agglomeration and settling.

– The stabilisation of dispersions and emulsions bases on the formation of high charge densities on the surface of the dispersed particles, respectively of the tiny oil droplets of an oil-in-water emulsion [11]. The repulsion of surface charges prevents coalescence of droplets. Reduction in charge density through oppositely charged polyelectrolytes or multivalent cations leads to destabilisation and increase in overall particle size. As a result, oil phases begin to coalesce, solid phases agglomerate. Such processes, for example, are found in waste water treatment to remove finely divided dispersed phases, as well as to support separation of oil phases during sedimentation, flotation, (ultra-)filtration processes.

As discussed above, the pH dependency of ζ-potential (Figure 4.8) is the result of a number of factors (e.g. dissociation of functional groups, sorption of ions, fibre swelling, structure of the fibre surface, fibre diameter and so on). Thus, the interpretation of ζ-potential changes on the basis of concentration of dissociating functional groups has to be undertaken with care.

It is important to consider that the attraction through opposite charge of a solute and the charges of a fibre surface represent only one element of the overall attractive forces activated in sorption of dissolved chemicals. Other forces, for example, hydrogen

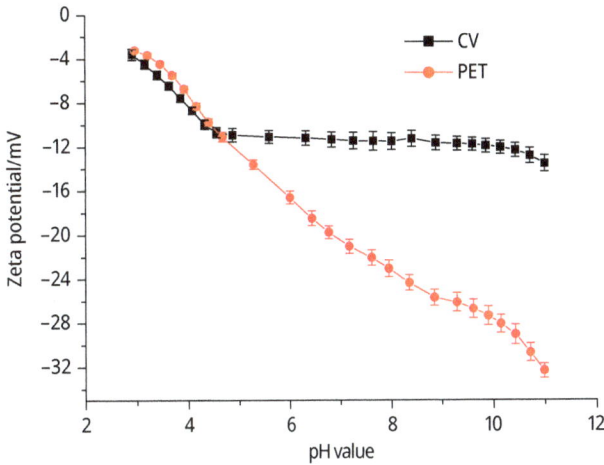

Figure 4.8: ζ-potential as function of pH value for different types of fibres (black: viscose fibre, CV; red: polyethylene terephthalate fibre) (Data: Avinash P. Manian, University of Innsbruck).

bonding, dipole–dipole attraction as well as hydrophobic interactions can exceed the effect of ionic attraction respectively repulsion. As an example, the sorption of an anionic group–containing reactive dye or direct dye would not occur on a cellulose fibre in alkaline solution, as these fibres usually exhibit a negative ζ-potential. Other binding forces overcompensate the charge repulsion between a cellulose fibre and the anionic dyestuff, and as a result spontaneous sorption is observed.

4.7 Donnan equilibrium

In the boundary layer of the surface of a fibre, an uneven distribution of ions is observed. At chemical equilibrium, the concentrations at the surface of the fibre and in the bulk are linked to each other. Two boundary conditions have to be fulfilled in the equilibrium state:

1. Electroneutrality: The sum of positive and negative charges in the bulk solution must add to zero.
2. Concentration differences of a certain species (e.g. Chloride) lead to diffusional transport, which, however, is compensated by an electrical potential difference. Thus, no net diffusional transport is observed in equilibrium state.

According to the Nernst equation, the concentration differences of species between bulk solution and boundary will lead to electrochemical potential differences, which, however, are compensated to zero in the thermodynamic equilibrium.

The principle can be demonstrated when distribution of salt (Na^+, Cl^-) between the bulk solution and the electrical double layer of a cellulose fibre is considered.

Due to oxidative processes, the cellulose fibre contains a distinct amount of carboxylic groups. In neutral state, the carboxylic groups dissociate $[-CO_2^-]_f$ and cations like sodium ions (Na^+) are present as counter ions.

Initially the sodium concentration in the stagnant layer is determined by the carboxylic group content of the fibre.

When sodium chloride diffuses into the stagnant layer, two different diffusion gradients have to be considered: the concentration gradient for sodium ions and the gradient for chloride ions.

In thermodynamic equilibrium, the condition of charge neutrality has to be fulfilled for the boundary layer and the bulk liquid (eqs. (4.13) and (4.14)).

Thus,

In solution:

$$[Na^+]_s + [Cl^-]_s = 0 \qquad (4.13)$$

In the boundary layer:

$$[Na^+]_b + [Cl^-]_b + [CO_2^-]_b = 0 \qquad (4.14)$$

In equilibrium, $[Na^+]_b$ will be higher than $[Na^+]_s$ and $[Cl^-]_b$ will be lower than $[Cl^-]_s$.

The Donnan equilibrium describes these concentrations in the state of thermodynamic equilibrium (eq. (4.15)):

$$K = [Na^+]_b[Cl^-]_b / [Na^+]_s[Cl^-]_s \qquad (4.15)$$

In an equilibrium state, the distribution of ions can be calculated according to eq. (4.16).

$$[Na^+]_b / [Na^+]_s = \left([CO_2^-]_b + \left(4c^2 + [CO_2^-]_b\right)^{1/2}\right)/2c \qquad (4.16)$$

with c being the concentration of $[Na^+]_s$ or $[Cl^-]_s$ in equilibrium which must be the same for reasons of electroneutrality.

In equilibrium state we will observe different concentrations of Na^+ and Cl^- in the boundary layer. The concentration of Na^+ in the boundary layer will be higher than in the bulk solution, while the concentration of Cl^- will be lower than in the bulk liquid. As a consequence of the concentration gradients a diffusion of Na^+ into the bulk would like to happen, which however is compensated by the tendency of Cl^- to diffuse from the bulk into the boundary layer. A dynamic equilibrium is established which maintains constant respective concentrations.

The concept of a Donnan equilibrium is of substantial importance for the interaction of fibres with aqueous solutions which contain electrolytes. This is the case for almost every process solution applied in textile dyeing and finishing.

For consideration of the concentration of a reagent which bears a charged group in the boundary layer of a fibre, the relations according to the Donnan equilibrium should be kept in mind as the other ions present in the system will also influence the diffusion of a certain ion.

References

[1] Atkins, P., de Paula, J. Atkins' Physical Chemistry, Oxford University Press, Oxford, UK, ISBN 0-19-879285-9, 2002.

[2] Komboonchoo, S., Bechtold, T. Sorption characteristics of indigo carmine as blue colorant for use in one-bath natural dyeing. Text. Res. J. 2010, 80, 734–743, doi:10.1177/0040517509342319.

[3] Glasser, WG., Atalla, RH., Blackwell, J., Brown, Jr. RM., Burchard, W., French, AD., Klemm, DO., Nishiyama, Y. About the structure of cellulose: debating the Lindman theory. Cellulose. 2012, 19, 589–598.

[4] Lindman, B., Karlström, G., Stigsson, L. On the mechanism of dissolution of cellulose. J. Mol. Liq. 2010, 156/1, 76–81.

[5] Matthews, JF., Skopec, CE., Mason, PE., Zuccato, P., Torget, RW., Sugiyama, J., Himmel, ME., Brady, JW. Simulation studies of microcrystalline cellulose I\ß. Carbohydr. Res.. 341 (2006) 138–152.

[6] Gallivan, JP., Dougherty, DA. A computational study of cation–π interactions vs salt bridges in aqueous media: Implications for protein engineering.. J. Am. Chem. Soc. 2000,122(5),870–874,DOI: 10.1021/ja991755c

[7] Mahmud-Ali, A., Bechtold, T. Aqueous thiocyanate-urea solution as a powerful non-alkaline swelling agent for cellulose fibres. Carbohydr. Polym.. 2015,116,124–130, DOI: 10.1016/j.carbpol.2014.04.084.

[8] Fauland, G., Constantin, F., Gaffar, H., Bechtold, T. Production scale plasma modification of polypropylene baselayer for improved water management properties. J. Appl. Polym. Sci. 2015, doi:10.1002/app.41294.

[9] Kongdee, A., Bechtold, T. Influence of ligand type and solution pH on heavy metal ion complexation in cellulosic fibre: model calculations and experimental results. Cellulose. 2009,16/1,53–63.

[10] Bechtold, T., Fitz-Binder, C. Sorption of alkaline earth metal ions Ca2+ and Mg^{2+} on lyocell fibers. Carbohydr. Polym. 2009,76/1,123–128.

[11] Holmberg, K., Jönsson, B., Kronberg, B., Lindman, B. Surfactants and Polymers in Aqueous Solution, Second Edition, John Wiley & Sons, Ltd, 2003, ISBN 9780471498834.

Take home messages

A number of attractive forces determine the interactions between fibre polymers. The same forces also regulate the interactions when a polymer comes in contact with solutions or gaseous media.

The interactive forces will lead to access of the solvent into a fibre structure and lead to swelling and adsorption.

At the border between fibre and solution a boundary layer is established. For chemical reactions such as dyeing processes or sorption of substances, the molecules have to pass the boundary layer to come in contact with the polymer surface.

The surface energy of the material also determines the wettability.

The presence of charges in the boundary layer leads to development of the zetapotential.

Quiz

Question 1. Why structural differences are detectable in the refractive index?

Question 2. Which main interactions are present between polymer chains?

Question 3. At what pH we will expect the transition between positive and negative zeta potential in case of bleached cellulose, an insoluble polyamine ($pK_a = 10$) and wool? (Compare with Chapter 2.)

Exercises

1. Estimate the equilibrium concentrations of Na^+ and Cl^- according to the Donnan equilibrium (eq. (4.16)) for a cellulose fibre (carboxyl-group content 70 mmol/kg, density 1.3 g/mL, K = 1.0) which is in contact with a solution of 100 mM NaCl.
2. Sort the materials/polymers with regard to refractive index and surface energy: water, PP, PAC, viscose, PET. Something remarkable? Any explanation for this?

Solutions

1. (Question 1): The refractive index is related to the relative permittivity, and the relative permittivity is related to the dipole moment and polarisability. Thus, structural influences which change dipole moment and polarisability will influence the refractive index.
2. (Question 2): Different attractive forces can be distinguished: ion–ion interactions, hydrogen bonds, Van der Waals forces, electron interactions, hydrophobic interactions.
3. (Question 3): The change in zeta potential depends on the protonation/dissociation of the functional groups. In case of cellulose the carboxylic groups will dissociate at pH 4–5, thus this will be the point of change in sign of surface charge. In case of wool this will be the isoelectric point (pH = 4.9), in case of a polyammonium ion dissociation of the alkylammonium group will be at pH 9–10.
4. (Exercise 1): From eq. (4.16):

$$[Na^+]_b / [Na^+]_s = \left([CO_2^-]_b + \left(4c^2 + [CO_2^-]_b \right)^{1/2} \right) / 2c$$

$[Na^+]_b = 100 \, (91 + (4 \times 100^2 + 91)^{1/2})/200 = (91 + 200)/2 = 146 \, \text{mmol/L } Na^+$

From eq. (4.15):

$[Cl^-]_b = K [Na^+]_s [Cl^-]_s/[Na^+]_b = 1.0 \ 100 \ 100/146 = 68$ mmol

5. (Exercise 2): Refractive index: water, viscose, PAC, PET, PP; Surface energy: water, viscose, PET, PAC, PP. Remarkable: The order is the same. Explanation: Both parameters are related to polarity as a core property.

List of abbreviations/symbols

EN	electronegativity
C	Coulomb
D	Debey
l	distance
HOMO	highest occupied molecular orbital
LUMO	lowest unoccupied molecular orbital
μ^*	induced dipole moment
E	electrical field
V	potential energy
α	polarisability
α'	polarisability volume
ε	vacuum permittivity
ρ	mass density of sample
M	molar mass
N_A	Avogadro number
k	Boltzmann constant
T	temperature
P_m	molar polarisation
n_r	refractive index
r	distance
E_{max}	potential energy at maximum attraction
PET	polyethylene glycol terephthalate
CV	viscose
γ	surface energy
CMC	carboxy methyl cellulose
ζ	zeta potential
dU/dp	slope of streaming potential vs. pressure
η	electrolyte viscosity
ε_0	permittivity
ε	dielectric constant of electrolyte
L	length of capillary system
Q	cross-sectional area of capillary system
R	AC resistance of cell using electrolyte solution
K	equilibrium constant
c	equilibrium concentraton

5 Thermodynamics and kinetics in fibre chemistry

5.1 Moisture sorption

All textile fibres absorb moisture from the ambient atmosphere. The level of moisture sorption behaviour is a characteristic property of a textile fibre. The amount of sorbed water depends on the partial pressure of water vapour in the atmosphere, which is usually expressed by the relative humidity (r.h.). For a given temperature, the relative humidity is defined as the ratio between the partial pressure of water in the ambient atmosphere (p_x) and the water vapour pressure water at the respective temperature (p_{sat}) (eq. (5.1)).

$$\text{r.h.} = 100\, p_x/p_{sat} \tag{5.1}$$

Often the vapour pressure p_x in eq. (5.1) is replaced by the respective masses: m_x is the mass of moisture present in the atmosphere and m_{sat} is the mass of moisture under conditions of saturation. The resulting values for r.h. are almost the same. Minor differences result from the non-ideal behaviour of water vapour, which is not an ideal gas. Only in the gas equation for an ideal gas, mass and pressure are related to each other according $pV = nRT$ (p, pressure; V, volume; n, the number of moles; R, gas constant and T, temperature).

The uptake of water by textile fibres through sorption is dependent on the type of fibre. Important factors determining the hygroscopy of a fibre are as follows:

- Presence of polar groups in the fibre polymer that are able to develop hydrogen bonds and Van der Waals forces to sorbed water molecules, for example, $-OH$, $-NH_2$, $-COOH$ and $-SO_3Na$.
- Accessibility of sorbing polar groups for water vapour through pores and voids. Thus, a high content of amorphous parts in the fibre structure will lead to higher moisture sorption, while a higher content of crystalline, thus less accessible parts, will reduce moisture sorption.
- Presence of small pores and voids, which can develop capillary condensation, in particular at high values of relative humidity.
- Ability of a fibre structure, for example, cellulose to expand and swell during uptake of moisture, thus increasing internal surface and space for further sorption of water molecules.

Moisture sorption is dependent on the ambient conditions (r.h. and temperature); thus, standard conditions have been defined for gravimetric assessment of moisture content of textile materials, for example, with 20 ± 2 °C and 65 ± 2% r.h. For comparison of moisture uptake, these conditions usually are taken as reference conditions for the determination and comparison of moisture content of a textile product. Determination and subtraction of moisture content is an important factor in trading of natural fibres, for

https://doi.org/10.1515/9783110795738-005

example, cotton and wool, where a significant part of the transported mass (10–20 wt%) results from sorbed moisture.

Similarly the "dry" state of a fibre is defined after a drying to constant weight at 105 °C. Sorption of water on fibres with low polarity, for example, polypropylene and polyester, is rather low (PP 0 wt%, PES 0.2–0.5 wt%, further values given in Chapter 2).

The moisture in a textile material are of two forms: the moisture regain (RG; eq. (5.2)) and the moisture content (MC; eq. (5.3)).

$$RG = 100 \, m_{mt} - m_{dry}/m_{dry} \tag{5.2}$$

$$MC = 100 \, m_{mt} - m_{dry}/m_{mt} \tag{5.3}$$

where m_{mt} is the mass of moist material and m_{dry} is the mass of dry material.

The important difference between both measures results from the different reference basis. In the case of the moisture regain RG, the moisture content is always related to the dry material, which thus allows direct comparison of results. In the case of moisture content, the actual weight of the sample forms the basis; thus, direct comparison of results at different moisture contents is more complicated.

For the determination and discussion of moisture sorption, usually the moisture regain is used, which then simplifies direct comparison of different materials.

5.2 Moisture sorption isotherms

An isotherm characterises a physical–chemical behaviour measured at a constant temperature. In the case of a moisture sorption isotherm, the uptake of moisture by a material under equilibrium conditions and under defined conditions with regard to temperature and water vapour pressure is described. Thus, moisture sorption isotherm describes the moisture regain of a material as a function of relative humidity.

A sorption isotherm can be recorded by stepwise acclimation of a sample to a given climate, which is kept constant until sorption equilibrium has been reached. The sample is weighed within certain time intervals, until constant weight is observed and sorption equilibrium thus has been confirmed. The change in weight is then a measure for the uptake/loss of moisture that occurred as a result of the climate change before and after acclimation. In a traditional experimental approach, concentrated salt solutions were used to establish a certain climate inside a desiccator. Special devices of glassware have to be used, which allow weighing without an inacceptable perturbation of the climate.

Modern devices utilise the technique of dynamic water vapour sorption. In an automated device, the samples are placed in a small chamber where a stepwise change of the climate can be achieved within a short time. After new climate conditions have been set, the samples are weighed within regular intervals of time until constant weight is observed. Then the conditions change to the next climate. As a result, moisture sorption/desorption curves are obtained.

When the experiment is started at 0% r.h. and the relative humidity is increased stepwise, a sorption isotherm is obtained. Often the experiment is stopped at 90 r.h.% to avoid condensation of water in the device and on the samples. Thus, a desorption experiment usually is started at 90% r.h. and then performed in the direction of lowered r.h.

Besides the observation of the sample weight at sorption equilibrium, the use of an automatic device also permits a recording of the rate of moisture uptake. Thus besides sorption and desorption isotherms, sorption kinetics can also be analysed.

In Figure 5.1 the time-dependent moisture uptake during the measurement of a moisture isotherm is shown. In the drawing the steps of climate change are visible and the weight gain of the sample as a result of moisture uptake is seen as a retarded reaction to the respective climate change. Form the weight gain with increasing value of r.h., the sorption isotherm is obtained. After a value of 90% r.h. had been reached, the direction of climate change reverses and the desorption isotherm is registered during the phase of stepwise reduction of r.h.

Representative examples for sorption isotherms of a regenerated cellulose fibre (lyocell-type fibre) and of cotton are shown in Figure 5.2 [1, 2]. A sorption isotherm exhibits following characteristics:

- Three characteristic parts of the curve can be distinguished (A, B, C in Figure 5.2).
- A hysteresis is observed between the sorption and the desorption curves of the experiment.
- The sorption and desorption kinetics can be observed in case the weight change between two neighbouring points of climate has been recorded (Figure 5.1).

The sectors A, B and C of the sorption isotherm can be attributed to different physical–chemical processes. The specific properties of the sorption curves depend on the fibre characteristics, for example, polymer type, porosity, accessibility and physical and chemical modification of the material.

5.2.1 Water activity

The activity of the sorbed water molecules is an important term to interpret the sectors A, B and C of a sorption isotherm with regard to their physical–chemical background [3]. The activity of a chemical is defined via the chemical potential μ (eq. (5.4)).

$$\mu = \mu_0 + RT \ln a_w \qquad (5.4)$$

where a_w is the activity of the substance (in this case sorbed water), R the gas constant for ideal gases, T the temperature in K, μ the chemical potential and μ_0 the chemical potential under standard conditions.

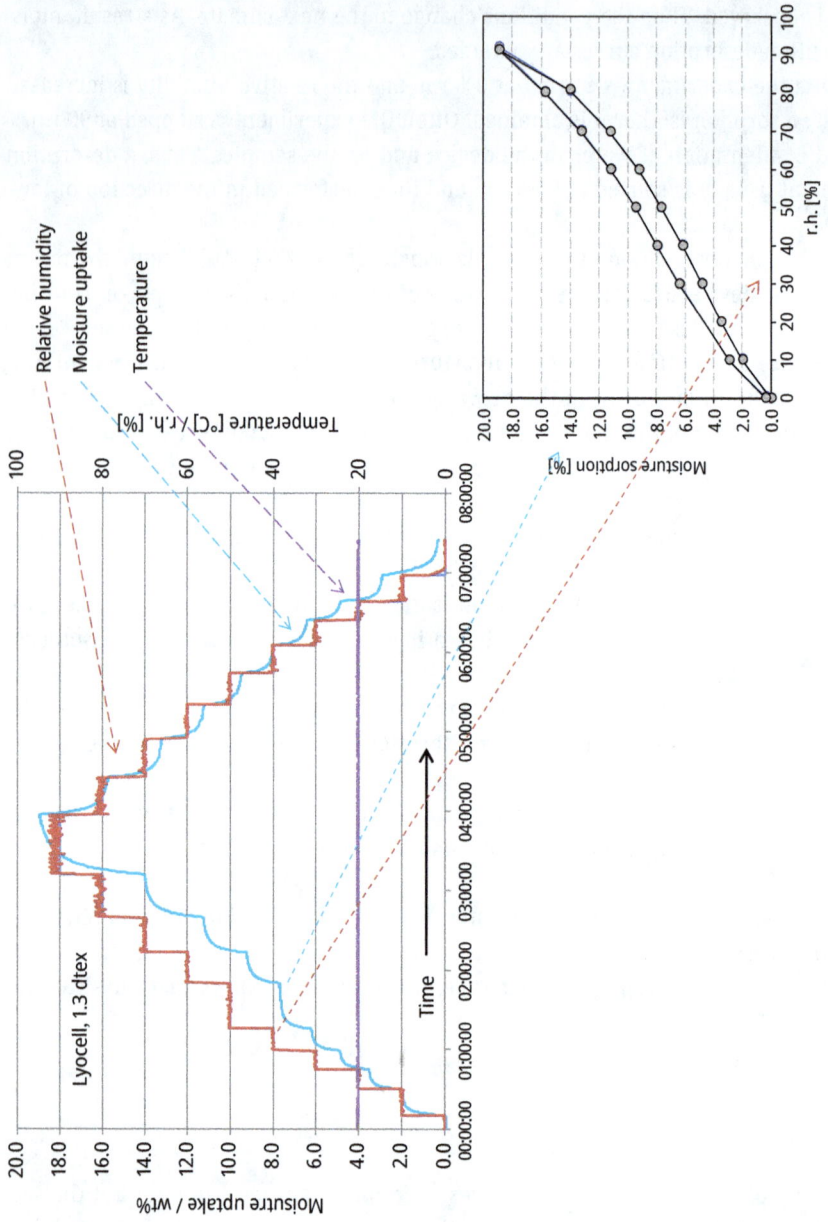

Figure 5.1: Dynamic water vapour sorption: (a) mass change as a function of relative humidity and time and (b) resulting sorption isotherm.

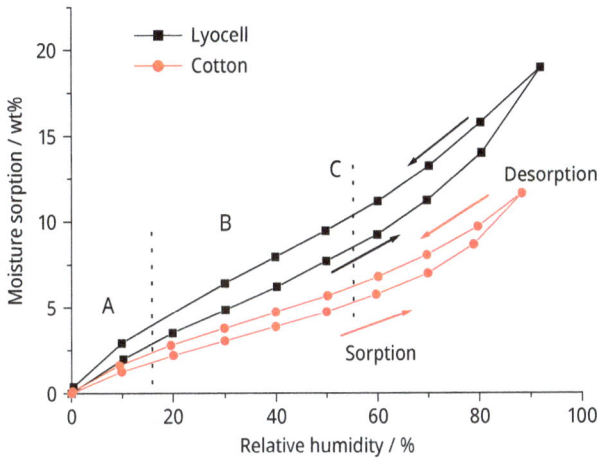

Figure 5.2: Sorption isotherm of a regenerated cellulose fibre.

In a pure system at standard conditions the activity a will be unity, while under real conditions, for example, aqueous solutions and sorbed water, the real activity will be reduced. The interaction with a sorbent reduces the activity of water.

Sector A: In this region direct molecular interaction with the sorbent occurs through interaction due to hydrogen bonding and Van der Waals forces. As a result hydrate-like structures are formed. Sorption of water to sites with high surface energy occurs as an exothermal process and as a result the activity of water is substantially lowered. This highly sorbed water does not exhibit freezing behaviour and thus is called non-freezing bound water.

The complete removal of non-freezing bound water requires temperatures above the boiling point of water. As a result of the presence of strongly bound sorbed water, in particular the definition of the dry state of a hygroscopic fibre, for example, cellulose- or protein-based fibres, is difficult. While the term "dry fibre state" has been defined as equilibrium state at 65% r.h. and 20 °C, the "oven dry" state is achieved after drying to constant weight, for example, at 105 ± 3 °C; none of these two states represents a fibre that is free of sorbed water molecules.

The saturation of these highly moisture absorbing sites in the fibre can be allocated to the region A of the sorption isotherm. Representative levels for the water activity are at $a_w < 0.2$.

Often the formation of a monomolecular layer of water on the fibre surface and on the walls of pores and voids is used as an instrument to describe the sorption on a molecular scale. Such models are valid for well-defined surfaces, for example, catalysts where such theory originally had been developed. However, the specific structure of a fibre polymer has to be considered before such an assumption is applied without any further concern.

Synthetic fibres with compact fibre structure, for example, polyester fibres, may exhibit a number of sorption sites with rather similar energy of sorption. A different situation is observed with hydrophilic fibres, for example, cellulose fibres or protein fibres, where a wide range of different sites for sorption is available and a gradual change in sorption energy will occur with increased coverage of the inner surface. Thus, the term of a monolayer should be used with care in these cases as a monolayer implies predominance of high number of sites with very similar sorption energy.

Sector B: In sector B of the sorption isotherm, the process of "monolayer" adsorption has been completed and multilayer adsorption begins to occur.

A typical range for the activity of water is in the interval of $0.2 < a_w < 0.6$.

For this part of the sorption isotherm, the Brunauer–Emmett–Teller (BET) model is used quite frequently [4]. The model is applied in the linear branch of the sorption isotherm, which usually is registered above 5% r.h. and below 50% r.h.

Several assumptions have to be made as prerequisite before the BET model can be applied:

- The adsorption energy of all molecules forming the first sorbed layer of molecules is equal. This means that all sorption places are the same or at least similar.
- All other layers adsorb with the binding energy of the pure adsorbate. Thus, the heat of adsorption of other layers corresponds to the heat of condensation of the pure substance.
- On the first layer of adsorbate both, the rate of condensation and evaporation of adsorbate are the same.

The adsorbed mass of a substance according the BET equation can be calculated according to eqs. (5.5) and (5.6).

$$m_{ads} = \frac{m_m\, C\, p_x}{(p_{sat} - p_x)\left[1 + (C-1)\,\frac{p_x}{p_{sat}}\right]} \tag{5.5}$$

$$C = e^{\frac{(E_1 - E_2)}{R\,T}} \tag{5.6}$$

where m_{ads} is the mass of adsorbed vapour at vapour pressure p_x in equilibrium, m_m mass of adsorbed gas in monolayer, p_{sat} vapour pressure of gas under conditions of saturation, E_1 heat of adsorption for the first adsorbed layer (monolayer) and

E_2 heat of condensation of the sorbate.

The BET equation can be presented in a linearised form as a function of the relative humidity (eq. (5.7)). In eq. (5.7) gas volumes have been introduced to replace masses (m_{ads}, m_m).

$$\frac{1}{V_{ads}}\left(\frac{p_x}{p_{sat} - p_x}\right) = \frac{C-1}{C V_{mono}}\frac{p_x}{p_{sat}} + \frac{1}{C V_{mono}} \tag{5.7}$$

V_{ads} is the adsorbed volume of gas, namely, water, and V_{Mono} is the volume of adsorbed water to form a monolayer.

In a diagram of $\frac{1}{V_{ads}} \left(\frac{p_x}{p_{sat} - p_x} \right)$ against p_x / p_{sat} a linear graph is obtained and the monolayer capacity can be calculated.

The interlinkage between adsorbed volume and mass of sorptive is given by the gas equation; in a simplified approach, the ideal gas equation can be used. Then V_{ads} can be replaced according eq. (5.8).

$$V_{ads} = n_{ads} R\ T / p_x \qquad (5.8)$$

From V_{Mono} and an assumption of the space requirement of one water molecule, the surface area of the sorbent can be calculated. For water the space covered by sorption of one molecule is assumed with $0.11\ nm^2$.

Sector C: For hydrophilic capillaries, the condensation of liquid occurs below the condensation temperature of the vapour. Thus, capillary condensation in narrow pores of a fibre occurs. The Kelvin equation correlates the pore size of a substrate with the vapour pressure for condensation (eq. (5.9)).

$$A_w = p_x / p_{sat} = \exp(-2\gamma M / r\rho R\ T) \qquad (5.9)$$

with γ being the surface tension, M the molar mass of the condensing liquid (water), r the radius of a capillary and ρ the density of liquid water.

A typical range of water activity for the process of capillary condensation is in the range of $a_w = \geq 0.6$. The effect of capillary condensation is dominant in Section C of the sorption isotherm and can be used to assess pore diameter and pore volumes when data about the mass change during condensation are available.

From the partial pressure of water vapour and the mass change of the sample, the diameter and volume of the capillaries can be calculated by using the Kelvin equation.

We need to consider the fact that these phenomena just represent models, and high complexity of a fibre structure makes a straightforward assignment to one simple process step difficult. However, the general models given are very useful to interpret and correlate the experimental observations with physical–chemical processes.

An important factor to be considered for an interpretation of surface phenomena, such as monolayer capacity or porosity and pore size distribution from sorption isotherms, is the type of fibre under investigation and the sorbent chosen to characterise the material.

Porosimetry studies by using nitrogen as sorbent allow the determination of pore characteristics without undesired dimensional changes of the sample material, for example, through swelling.

For many studies, the sorbent of interest, however, is water vapour. In this case, fibres that already exhibit low sorption and do not swell in the presence of water are less dependent on the presence of moisture in the structure of the porous fibre. Water uptake in cellulose-based fibres and protein fibres goes ahead with substantial dimensional changes of the 3D fibre structure. Thus, the sorption process itself changes the polymer order, porosity and accessibility of the fibre for the sorbent. In such cases, the fibre reorganises and changes its molecular order during the sorption process as a function of relative humidity present in the ambient atmosphere. The concept of a well-defined surface for adsorption and the assumption of stable polymer structure then both are not strictly valid. However, the concept of BET sorption is widely used because of the relatively simple experimental approach and the useful data that can be extracted, which at least allow comparison of similar material samples [5].

5.3 Sorption kinetics for adsorption from the gas phase

Besides the analysis of the equilibrium sorption, kinetic data can also be obtained by applying dynamic water vapour sorption methods. Following to a rapid change in ambient climate conditions, the change in sample weight through sorption or desorption is registered by repetitive measurement of the sample weight until equilibrium has been achieved. Through the stepwise change in moisture regain, the uptake/release of moisture can be observed as a function of ambient climate. Thus, sorption/desorption kinetics can be observed (Figure 5.3).

The kinetics of the moisture sorption depends on the accessibility of the sorbent for moisture and on the mechanism of sorption. In a textile substrate, a sorption process consists of a series of consecutive steps that determine the overall kinetics. While access to the fibre surface and sorption on the surface can be assumed to be rapid processes, the diffusion inside the fibre and sorption on sites at inner parts of a fibre depend on the accessibility of the fibre structure and thus can be comparatively slow processes.

Different models that describe the kinetics of moisture sorption in fibres and textile structures have been proposed in the literature. The complexity of the models increases with the number of individual steps that have been included into the model. A basic model to analyse the sorption kinetics in textile fibres has been proposed as the so-called parallel exponential kinetics (PEK) model [6] (Figure 5.4).

The PEK model simplifies the complex sorption process by an approximation of the sorption curve with only two independent sorption reactions. Both sorption reactions obey first-order kinetics. One reaction represents a fast sorption process that describes, for example, sorption on the surface and on easy accessible sites. The second process summarises slow sorption reactions, which, for example, occur as a result of a slow diffusion of the sorbent inside less accessible parts of the fibre structure. Both processes are independent of each other; thus, the reactions proceed in parallel. As the basic formulations use an exponential relation between the moisture regain and the time, the model is called "parallel exponential kinetics" (PEK) model.

The respective equations for the sorption and desorption processes according to the model are given as eqs. (5.10) and (5.11).

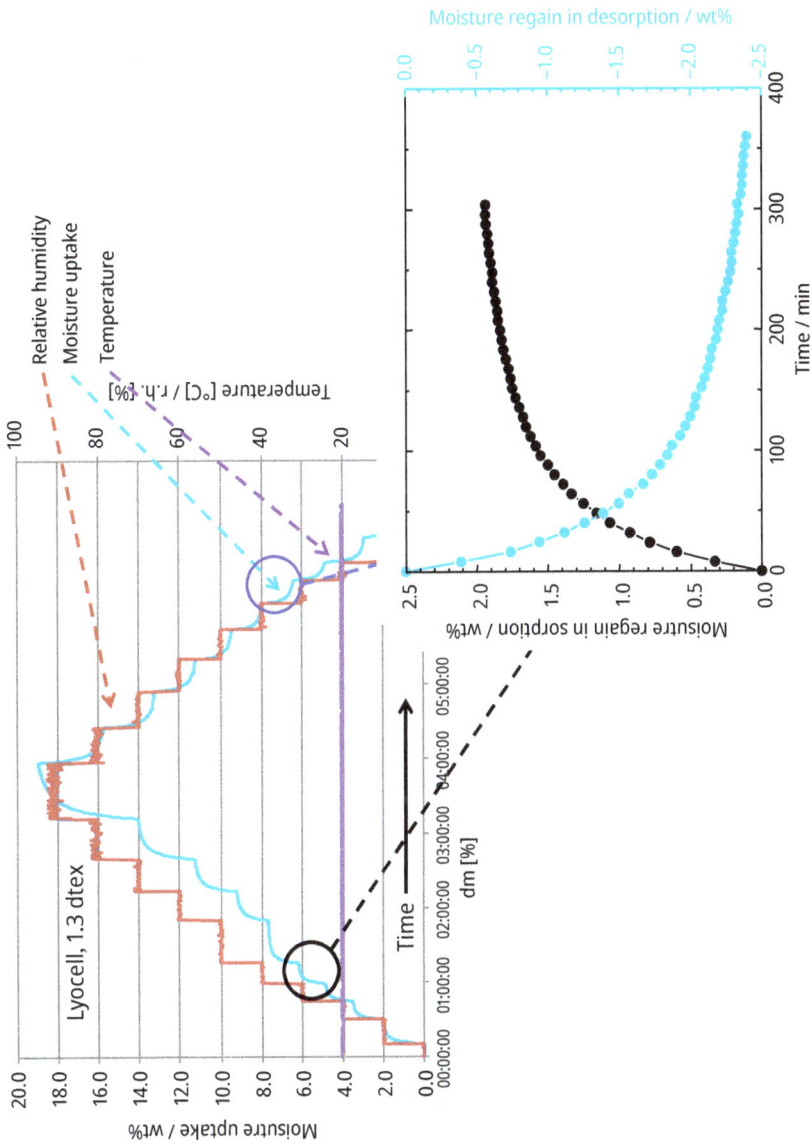

Figure 5.3: Dynamic water vapour sorption. Left: Primary data recording the mass change following to a rapid change in ambient climate; right: extracted data of mass change as a function of time for a distinct change in climate.

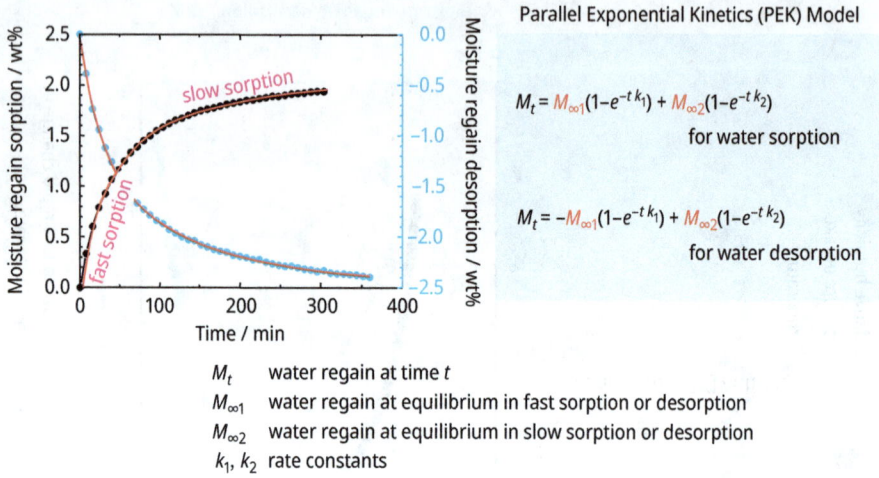

Figure 5.4: The PEK model: sorption/desorption of moisture and separation of the sorption kinetics into a fast and a slow sorption process.

$$M_t = M_{\infty 1}(1 - e^{-tk1}) + M_{\infty 2}(1 - e^{-tk2}) \tag{5.10}$$

$$M_t = -M_{\infty 1}(1 - e^{-tk1}) - M_{\infty 2}(1 - e^{-tk2}) \tag{5.11}$$

Each sorption/desorption step is characterised by the moisture regain M_i and the rate constant k_i. Instead of the rate constant, often the characteristic time τ is used, which is related to the rate constant k by eq. (5.12).

$$k_i = 1/\tau_i \tag{5.12}$$

The kinetics of moisture sorption or desorption processes in a textile fibre can be described by a set of four characteristic constants. The first two parameters ($M_{\infty 1}$ and k_1) describe the rapid sorption process in terms of maximum sorption capacity and rate of sorption. The second pair of parameters ($M_{\infty 2}$ and k_2) describes the slow sorption process.

A fibre with high M_1 thus exhibits a high sorption capacity at sites that can be accessed rapidly. A fibre with a high value of M_2 offers substantial capacity for sorption at sites that will equilibrate much slowly.

Thus, an analysis of the respective capacities and rates of sorption delivers valuable information about the capacity and accessibility of internal structures of a material, as well as changes of the internal structure, for example, during fibre modification. Incorporation of a fibre material as reinforcement into a composite structure changes the accessibility of the fibre structure, which can thus be characterised by comparison of the sorption characteristics [7].

Hysteresis: When the sorption and desorption curves obtained from one sorption and desorption cycle are compared with each other, a hysteresis is observed. At a given relative humidity of the ambient atmosphere, the respective moisture regain of the material measured during the desorption process is higher than the corresponding value observed during the sorption experiment.

From the models for sorption kinetics as well as from thermodynamic considerations, there would be no explanation for the direction dependence of the result. The sorption behaviour of the material should be independent of the direction when a certain state of equilibrium had been obtained.

Hysteresis behaviour in sorption experiments is an indicator for structural differences in the sorbent material, which had resulted from a reorganisation of the polymer structure during the phase of moisture uptake.

Such phenomena are of particular relevance in cellulose fibres. At relative high humidity, substantial amounts of water become absorbed in the fibre structure, which begins to reorganise due to local "plasticising" effects of water. The hysteresis thus will also be dependent on the maximum value of relative humidity reached, which had been chosen as the point of reverse in the sorption experiment. Fibres that exhibit a small hysteresis are only subject to minor moisture-dependent reorganisation effects. Thus, the magnitude of a hysteresis in sorption experiments and possible changes as a result of fibre processing can deliver useful information about an achieved modification of the internal fibre structure.

(Figure 5.5) shows representative sorption isotherms for important fibre materials. High moisture regain is observed for regenerated cellulose fibres and wool. A lower sorption capacity is determined for cotton, which also exhibits higher fibre crystallinity. As sorption is assumed to occur mainly in the amorphous parts of a fibre, the reduced capacity of cotton for uptake of humidity can be explained by its higher crystallinity. A modification of a cellulose fibre, for example, through pad-dry crosslinking in finishing steps will reduce the moisture uptake further, which will also influence the wear properties of the fabric.

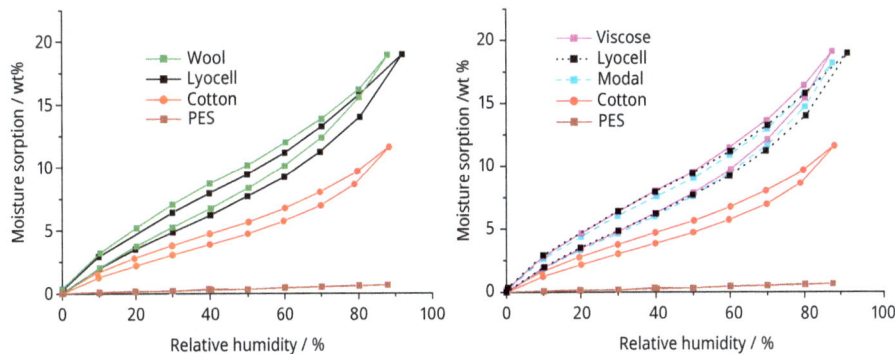

Figure 5.5: Sorption isotherms of important fibre material.

Polyester (polyethyleneterepthalate) represents a fibre that exhibits only low uptake of moisture and also a very small hysteresis. Thus, only very minor fibre reorganisation is expected to occur during a sorption/desorption cycle.

5.4 Sorption from the liquid phase

5.4.1 General considerations

An understanding of adsorption isotherms for sorption of dissolved molecules from the liquid phase is of high relevance for the interpretation of many textile chemical processes, which are performed by means of liquid process baths. In physical chemistry adsorption, isotherms preferably consider sorption from the gas phase to the surface of a solid substrate or into a liquid phase. Sorption isotherms describe the conditions in the equilibrium state; thus, from sorption equilibria and the respective equilibrium constants, thermodynamic data describing the sorption process can also be derived.

In the case of a gas/solid sorption, the isotherm describes the amount of sorptive n per mass of sorbent m, as a function of partial pressure p of the respective component in the ambient atmosphere.

For the process, $A_{(gas)} \rightleftarrows A_{(sorb)}$ in equilibrium state, a simple sorption isotherm can be formulated, for example, according to eq. (5.13).

$$n_{ads}/m = K p^x \tag{5.13}$$

with K being the respective equilibrium constant and p being the partial pressure, and x represents a factor that is dependent on the type of sorption isotherm and can take values between 0 and 1.

In liquid phase, the sorption isotherms have to use other approaches to describe the interrelation between the concentration of compounds sorbed in the solid phase (the textile fibre) and the respective concentration in liquid phase (the dyebath).

For description of the concentration in the solid fibre, an expression in terms of molality is used (units mol/kg; moles of sorptive per kilogram of sorbent) and for the description of the concentration in solution the usual unit mol/L is used (moles of sorptive per litre of the process liquid).

For a critical analysis of thermodynamic data obtained from the calculations using sorption isotherms for liquid/solid sorption, several assumptions have to be kept in mind. Molality is used as a substitute for the "concentration" in the solid phase, which is based on the assumption that the full volume of fibre is equally accessible to the sorbate. Such an assumption means that the volume of the crystalline domains of the fibres is ignored, which however is known to be not accessible for the major part of substance applied in textile chemistry.

A fibre with a crystallinity of 50% thus offers a significantly lower volume for sorption than a fibre with mainly amorphous structure. The phenomenon of swelling of fibres in aqueous phases or other liquid systems is not considered in such an approach, as the mass of fibre is taken as the basis and a change in volume through swelling and expansion of the fibre structure would also change the accessible volume substantially. Despite these substantial simplifications, sorption isotherms are a widely used concept to describe the uptake of dissolved compounds into a fibre structure. Three types of sorption isotherms are of relevance for the analysis of liquid/solid sorption, for example, in dyeing operations [8]: the Nernst-type isotherm, the Freundlich-type isotherms and the Langmuir-type isotherm.

5.4.2 The Nernst isotherm

The Nernst isotherm represents cases in which the observed distribution between the two phases follows the principles of an extraction. The classical examples are gas/liquid sorption, extraction and distribution of a substance between the two phases of two immiscible liquids.

In the case of a textile operation, the distribution between a solid fibrous material and a liquid phase, for example, the dyebath, is considered. The concentration of a substance in the fibre $[D_f]$ is related to the concentration of the substance in solution $[D_s]$ by a constant K (eq. (5.14), Figure 5.6).

$$[D_f] = K [D_s]$$ (5.14)

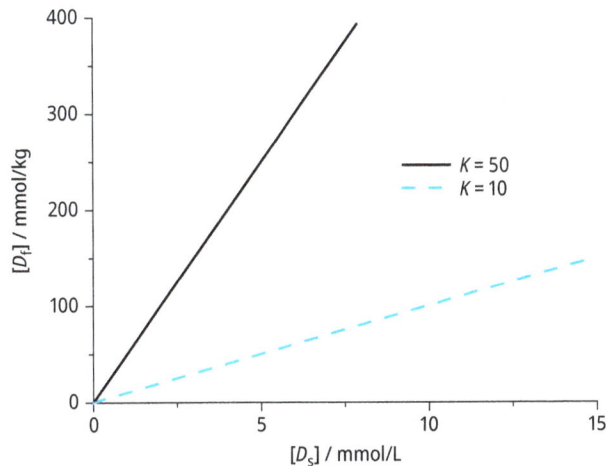

Figure 5.6: Nernst-type sorption isotherm.

The sorption behaviour of a Nernst sorption isotherm is observed for limited concentrations of the dissolved substance $[D_s]$. Theoretically, there is no limitation for $[D_f]$ at high concentrations of $[D_s]$, which indicates a restriction for the applicability of the Nernst-type isotherms at high concentrations where saturation effects or solubility limits have to be taken into account.

The Nernst equation represents the physical case of an extraction and distribution of a dissolved substance between two immiscible liquids. In textile chemical operations, the behaviour of disperse dyes in the exhaust dyeing of polyester fibres follows the Nernst isotherm. Using the free-volume model to describe the polyester fibre, we can imagine the fibre above the respective glass transition temperature T_g to exhibit accessible domains that behave similar to a liquid phase. These parts of the fibre exhibit higher molecular flexibility and permit extraction and dissolution of the disperse dye into the amorphous parts of the fibre structure.

The equilibrium constant K is related to the free energy ΔG of the sorption process according eq. (5.15).

$$\Delta G = -\,RT \ln K \tag{5.15}$$

From the temperature dependence of the equilibrium constant, the enthalpy ΔH and the entropy ΔS of the sorption process can be calculated according eqs. (5.16) and (5.17).

$$\ln K_2/K_1 = -(\Delta H/R)(1/T_2 - 1/T_1) \tag{5.16}$$

$$\Delta G = \Delta H - T\Delta S \tag{5.17}$$

5.4.3 The Freundlich isotherm

A sorption equilibrium that follows the Freundlich isotherm exhibits a non-linear dependency of the content of sorbate in the sorbent on the concentration of sorptive in solution (Figure 5.7).

The general formulation of a Freundlich-type sorption isotherm is given in eq. (5.18).

$$[D_f] = K\,[D_s]^{1/n} \tag{5.18}$$

The concentration of the sorbate $[D_f]$ depends on the concentration of sorptive in solution $[D_s]$; however, the relationship is non-linear. Besides the equilibrium constant K, an exponential factor, the so-called heterogeneity factor $1/n$ is also introduced. The factor $1/n$ can take values between unity and zero. In its logarithmic form, a linear relationship of eq. (5.18) is obtained (eq. (5.19)), which permits calculation of the equilibrium constant K and the exponent $1/n$.

$$\log [D_f] = \log K + 1/n \log [D_s] \tag{5.19}$$

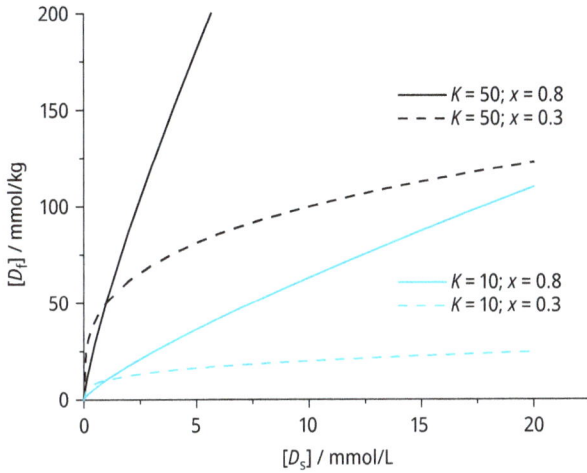

Figure 5.7: Freundlich-type sorption isotherm (x = 1/n).

Physically the Freundlich isotherm is based on the finding that for many sorption processes the assumption of a defined number of identical sites available for sorption of molecules up to monolayer coverage is not correct. The sorption energy for the first sorbed molecules is substantially higher than for molecules that adsorb in a later phase of the equilibration. This behaviour is considered by the introduction of the heterogeneity factor n in the Freundlich isotherm. The magnitude of the factor n is an indication for the non-uniformity of the sorption sites of a sorbent.

The physical background of the heterogeneity factor is based on several reasons: The interaction energy of the places where the molecules sorb is not uniform, which thus indicates structural differences of the adsorbing sites available on the surface of the sorbent. Repulsion and interaction between already sorbed molecules and molecules that intend to interact with a free site will also lower the sorption energy for molecules with increasing coverage of the surface. As a result, the binding energy of sorptive molecules reduces with the increasing surface coverage.

In case $n = 1$, the Freundlich isotherm converts into the Nernst isotherm. This means that the equilibrium constant now will be no longer dependent on surface coverage of the sorbent or the concentration of sorptive present in the solution.

The more n deviates from unity, the higher will be the influence of surface coverage on the sorption energy of molecules. Thus, the sorption isotherm exhibits a curved shape. With increase in surface coverage by the sorptive over proportionally higher concentrations of sorbent will be required in solution to achieve further sorption.

A clear limitation of the Freundlich isotherm appears when the concentration of sorbent in solution is increased to very high values. In this case, unrealistic concentrations of sorbate on the sorbent will result as no saturation effects have been considered.

Sorption behaviour according to the Freundlich-type isotherm is often observed in sorption processes of chemicals from aqueous solution on cellulose fibres. The heterogeneity of the surface of a swollen cellulose fibre leads to a gradual change in sorption properties as a function of substrate coverage. The interactions for sorption are based on hydrogen bond formation and Van der Waals forces. The sorption thus is not dependent on a stoichiometric interaction with a certain functional group available on the sorbent. Typical cases for Freundlich-type isotherms are observed in dyeing operations with direct dyes and during the sorption phase of reactive dyes. In reactive dyeing, follow-up reactions have to be considered as the formation of a reactive bond between sorbed dye and cellulose fibre will alter the effective concentration of sorbed dye. Thus, often sorption experiments with reactive dyes are performed with hydrolysed anchor groups to avoid disturbing side reactions. However, it has to be considered that the sorption isotherm of a hydrolysed reactive dye differs from the isotherm of the respective hydrolysate.

A special case for sorption processes on cellulose fibres is observed when the carboxylic groups present in the cellulose participate in a stoichiometric reaction, for example, during sorption of the cationic dye methylene blue. In such a case, the Langmuir-type isotherm is valid. The quantification of the sorbed methylene blue then allows the determination of the carboxylic groups of cellulose fibres.

5.4.4 Langmuir adsorption isotherm

Langmuir-type adsorption isotherms are observed for processes that fulfil a number of conditions:
- The sorption process follows a stoichiometric reaction pathway.
- There exists a number of chemically well-defined sites that serve as binding centres for sorptive molecules.
- A defined reaction ratio, for example, 1:1, exists between a sorption site and the sorptive, which in case of a 1:1 stoichiometry means that one sorption site binds one molecule of sorptive.
- When all sites are occupied by sorptive molecules, no further adsorption takes place. The saturation concentration $[S]$ describes this maximum value.

The relation between the concentration of sorptive on the sorbent $[D_f]$ and the concentration of sorptive in solution follows eq. (5.20).

$$[D_f]/([S] - [D_f]) = K\,[D_s] \tag{5.20}$$

A representative graph for the Langmuir type behaviour is shown in Figure 5.8.

By rearrangement of eq. (5.20) a linear function can be formulated (eq. (5.21)), which permits the determination of the equilibrium constant K from the slope and the saturation concentration $[S]$ from the intercept of the graph.

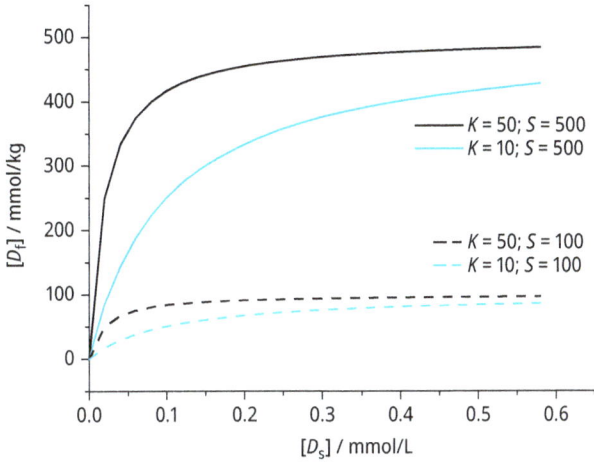

Figure 5.8: Langmuir-type sorption isotherm.

$$[D_f]/[D_s] = -K[D_f] + K[S] \tag{5.21}$$

In considerable number of textile operations, sorption of chemicals obeys the Langmuir-type sorption isotherm. In particular, when ion-exchange properties of the fibrous sorbent will determine the maximum sorption capacity, this will define the point of fibre saturation [S]. Representative examples are as follows:

– Sorption of the cationic dye methylene blue on negatively charged carboxylate groups of cellulose, which serve as binding sites. As the stoichiometry of the reaction is clearly defined, such a method is used to determine the carboxylic group content of cellulose fibres.
– Sorption of cationic dyes on the anionic groups in polyacrylonitrile copolymers. Such copolymers represent the major part of technically produced polyacrylonitrile fibres. Careful calculation of fibre saturation with the cationic dye molecules is required to avoid deviations in shade and low levelness as a result of unwanted fibre saturation by competing dyestuff molecules.
– The ammonium groups in polyamide and wool can serve as selective binding sites for small anionic dyes, for example, indigocarmine, which then exhibit a Langmuir-type sorption isotherm [9].
– The sorption of cationic ammonium ion-based softeners also requires the presence of negatively charged groups on the fibre surface. While in the case of cotton, carboxylic groups will serve as binding sites, the negative charged sulphonate groups of sorbed direct dyes and bound reactive dyes can also bind cationic softeners.

The type of sorption isotherm on wool fibres changes as a function of the dyebath pH. At pH 4 the sorption sites are formed by ammonium groups ($-NH_3^+$) present in the protein, while at pH 6 a less specific sorption through hydrogen bond interaction and

Van der Waals forces is observed. Thus, the sorption of indigo carmine on wool at pH 4 follows a Langmuir-type isotherm, while at an increased pH value of 7 the sorption of the dye follows the Freundlich isotherm. In parallel the overall sorption of the dye is reduced. At pH 4 and in the presence of a concentration of 100 mg/L dyestuff, approximately a mass of 8–9 mg of dye sorbs on 1 g of wool, while at pH 6 for the same concentration of dye in solution only 1–1.5 mg/g dye sorbs on the fibre (Figure 5.9).

5.4.5 Special aspects of sorption isotherms

On basis of sorption isotherms, a number of observations in textile chemical operations can be explained [8]. In many dyeing operations salt, for example, sodium chloride or sodium sulphate is added to the dyebath. The effect of this addition, however, is dependent on the type of fibre to be treated.

A. *Wool and polyamide*: The addition of salt in dyeing of wool and polyamide with acid dyes leads to a levelling effect and reduced dyestuff sorption. In the case of protein fibres, the fibre charge depends on the pH of the dyebath (eq. (5.22)).

$$H_3N^+ - Polym - COOH \rightarrow H_3N^+ - Polym - COO^- \rightarrow H_2N - Polym - COO^-$$

$$H_2N - Polym - COOH$$

| Acidic pH | Isoelectric pH | Alkaline pH | (5.22) |

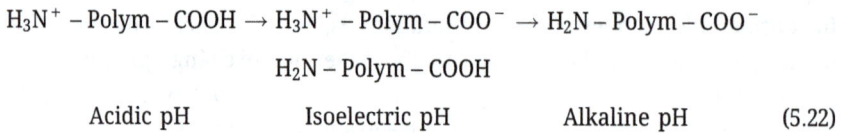

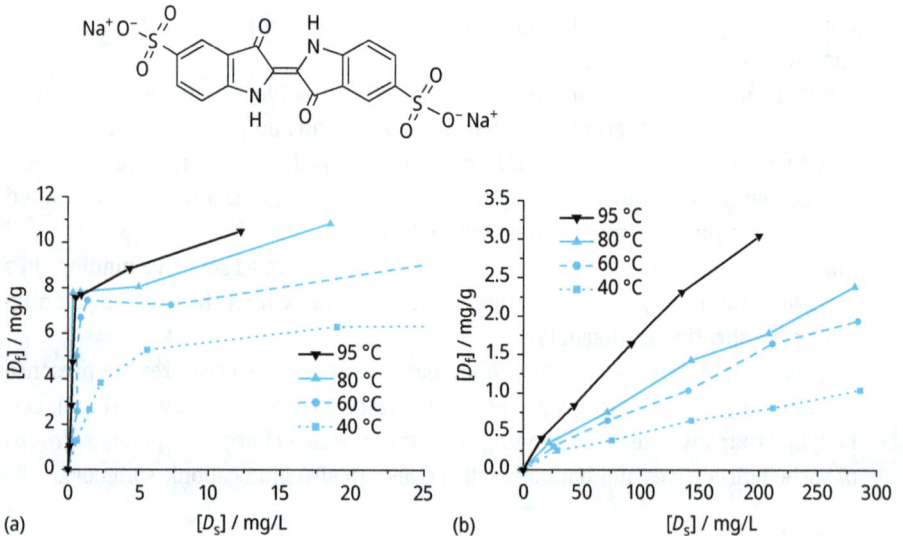

(a) $[D_s]$ / mg/L

(b) $[D_s]$ / mg/L

Figure 5.9: Sorption isotherm of indigocarmine on wool at (a) pH 3 and (b) pH 6 (in this example the mass concentrations were used instead of molality and molarity). From Komboonchoo and Bechtold [9].

Typical values for the pH in wool dyeing with acid dyes are at pH 3–4. Under these conditions, the fibre is present in its cationic form, the carboxylic groups being not dissociated and the amino groups being present in the protonated ammonium form.

The binding of an anionic dyestuff molecule then occurs through an anion-exchange process (eq. (5.23)). The sorption equilibrium then can be formulated according eq. (5.24) with K_{DX} as the equilibrium constant.

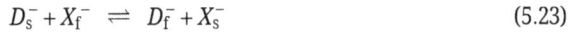

$$D_s^- + X_f^- \rightleftharpoons D_f^- + X_s^- \tag{5.23}$$

$$[D_f^-][X_s^-]/[D_s^-][X_f^-] = K_{DX} \tag{5.24}$$

At the same time, the condition of charge neutrality has to be observed. The concentration of positive charges (ammonium groups) $[S^+]$ and the sum of negative counter ions present, which can sorb dyestuff anions $[D_f^-]$ or salt anions $[X_f^-]$, for example, sulphate and chloride, have to be equal (eq. (5.25)).

$$[S^+] = [D_f^-] + [X_f^-] \tag{5.25}$$

From eq. (5.25) the concentration of salt anions $[X_f^-]$ can be formulated to be equal to $[S^+] - [D_f^-]$, which can be substituted in eq. (5.24), and after some rearrangement results in eq. (5.26).

$$[D_f^-]/([S^+] - [D_f^-]) = K_{DX}([D_s^-]/[X_s^-]) \tag{5.26}$$

$$[D_f^-]/([S^+] - [D_f^-]) = K[D_s^-] \tag{5.27}$$

Equation (5.26) represents a Langmuir-type isotherm (compare eq. (5.27)), with the difference that the general equilibrium constant K has been separated into a constant part K_{DX} and a term $1/[X_s^-]$, which considers the concentration of salt anions in the solution.

The addition of salt into a dyebath increases the concentration of salt anions in solution $[X_s^-]$ and thus leads to a reduction of dye sorption. For a given concentration of dye sorbed on the fibre, the dyestuff concentration in the bath has to be increased to compensate for higher concentrations of negative salt ions in solution $[X_s^-]$.

The same behaviour is observed when certain ions, which had been bound to an ion-exchanger, become eluted in the presence of a high concentration of competing ions with the same charge. The physical–chemical principle behind this phenomenon is the controlled displacement of sorbed ions by other ions of the same charge that compete to the ion-exchange equilibrium (eq. (5.23)).

The principle of ion-exchange processes in sorption of dyes and auxiliaries has to be considered in a number of applications: The behaviour of wool is similar to a weak anion exchanger. The sorption of acid dye is based on ion-exchange effects on positively charged ammonium groups in the wool fibres; thus, the deprotonation of ammonium groups in an alkaline wash bath leads to a loss of anchor groups for the acid dye molecules. This explains the unwanted release of dyestuff into alkaline wash baths and the observed lowered wash fastness of acid dyes under alkaline wash conditions. In dyeing of polyacrylic fibres, which actually contain anionic groups for binding of cationic dyes, colourless organic cations are used as retarders. These molecules sorb on the negative sites of the polyacrylonitrile co-polymer and compete with the coloured dye molecules. As the binding constant for the dye K_{DX} is higher than the constant for the retarder K_{RET}, the dyestuff molecules displace the sorbed retarder molecules. As an overall effect, the dyeing rate of the cationic dye is reduced that supports the formation of equal dye distribution and levelness of dyeing. High amounts of retarder, however, lead to a substantial competition to the intended dyestuff sorption and lighter dyeings are obtained. Thus, in polyacrylic fibre dyeing with retarder systems, careful consideration of dyestuff concentrations and dyestuff binding constants, fibre binding capacity and retarder concentration and retarder binding constant to the fibre is required.

B.   *Addition of salt in direct dyeing and reactive dyeing of cellulose fibres*: In exhaust dyeing, the addition of salt in the dimension of 10–50 g/L, salt is used to increase the affinity of the dyestuff to the fibre. (In this book we follow the diction of Zollinger [8], page 393, who recommends to use the term affinity in a semi-quantitative way and to describe the tendency of a substance to exhaust from a bath to a fibre.)

The influence of the addition of salt on the sorption of the negatively charged dyes can be explained by the formulation of a few basic relations: At first the requirement of charge neutrality on the fibre surface has to be fulfilled (eq. (5.28)).

$$[S^-] + [D_f^-] = [Na_f^+] \tag{5.28}$$

In the fibre, the sum of negative charges present due to dissociation of carboxylic groups $[S^-]$ and sorbed anionic dyes $[D_f^-]$ (direct dye or non-covalently bound reactive dye) has to be compensated by sodium ions $[Na^+]$. For simplification, only sodium ions are assumed to be present in the dyebath as positively charged ions.

As a second condition, a Donnan-type equilibrium (compare Chapter 4) between the ions present in bulk solution and the ions in the boundary layer of the fibre has to be formulated (eq. (5.29)).

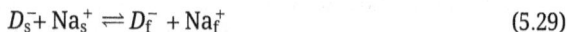

$$D_s^- + Na_s^+ \rightleftharpoons D_f^- + Na_f^+ \tag{5.29}$$

In the equilibrium state of sorption, the adsorbed dye present on the fibre surface $[D_f]$ and the counterions $[Na_f^+]$ (in this case only sodium ions) present in the boundary layer establish a dynamic equilibrium with the corresponding ions of the bulk solution $[D_s^-]$ and $[Na_s^+]$. The equilibrium constant K_{NaD} quantifies this relationship (eq. (5.30)).

$$[D_f^-][Na_f^+]/[D_s^-][Na_s^+] = K_{NaD} \tag{5.30}$$

When $[Na_f^+]$ is expressed according eq. (5.28) and is substituted into eq. (5.30), we obtain eq. (5.31).

$$[D_f^-]([S^-] + [D_f^-]) = K_{\text{NaD}}[D_s^-][Na_s^+] \tag{5.31}$$

In eq. (5.31) for a given fibre substrate, the concentration of sorbed dye $[D_f^-]$ depends on the concentration of dye in bulk solution $[D_s^-]$ and the concentration of sodium ions $[Na_s^+]$.

Two border cases can be analysed with regard to the concentration of bound negatively charged ions $[S^-]$, which in the case of cellulose mainly are carboxylate groups.

a. Under usual conditions the number of carboxylic groups in cellulose $[S^-]$ is small and the influence of sorption of a direct dye, a reactive dye or a reduced vat dye can be neglected. In this case $[S^-]$ is substantially smaller compared to $[D_f^-]$. Under this circumstance, $[S^-]$ can be put to zero and eq. (5.31) rearranges to the form shown in eqs. (5.32) and (5.33).

$$[D_f^-]^2 = K_{\text{NaD}}[D_s^-][Na_s^+] \tag{5.32}$$

$$[D_f^-] = (K_{\text{NaD}}[D_s^-][Na_s^+])^{1/2} = K[D_s^-]^{1/2} \tag{5.33}$$

Equation (5.33) describes an exponential dependency of the concentration of sorbed dye from the concentration of sodium ions in solution $[Na_s^+]$. In the presence of high salt concentrations, the concentration of sodium ions $[Na_s^+]$ can be assumed as a constant as the contribution of the positive counter ions of the dyestuff on $[Na_s^+]$ will be low. In this case, the formulation of eq. (5.33) is identical to the Freundlich isotherm, with a heterogeneity factor $1/n = 0.5$.

An increase in sodium concentration by a factor of 2 will modify the equilibrium constant K by a factor of $2^{1/2} = 1.41$. The dependency of K with regard to the concentration of $[Na_s^+]$ thus is not linear. As a rough estimate, an increase of $[Na_s^+]$ by a factor of 4 should be required to increase the exhaustion of a negatively charged dye on cellulose fibres by a factor of 2. In reactive dyeing, concentrations of up to 50 g/L NaCl or Na_2SO_4 are quite common for exhaust processes. A further increase in concentration of salt often is not efficient due to limited solubility of the dyes in highly concentrated salt solutions and due to the already high salt concentration in the dyebath.

b. A second border case can be formulated for a cellulose fibre that exhibits a high number of carboxylic groups. This case is not usual for normal cellulose fibres; however, it can be observed in highly oxidised pulp fibres or after selective oxidation of cellulose fibres, for example, with the use of tetramethylpiperidine-N-oxide (TEMPO) in combination with halogen [10].

In this case the concentration of negatively charged groups on the fibre $[S^-]$ already exceeds the concentration of sorbed dye $[D_f^-]$. In eq. (5.31) the factor $[D_f^-]$ can be neglected and the equation then rearranges to eqs. (5.34) and (5.35).

$$[D_f^-][S^-] = K_{\text{NaD}}[D_s^-][Na_s^+] \tag{5.34}$$

$$[D_f^-] = K_{\text{NaD}}\,[D_s^-][\text{Na}_s^+]/[S^-] \tag{5.35}$$

In this case, a linear dependency of the concentration of dye sorbed on the fibre $[D_f^-]$ with regard to both the concentration of dye in solution $[D_s^-]$ and the sodium concentration in solution $[\text{Na}_s^+]$ is obtained.

An interesting output of eq. (5.35) is the indication of an influence of the concentration of bound negatively charged groups on the dyestuff sorption. The higher the concentration of negatively charged groups in the fibre $[S^-]$, the lower will be the sorption of the dye on the fibre. Usually this observation is explained with repulsion between the negatively charged surface and the anionic dye. However from the basic formulations given in eq. (5.35), we can recognise that the changed sorption behaviour is a result of the Donnan equilibrium between boundary layer and bulk solution (eq. (5.29)).

Critical limitations for the discussion of sorption isotherms

Purity of sorbates (dyes, chemicals): It is worth to mention that sorption isotherms should be investigated preferentially with the use of pure substances. While sorption studies on natural colorants may deliver useful information about the dyeing properties of a certain plant extract, the calculation of thermodynamic data is not possible. A detailed interpretation of the sorption behaviour of plant extracts requires careful consideration and analysis of the actually exhausted chemical compounds of the extract and the non-exhausted coloured substances that do contribute to the colouration process.

Molecular dissolution: For dyeing experiments, the theory of adsorption isotherms is based on the assumption of a molecular disperse solution of the dye in the dyebath. Aggregation of dye molecules as well as precipitation leads to more complicated sorption schemes as preceding equilibria between agglomerates forming dispersed phases and the dissolved dye molecule have to be considered.

In case of polyester dyeing with disperse dyes, the major part of the dye in the dyebaths is present in dispersed solid form. The dyeing process proceeds via the molecular dissolved form of the dye. According to Zollinger [8], representative concentrations of molecular dispersed, dissolved dye are in the dimension of 1–20 mg/L dyestuff at 80 °C dyebath temperature. In a medium colour depth of 3%, an amount of 30 g dye is required for dyeing of 1 kg of material. When a liquor ratio of 1:15 is used for dyeing, a total concentration of 2 g/L dye is present in the dyebath as initial dye concentration. This value exceeds the solubility of the dye in the aqueous dyebath by far.

The overall dyeing process thus includes a preceding step of dissolution of disperse dye according eq. (5.36).

$$\text{Dye}_{\text{disp}} \overset{K_1}{\rightleftharpoons} \text{Dye}_{\text{sol}} \overset{K_2}{\rightleftharpoons} \text{Dye}_{\text{fib}} \tag{5.36}$$

For an analysis of the dyeing process and the corresponding isotherms, only the concentration of dye in dissolved form is relevant. At very low concentrations of dispersed dye and in the presence of molecular dissolved dye, the isotherm follows the Nernst-type behaviour. At higher concentration of dye, the presence of a dispersed phase stabilises the concentration of molecular dissolved dye in the dyebath at its saturation level. This prevents the development of a sorption isotherm, as the sorption step proceeds from an almost constant dye concentration in the dyebath. The concentration of disperse dye in solution decreases first when the reservoir of dispersed dye has disappeared.

Similar effects result from dyestuff aggregation in other dye systems. Many dyes tend to form associates, for example, dimers, trimers and higher aggregates both in solution (Dye_{sol}) and sorbed on the fibre (Dye_f) (eq. (5.37)).

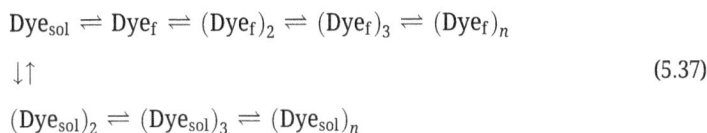

$$Dye_{sol} \rightleftharpoons Dye_f \rightleftharpoons (Dye_f)_2 \rightleftharpoons (Dye_f)_3 \rightleftharpoons (Dye_f)_n$$

$$\downarrow\uparrow \qquad\qquad\qquad\qquad\qquad\qquad\qquad\qquad\qquad (5.37)$$

$$(Dye_{sol})_2 \rightleftharpoons (Dye_{sol})_3 \rightleftharpoons (Dye_{sol})_n$$

The formation of aggregates in a solution can be detected by photometric measurement of dyestuff absorption. Concentration-dependent changes in the shape of the absorption spectra indicate the formation of aggregates. Aggregates also can be detected by other spectroscopic methods, for example, NMR techniques or by conductometry and polarography. The presence of aggregates has substantial influence on the sorption equilibrium of the dye, as preceding and following equilibria have to be considered.

Another important case where application of sorption isotherms has to be considered with care is the dyeing of cellulose or wool fibres with reactive dyes. In this case, the following reaction between the sorbed reactive dye and the fibre polymer leads to the formation of a covalent bond between the dye and the polymer. The covalently bound reactive dye molecules are no longer involved into the sorption equilibrium. As a result, compensation for the covalently bound reactive dye molecules through sorption of further dye molecules from solution to the fibre surface occurs. The overall equilibrium apparently shifts towards sorbed reactive dye, which however is not the case. The equilibrium constant remains a constant. The sorption of further reactive dye, after the dye fixation has been initiated by the addition of alkali, is a result of the reaction of sorbed dye with the fibre polymer (eq. (5.38)).

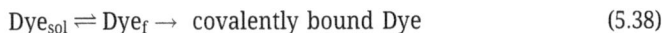

$$Dye_{sol} \rightleftharpoons Dye_f \rightarrow \text{ covalently bound Dye} \qquad (5.38)$$

An important application of the controlled blocking of the sorption equilibrium is observed in indigo dyeing. The sorption equilibrium of reduced indigo on the cellulose fibre would permit only the production of light dyeings as the affinity of the reduced indigo to the cellulose fibre is rather limited [11]. Dark shades of indigo have to be dyed in repetitive steps of dyeing followed by air oxidation. During the dyeing step, the sorption equilibrium between reduced indigo (leuco-indigo) present in dissolved state in the dyebath and adsorbed leuco-indigo on the fibre is established. During air oxidation, the adsorbed leuco-indigo is oxidised to indigo and thus eliminated from the sorption equilibrium. During the next dyeing step, the sorption equilibrium between fibre and dyebath establishes again; however, the time of immersion into the dyebath is kept short enough to prevent the reduction of indigo present on the fibre in the oxidised form. Thus in every repetitive dyeing/oxidation step, the sorption equilibrium is established again (eq. (5.39)). The total amount of indigo present on the denim yarn at the end of a series of dyeing/oxidation steps is much higher than that expected from the sorption isotherm of leuco-indigo (eq. (5.24)). For further information about indigo dyeing, see Chapter 10.

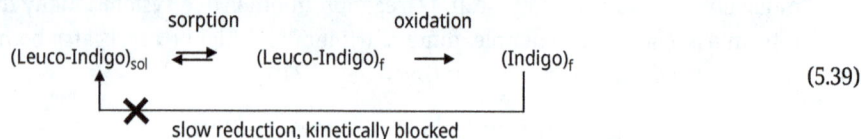

$$
\begin{array}{ccc}
& \text{sorption} & \text{oxidation} \\
(\text{Leuco-Indigo})_{sol} & \rightleftarrows \ (\text{Leuco-Indigo})_f & \longrightarrow \ (\text{Indigo})_f
\end{array}
\tag{5.39}
$$

slow reduction, kinetically blocked

References

[1] Okubayashi, S., Griesser, UJ., Bechtold, T. A kinetic study of moisture sorption and desorption on lyocell fibers. Carbohydr. Polym. 2004,58,293–299.

[2] Okubayashi, S., Griesser, UJ., Bechtold, T. Moisture sorption/desorption behaviour of various manmade cellulosic fibres. J. Appl. Polym. Sci. 2005, 97/4, 1621–1625.

[3] N. Pan, and Z. Sun,, Chapter 3, Essentials in psychrometry and capillary hydrostatics, Thermal and Moisture Transport in Fibrous Materials, Ed. N. Pan and P. Gibson, 2006, Woodhead Publishing Ltd, Cambridge, England

[4] Peter, Atkins., Julio, de Paula., Atkins' Physical Chemistry, Oxford University Press, Oxford, UK, 2002, ISBN 0-19-879285-9.

[5] Barbora, Š., Avinash, PM., Noisternig, MF., Henniges, UC., Kostic, M., Potthast, A., Griesser, UJ., Bechtold, T. Wash-dry cycle induced changes in low-ordered parts of regenerated cellulosic fibers. J. Appl. Polym. Sci. 2012, 126/S1, E397–E408.

[6] Kohler, R., Dück, R., Ausperger, B., Alex, R. A numeric model for the kinetics of water sorption on cellulosic reinforcement fibres. Compos. Interfaces. 2003, 10(2–3), 255–276.

[7] Cordin, M., Griesser, U., Bechtold, T. Analysis of moisture sorption in Lyocell-Polypropylene Composites. Cellulose. 2017, 24(4), 1837–1847, DOI: 10.1007/s10570-017-1227-8.

[8] Heinrich, Zollinger. Color Chemistry, 2003, Wiley-VCH, Weinheim, Germany.

[9] Komboonchoo, S., Bechtold, T. Sorption characteristics of indigo carmine as blue colorant for use in one-bath natural dyeing. Text. Res. J. 2010, 80, 734–743. doi:10.1177/0040517509342319.

[10] Fitz-Binder, C., Bechtold, T. One-sided surface modification of cellulose fabric by printing a modified TEMPO-mediated oxidant. Carbohydr. Polym. 2014, 106, 142–147.

[11] Etters, JN., Hou, M. Equilibrium sorption isotherms of indigo on cotton denim yarn: effect of pH. Text. Res. J. 1991, 61, 773–776.

Take home messages

The presence of polar groups in fibre polymers lead to a sorption of moisture. Based on polymer structure and fibre porosity, different amounts of moisture are sorbed in fibres.

Sorption isotherms describe the situation in equilibrated state.

The sorption behaviour of a substance on a fibre can be described by three different sorption isotherms: the Nernst-type isotherm, the Freundlich-type isotherm and the Langmuir-type isotherm. The mechanism of sorption is directly related to the characteristics of the sorption isotherm.

From an analysis of sorption isotherms, structural characteristics of the fibre polymer (e.g., internal surface and pore size) can also be obtained.

Sorption behaviour of charged substances from liquid phase has to consider also the overall charge balances and the establishment of a Donnan equilibrium. Thus, the presence of salt can have a direct influence on the sorption of a charged dyestuff.

Careful consideration of the sorption process with regard to disturbing aggregate formation or follow-up reactions is important.

Quiz

1. Why do we have to carefully consider the present climate conditions for weighing of textile fibres? Explain on the basis of a sorption isotherm.
2. Explain the characteristics of a moisture sorption isotherm.
3. What is the principle of the PEK model.
4. Why the Freundlich-type isotherm is curved? What represents the saturation point in the Langmuir-type isotherm?
5. What is the effect of a Donnan-type equilibrium in sorption processes?
6. Give examples where sorption isotherms become disturbed by preceding or following reaction steps.

Exercises

1. Calculate the amount of $[D_f]$ for the Nernst-type isotherm, the Freundlich-type isotherm and the Langmuir-type isotherm for the following conditions:
   $K = 50$ or $K = 10$; $[D_s] = 0.2 \times 10^{-3}$ mol/L or 2×10^{-3} mol/L. Freundlich (n = 1.25); Langmuir $[S] = 100 \times 10^{-3}$ mol/kg. Explain the results.

Solutions

1. (Question 1): Based on the molecular structure of textile fibres, considerable amounts of moisture are absorbed in the fibre matrix. The amount is dependent on the ambient climate. The sorption curves describes this behaviour.
2. (Question 2): The sorption isotherm describes the uptake of moisture as a function of relative humidity. Thus, a sorption isotherm is valid for a certain temperature. Three different sectors can be distinguished: Formation of a monolayer, multi-layer

adsorption and capillary condensation. During a sorption/desorption cycle, a hysteresis is observed, which results from structural changes at higher humidity.

3. (Question 3): In a simplified approach, the sorption kinetics are separated into two processes: a rapid sorption process and a slow sorption process. The model helps to describe a sorption process and indicates capacity and rate of sorption for the two processes, thus also leading to description of fibre characteristics.

4. (Question 4): The Freundlich isotherm is curved as a result of different adsorption energies on the surface of the fibre. The heterogeneity of the surface is characterised by the curvature of the isotherm. In a Langmuir-type sorption, the saturation [S] is reached when all places for sorption have been occupied. Fibre saturation is reached only at infinitely high concentration of sorptive in solution [D_f].

5. (Question 5): As a result of the formation of Donnan-type equilibria in the boundary layer of a fibre, the presence of any charged substance in the solution (e.g., salt) will contribute to the sorption behaviour of charged dye molecules.

6. (Question 6): Disperse dyes dissolve only at very low concentration; thus, a preceding equilibrium to the dispersed dye particles is formed. In vat dyeing and in reactive dyeing, follow-up reactions can interrupt the sorption equilibrium.

7. (Exercise 1): Substitute the values given into the respective formulas for the isotherms and calculate [D_f].

	[D_f] Nernst mol/kg	Freundlich mol/kg	Langmuir mol/kg
$K = 50$ [D_s] $= 0.2 \times 10^{-3}$ mol/L	0.010	0.055	0.0010
$K = 50$; [D_s] $= 2 \times 10^{-3}$ mol/L	0.100	0.347	0.0091
$K = 10$; [D_s] $= 0.2 \times 10^{-3}$ mol/L	0.002	0.011	0.0002
$K = 10$; [D_s] $= 2 \times 10^{-3}$ mol/L	0.020	0.069	0.0020

Discussion: The Nernst equation leads to a linear relationship with concentration. Compared to the Nernst equation, the Freundlich isotherm indicates higher sorption at [D_s] < 1 mol/l and lower sorption at [D_s] > 1 mol/l. Nernst and Freundlich isotherm will cross at [D_s] = 1 mol/L. The Langmuir isotherm leads to substantially lower sorption when the same value for K is used in the calculations.

List of abbreviations/symbols

ρ	density
γ	surface tension
ΔG	Gibbs energy
ΔH	enthalpy
ΔS	entropy
[D_f]	concentration of sorptive (= sorbate) in fibre (molality)
[D_s]	concentration of sorptive in solution (molarity)
[S]	value for saturation

μ	chemical potential
μ_0	chemical potential at standard conditions
a_w	activity of water
E_1	heat of adsorption for the first adsorbed layer (monolayer)
E_2	heat of condensation of the sorbate.
K	equilibrium constant
k_i	rate constant in the PEK model
M	molar mass
m_{ads}	mass of adsorbed gas in equilibrium
MC	moisture content
m_{dry}	mass of dry fibres
M_i	moisture regain in the PEK model
m_m	mass of adsorbed gas in monolayer
m_{mt}	mass of moist fibres
n	number of moles
p	pressure
p_{sat}	vapour pressure of gas (sorptive, water vapour pressure) under conditions of saturation
p_x	pressure of sorptive (water vapour pressure) in equilibrium
r	radius of a capillary
R	gas constant
r.h.	relative humidity
RG	moisture regain
T	temperature
V	volume
V_{ads}	adsorbed volume of gas
V_{Mono}	volume of adsorbed gas to form a monolayer
τ	half life time in the PEK model

6 Kinetics of textile chemical processes

The models and practical aspects presented in Chapter 6 consider a non-equilibrium state. Through differences in concentration of chemicals, transport processes occur. A negative free energy can lead to spontaneous chemical reactions. Based on the type of reaction and the time given, an equilibrium state can be established, which represents the end of a dynamic process. In the preceding Chapter 5, the presence of an equilibrium state is a central precondition to apply the models given there for example sorption isotherms. This chapter describes that the presence of a non-equilibrium state is a precondition to trigger transport phenomena and chemical reactions.

6.1 Elementary steps in polyester dyeing

Compared to other dyeing processes, the dyeing of polyester fibres with disperse dyes appears to be more straightforward than other dyeing techniques.

Above T_g, the dissolved dye molecules become extracted by the solid polyester fibre. The apparently uncomplicated process ends with an equilibrium state between dispersed dyestuff dissolved in the fibre and dyestuff remaining in the dyebath, which follows Nernst-type isotherm.

Dyeing of wool includes the formation of ionic bonds and hydrogen bonds; in reactive dyeing processes, we have to consider the formation of a covalent bond between a dye and a polymer. In vat dyeing, reduction and dissolution of the dispersed vat dye and a reoxidation of the reduced dye occur. In sulphur dyeing, the creation of disulphide bridges leads to formation of a nearly polymeric dye as the final product.

A polyester dyeing process does not include chemical reactions, thus, the overall process scheme consists of a set of heterogeneous transport processes (Figure 6.1).

6.2 Step A – dissolution of dispersed dye

Despite the fact that disperse dyes are almost insoluble in aqueous phase at least a certain minimal solubility in water is a necessary precondition to achieve reasonable dyeing rates.

Through its limited water solubility, the dyestuff present in the dyebath does not dissolve completely as the concentrations used for dyeing exceed the solubility limit. At the beginning of the dyeing process, the major part of the dyestuff is present in the dispersed form and only a minor amount is available as the dissolved state. As a result of the dye uptake by the fibre, the concentration of dissolved dye lowers and dissolution of disperse dye from the surface of the disperse dye particles begins. The

https://doi.org/10.1515/9783110795738-006

Figure 6.1 content:

D	C		B		A
					c_S
Solid fibre	c_{Fsurf}		Dyebath		Dispersed dye
			c_B		
c_F		δ			
	c_{Surf}			δ'	

Dye concentration   Boundary layer   Dissolution

$$dm/dt = AD_{aq}(c_S - c_B) / \delta'$$

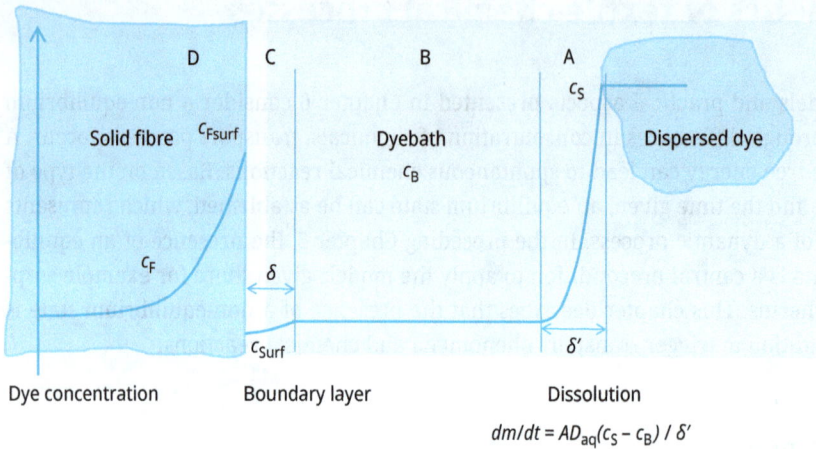

Figure 6.1: Elementary steps and respective dye concentrations in dyeing of polyester with disperse dyes. From Peters and Freeman [1].

driving force for this process results from the objective to restore the solution equilibrium between dispersed and dissolved phases of the dye (eq. (6.1)):

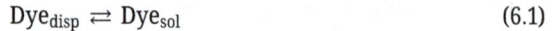

$$\text{Dye}_{disp} \rightleftharpoons \text{Dye}_{sol} \tag{6.1}$$

The dissolution process of the disperse dye can form the bottleneck of the dyeing process in case the rate of dissolution is not sufficiently high to maintain the concentration of dissolved dye near the saturation solubility c_s.

For dispersed organic compounds the dissolution rate can be estimated by the diffusion layer theory (eq. (6.2)). In this approach, the diffusion dye through the boundary layer around the dye particle determines the dissolution rate [1]:

$$dm/dt = A\,D_{aq}(c_s - c_B)/\delta' \tag{6.2}$$

The rate of dissolution dm/dt depends on the surface area of the particle A, the apparent diffusion coefficient of the dye in water D_{aq}, the saturation solubility and the concentration in the bulk solution c_B and the thickness of the diffusion layer δ'.

The dissolution rate of the disperse dye will become rate determining when the area A of the particles is low; thus, dispersions with particle size in micrometre and sub-micrometre dimensions are in use to offer a high area A for dissolution. In addition, dispersing agents are used to prevent agglomeration. A high concentration gradient $(c_s - c_B)$ through the boundary layer supports the dissolution; thus, dyes with higher solubility c_s can be expected to exhibit a higher dissolution rate.

During the dyeing step the size of the particles reduces, which also influences the saturation solubility c_s. An estimate for this effect for particles with size of submicron can be obtained from the Ostwald–Freundlich equation, which compares saturation solubility c_s at different radii of the particles (r_1 and r_2) (eq. (6.3)):

$$\ln(c_s(r_1)/c_s(r_2)) = (1/r_1 - 1/r_2)2M\,\sigma/RT\,\rho \qquad (6.3)$$

with $c_s(r_1)$ and $c_s(r_2)$ being the saturation solubilities at the respective particle radii r_1 and r_2, M the molecular weight of the dye, σ represents the specific free energy at the interface between the particle and the liquid, and ρ represents the density of the solid.

The diffusion coefficient in aqueous solution D_{aq} can be determined experimentally or estimated from the molecular weight considerations (eq. (6.4)) [2]:

$$D_{aq} = 7.4 \times 10^{-8}(yM)^{1/2}T/\eta\,V_m^{0.6} \qquad (6.4)$$

where η represents the viscosity of the solvent (water), y represents a factor for hydrogen bonding in the solvent (e.g. for water $y = 2.6$) and V_m is the volume of 1 mole (e.g. cm^3/mol).

For a given disperse dye, technical measures to maximise the dissolution rate of a dye can be derived from eqs. (6.2)–(6.4)

- Finely divided particles offer a higher surface A for dissolution
- Agitation of the dyebath reduces the thickness of the diffusion layer δ'
- Temperature increases D_{aq} the diffusion coefficient in water and in parallel reduces the viscosity η

Besides the theoretical considerations given above, also the presence of different crystal polymorphs of a dye and the addition of auxiliaries that apparently change the concentration in solution e.g. through formation of micelles or better dispersion of particles will influence the observed rate of dissolution.

6.3 Infinite and finite dyeing kinetics

In an idealised case, a steady state is established between the dissolution of dyestuff from its dispersed phase into the dyebath and the dye uptake by the fibres. As long as the rate of dissolution (step A) is not the rate-determining step, the concentration of dye in solution c_B remains constant. The dyebath then behaves as it would be of infinite size. The solid reservoir of dispersed dye stabilises c_B the concentration of dye in solution on a constant level. The observed kinetics are then called "infinite dyebath" kinetics as the uptake of colour proceeds from an apparently infinite dyebath which exhibits a stable dye concentration c_B (of course on the costs of the dispersed dye present in the dye bath).

At the moment when the dyestuff concentration in the dyebath reduces as a result of the dye uptake by the fibre, "finite" dyebath kinetics will be observed. The rate of dyestuff uptake is then dependent on time, as the concentration of dye c_B decreases with time.

An example for an estimate when infinite conditions can be expected is (eq. (6.5))[3]:

$$c_0 \geq c_s(1 + K/\text{LR}) \tag{6.5}$$

where c_0 is the initial total concentration of dye (dissolved and dispersed dye), LR is the liquor ratio (volume of dyebath in dm^3 per mass of goods in kg) and K the partition coefficient for the dissolved dye between polyester fibre and dyebath.

Thus infinite dyebath kinetics is favoured by dyestuff systems that exhibit low dye solubility c_s in dark dyeings (high c_0) and in dyeings with short liquor ratio as the amount of dye to maintain c_s in the full dyebath volume is lower compared to the same dyeing at higher liquor ratio.

The total amount of dye required for a 2 wt% dyeing is 20 g/kg, which corresponds to c_0 of 2 g/L at an LR of 1:10 and of 0.5 g/L at an LR of 1:40. For a dyestuff solubility of $c_s = 0.05$ g/L, the "infinite" dyebath conditions will change to finite dyebath conditions when c_0 drops below 0.05 g/L. From this point, the rate of dyeing will be dependent on the constantly decreasing dyestuff concentration in solution. The final concentration in the dyebath will depend on the respective partition coefficient K. However, in any case the residual amount of dye will be four times higher at the dyeing with LR = 1:40 compared to LR = 1:10.

6.4 Step B – hydraulic transport in the dyebath

Intensive agitation of the dyebath has to be achieved as a condition to obtain equal uptake of dyestuff at all sites of the material (eq. (6.6)):

$$\text{Dye}_{sol}(\text{bulk}) \rightleftharpoons \text{Dye}_{sol}(\text{near boundary layer of fibre}) \tag{6.6}$$

Equal conditions near the boundary of the textile fibres have to be maintained throughout the full dyeing process, in particular during the critical phase of dyestuff sorption. This can be achieved by careful control of the dyeing conditions, for example, temperature, pH with the aim to retard the dyestuff sorption to such an extent that the hydraulic transport of dye near the boundary layer of the fibres is substantially more rapid than the diffusion through the boundary layer and the following sorption process.

A negative influence of insufficient exchange of dyebath can be avoided by technical measures, for example,

– well-controlled sorption process through slow heating rates near T_g of synthetic fibres, dosage of acid in wool dyeing;
– intensive circulation of dyebath in dyeing apparatus and rapid transport of goods in jet-dyeing.

While these measures appear as straightforward strategies, the structure of many textile materials adds complications. The architecture of a textile consists of several levels. Penetration of a dyebath through fabric can be achieved by appropriate pumping or movement of the goods. The penetration of yarn forming the fabric is more challenging. The highly twisted fibres form a dense inner part in which a stagnant volume of liquid is trapped. As a result of the insufficient liquor exchange to the core of a yarn, this part remains lighter or not dyed. Besides the high rates of bath circulation also higher dyeing temperatures support the dye penetration into the yarn core, as the viscosity of water reduces with temperature and diffusion transport is higher.

However, requirements with regard to a better controlled dye uptake through slow increase in dyebath temperature will require a compromise between better levelness of dyeing through slower temperature-controlled fixation and better penetration of yarn core at high dyebath temperature and long dyeing time. Dye classes that can establish a sorption equilibrium during the exhaust phase permit introduction of a levelling period after the major part of dye had been exhausted. Examples are disperse dyes, direct dyes, acid dyes and vat dyes. On the contrary, levelling of reactive dyes is only possible before dyestuff fixation has been initiated by the addition of alkali.

In case of indigo dyeing, for the production of denim yarn low dye penetration into the yarn is desired. Thus, the process design utilises all "negative" effects that prevent uniform dye distribution in the cross section of the yarn. The desired "ring dyeing" is achieved by a number of measures:

- pre-wetting of the yarn with water to fill the yarn core with liquid and reduce accessibility;
- short dip into the dyebath to limit the time for exchange of the liquid present in the yarn core with the dyebath;
- low dyebath temperature; and
- low concentration of dye in the dyebath and repetitive dyeing.

6.5 Step C – diffusion through the boundary layer of the fibre

As a third step in the dyeing process, the transfer of dissolved dye through the boundary layer of the fibre occurs. Similar to step A, the transport process through this stagnant layer is driven by diffusion.

In a very general approach, the transport through the boundary layer can be formulated according to Fick's law (eq. (6.7)):

$$J = -D_{aq}(c_{surf} - c_B)/\delta \tag{6.7}$$

where J represents the flux of mass per unit area (g/cm^2/s) and δ is the thickness of the boundary layer of the fibre. For a given dye, the transport through the boundary fibre thus depends on the concentration gradient between solution and fibre surface $(c_B - c_{surf})$ and the thickness of the diffusional boundary layer δ. High shear rates and intensive dyebath circulation reduce the thickness of the boundary layer and thus directly intensify the diffusion transport of dyestuff molecules to the fibre surface.

The concentration difference $(c_B - c_{surf})$ depends on the process conditions (e.g. colour depth and degree of dyebath exhaustion).

In case of infinite dyebath conditions, the concentration gradient $(c_B - c_{surf})$ remains almost constant.

In the equilibrium state, the concentration of dyestuff in the liquid at the fibre surface c_{surf} is linked to the concentration in the polymer surface c_{Fsurf} through the partition coefficient K (eq. (6.8)):

$$K = c_{Fsurf}/c_{surf} \tag{6.8}$$

Thus, step C actually consists of two separate processes: diffusion and adsorption (Figure 6.2).

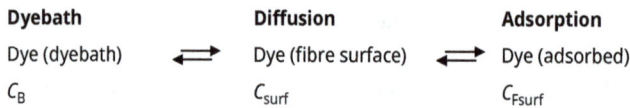

Dyebath		**Diffusion**		**Adsorption**
Dye (dyebath)	$\rightleftharpoons$	Dye (fibre surface)	$\rightleftharpoons$	Dye (adsorbed)
c_B		c_{surf}		c_{Fsurf}

Figure 6.2: Step C – dyestuff diffusion through boundary layer of the fibre and dyestuff adsorption.

Usually the rate of adsorption is assumed to be substantially higher than the diffusion process; thus, in the series of these two consecutive processes, the transport through the boundary layer by diffusion (eq. (6.7)) will be rate determining.

However, from the point of a dyeing process, it is important to consider that the equilibrium concentration of the dye in the fibre c_{Fsurf} is substantially higher than c_{surf}, and the proportionality factor is the partition coefficient K. As a consequence, short time after the dyestuff uptake has been initiated, a high concentration c_{Fsurf} will be reached. Thus the sorption process in the fibre will remain far from the equilibrium state for a long time and an increase of c_{surf} due to saturation effects in the fibre will appear only during a later phase of the dyeing process.

Saturation effects should not appear as long as $c_{surf} K \geq c_{Fsurf}$. In addition, the diffusion of the sorbed dye into the core of the fibre contributes to a reduction in c_{Fsurf}.

6.6 Step D – dyestuff sorption in the fibre

The dyestuff initially adsorbed in the surface layer of the fibre diffuses into the core of the fibre according to Fick's first law:

$$J = D_f dc_F/dx \qquad (6.9)$$

where D_f is the diffusion coefficient of the dyestuff in the fibre polymer and dc_F/dx is the concentration gradient along the radius of the fibre. The basic eq. (6.9) assumes that the coefficient of diffusion D_f in the fibre is independent of the dyestuff concentration and is also strictly valid only for a linear diffusion, which for example is obtained in experiments with films. In a round fibre, a radial diffusion to the core adds an additional complication.

Typical values of D_f for disperse dyes in polyester fibres at a dyebath temperature of 130 °C are in the magnitude of 10^{-10} cm²/s.

It is important to consider that the basic concepts discussed here always include the assumption of a uniform fibre structure, in which the dye molecules absorb and diffuse. In real fibres, differences in the molecular order, crystallinity and core–shell properties have to be considered.

The given model represents a quite acceptable approach for fibres where the free volume model is appropriate, for example, polyester fibres and for dyes, which do not exhibit follow-up reactions (e.g. disperse dyes).

For cellulose fibres, the pore model represents the more appropriate description of the fibre structure, which however adds substantial complications through diffusion processes of dyes in the voids and pores.

Another effect that must be kept in mind is fibre swelling, which alters the polymer accessibility in cellulose fibres and the available free volume in polyester fibres, for example, in the presence of carrier molecules.

6.7 Levelling

In case the equilibria through stages A–D are not interrupted by the follow-up reactions, an initially uneven distribution of dye can be levelled out during the last phase of a dyeing process.

During the levelling phase, the concentration of dyestuff in solution remains almost constant at a certain equilibrium value of c_B. The concentration of dye in bulk solution c_B, however, is not in equilibrium with every part of the dyed fibre, for example, some sections of a fibre may have a higher/lower concentration than c_B would predict as the equilibrium concentration. Thus, from sites with high dye concentration diffusion into the bulk solution occurs with the aim to increase c_B, which however leads to an increased adsorption at sites where the concentration of dye in the fibre is too low. As a result, an overall transport of dye from sites of higher dyestuff concentration to sites of low dyestuff concentration occurs, and differences in colour depth and shade level out (eq. 6.10)):

$$\text{dyestuff} \atop \longrightarrow$$

$$\underset{\text{dark}}{c_{F-\text{high}}} \rightleftarrows c_B \rightleftarrows \underset{\text{light}}{c_{F-\text{low}}} \tag{6.10}$$

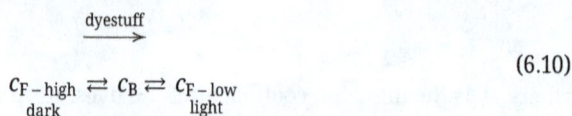

where $c_{F-\text{high}} > c_{F-\text{low}}$.

The main characteristic of a dye with high levelling capacity is the diffusion coefficient of the dye inside the fibre D_f. Dyes with high D_f achieve more rapid equilibration in an highly agitated dyebath, as the major rate-determining step then will be the diffusion transport of dye in the fibre structure.

Apparently, the concentration of c_B is of minor influence on the levelling properties of a dye, which is the case as long as the hydraulic transport of dye in the agitated dyebath is sufficiently high to avoid local concentration differences in the bulk phase.

6.8 Follow-up reactions – kinetics in reactive dyeing

In the application of reactive dyes, the formation of the covalent bond between dye and fibre polymers leads to a formal reduction of the concentration of sorbed dye c_F, which thus influences the diffusion gradient between the bulk and fibre surface. In case of reactive dyeing of cellulose fibres, substantial amounts of alkali are added to the dyebath to initiate the reactive coupling reaction with the hydroxyl groups of the cellulose polymer. Besides the desired fixation reaction also hydrolysis of the reactive dye occurs as a competitive process through the reaction with the alkali present in the dyebath (Figure 6.3).

The hydrolysis reaction in the dyebath is proportional to the concentration of dye in bulk c_B and to the concentration of hydroxyl ions c_{OH^-}. In a simplified approach, the reaction rate for dyestuff hydrolysis thus corresponds to a second-order reaction. In eq. (6.11), the hydrolysis reaction has been split into a bulk reaction and a reaction inside the fibre, which thus considers differences in concentrations and reactivities between the bulk solution and fibre:

$$dc_{\text{Hydrol}}/dt = k_{\text{Hydr}}\, c_B\, c_{OH^-} + k_{\text{FHydr}}\, c_F\, c_{FOH^-} \tag{6.11}$$

where c_{Hydrol} is the concentration of hydrolysed reactive dye, k_{Hydr} the rate constant for hydrolysis of reactive dye in bulk solution, k_{FHydr} the rate constant for hydrolysis of the reactive dye in the fibre matrix. c_{OH^-} is the concentration of hydroxyl ions in bulk and c_{FOH^-} the concentration of hydroxyl ions in the fibre.

The transport process into the fibre can be described by a similar approach as given earlier, using Fick's second law for the mass transport, with consideration of a parallel chemical reaction:

$$dc_F/dt = D(d^2 c_F/dx^2) - k\, c \tag{6.12}$$

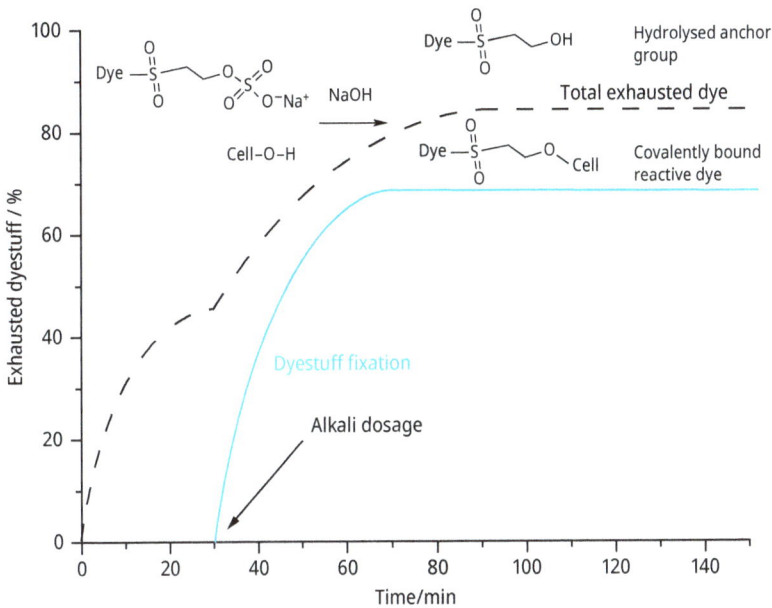

Figure 6.3: Fixation and hydrolysis of a reactive dye on cellulose.

In eq. (6.12) the rate constant k includes all reactions that lead to disappearance of the dye, for example, through reaction with the fibre or through hydrolysis reactions.

According to literature, a simplified model of fibre surface and diffusion processes can be used to formulate the fixation rate of the reactive dye per unit area of fibre based on eq. (6.13). In eq. (6.13) the rate of dyestuff fixation dF/dt is formulated as a function of fibre surface concentration of dye c_{Fsurf}, the observed diffusion coefficient D_{obs} and the pseudo-first-order rate constant for dyestuff fixation reaction k_F per area of fibre surface [4]:

$$dF/dt = c_{Fsurf}(D_{obs}\ k_F)^{0.5} \tag{6.13}$$

It is important to consider that in these models, the fixation rate constant k_F has been introduced as a pseudo-first-order reaction rate constant. Actually the fixation reaction is a heterogeneous reaction and also depends on the hydroxyl ion concentration and the surface properties of the fibre material (eq. (6.14)):

$$dF/dt = f(c_{Fsurf},\ c_{FOH^-};\ \text{area and reactivity of fibre surface}) \tag{6.14}$$

Thus, modification of dyeing conditions in terms of type of alkali and concentration used will also influence the fixation rate. For cellulose type of fibres, the surface reactivity of the material will influence fixation rate; thus, pre-treatment, alkali activation and hornification due to high drying temperatures all will modify the observed reaction rate.

By combination of the process of dyestuff fixation (eq. (6.13)) with the parallel running hydrolysis reaction of the dye (eq. (6.11)), an overall formulation for the consumption of reactive dye dc/dt can be formulated as follows:

$$-dc/dt = dF/dt + dc_{\text{Hydrol}}/dt = c_{\text{Fsurf}}(D_{\text{obs}}\ k_{\text{F}})^{0.5} + k_{\text{Hydr}}\ c_{\text{B}}\ c_{\text{OH}^-} + k_{\text{FHydr}}\ c_{\text{F}}\ c_{\text{FOH}^-}$$

$$(6.15)$$

The overall consumption of reactive dye during a dyeing process thus will depend on the rate of several competitive processes (eq. (6.15)): the desired reaction with the cellulose fibre and unwanted losses through hydrolysis of anchor groups in the bath and inside the fibre.

A representative application of eq. (6.15) is observed in exhaust dyeing processes using reactive dye and alkali dosage. During the first phase of the dyeing, dyestuff exhaustion is achieved in the presence of salt. Thus during this phase uniform distribution of dyestuff on the material is intended. The reaction in eq. (6.15) is then initiated by the dosage of alkali solutions. Small additions of alkali lead to dyestuff fixation, while the competitive hydrolysis reactions according to the second and third terms of eq. (6.15) can be retarded.

6.9 Preceding and following reactions – vat dyeing

Another complex series of preceding and consecutive reactions is observed in vat dyeing processes. Initially the dye is present in a dispersed oxidised form, which then is transferred into the corresponding alkali-soluble reduced form. The reduced form sorbs on the cellulose fibre and equilibrates during the levelling phase. At the end of the dyeing process, the reduced form of the adsorbed vat dye is transferred into the insoluble oxidised form and thus immobilised again (compare Chapter 10). The dyeing process can be formulated based on Figure 6.4.

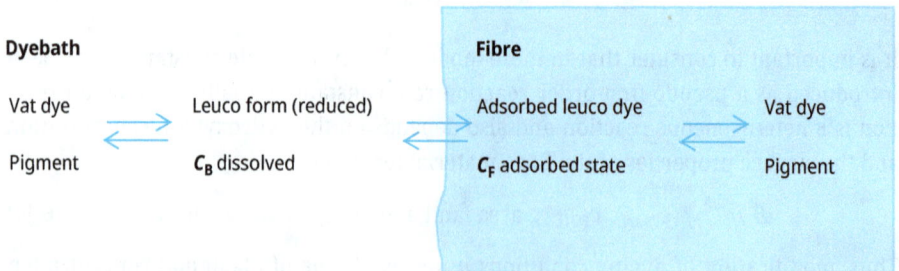

Dyebath

Vat dye   Leuco form (reduced)

Pigment C_{B} dissolved

Fibre

Adsorbed leuco dye   Vat dye

C_{F} adsorbed state   Pigment

Figure 6.4: Dyeing kinetics in vat dyeing.

Reduced vat dyes exhibit rather high rate of dye exhaustion; thus, uniform dyeing is supported by different measures:

The rate of dyestuff sorption can be steered through addition of the reducing agent. As only the reduced form adsorbs rapidly on the fibre, appropriate control of concentration of leuco dye in the dyebath permits steering of the adsorption rate.

A levelling temperature above 80 °C is required in exhaust processes. At such temperatures, the equilibrium concentration c_B increased to a level, which permits effective redistribution of unevenly adsorbed leuco dye.

Stable reducing conditions are essential to maintain the dye present in the dyebath in its reduced form. Unreduced or partially reoxidised vat dye does not contribute to the dyeing and exhaust equilibrium, however, will contribute to the final distribution of dye on the textile goods.

Reoxidation of the leuco dye has to be achieved under conditions that prevent unwanted redistribution of leuco dye in the fibre.

6.10 Aggregation

Similar to the dissolution of dispersed dye in polyester dyeing (step A, Figure 6.1), aggregation of dissolved dyestuff molecules leads to a reduction in c_B, the effective concentration of dye in solution ($c_0 > c_B$).

Dyestuff aggregation can be observed with all classes of dyes.

Driving forces for aggregation can be:

- Hydrophobic interactions, for example, in case of presence of longer aliphatic groups in the dye molecule
- Hydrogen bonding between dye molecules and dye–water molecules
- Aromatic interactions (π–π interactions) between benzene rings and annealed rings
- Van der Waals forces in particular when dye molecules arrange parallel or in layered structure

Electrostatic repulsive forces: Charged ionic groups (mainly sulphonate groups) that are present in the molecules to improve dyestuff solubility also lead to a electrostatic repulsion between dyestuff molecules. Thus the number of sulphonate groups and their position in the dye molecule will be of influence for tendency of dye molecules to aggregate.

References

[1]   Peters, AT., Freeman, HS. Advances in color chemistry series – Volume 4, Physico-chemical Principles of Color Chemistry, Blackie Academic & Professional, London, 1996.
[2]   Wilkie, CR., Chang, P. Correlation of diffusion coefficients in dilute solutions. AIChE J. 1955, 1, 264–270.
[3]   McGregor, R., Etters, JN. Transitional kinetics in disperse dyeing. Text. Chem. Color. 1979, 11, 202–206.
[4]   Sumner, HH., Weston, CD. Pad-dyeing methods for reactive dyes on cotton: the practical implications of a theoretical study. Am. Dyest. Rep. 1963, 52, 442–450.

Take home messages

The apparently simple process of dyeing consists of a series of transport processes. Often chemical reactions and preceding or consecutive reactions have to be considered.

Hydraulic transport is usually not rate determining, as agitation of dyebath and movement of good are high enough to achieve equal distribution of the dyebath in the goods.

Often diffusion through the boundary layer between dyebath and fibre and diffusion inside the fibre polymer are rate determining.

The complex architecture of a textile (fibre polymer, yarn and fabric structure) makes theoretical considerations complicated.

Levelling of a dyeing can be achieved in processes that permit equilibration. Thus, covalently bound reactive dyes cannot equilibrate any more. Also the oxidised form of a vat dye and the dispersed form of a disperse dye do not participate directly to a levelling process, as equilibration only proceeds through the molecular dispersed form (c_B).

Simplified reaction models, however, can be very useful to identify the rate determining steps and critical conditions in dyeing processes.

Quiz

1. Why agitation of the dyebath contributes to the dyeing rate?
2. What is the physical background for milling of dye particles to the dimensions of submicrometre?
3. Why characteristics of the fibre structure determine the rate of dyeing and the levelness?
4. In reactive dyeing higher dyestuff fixation could be obtained by use of double anchor dyes (reactive dyes that bear two anchor groups). Can you explain this on the basis of the dyeing kinetics (eq. (6.15))?

Exercises

1. A polyester dyeing (assume density of fibre material with 1 g/mL) is performed with a dispersed dye of 2% o.w. colour depth, at a liquor ratio of LR = 1:20. The partition coefficient between dyebath and polyester fibre (a) K = 50 and (b) K = 200. The bulk solubility during dyeing c_B of the dye at 110 °C is measured with 0.1 g/L. What is the total concentration of dye in the dyebath at the beginning of dyeing? At which total concentration of dye in the dyebath will finite dyebath kinetics would begin? Which amount of dye will be left in the exhausted dyebath?

Solutions

1. (Question 1): Agitation reduces the thickness of the stagnant layer δ between dyebath and fibre surface, and thus supports the diffusion transport from the dyebath to the fibre surface.
2. (Question 2): The dissolution of a dispersed dye and the reactivity of a dispersed vat dye will depend on the surface of the particle A. The rate of the heterogeneous reaction thus can be increased by milling of the particles.
3. (Question 3): The rate of diffusion of a dye inside the fibre depends on structural characteristics of the fibre (crystallinity, porosity, accessibility and internal surface) and thus will influence the dyeing characteristics.
4. (Question 4): The use of a double anchor dye leads to a higher probability for a covalent linkage between the dye and the cellulose fibre. Let us assume the probability for a reaction of the one anchor group with the fibre lies at 60%, and then 40% of the dye will be hydrolysed. As these molecules bear another active anchor group, 60% of the remaining unfixed 40% of dyestuff get a second chance for fixation; thus, approximately 40×0.6 = 24% still could react with the fibre. A rough estimation of the fixation rates will be at approximately 60% for a mono-functional dye and 84% for a bi-functional dye, thus indicating the substantial increase in dye fixation.
5. (Exercise 1): The total concentration at start of dyeing: A 2 wt% dyeing corresponds to 20 g dyestuff per 1 kg of goods, which are dyed at LR = 1:20 in 20 L of dyebath. Thus, the total concentration of dye in the bath will be 1 g/L.

    Finite dyebath conditions would begin at c_B = 0.1 g/L. As the initial concentration of dye was 1 g/L the finite dyebath conditions will start after 90% of dyebath exhaustion.

    The total amount of dye remaining in the dyebath can be calculated from two equations: the partition coefficient $K = c_F/c_B$; thus, $c_B = c_F/K$ and from the total mass balance of the dyestuff in fibre and solution $c_F V_f + c_B V_B = m_{dye}$.

    For K = 20 c_F can be calculated directly from $c_F = K c_B$ = 50 × 0.1 = 5 g/L, respectively, 5 g/kg. In this case, fibre saturation is achieved and the dye is not fully exhausted. About 15 g of dye remains partially dispersed in the dyebath at a total concentration of $c_{dyestuff}$ = 15/20 g/L = 0.75 g/L.

For $K = 200$: $c_F V_f + c_B V_B = m_{dye}$; thus, $c_F 1 + c_B 20 = 20$. When c_F is a function of c_B inserted in the partition equation $c_B = 0.09$ g/L, finite dyebath conditions will be reached. c_F is calculated as 18.18 g/kg fibre (density = 1 g/mL).

List of abbreviations/symbols

T_g	glass transition temperature
dm/dt	rate of dissolution
A	surface area of particle
D_{aq}	diffusion coefficient in aqueous solution
D_{obs}	observed diffusion coefficient
c_s	saturation concentration
c_B	concentration in bulk solution
c_0	overall concentration of dye present in the dyebath (incl. dispersed dye)
c_{surf}	concentration immediately to the fibre surface
c_{Fsurf}	concentration in the fibre surface
c_F	concentration in the fibre
δ	thickness of diffusion layer
σ	specific free energy of the interface between particle and solution
R	ideal gas constant
T	temperature
x_i	radius of particle i
ρ	density
V	molal volume (cm^3/mol)
y	association factor
η	viscosity
LR	liquor ratio
K	partition coefficient/equilibrium constant
c_{Hydrol}	concentration of hydrolysate
k_{Hydr}	rate constant for hydrolysis of reactive dye in bulk solution
k_{FHydr}	rate constant for hydrolysis of reactive dye in fibre
c_{OH^-}	concentration of hydroxyl ions in bulk
c_{FOH^-}	concentration of hydroxyl ions in the fibre
k_F	pseudo-first-order reaction rate constant for dyestuff fixation in the fibre

7 Basics of colour development

7.1 The phenomenon of colour – how to approach?

The concept of colour can be approached from different points of view.

The physical chemistry of a dyeing process with regard to sorption equilibrium and heterogenic reaction kinetics has been discussed in Chapters 4–6.

In Chapters 7–10, 13 and 15, concentration on one major aspect is made at a time, which permits a disentanglement of the complex and heterogeneous situation.

The physical basis of colour development can be explained on physical-chemical processes, including quantum mechanics. This will be highlighted in Chapter 7 of this book.

In textile chemical approaches the majority of colourants base on larger organic molecules which are applied in form of dyes and pigments. The general principles of dyes and pigments will be discussed in Chapter 8, which thus focusses on organic chemistry of colourants.

Colour perception as a sensorial aspect and methods to quantify a physiological response will be discussed in Chapter 9.

The chemistry of dye application, processes and process technology will be discussed in Chapters 10 and 13, process balances and environmental aspects will be summarized in Chapter 15.

7.2 Physical aspects of colour development

Under the term 'visible light' usually the wavelength region between 400 and 700 nm is considered, as this is the range of electromagnetic irradiation where perception by the human eye is possible. This leads to the sensation of colour.

The physical explanation of colour thus considers the intensity distribution of electromagnetic irradiation in this wavelength region of the full spectrum of electromagnetic irradiation (Figure 7.1).

Frequency v (Hz, s^{-1}) and wavelength λ (m) of electromagnetic irradiation are linked together through the speed of light c (300,000,000 m/s) (eq. (7.1))

$$c = \lambda\, v \qquad\qquad (7.1)$$

The amount of energy transported by the irradiation follows Planck's distribution of energy (eq. (7.2)).

$$E = n\, h\, v \qquad\qquad (7.2)$$

with h being Planck's constant (6.626×10^{-34} J s). As a result, short wavelength irradiation such as violet light or ultra violet irradiation (UV) transports higher energy in a

https://doi.org/10.1515/9783110795738-007

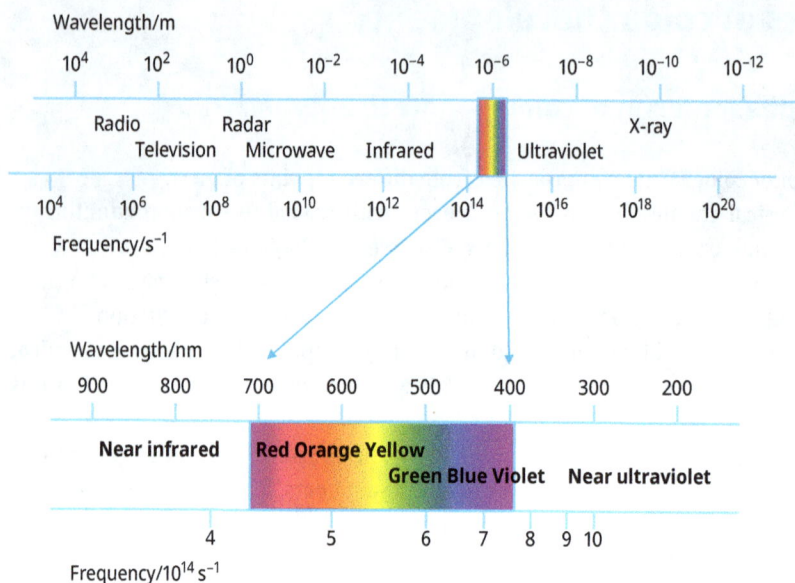

Figure 7.1: The range of visible light in the full spectrum of electromagnetic irradiation.

single photon than irradiation of longer wavelength does, for example, red light or infrared (IR) irradiation. IR irradiation mainly leads to a warm up of a sample through excitation to higher levels of rotational and vibrational states, while UV irradiation will cause electron excitation to higher states.

Despite the fact that colour is actually always bound to a perceived sensation, a purely physical approach is possible based on the intensity distribution of light in the wavelength range of 400–700 nm only.

The sensation of colour is always dependent on the intensity of perceived light. Thus in a physical assessment of colour- and wavelength-dependent intensity of light, the intensity of the non-absorbed part of irradiation will always be of relevance.

The transmission of a transparent solution T is defined by comparison of the intensity of light after passing through a solution I with the initial intensity of the light I_0 (eq. (7.3)).

$$T = 100 \; I/I_0 \tag{7.3}$$

For chemists the Lambert–Beer's law is a well known relation, which describes the intensity loss when light passes through a coloured transparent solution (eq. (7.4)). The absorbance E is then the logarithm of $100 \; I/I_0$ and related to the concentration of a coloured substance in solution.

$$E = \varepsilon \; c \; d = \log 100 \; I/I_0 = \log 100/T \tag{7.4}$$

The corresponding properties of a non-transparent material have to be described by the reflectance β (eq. (7.5)). For a given wavelength the reflectance is defined as the quotient of the intensity of reflected light I_R and the initial intensity of irradiation I_0.

$$\beta = I_R/I_0 \tag{7.5}$$

$$\alpha = 1 - \beta \tag{7.6}$$

The concept of reflectance allows a physical description of colour by use of reflectance curves, which correspond to the absorption spectra of transparent liquids. Often absorption curves are also shown instead of reflectance curves, the absorption α being defined as $(1 - \beta)$ (eq. (7.6)). In many cases the reflectance or absorption values are multiplied with 100 and given in percentage.

In a simplified approach we can assume that the energy distribution in daylight is almost uniform between 400 and 700 nm. A body which then reflects daylight in a diffuse form without absorption appears white.

Thus in the diffuse reflectance curves given in Figure 7.2, the reflectance of a perfect white body appears as a line at 100% reflectance and 0% absorption. A black body theoretically absorbs any wavelength of incident light to 100% ($I_R = 0$, $\beta = 0$).

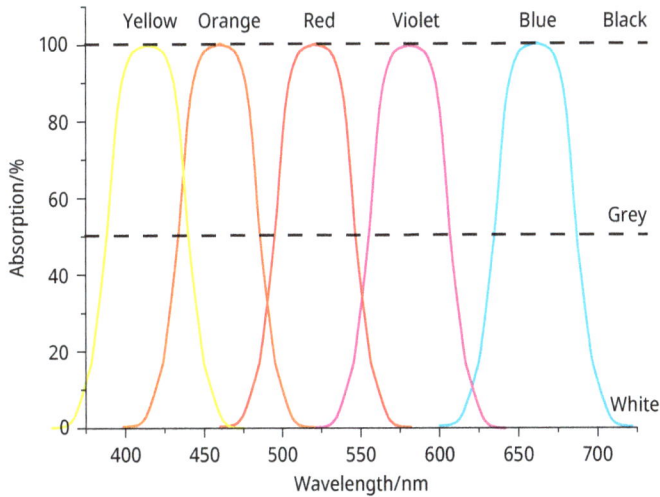

Figure 7.2: Diffuse reflectance curves of coloured matter.

While white bodies can be simulated quite well with use of white pigments such as $BaSO_2$ or MgO, the simulation of an ideal black body which absorbs any light is difficult. Thus, in theory only the 'colour' of black exists because of its definition by 100% absorbance. In practical textile dyeing such conditions cannot be fully realized, as a minimum part of the incident light is always reflected. As a result a wide range of 'blacks' exist, from greenish, blueish to more reddish, which actually cause substantial problems in colour matching of black dyeings on different textile fibres.

A surface which reflects all wavelengths more or less to the same degree, for example 50%, will appear grey. White, grey and black are considered achromatic colours.

When only a certain range of the spectrum is absorbed/reflected more, a chromatic colour appears.

An estimation of the perceived colour based on the reflectance curves has to be made with care as the perceived colour not always can be derived from the absorbance curves. In particular, dark colours will be difficult to identify.

However, as a general rule the main range in the spectrum where visible light is absorbed represents the complementary colour of the light perceived.

Thus, it is often more simple to predict the colour of a body on the basis of colours which were absorbed and now are no longer present in the reflectance/absorption curves.

A coloured body which mainly absorbs violet light in the wavelength range of 400–430 nm will thus appear in yellow, a surface which absorbs preferentially green light in the range of 480–550 nm will appear in red.

The curves given in Figure 7.3 represent the reflectance curves of a cochenille dyeing (red natural dye) and of synthetic indigo (blue).

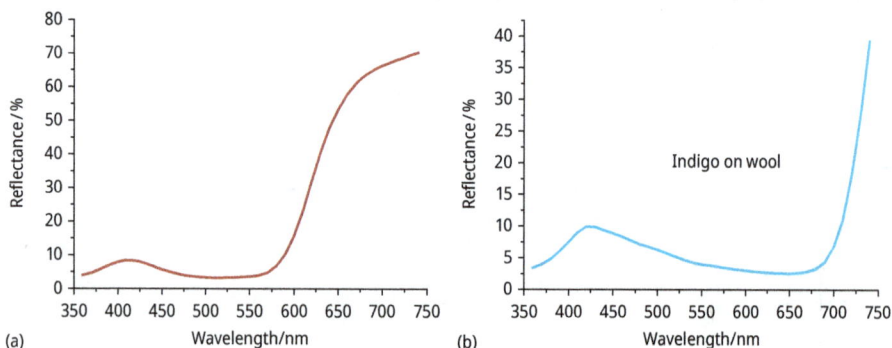

Figure 7.3: Absorbance curves of (a) the natural dye Cochenille on silk (red) and (b) synthetic indigo (blue) on wool.

The indigo dyeing exhibits the strongest absorbance in the wavelength region near 600 nm (yellow –orange), which will thus appear as dark blue with a reddish shade. The cochenille dyeing mainly misses the wavelengths of green and blue and has a higher reflectance at 600–700 nm, thus appears to us as dark red.

7.3 Additive and subtractive colour

Perceived colour always bases on the spectral intensity of light which is absorbed by the light sensitive photoreceptors available on the retina.

The formation of colour thus can be achieved by two different principles:

The subtractive mixture of colours: In this case the modification of the intensity of white light which impacts the surface of a body is achieved through absorption, interference and so on. As a predominance of light in a certain range of wavelengths occurs, colour is developed. In case two different substances which can absorb light are available at the surface at the same time (e.g. a mixture of two dyes or two pigments), the intensity is modulated by both substances. As a result, the absorption of both dyes adds and only the remaining light which has not been absorbed by one of the colourants will be reflected and thus will be visible.

Each dye subtracts a certain amount of light from the initially equally distributed level of intensity in white light.

Mixing red and blue colours will thus result in violet, and mixing yellow and blue colours will leave light mainly in the range of green.

In a subtractive colour mixture every colourant increases absorbance thus intensity of reflected light reduces. In its maximum expression the subtractive colour mixture will result in a black colour (no light left for perception through the eyes of the observer).

Textile chemists usually work with dyes and thus utilizes the principle of a subtractive colour mixture.

A black dyeing is thus generated by mixing complementary colours to such an extent that only a minimal and more or less uniformly low reflection is obtained. β is almost zero. The fact that equally low values of β throughout the full spectrum cannot be obtained in practice leads to the already mentioned difficulty of 'many' black shades, only one theoretical black exists.

It is important to consider that brilliant colours are required to realise a wide range of colours in subtractive mixing. Dull colours will become even more dull when mixed with other colours.

Additive mixtures are formed when light from sources with different colour is superimposed. Thus with each addition of a light source with an increased intensity, the colour gets lighter and in a perfect superimposition of appropriate sources a white colour will be obtained. Black then represents the absence of any light.

The basic colours in additive colour mixing are red (R), green (G) and blue (B). Addition of red and green will deliver orange colour, addition of green and blue will form the colour magenta (a blue-green shade) and addition of red and blue will form cyan (a violet shade).

Important examples for additive mixtures of colours are found in the screens of computers and in LED lamps. In computers, colour of items displayed can be adjusted using the RGB system. By mixing of red, green and blue light in different intensities almost any colour can be realised on the screen.

Additive colours are of increasing importance in textile chemical application, not in direct application for dyeing and colouration of goods, but to a growing extent in the field of illumination. In modern illumination concepts, energy saving LED lamps are available in stores, which allow an adjustment of light colour and intensity distribution of light of different wavelength. In this way, an additive combination of light is obtained and different colours of surfaces can be emphasised.

The perceived colour of textile goods is dependent on the light colour chosen, which can lead to difficulties through illumination-dependent colour variations and metamerism (see Chapter 9 for colour measurement).

7.4 Development of colour

Reflection: A combination of microstructured elements in the wings of the *Morpho* butterfly leads to a colour development as a result of reflectance and interference of light on submicron structures. The observed colour depends on the angle of the light as the conditions for interference depend on the geometric conditions between light and the submicron structure [1]. Through the lamellar structure of ridges present in the wings of the butterfly, a constructive interference for distinct reflected wavelengths is obtained. As a result, colours with high intensity are observed, because the colour is not a result of absorptive effects (Figure 7.4).

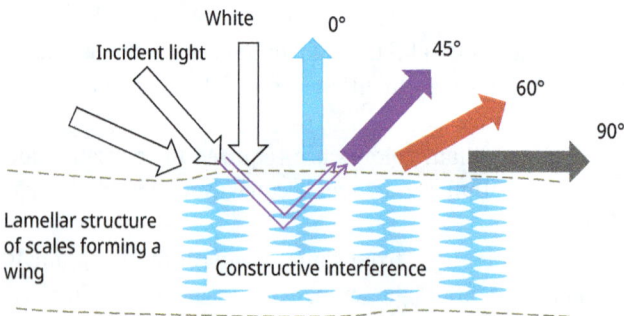

Figure 7.4: Principle of the lamellar structure of ridges in the wings of butterfly [1].

Refraction (Rainbow): When light passes from a medium of given refractive index n_1 into a medium with different refractive index n_2, the direction of the beam changes. The light beam is bent or refracted. In case of $n_1 < n_2$ the exit angle of the beam is directed towards the 90° vertical, in case of $n_1 > n_2$ the exit angle is higher than the entrance angle.

Refraction is a wavelength-dependent phenomenon, thus refraction of white light separates the incident beam according to the respective wavelength or the colour of the light. This effect is called dispersion or dispersive refraction.

Typical examples for colour development through refraction are the colours of a rainbow, or the colours of a faceted diamond.

Interference: Interference at thin layers with repetitive structural elements: Repetitive structural elements in dimension of the magnitude of a wavelength of the visible light can cause colouration effects. Typical examples are observed in feathers, wings of butterfly and pearlmut. As an example of such a structural interference, recently the formation of blue colour in the wings of *Hypolimnas salmacis*, a butterfly, has been studied in detail [2]. The colour is the result of interference at a thin lamina in the scales, which acts as a photonic crystal. The colour of the wings shifts when the wing is tilted as the conditions for interference are altered.

The changing colour of a soap bubble is another example for colour development through interference.

Scattering: Spongy structures found in feathers of birds can also lead to development of colour [3]. Scattering follows the theory of Rayleigh as long as the particles are in the dimension of 1/10 of the wavelength of light. Scattering s is dependent on the wavelength of light to the fourth power (eq. (7.7)).

$$s = 1/\lambda^4 \qquad (7.7)$$

In feathers, the spongy structures can exceed this limit and thus the Mie theory for scattering has to be applied. In the Mie theory, the factor 4 in the exponent lowers depending on the size of the scattering element [4]. In case the dimension of a spherical scattering element increases from 20 nm to 80–150 nm, the exponent in eq. (7.7) becomes lower than 4. Also, a shift of the spectral range towards longer wavelength occurs where the strong scattering effects are observed.

An application and combination of such effects are utilised in nature by animals which can adapt their colour, such as for camouflage effects. Representative examples are chameleon, cuttlefish and some fish such as the Atlantic cod [5].

Chromatophores are cells which can change colour either through modification of light absorption of pigments and pigment translocation, through scattering, diffraction or reflection.

Light absorption: The most important principle for development of colour in textile chemistry bases on a change in intensity of reflected and scattered light due to absorption of electromagnetic irradiation in the range of visible light (400–700 nm).

Higher absorption of irradiation in certain regions of the visible spectrum leads to a modulation of the intensity and to perception of colour.

The absorption occurs through excitation of electrons, which are excited to energetically higher molecule orbitals. When the exited state releases its energy, for example, through a radiationless transition, the absorbed light is transformed into vibrational and rotational energy, and thus remains absorbed by the matter.

The wavelength of light absorption in an organic molecule depends on the electron structure of the molecule. For an excitation of electrons in the visible range of light, the energy difference between the ground state and the excited state has to be in the order of 170–300 kJ/mol [6]. In a larger delocalised electron system, a higher number of electrons are available for excitation through absorption of light. The energy difference between the ground state and the first available excited state then defines the wavelengths of irradiation which can be absorbed (eq. (7.2)).

In a simplified model, the molecular orbital (MO) theory can be used to provide a basic understanding of colour development in dye molecules. The molecular orbitals are formed by Linear Combination of the respective Atomic Orbitals with respect to their wavefunctions (LCAO-MO model).

In a very simplified explanation of the MO theory, the atomic orbitals of atoms participating to a bonding in a molecule form molecule orbitals. The linear combination of the wavefunctions of the atoms leads to formation of molecular orbitals with different energy levels. Two electrons with opposite spin can occupy one MO. The electrons occupy the molecular orbitals beginning with the energetically lowest MO. Then orbitals with higher energy become occupied. Molecular orbitals which are formed by overlap of atomic orbitals and where a significant share of electron density is allocated to each atom are responsible for bond formation.

Depending on the symmetry of the atom orbitals (s, p, d, . . .) and the hybridisation of the atomic orbitals, a tetrahedral geometry (sp^3 in aliphatics), a trigonal geometry (sp^2 double bonds, on olefins and aromatics) and linear geometry (sp, in acetylenes) are obtained. In case of a sp^3 hybridisation, one s orbital and 3 p orbitals are combined and 4 σ-bonds are formed. In case of a sp^2 hybridisation, one s orbital and 2 p orbitals are combined to form three independent σ-bonds, one p-orbital remains which then can form a double bond, a π-bond, with a neighbouring atom. In case of a sp hybridisation, a linear geometry of two σ-bonds results and two π-bonds can be formed which either contribute to a triple bond or to a cumulene (e.g. propadien, allene). The different forms of hybridisation are shown in Figure 7.5.

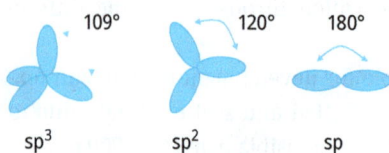

Figure 7.5: Hybridisation of sp3, sp2, sp hybrid orbitals.

As a result of bond formation, σ-bond orbitals and π-bond orbitals are obtained.

Generally the energy of σ-bond orbitals is lower than of π-bond orbitals.

As a result of the linear combination for every binding, σ and π orbitals are formed with an anti-bonding orbital. These σ* and π* orbitals are energetically higher and will be occupied by electrons only under circumstances where all lower-binding orbitals have already been filled with electrons. If too many anti-bonding orbitals are filled with electrons, the bonding will be destabilised and break.

A molecular orbital which is mainly concentrated on a single atom does not contribute directly to the bonding system. Such an orbital is called a non-bonding orbital (*n*-orbital).

As a result of these models, the MO energy diagram of a dye molecule can be drawn as combination of bonding σ and π orbitals, anti-bonding σ* and π* orbitals and non-bonding *n*-orbitals.

This highly simplified approach of molecular orbitals in a molecule helps to understand the principle of light absorption. The absorption of electromagnetic irradiation will be possible in case the energy of the photon ($E = h v$) is sufficient in height to lift an electron from the highest occupied molecular orbital (HOMO) into the lowest unoccupied molecular orbital (LUMO). Thus, the energy levels of the molecular orbitals present in a chemical compound will determine which minimal energy of a photon will be able to excite an electron. This will determine the wavelengths which will be absorbed and thus will form the molecular background for development of colour.

Another complication arises from the spin of the electrons. As indicated in Figure 7.6, in most cased the two electrons which occupy one bonding orbital will be present as an electron pair with opposite spin (Ground state, singlet state S_0). When an electron absorbs energy and changes from the ground level into an excited state, the spin of the electron does not change to the first (or second) excited state. Thus, initially the excited state will be a singlet state (S_1, S_2). In case the spin of the electron in the excited state changes its direction, two unpaired electrons with same direction of spin are present in the molecule and triplet state (T_1) is formed.

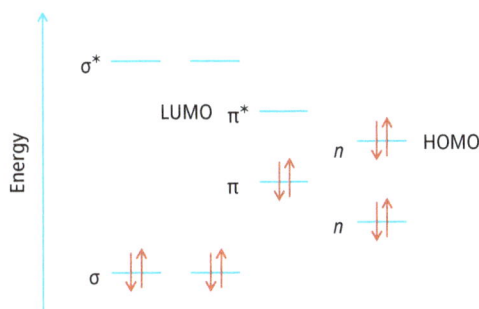

Figure 7.6: Simplified MO scheme for azobenzene (from Zollinger [6]).

In dyestuff molecules the light absorbing system usually consists of a large conjugated electron system. The high number of electrons on similar energy levels leads to a high number of occupied and unoccupied molecular orbitals which are in short distance of energy. Thus, absorption of irradiation in the visible region of the electromagnetic spectrum is already sufficient to excite electrons into a higher unoccupied molecular orbital. As a result of this absorption less light of this wavelength is reflected and colour develops.

Any substituent which changes the energy levels of the electrons in a dyestuff molecule will thus alter the shade of a dye.

Usually dyes contain multiple conjugated double bonds, including aromatic ring systems. These conjugated systems are often called chromophores. Substituents to the ring are classified into auxochromes, which are electron donating groups (e.g. –OH or –NH$_2$ groups) and antiauxochromes, which are electron withdrawing groups (e.g. –COOH, –NO$_2$).

Depending on the electron structure of a substituent, it functions as electron donating (+) or subtracting (–) group. The effect can be mesomeric +M (e.g. a phenolate group: Aromat –O$^-$) or –M (e.g. –NO$_2$) which are described by mesomeric structures. The second principle is the inductive effect, which is due to shift in electron distribution as result of differences in electronegativity or electron density. A +I effect can be due to electron donation by a saturated hydrocarbon group, a –I effect can be due to an electronegative halogen substituent. Often both effects are observed for a single substituent, their extent being dependent on the type of substituent.

Changes in the molecular structure of a dye lead to a modification in the absorption behaviour (Figure 7.7). The height of absorbance can change and the wavelength of maximum absorbance can shift.

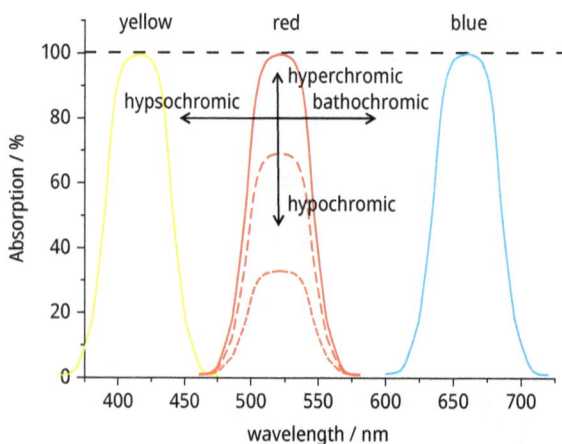

Figure 7.7: General definitions for modification of an absorption curve.

A shift in wavelength of maximum absorbance indicates a change in the energy levels of the light-absorbing electrons. A hypsochromic shift indicates that the energy difference between the ground state and the first exited state of the electrons has increased as light with shorter wavelength and vice versa with higher energy is required to excite an electron. A bathochromic shift indicates that the energy difference between occupied and unoccupied energy levels has become smaller, photons with lower energy will be sufficient to push electrons to the excited state.

The absorbance of a dye is dependent on the probability of an electron to become excited and to reach an exited state. Thus, for a given shade, a dye with high colour strength exhibits a higher probability that electrons reach an excited state, compared to a dye with lower colour strength. An increased absorbance is called a hyperchromic effect, which indicates an increased probability of electron excitation, while a hypochromic effect indicates that the probability to absorb light by electron excitation is reduced. For technical applications, a dye with high colour strength is favourable as lower amount of product would be required to reach a desired colour depth.

Depending on the design of a dyestuff molecules, solubilising groups are present in the structure. Furthermore, anchor groups are present in reactive dye, which enable the molecule to establish a covalent bonding with the hydroxyl groups of cellulose or, in case of wool, with the amino groups and hydroxyl groups of the protein structure.

C.I. Reactive Black 5 can be used as an example for a widely used reactive dye which exhibits all relevant structural elements discussed so far (Figure 7.8).

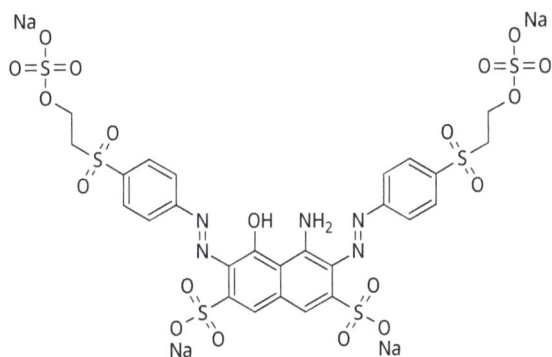

Figure 7.8: C.I. Reactive Black 5, a representative of a widely used synthetic dye containing a chromophore, auxochromes, solubilising groups and anchor groups.

In the dyestuff molecule, all relevant elements discussed are present (a large delocalised electron system containing two azo bridges, auxochromes and solubilising groups, as well as two reactive anchor groups).

During the reaction of the dye with the fibre polymer, the sulphatoethyl-sulphonate groups (vinyl-sulphone groups) react to form an ether bridge with the polymer.

As a side reaction also, hydrolysis of the sulphatoethyl groups occurs and the corresponding hydrolysed dye is formed (Figure 7.9).

From the point of colour formation, the sites where hydrolysis of the anchor groups occurs is far from the chromophoric system, thus no strong influence on the colour of

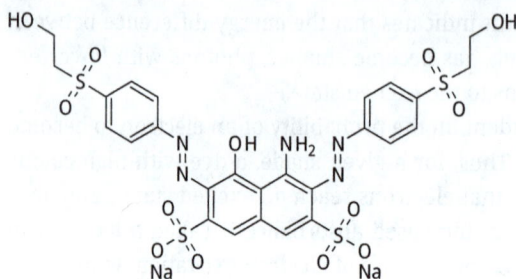

Figure 7.9: Hydrolysed C.I. Reactive Black 5.

the dye is expected. However, in practice the hydrolysis of the anchor groups changes the absorption spectrum of the dye. As a result, it is important to consider that any group present in a dye molecule will contribute to the final shade and also to the technical properties of a dye (e.g. lightfastness property).

The removal of a solubilising group in a chromophore does not only change the technical properties of the dye, but also changes the shade of the dyestuff.

Sulphonation of indigo to indigo carmine adds two sulphonate groups to the indigo dye molecule, which transfers the insoluble indigo into a water-soluble anionic dye. As a result of the substitution also the shade of the dye changes. Indigo carmine (Acid Blue 74) is a completely different blue dye when compared to indigo (Vat Blue 1)(Figure 7.10).

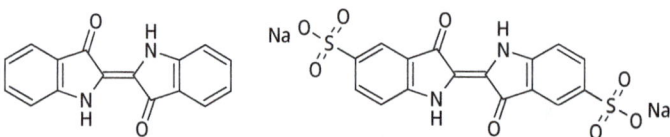

Figure 7.10: Indigo (Vat Blue 1) and indigo carmine (Acid Blue 74).

Besides the evident effects of molecular structure and substituents on the colour of a dye, many other effects also influence the final appearance of a colour.

7.5 Colour variations independent of molecular structure

The chemical surrounding of the dyestuff also influences the absorbance, thus the polarity of the fibre polymer surrounding a dye as well as the solvent in which a dye is directly dissolved will influence the absorption curve (solvatochromy). Colour changes of dyes are sometimes used to probe the polarity inside a fibre structure [7]. Any change in the chemical surrounding leads to a shift of the energy levels in the orbitals of the dye molecule. As a result, the colour of a dyeing with the same dye shows more or less visible differences which depend on the fibre substrate to be dyed. Also, the moisture content of a textile sample leads to changes in the absorption curve. Thus, in practice a sample taken to assess the shade of the goods during an exhaust dyeing should always be humidified before the measurements. After rapid drying at elevated temperature, moisture has to be absorbed by the sample to avoid colour changes due to overdrying and low moisture content.

Temperature also can lead to a reversible colour change; dyes which exibit a pronounced colour change are called thermochromic dyes (thermochromy).

Irradiation can cause reversible rearrangement of the molecular structure through isomerism which thus leads to photochromic effects (photochromy). Photochromy can cause problems during colour measurement as storage and handling of a sample under light also influences the observed colour. Photochromy leads to measurable irradiation-dependent changes in the absorption curve of a dyeing. This phenomenon must not be messed up with the observation of metamery, where illumination-dependent changes in observed colour occur. In case of metamery, the perceived change in colour is due to the differences in the energy distribution of light source and not due to changes in the absorption curves of the dyes.

Any chemical change in dyestuff structure, such as reduction of a vat dye (indigo), change in the pH of a solution (indicator dye), complexation in case of mordanting of a natural dye, will lead to substantial changes in colour. However, this behaviour is due to a direct change in the molecular structure of the chromophore, thus is similar to the effects of different substituents bound to a chromophore.

A related case of colour development is observed in case of the red colour of ruby, which is actually composed of colourless aluminium oxide. The red colour develops due to the presence of small amounts of other metal ions (e.g. the green Cr_2O_3). According to Ligand field theory, the electron shells of the dope metal ions change their energy levels and the observed red colour appears. Upon heating, ruby turns from red to green as the atomic orbitals of the chromium atoms change their energy levels and the absorption spectrum alters.

7.6 Fluorescence and phosphorescence

The Jablonsky diagram describes in a simplified model the different electronic states that can appear during absorption of light by electrons (Figure 7.11).

In the ground state (S_0) the electrons are present in paired form with antiparallel spin, thus a singlet state is observed. Different vibrational energy levels are indicated by lines at short energy distance to the vibrational ground state. The short distance between the vibrational levels indicates that electromagnetic irradiation, which is insufficient in energy to excite an electron to the first excited level (S_1), will only cause the molecule to achieve an energetically higher vibrational state.

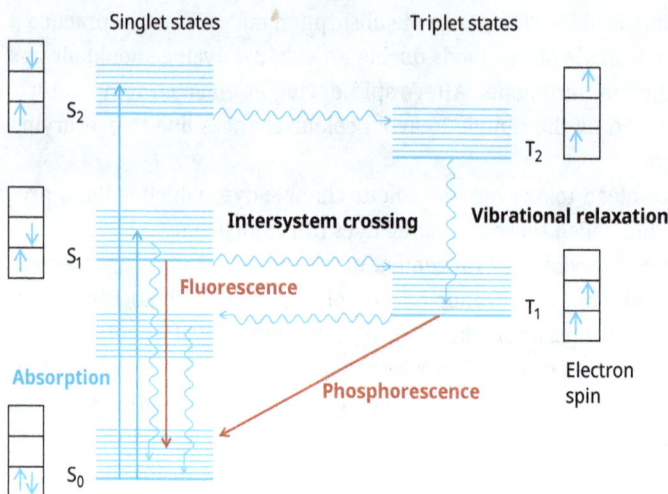

Figure 7.11: Jablonsky diagram.

Absorption of IR irradiation thus changes the vibrational state of a molecule, which will return into an energetically lower state through radiationless energy transfer (dissipative processes), for example, collision with neighbouring molecules.

Absorption of electromagnetic irradiation which is sufficient in energy will lift the electron into its first excited state (S_1). Possibly, a higher vibrational state (S_2) can also be occupied.

As long as the direction of the electron spin does not change, all excited states are singlet states. The excited state can then release the absorbed energy to the environment though several pathways.

Higher vibrational states can transfer energy to neighbouring molecules, through collision, without emission of any irradiation.

Interconversion is a radiationless transition from an excited electronic state to a lower state. The absorbed energy is then transferred to the neighbouring molecules in the form of vibrational/thermal energy (heat). For such processes a first-order rate constant can be formulated. The respective rate constant for vibrational relaxation has the dimension of $10^{12}\,s^{-1}$, which is substantially smaller than the rate constant of absorption ($10^{15}\,s^{-1}$).

A release of absorbed energy can also be achieved through emission of radiation. These phenomena are summarised under the term luminescence. Two important pathways can be distinguished depending on the respective rate constant.

Fluorescence: After having reached the vibrational ground state of an electronically excited state, the release of energy and a jump back into the ground state can be achieved through emission of radiation. As part of the absorbed energy had already been dissipated in the form of thermal energy, the energy of the emitted radiation is lower than the energy of the absorbed irradiation. As a result, the wavelength of the fluorescent light is longer compared to the absorbed light $\lambda_{\text{fluoresc.}} > \lambda_{\text{ads}}$.

The rate constant for fluorescence is high with $10^6 - 10^9 \, \text{s}^{-1}$, which indicates that fluorescence is coupled with absorption of radiation and disappears almost immediately when incidence of radiation stops.

Many fluorescent molecules exhibit a rather rigid chemical structure. As an example, fluorescein is a highly fluorescent dye, while in case of phenolphthalein the less rigid molecule structure permits more rapid dissipation of absorbed energy by internal conversions and energy dissipation, thus without emission of radiation (Figure 7.12) [6].

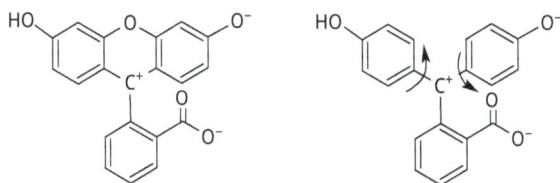

Figure 7.12: Molecular structure of fluorescein and phenolphtalein.

Phosphorescence: Despite being a quantum chemically forbidden transition, a change in direction of the electron spin can occur from an excited state. The spin of both the unpaired electrons is parallel and a so-called triplet state is formed. This process is called intersystem crossing. The energy level of a triplet state is lower, compared to the respective singlet state. Part of the absorbed energy is dissipated to the environment in form of thermal energy. The transition to the ground state again requires a forbidden change in direction of the electron spin, which is a substantially slower process compared to fluorescence. In case the transition to the ground state is achieved by emission of irradiation, this phenomenon is called phosphorescence. The wavelength of phosphorescence is longer compared to the fluorescence, as the energy difference between the first excited triplet state T_1 and the ground state S_0 is smaller than the difference between the first excited singlet state S_1 and the ground state S_0 ($E_{\text{fluoresc}} > E_{\text{phosphor.}}$).

The first-order rate constant for phosphorescence is in the dimension of $10^{-2} - 10^4 \, \text{s}^{-1}$, which indicates a substantially longer lifetime of the excited triplet state. Thus, phosphorescence can also be observed during minutes to hours after illumination of a sample has stopped.

7.7 Textile chemical relevance

In an exited state, at least one electron is present in a higher energy level, often in an anti-bonding orbital. The presence of an electron in an anti-bonding orbital lowers the bonding energy substantially. In case of a double bond, approximately one binding equivalent to the double bond is lost. In case of a single bond, the bonding energy is practically zero.

As a reaction on this situation, the distance between the linked atoms increases due to repulsive forces. The physical movement of the atoms, however, requires a substantially longer time as the absorption process does.

The exited state is the origin for following photochemical reactions. The average lifetime of a singlet state is also very short (1–100 ns). Thus, the probability for a chemical reaction to occur within this short period is low. Singlet states thus stabilise either through radiationless transition or through fluorescence.

Due to longer lifetime of the triplet state (100 ns–10 s), chemical follow-up reactions are likely to originate from this state. Possible reactions can be dimerisation, cleavage of bonds and reactions with neighbouring molecules.

In case of textile dyes the existence of triplet states, which do not stabilise rapidly through phosphorescence or dissipative processes, can be the physical basis for photochemical fading, resulting in low fastness to light. A dyestuff with high light fastness is thus a molecule which either exhibits a low tendency to enter in triplet states or which exhibits a chemically stable structure which resists degradation during phases of being in an excited triplet state. In this case a triplet state dissipates energy either to the surrounding or emits part of the absorbed light as phosphorescence. In an undesirable variant, the excited dye transfers its absorbed energy to a neighbouring dye molecule and thus initiates possible degradation of another colourant. Such phenomena can be observed when inappropriate combinations of dyes are used in a dyeing. The combination of two dyes then exhibits lower light fastness as each of the dyes alone.

Fluorescence quenching: The radiationless deactivation of an excited state can be supported by a number of factors, which can cause a partial reduction or complete quenching of the fluorescence.

Relevant factors which can support fluorescence quenching are:

- Certain substituents present in the molecule can support deactivation: $-NO_2$, $-SH$, $-NH_3^+$
- Increased temperatures support vibrational energy transfer from excited states
- The presence of a solvent: The fluorescence of a dye in dissolved state can differ from the situation in adsorbed state.
- The concentration of the fluorescent substance itself can support quenching effects. At higher concentration the life time of the excited state is reduced by collisions with non-excited molecules, thus leading to energy transfer and dissipation. As a result, the probability for emission of irradiation is lowered.
- Also, the presence of heavy metals can lead to fluorescence quenching. A similar effect is observed with mordants used in the application of natural colourants, where an improvement in light fastness is observed in many cases in presence of iron mordants. However, in this particular case it must be considered that after mordanting also a new chemical individual, a metal complex, is formed, which cannot be compared directly with the non-mordant dye [8].

Optical brighteners: Examples for fluorescent dyes are found in highly reflective clothing as elements to increase visibility at conditions of reduced illumination. Another important class of fluorescent substances with textile chemical relevance are optical brighteners. These are more or less non-coloured substances which absorb irradiation in the near UV-range and emit irradiation in the visible range of the spectrum.

The reflectance curves of a raw cotton fabric and the same after different treatments to obtain a perfect white are shown in Figure 7.13. The reflectance curve of raw cotton remains substantially below the 100% reflectance of a theoretical white. As the curve exhibits lowest reflectance in the violet range of the visible spectrum of light, the sample appears yellow-beige. After a bleach procedure, the naturally coloured substances are removed; however, the reflectance curve is still below the 100% line. The addition of a minimal concentration of a blue dye changes the yellowish shade into a more neutral light grey, which is perceived as 'better' white compared to the bleached cotton.

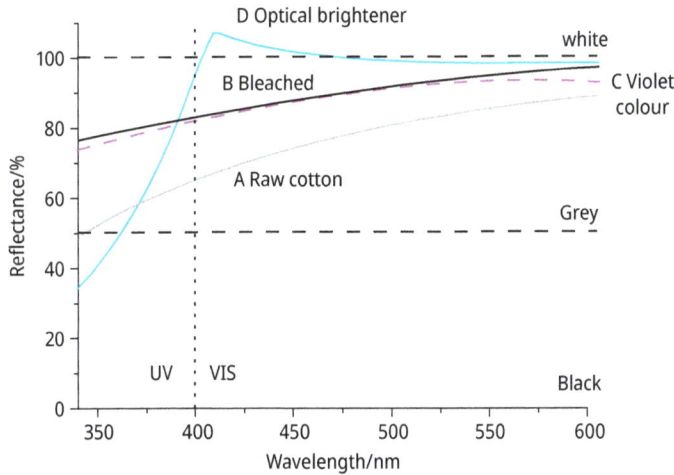

Figure 7.13: Diffuse reflectance of co-fabric (A) raw cotton (grey line), (B) bleached (black line), (C) violet tinged (pink, dashed line), (D) with optical brightener (blue line).

The addition of an optical brightener changes the reflectance curves substantially. A high absorption is observed in the near UV range, while the fluorescent light emitted by the sample in the visible range adds to the reflectance of light. As a result the intensity of light reflected in the range between 400 and 500 nm is even higher than the theoretical value of a white sample. Thus, the old marketing slogan for washing powder to wash 'whiter than white' is physically true due to the function of the optical brightener, which adds fluorescent light in addition to the reflected light.

References

[1] Kinoshita, S., Yoshioka, S., Kawagoe, K. Mechanisms of structural colour in the *Morpho* butterfly: cooperation of regularity and irregularity in an iridescent scale. Proc. R. Soc. B. 2002,269,1417–1421, https://doi.org/10.1098/rspb.2002.2019.

[2] Siddique, RH., Vignolini, S., Bartels, C., Wacker, I., Hölscher, H. Colour formation on the wings of the butterfly Hypolimnas salmacis by scale stacking. Sci. Rep. 20166, 36204, doi:10.1038/srep36204.

[3] Finger, E. Visible and UV coloration in birds: Mie scattering as the basis of color in many bird feathers. Naturwissenschaften. 1995, 82, 570–573.

[4] Mie,, G. Beiträge zur Optik trüber Medien, speziell kolloidaler Metallösungen. Ann. Physik. 1908, 25, 377–445.

[5] Aspengren, S., Sköld, HN., Quiroga, G., Mårtensson, L., Wallin, M. Noradrenaline- and melatonin-mediated regulation of pigment aggregation in fish melanophores. Pigm. Cell Res. 2003, 16(1), 59–64.

[6] Zollinger, H. Color Chemistry Wiley-VCH, Weinheim, Germany, 2003.

[7] Fidale, LC., Ißbrücker, C., Silva, PL., Lucheti, CM., Heinze, T., El Seoud, OA. Probing the dependence of the properties of cellulose acetates and their films on the degree of biopolymer substitution: use and solvatochromic indicators and thermal analysis. Cellulose 2010, 17, 937–951.

[8] Manian, AP., Paul, R., Bechtold, T. Metal Mordanting in Dyeing with Natural Colourants: A Review. Color. Technol. 2016,132,107–113.

Take home messages

Colour bases on a perceived sensation. The physical basis for colour is a modification of the light intensity of reflected light.

Reflectance curves of surfaces permit a physical description of the absorption/reflectance characteristics of a sample, which is related to the development of colour.

Different processes lead to modulation of the intensity of reflected light: reflection, scattering, interference, refraction and absorption.

Absorption of light leads to excitation of electrons in chromophoric systems of a dye. Stabilisation of the excited state can occur through fluorescence, phosphorescence and radiationless transitions (dissipative processes).

Light-induced degradation of molecules occurs mainly from the triplet state due to the longer lifetime compared to singlet states.

Quiz

1. Which colour will be exhibited the maximum in the absorption curve at shorter wavelength: yellow or violet?
2. What kind of change in the absorption curve would you expect if the energy distance between a HOMO and LUMO is lowered and the probability of a transition by absorption of light is increased?

3. Why the lifetime of fluorescence is shorter compared to phosphorescence?
4. Why the wavelengths of fluorescence and phosphorescence irradiations are longer than the wavelength of absorbed light?

Exercises

1. Draw a square in any computer program available and colour the square in the following RGB colours: 1) R = 236, G = 252, B = 86; 2) R = 0, G = 255, B = 255; 3) R = 255, G = 0, B = 255; 4) R = 255, G = 255, B = 0.

   Discuss the results in terms of colours you are mixing. What coordinates result in black, white, grey?

2. Identify in the following dyestuff the chromorpore system, auxochromes/antiauxochromes, anchor groups, solubilising groups.

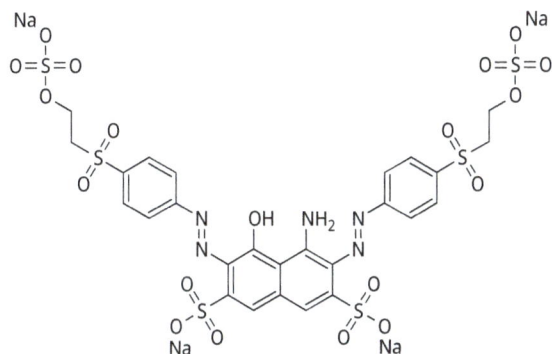

3. Calculate the energy for excitation of electrons in case of colours such as blue (650 nm), red (525 nm) and yellow (410 nm). (h = 6.63 × 10^{-34} J s; N_a = 6.02 × 10^{23}, c = 3 × 10^8 m/s).

Solutions

1. (Question 1): The absorption maximum of yellow colour will be near 400 nm, the absorption maximum of violet will be near 600 nm.
2. (Question 2): If the energy distance between HOMO and LUMO is lowered we expect a bathochromic shift, if the probability of a transition is increased we expect a hyperchromic shift.
3. (Question 3): Fluorescence does not require a change in direction of electron spin, thus transition into the ground state is more rapid.

4. (Question 4): Part of the energy of the excited state is dissipated from higher vibrational states, the residual energy available for emission of fluorescence light is thus lower and the wavelength will be longer.

5. (Exercise 1): RGB colours: 1) R = 236, G = 252, B = 86 yellow; 2) R = 0, G = 255, B = 255 turquoise; 3) R = 255, G = 0, B = 255 violet; 4) R = 255, G = 255, B = 0 yellow Coordinates of black R = 0, G = 0, B = 0; white R = 255, G = 255, B = 255; grey R = 120, G = 120, B = 120.

6. (Exercise 2):

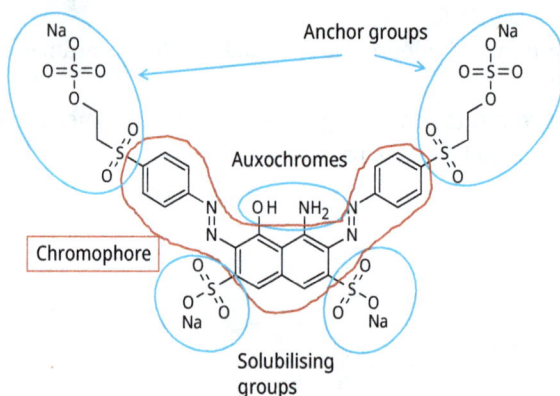

7. (Exercise 3):

	Wavelength	Frequency	Energy		
nm	m	s^{-1}	E=hv	E (J/mol)	E (kJ/mol)
650	6.50×10^{-7}	4.62×10^{14}	3.06×10^{-19}	1.84×10^{5}	184.21
525	5.25×10^{-7}	5.71×10^{14}	3.79×10^{-19}	2.28×10^{5}	228.07
410	4.10×10^{-7}	7.32×10^{14}	4.85×10^{-19}	2.92×10^{5}	292.04

List of abbreviations/symbols

c	speed of light
λ	wavelength
v	frequency
s	scattering
E	energy
n	integer number 0, 1, 2, . . .
h	Planck's constant
T	transmission
I	intensity of light
I_0	initial intensity of light

E	absorbance (extinction)
ε	coefficient of absorbance
c	concentration
d	path length of light
β	reflectance
I_R	intensity of reflected light
a	absorbance
N_a	Avogadro constant
S_i	singlet states
T_i	triplet states

8 General principles of dyes

8.1 The molecular structure of dyes

The dyes are classified with regard to different aspects:
- The chemical constitution permits a classification based on a major structural element in the chromophore, for example, an azo group, a carbonyl or quinoid group, disulphide sulphur bond, a phthalocyanine group and metal complexes.
- Another principle of classification uses the field of application, for example, for cellulose fibres, wool, synthetic material and for screen printing or ink-jet printing. Other classes of dyes could cover food application, colouration of plastics, paints, cosmetics, printing of paper and so on.
- For practical applications, different dyes have been grouped into a colour gamut that exhibits comparable applicatory properties and thus can be used in mixtures to dye a material to a certain shade. In such colour gamuts, we often find many different classes of dyes, for example, in reactive dyes the chromophore can be a mono-azo type, a diazo compound, an anthraquinone and a phthalocyanine.
- In addition, the origin of a dye can be used to form a group: synthetic dyes versus natural colorants. Here even the same molecule, for example, indigo can be a representative of both classes, dependent on the route that had been used to isolate the product.

In this chapter, relevant types of chromophores, irrespective of the applicatory properties, are discussed, to present an overview about the wide range of possible structures that can be elements in technically used dyes.

Detailed descriptions of dyestuff chemistry including synthetic aspects are given by monographies with emphasis on dyestuff chemistry [1, 2].

8.2 The azo chromophore

8.2.1 Mono-, di- and polyazo dyes

The majority of dye molecules used at present for textile dyeing processes contain an azo-group –N=N– in their chromophore. Depending on the number of azo groups present in the molecule, we can distinguish between mono-, di-, three- and so on poly-azo dyes (Figure 8.1).

Often both organic substituents that are directly linked to the azo group are aromatic ring systems. As a consequence of the synthesis strategy, one of the aromatic ring systems of a monoazo dye usually bears electron-withdrawing substituents and on the other ring system electron-donating substituents are present.

https://doi.org/10.1515/9783110795738-008

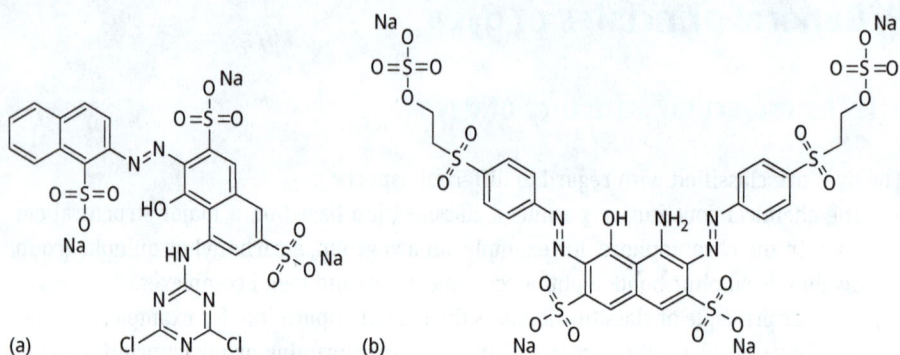

Figure 8.1: Structures of a monoazo mono-functional reactive dye (C.I. Reactive Red 11) and a diazo bifunctional reactive dye (C.I. Reactive Black 5).

The general synthesis follows the so-called azo coupling between a diazonium salt and an electron-rich coupling component. In terms of reaction type, the azo coupling is an example of an electrophilic aromatic substitution. The diazonium salt represents the electrophile that attacks the aromatic ring of the activated coupling component. As the diazonium salt is a poor electrophile, intensive activation of the coupling component is necessary to achieve successful coupling reaction.

8.2.2 The diazotisation

As a first step the diazonium salt has to be synthesised by the reaction of nitrous acid with a primary aromatic amine. The first product formed is the nitrosamine, a *N*-nitroso compound, depending on the pH of the solution, nitrosamine tautomerises (Figure 8.2).

Possible products are the diazo acid, which can dissociate into the corresponding anion, the diazotate R–N=N–O⁻. The diazotate is not longer active as a coupling component; thus, high pH values need to be avoided to prevent deactivation of the diazonium salt.

In case the solution pH is low enough (e.g., pH 1–5, depending on the diazonium salt used), the diazo acid is protonated and water is split off. Thereby the diazonium salt formed is a temperature-sensitive compound. At elevated temperature, the diazonium salt decomposes and nitrogen is released.

In traditional naphtol dyeing, the diazotisation reaction was performed in the dyehouse by direct reaction of the naphtol base (the aromatic amine) with sodium nitrite and hydrochloric acid in the presence of ice. Thus in practice, these colours were often also called "ice colours". The health hazards due to deliberation of nitrous oxides lead to a replacement of the direct diazotisation by using stabilised solid diazonium salt containing products, which could be dissolved directly in water. In concentrated form, the diazonium salts were very instable, even explosive substances, which required appropriate

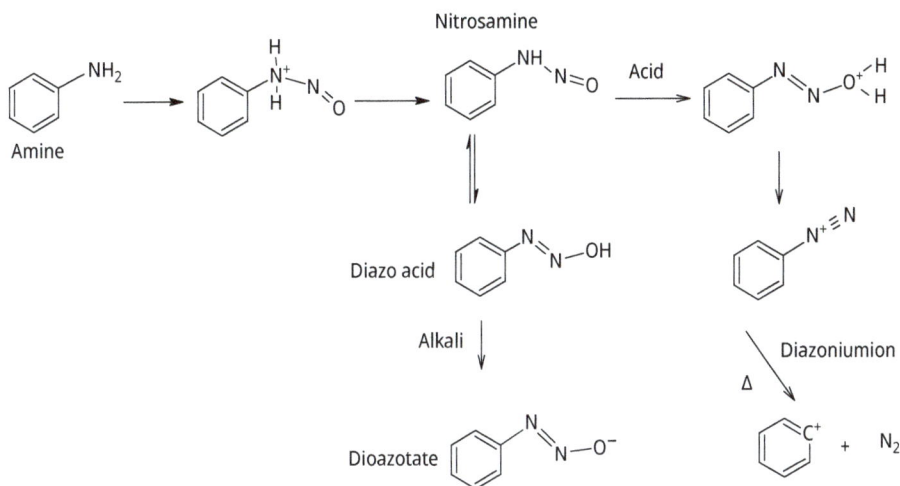

Figure 8.2: The diazotisation reaction and side reactions.

stabilisation to deliver safe and time-stable products. Important techniques to formulate a commercial solid form of the diazonium salt utilise the addition of large counter ions as anions to the diazonium ion (e.g., tetrafluoroborate BF_4^-, tetrachlorozincate $[ZnCl_4]^{2-}$ and naphthalene sulphonate) and addition of larger amounts of inorganic salt (e.g., Na_2SO_4, NaCl and $MgSO_4$), which then allows the preparation of dye formulations with a base content of up to 20 wt%.

In addition, reversible trapping of the diazonium salt through the formation of intermediate products is a possible method (Figure 8.3).

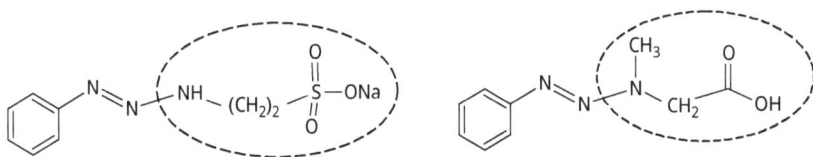

Figure 8.3: Stabilisation of a diazonium salt through the formation of amino-azo compounds with taurine (left) or sarcosine (right).

Primary aromatic amines represent the major amines for diazotisation as stabilisation of the diazonium ion through the aromatic ring system is possible (Figure 8.4). Aliphatic amines are not suitable for these purposes. In case of a primary aliphatic amine, the formation of a diazonium salt leads to the formation of unstable products and rapid deliberation of nitrogen.

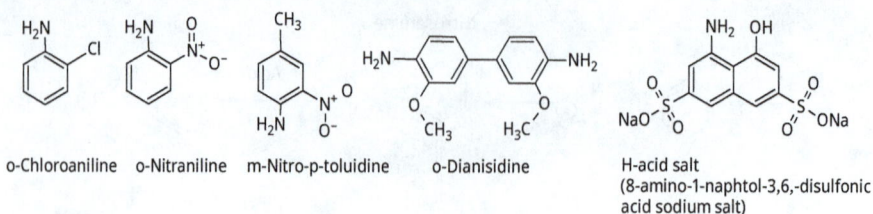

Figure 8.4: Representative structures of amino components.

In the case of secondary amines, the reaction ends with the formation of the nitrosamine (RR'N–N=O) as no tautomerism to the diazo acid is possible.

Electron-withdrawing substituents, for example, nitro groups and chloro substituents, increase the stability of the diazonium salt.

8.2.3 The coupling component

Strong activation of the aromatic ring system with electron-donating substituents is necessary to achieve a rapid coupling reaction.

Important substituents (electron donating) to activate aromatic ring systems are as follows: amino groups (mono- and diamines) and phenolic hydroxyl groups. A particular form of activation is achieved in the case of phenolic hydroxyl groups, where dissociation to the corresponding phenolate (naphtholate) is achieved by an increase in pH of the solution (Figure 8.5).

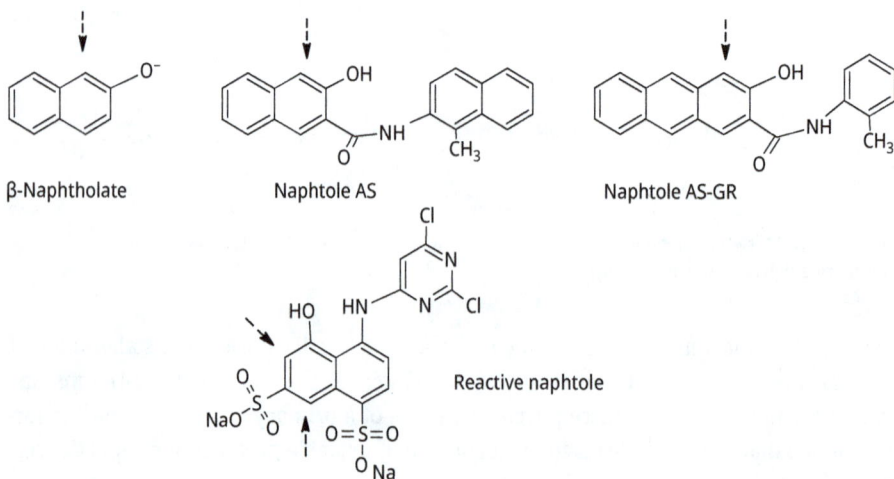

Figure 8.5: Representatives for activated aromatic ring systems for azo coupling (arrow indicates the site of coupling).

The formation of the azo bridge follows the usual substitution pattern according the substitution rules for electrophilic substitution on aromatic ring systems in the presence of electron-donating substituents. Thus for example, a phenolate ion will direct the substitution into the ortho or para position of the aromatic ring.

The selection of pH value for the coupling process seems rather straightforward; however in practice, careful selection and buffering is required. The optimum range depends on both of the two chosen coupling components. In case pH value is too high, deactivation of the diazonium salt through the formation of diazo acid/diazotate will occur. In case pH is adjusted too low, the activation of the electron-rich component is reduced, for example, through protonation of aromatic amino groups or of the phenolates present in the activated aromatic coupling component (Figure 8.6).

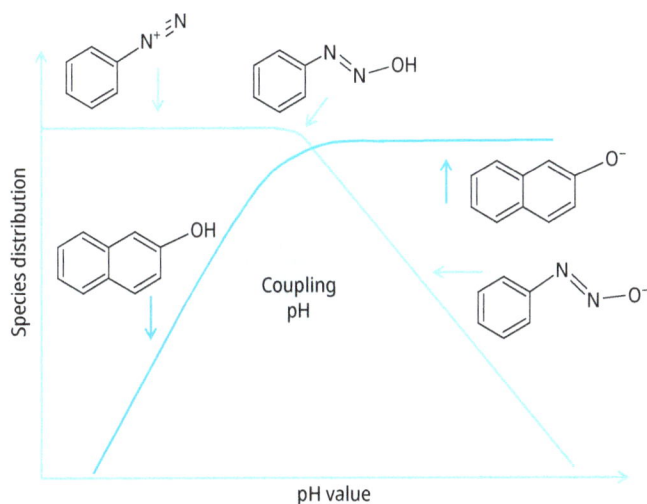

Figure 8.6: Influence of pH on azo-coupling reaction.

Based on the substituents on the aromatic rings that are attached to the azo bond, tautomerism can be observed. In case of hydroxyazo dyes, tautomerism between the azo form and the corresponding hydrazone form is possible (Figure 8.7). For amino-azo dyes, the corresponding tautomer form is the imino form. Tautomerism is important due to several facts:

The two tautomeric forms of a dye differ in a number of relevant properties: shade, colour strength, fastness (e.g., light fastness) and also toxicological properties.

From the point of textile chemical application, a dye with high tinctorial strength, namely, high coefficient of extinction, is desirable. Interestingly the hydrazone form usually exhibits the highest absorbance; however, often the azo form is the predominant isomer.

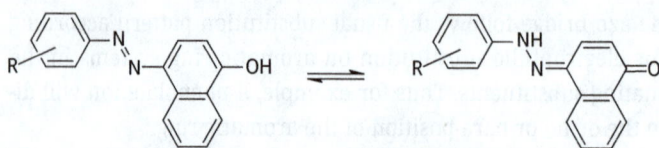

Figure 8.7: Tautomerism of an azo dye. From Hunger [2].

For example, 4-phenylazo-1-napthol exhibits in the azo form (yellow), with a coefficient of extinction of ε_{max} = 20,000 (λ_{max} = 410 nm), and in its orange hydrazone form, with a value of ε_{max} = 40,000 (λ_{max} = 480 nm).

In practice, the molecular structure of the dye determines which form is the predominant one; for example, azophenol dyes and aminoazo dyes preferentially are present in their azo form.

In disazo dye that is based on aminonaphthols (e.g., H-acid) as a coupling component (e.g., C.I. Reactive Black 5), one of the two azo groups is present in azo from, while the second azo group tautomerises into its hydrazone form.

The protonation of a dye can also shift the equilibrium between the azo and hydroazone form, a principle that is used in indicator dyes, for example, methyl orange. For textile chemical applications a strong dependency of the colour of dyes on external conditions e.g. changes in pH is undesirable. As an example pH sensitive dyes could exhibit colour changes during periods sweating, or as a result of alkaline pH applied in laundry (Figure 8.8).

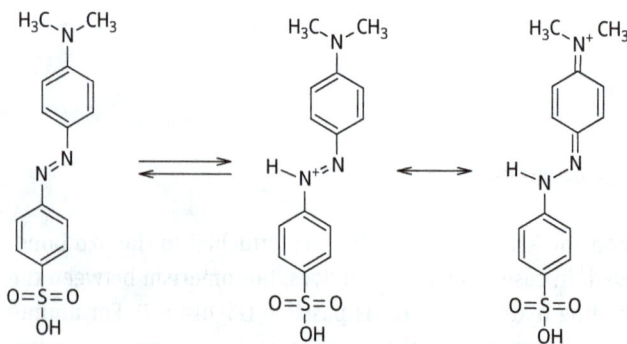

Figure 8.8: pH-dependent change in the structure of a dye as the basis reaction for an indicator dye (methyl orange). From Hunger [2].

Monoazo dyes are found in many applicatory classes of dyes, for example, direct dyes for cellulosic fibres, reactive dyes, acid dyes for protein fibres and polyamide, disperse dyes for polyester and polyamide, cationic dyes for polyacrylic fibres, as organic ligands in metal complex dyes and as insoluble pigments.

The respective applicatory properties are introduced with the respective functional groups, which are listed as follows:

- sulphonate groups to increase water solubility (direct dyes and acid dyes)
- cationic groups to attach a positive charge to the molecule (cationic dyes)
- reactive anchor groups to permit covalent bond formation to functional groups of the fibres (reactive dyes)
- formation of a metal complex as a characteristic element of the dyestuff (metal complex dyes).

Disazo dyes usually are formed by coupling of a bifunctional diazo compound with two-donor coupling components (Congo red) or by a reaction of two diazo compounds with two sites at a common donor-coupling component (C.I. Reactive Black 5) (Figure 8.9).

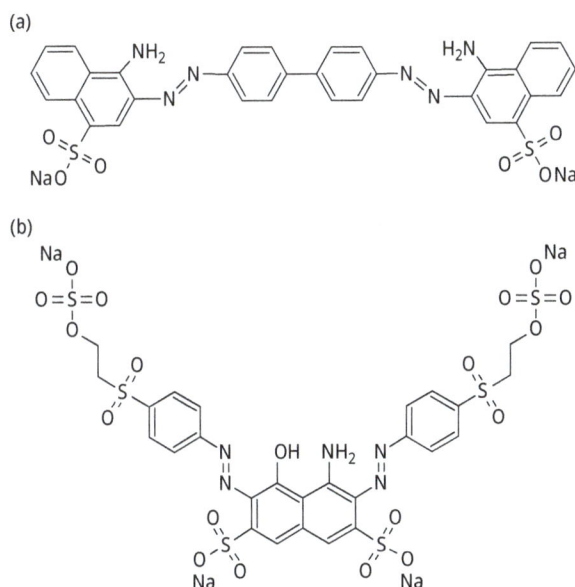

Figure 8.9: Structure of (a) Congo Red and (b) C.I. Reactive Black 5.

In the case of C.I. Reactive Black 5, the aminophenolate (H-acid) was coupled with two diazonium components.

8.3 Anthraquinone chromophores

Besides the azo chromophores, anthraquinoid structures represent the second most important chemical structure of dyes. Historically the classical red natural dyes

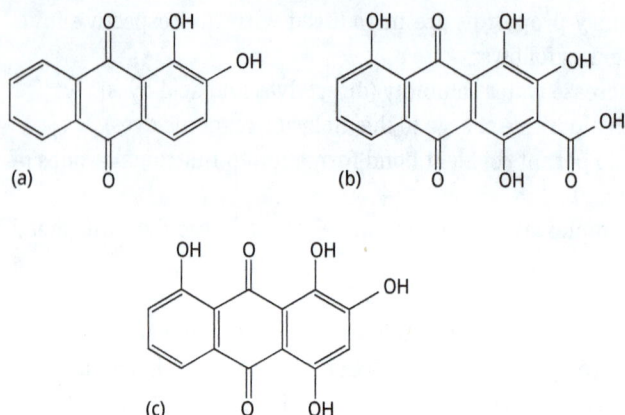

Figure 8.10: Structures of (a) alizarin, (b) pseudopurpurin and (c) purpurin.

extracted from madder were anthraquinones, for example, alizarine, pseudopurpurin and purpurin (Figure 8.10) [3].

The representative structural element is based on 9,10-anthraquinone. The majority of vat dyes also contain an anthraquinoid chromophore.

Anthraquinoid dyes exhibit several characteristics: The aromatic ring structure is rather stable to light exposure; thus in many cases, fastness to light is good. Unfortunately the tinctorial strength is comparatively low and the production of the dyes is more expensive. A chemical characteristic is the ability of many anthraquinoid dyes to undergo reversible reduction. Azo group containing dyes can be discharged through a reductive process. In the case of the Remazol Brilliant Blue R (C.I. Reactive Blue 19), the reversibility of the reduction of the anthraquinoid chromophore makes this dye difficult to discharge with reductive standard discharge chemicals (Figures 8.11).

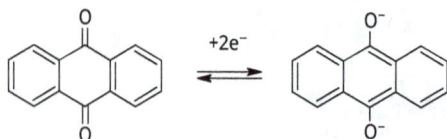

Figure 8.11: Reduction of an anthraquinoid group.

Similar to the azo dyes, the applicatory class of these dyes is dependent on the functional groups present in the molecular structure.

When solubilising groups are absent, a disperse dye is obtained, the introduction of sulphonate groups will deliver an acid dye (Figure 8.13) and reactive groups are introduced to synthesise reactive dyes (Figure 8.12).

In the application of vat dyes, the reduction of the anthaquinoid structure is a decisive step to achieve solubility of the dye. While the oxidised form of the vat dye is

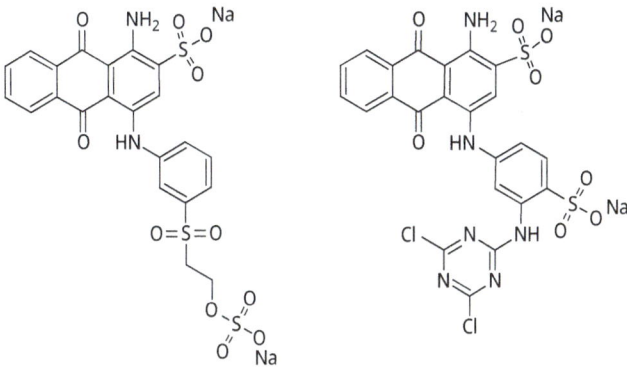

Figure 8.12: Reactive dyes with an anthraquinoid group: C.I. Reactive Blue 19 and Reactive Blue 6.

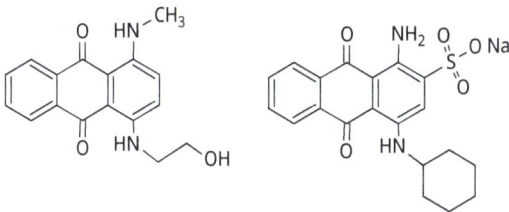

Figure 8.13: Structures of anthraquinoid disperse dyes C.I. Disperse Blue 3 and C.I. Acid Blue 62.

not soluble in alkaline solution, the reduced form dissociates under formation of a leuco-dianion, which is soluble in the alkaline dyebath. The leuco form adsorbs on the cellulose fibres and then is reoxidised to the insoluble anthraquinoid form at the end of the exhaustion process.

Vat dyes often exhibit larger systems of annelated rings; two anthraquinoid building blocks can also be observed in the chemical structure of vat dyes, for example, the indanthrone structure of (C.I. Vat Blue 4, C.I. Vat Blue 6, C.I. Vat Blue 14) (Figure 8.14).

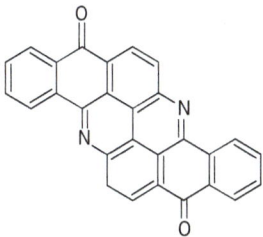

Figure 8.14: Structure of C.I. Vat Yellow 1 (Indanthrene Yellow G).

8.4 Indigoid chromophores

Indigo (Vat Blue 1) is the historical blue dye, which has been extracted from different plant sources all over the world. Another indigoid dye derived from natural sources is the Tyrian Purple (6,6'-dibromoindigo).

Despite the high attention indigo has gained both as a natural dye extracted from plant sources and today as synthetic dye for denim products, the range of technically relevant examples of dyes with indigoid structure, with exception of indigo, is very limited (Figure 8.15).

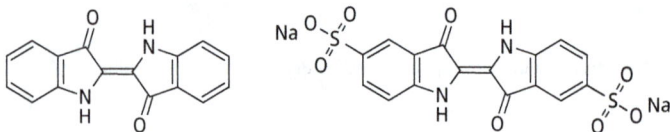

Figure 8.15: Structure of indigo (C.I. Vat Blue 1) and indigocarmine (indigodisulphonic acid, C.I. Acid Blue 74).

Technically indigo belongs to the class of vat dyes and thus requires a reduction step to achieve its water-soluble form. By sulphonation a water-soluble dye, indigocarmine (C.I. Acid Blue 74, C.I. Natural Blue 2, indigo-5,5'-disulphonic acid), is obtained, which can be used as an acid dye for wool [4, 5].

8.5 Cationic dyes

The presence of a cationic charge either in localised or in delocalised form is the characteristic sign of a cationic dye. In the vast majority of cationic dyes, the positive charge is carried by a nitrogen atom, which usually is a fully substituted ammonium ion.

Delocalised structures can be based on the methine dye system, which is vinylogous to an amidinium ion. In addition, aromatic ring systems can participate in the delocalisation (Figure 8.16).

Another condensed system in which the positive charge can be distributed is found in the azine dyes. A representative example for a thiazine is methylene blue (Figure 8.17).

In case a positive charge is present as a localised structural element, the charge does not contribute to the colour development to such an extent as it is the case for dyes with delocalised positive charge. The influence of a localised quaternary ammonium group on the chromophore is limited and the cationic charge is mainly present to provide the required sorption behaviour for substrates that bear negatively charged sites (e.g., polyacrylic fibres, leather or paper).

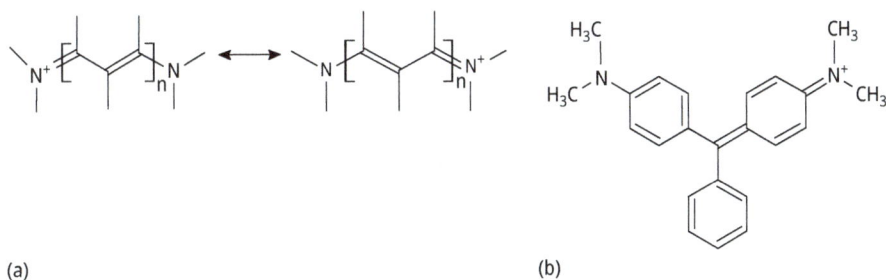

(a)

(b)

Figure 8.16: Representative structures of cationic dyes with delocalised structure: (a) methine dye and (b) triphenylmethane dye (malachite green).

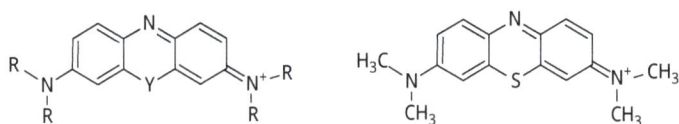

Figure 8.17: General formula of azine dyes (Y = –O–, –S–, –NR–) and structure of methylene blue.

8.6 Polymethine dyes

Methine and polyene dyes contain a conjugated double bonds built of methine groups. An odd number of methine groups forms the conjugated system, which then bears two terminal groups X and Y. In many cases, the terminal atoms are nitrogen or oxygen. The terminal group can also be a ring system, which then contains the heteroatom (Figure 8.18).

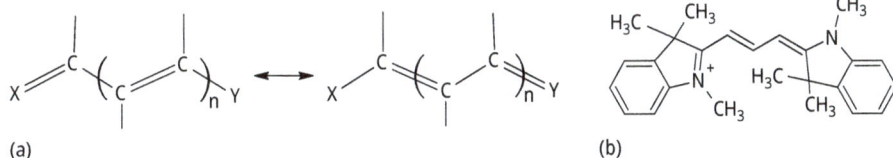

(a)

(b)

Figure 8.18: General formula of methine dyes and a representative example of a polymethine dye.

Important representatives for these dyes are β-carotene, which contains 22 methine groups and cyanine dyes. Important applications of cyanine dyes are their use as sensitising dyes in silver halide photography and as IR-absorbing dyes for optical data storage (Figure 8.19).

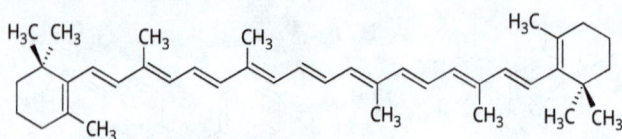

Figure 8.19: Structure of β-carotene.

8.7 Phthalocyanine dyes – Aza[18]annulenes

The structure of phthalocyanine dyes is related to the porphyrin structure and can be seen as tetrabenzo–tetraaza–porphyrin (Figure 8.20). The phthalocyanine structure forms stable complexes with numerous metal ions, for example, copper, nickel and iron.

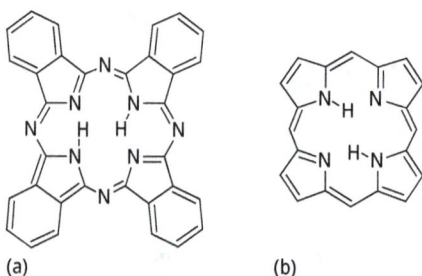

(a)   (b)

Figure 8.20: General structure of phthalocyanine and porphyrin.

Haemoglobin, chlorophyll and vitamin B_{12} are important naturally occurring porphyrins. Through exchange of the magnesium centre ion in chlorophyll by copper, a stable green dye is obtained based on natural resources, which however is no longer regarded as a natural colorant.

The most important representative for metal complexes on the basis of phthalocyanines is copper phthalocyanine, which is a very stable green dye with high coefficient of absorption. By the introduction of solubilising groups, for example, sulphonate groups, a green direct dye is obtained: C.I. Direct Blue 86. When reactive anchor groups are attached to the phthalocyanine dye system, a reactive dye is obtained (Figure 8.21).

Phthalocyanine dyes are highly important blue, green and turquoise dyes with excellent stability and are thus found in many colour gamuts.

The fact that the phthalocyanine dyes are metal complexes has to be considered when heavy metal concentrations in waste water of a dyehouse are analysed. Considerable amounts of copper or nickel can arise from the presence of the respective complexes, for example, in form of green and turquoise dyes in the effluent.

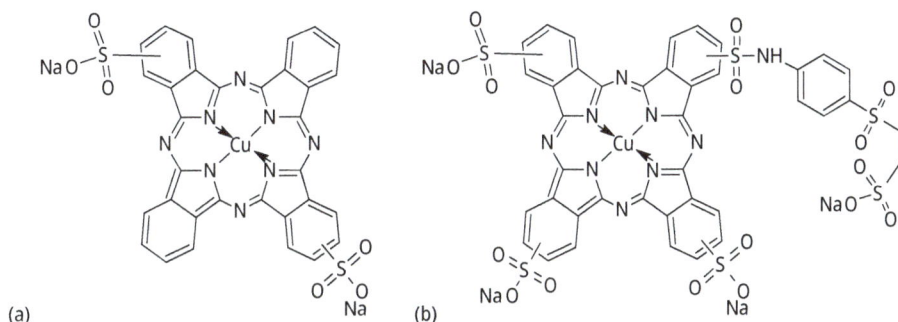

(a) (b)

Figure 8.21: Formula of C.I. Direct Blue 86 and C.I. Reactive Blue 21 (Remazol Turquoise Blue G 133).

8.8 Sulphur-based chromophores

In contrast to the majority of all other textile dyes, the structure of the sulphur dyes is chemically not uniform. This is a result of their synthesis in which organic precursor molecules, for example, 2,4-dinitrochloro-benzene ore 2,4-dinitrophenol, were heated in the presence of sulphides. Based on the route of synthesis, we can distinguish between

– bake dyes in which, for example, toluidine is heated in the presence of molten sulphur
– melt dyes in which the thionation is performed in solution.

As a general sign, sulphur is bound into a heterocyclic structure or is present as thiophenolic sulphur. In many cases, the dyestuff exhibits a polymeric structure with disulphide and polysulphide bridges between heterocyclic blocks.

During synthesis the dyestuff molecules are present in the reduced form. To isolate the insoluble dye, the crude dye is precipitated through oxidation, for example, by air oxygen. The product is filtered and rinsed before the commercial dyestuff is prepared.

Only few colours mainly blue, brown and black tones are available, which are used for dyeing of velvet and working cloths. The most important application of sulphur black is for dyeing of black shades in the production of black denim and for topping and bottoming of indigo dyed denim, for example, with Sulphur Black 1. Because of its high colour yield and cost-effective synthesis, large amounts of Sulphur Black 1 and its variations are used, as this dye offers "the highest amount of black per dollar", which places this dye as number one in terms of the yearly synthesised amount (Figure 8.22).

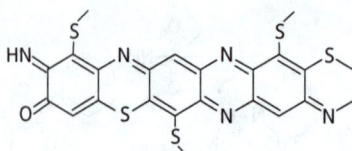

Figure 8.22: Representative structure of C.I. Sulphur Black 1. From Bechtold et al. [6].

8.9 Metal complexes

Traditional mordant dyeing with natural colorants and the addition of metal salt–based mordants utilise the principle of metal complex formation since the early stages of textile coloration. The addition of a metal ion and the formation of a dye complex led to substantial changes in hue of the chromophore and in many cases also to an improvement of the colour fastness to light and washing [7].

The organic dye molecule takes the function of the ligand that forms stable metal complexes with the metal ion. The metal complex formation can be undertaken in situ in the dyebath through the reaction between the ligand molecules and metal salt. In dyeing with natural colorants, the metal ions can be introduced first (pre-mordanting), directly during the dyeing step (meta-mordanting) or as a separate step after the dyebath had been released (after mordanting). In modern dyehouses, the direct application of metal salt solutions to form metal complex dyes in situ is no longer in use. Major reasons being the difficulty to apply stoichiometric amounts of metal ions to form the complex. As a result, substantial concentrations of heavy metals (e.g., Cu, Cr, Co and Ni) then are found in the waste water from such processes.

Metal complex dyes are widely used as acid dyes for wool and polyamide; however, the phthalocyanine dyes (e.g., reactive dyes and direct dyes for cotton) also belong to this class.

An present, mainly commercially available metal complex dyes are used, which thus release only minor amounts of metal ions into the waste water in the form of not exhausted unfixed dye remaining in the spent dyebath and the rinsing water.

Major metal ions used in metal complex dyes are copper, chromium and cobalt. Nickel complexes are used as pigments, where the mobility of nickel ions is prevented by the insolubility of the pigment and the stability of the complex. Iron and aluminium complexes are found as mordants in the application of natural dyes [8].

Based on the number of ligands that form the complex with the centre ion, we can distinguish between 1:1 metal complex dyes and 1:2 metal complex dyes. In 1:2 metal complex dyes, usually both organic molecules are the same. The ligand molecules form chelate complexes, which means that one ligand coordinates with two or more linkages to the metal ion. The ligands contain solubilising groups to achieve sufficient water solubility for dyeing from an aqueous phase.

The formation of complexes with high stability is the basis to obtain dyeings that are stable enough to resist to commercial detergent formulations, which contain complexing

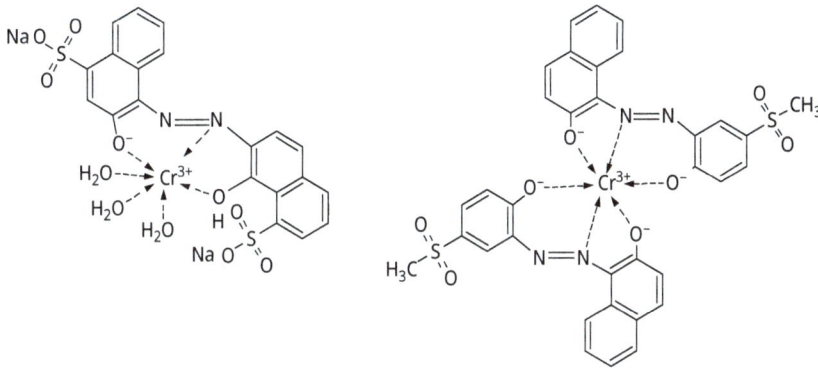

Figure 8.23: Metal complex dyes: 1:1 metal complex dye (C.I. Acid Blue 158; left) and 1:2 metal complex dye (C.I. Acid Violet 78; right).

agents. Removal of the centre ion through complexing agents during wash would destroy the dyestuff and lead to unacceptably low wash fastness (Figure 8.23).

8.10 Formazan dyes

The structure of formazan dyes is related to azo dyes (Figure 8.24). The synthesis often uses the coupling between a hydrazone and a diazonium salt in an alkaline medium. An important feature is the formation of a six-membered ring with a hydrogen linkage. Formazan dyes can be metallised to form metal complex dyes with high colour strength and are important reactive dyes today.

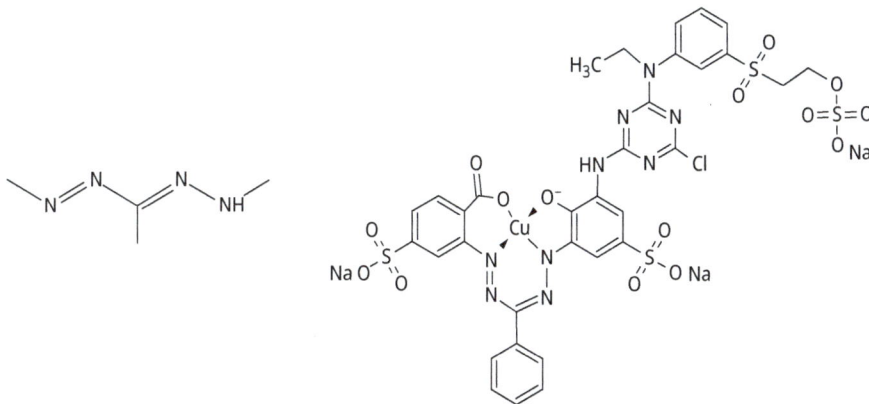

Figure 8.24: General structure of a formazan (left) and structure of a copper formazan dye (C.I. Reactive Blue 221; right).

8.11 Fluorescent brighteners/fluorescent whitening agents

While dyes absorb radiation in the visible range of the spectrum of electromagnetic radiation, the fluorescent dyes absorb mainly in the UV region and emit irradiations in the wavelength range of 400–500 nm.

The vast majority of fluorescent whitening agents contain a (E)-configured double bond (olefin –CH=CH– or azomethine –N=CH–) or a carbonyl group, which are conjugated to an aromatic system (e.g., benzene, naphthalene and heteroaromatic rings). Usually the molecules contain a flat and rigid structure. This hinders the electronically excited molecules to release their energy in radiationless processes, for example, in the form of vibrational energy to neighbouring molecules. Thus, return to the ground state requires emission of fluorescent light.

The characteristics of a fluorescent whitening agent are given by the molar extinction coefficient ε and the quantum yield Φ. The quantum yield is defined as ratio between the intensity of emitted light I_e and the absorbed light I_a . Values for Φ can range between 0.48 and 0.85, which means that more than 50% of the absorbed light is emitted in the form of fluorescent light.

$$\Phi = I_e/I_a \tag{8.5}$$

The wavelength of the emitted light is not monochromatic. Each optical brightener also contributes to the development of the colour; thus optical brighteners can also be classified with regard to their shade (e.g., red, blue and green; Figure 8.25).

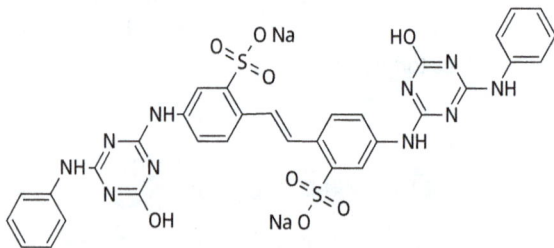

Figure 8.25: Example for a fluorescent brighteners (fluorescent whitening agent) – C.I. Fluorescent Brightener 40.

8.12 Photodegradation of dyes

The photodegradation of dyes leads to colour change and fading. The chemical processes are complex and thus still many questions are unclear.

Major factors that are of significant influence to light fastness are as follows [9]:

- structural factors – constitution of the dye molecule
- chemical environment of a dye molecule, for example, adsorbed in monomolecular state, associated, chemical interaction with substrate
- substrate (e.g., cellulose fibres, protein fibres and polyamide)
- presence of moisture and gases (e.g., oxygen)
- energy (wavelength) and intensity of irradiation
- probability of the formation of reactive intermediates during absorption of irradiation.

Absorption of light leads to promoting molecules to an exited state. An electron is moved from the highest occupied molecular orbital to the lowest unoccupied molecular orbital. As discussed in Chapter 7, the return to the ground state requires a release of energy, which can occur through different pathways. Emission of radiation of longer wavelength than the absorbed irradiation in the form of fluorescence and phosphorescence can occur. Radiationless transitions also are possible, thus leading to dissipation of the absorbed energy as heat. All three processes, to return to the ground state, have in common that the molecular structure is not altered and the molecule remains intact. A general presentation of these processes is given in Figure 8.26.

The excitation of an electron begins from the ground state, which is a singlet state with paired electrons (S_0). The first exited state again is a singlet state (S_1), which can be converted into a triplet state (T_1) through intersystem crossing.

The potential curves in Figure 8.26 indicate that the potential curves of the exited states have their minimum at longer internuclear distance compared to that of the ground state. In the triplet state, the internuclear distance has increased considerably

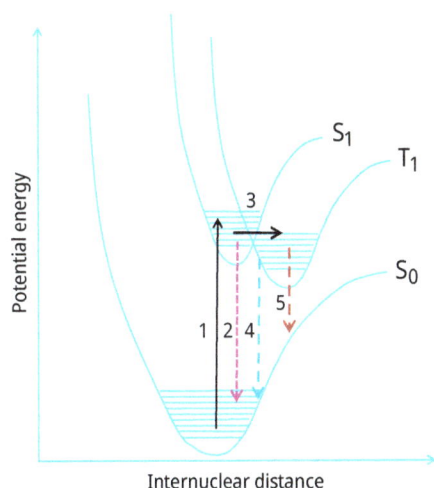

Figure 8.26: Potential energy diagram of ground state, singlet state and triplet state of a molecule (from Peters and Freeman [9]):1 absorption; 2 fluorescence; 3 intersystem crossing and 4 and 5 phosphorescence.

and the lifetime of the state is with 10^{-3}–10 s substantially longer than for a singlet state (10^{-9}–10^{-6}). Thus, a certain probability exists that the length of a bonding increases to such a distance that bond breakage and follow-up reactions may occur.

Through this pathway photochemical reactions lead to bond dissociation, redox reactions, molecular rearrangements and so on.

Photofading of dyes leads to unsatisfying fastness to light and degradation of fibre material. In many important technical applications the photochemical initiation of chemical reactions is utilised by using appropriate dyes as the photosensitiser and photocatalyst. As shown in Figure 8.26, the majority of photochemical processes begin from the more time stable triplet state. Depending on the chemical surrounding, for example, the presence of oxygen, different mechanisms have been proposed.

Photochemical reactions in the absence of oxygen can involve the formation of radical species (eq. (8.1)).

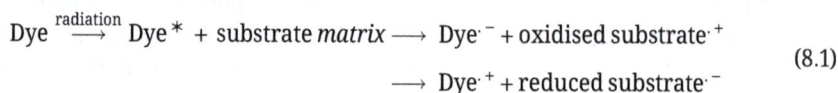

$$\text{Dye} \overset{radiation}{\longrightarrow} \text{Dye}^* + \text{substrate } matrix \longrightarrow \text{Dye}^{\cdot -} + \text{oxidised substrate}^{\cdot +}$$
$$\longrightarrow \text{Dye}^{\cdot +} + \text{reduced substrate}^{\cdot -} \tag{8.1}$$

In the presence of oxygen (ground state is a triplet state $^3O_2$), the oxygen molecule can participate in the first reaction step, thus leading to singlet oxygen (eq. (8.2)):

$$\text{Dye} \overset{radiation}{\longrightarrow} \text{Dye}^* + {}^3O_2 \longrightarrow \text{Dye} + {}^1O_2 \tag{8.2}$$

The lifetime of the singlet oxygen formed is dependent on the chemical environment, for example, solvent. The high reactivity of the singlet oxygen then leads to follow-up reactions.

The chemistry of the follow-up reactions is dye specific and also depends on external factors, for example, humidity, fibre type and temperature. Thus, the prediction of light fastness is still a challenging task. Some more general predictions for changes in light fastness are as follows: Increasing humidity often leads to a decrease in light stability of dyes, thus results in higher fading rates. Increasing dye concentration improves light fastness of dyes; thus, darker shades exhibit better stability.

8.13 Photodegradation of dye classes

Based on the chemical basis of a chromophore, different reaction mechanisms can occur.

Azo dyes: These can undergo both reductive and oxidative degradation and also photoisomerisation.

In the case of a photoreduction, the azo group is reduced to the hydrazo group via radicals as intermediates, which is then reduced further to the corresponding amino groups (eq. (8.3)).

$$R - N = N - R' \longrightarrow R - NH - NH - R' \longrightarrow R - NH_2 + R' - NH_2 \qquad (8.3)$$

For oxidative fading, different mechanisms have been proposed, where some of these involve singlet oxygen [9]. The reaction is formulated to occur at the azo group, which cleaves under the formation of a diazonium ion and decomposes to a hydrocarbon and an oxidised counterpart (eq. (8.4)).

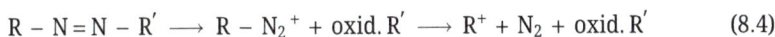

$$R - N = N - R' \longrightarrow R - N_2{}^+ + \text{oxid.} R' \longrightarrow R^+ + N_2 + \text{oxid.} R' \qquad (8.4)$$

Another photochemical effect that leads to a change in colour of an azo dye is the photoinduced cis–trans isomerisation of the azo group. The azo group usually is present in its trans form. On the basis of the dye structure, the chemical environment and physical conditions, for example, temperature under irradiation, an equilibrium between trans and cis form of the azo dye is obtained. As the absorption spectra of both isomers are different, an irradiation-dependent change in colour is observed.

Fluorescent dyes: The photodegradation of fluorescent dyes can follow the mechanisms already discussed, for example, photoreductive cleavage, oxidative processes and attack of singlet oxygen.

In the case of stilbene-based materials, cis–trans isomerisation can also occur, followed by decomposition of both isomeric forms.

A scheme describing the isomerisation of the trans form to the cis form and the formation of a dioxetan as a result of the addition of singlet oxygen on the double bond is shown in Figure 8.27.

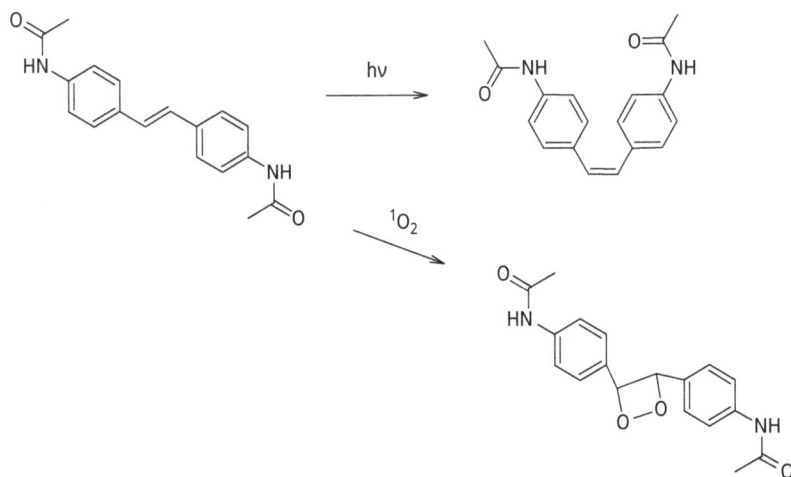

Figure 8.27: Photoisomerisation of a stilbene-type fluorescent whitening agent from the trans form to the cis form (top) and to the dioxetan (lower formula) (from Peters and Freeman [9]).

Anthraquinoid dyes: In general anthraquinoid dyes, for example, vat dyes, the natural colorant extracted from madder and other dyes that contain an anthraquinoid chromophore exhibit high fastness to light.

As a first step, excitation of the anthraquinoid system occurs that then leads to a series of possible follow-up reactions.

– *N*-alkyl substituents on the aromatic ring system can be dealkylated.
– Introduction of oxygen to the aromatic rings, for example, in the form of a hydroxyl group can occur.
– Photoreduction of the 9- and 10-position of the anthraquinone to the corresponding phenols can be observed in the presence of oxidable substances that serve as reducing agents (Figure 8.28).

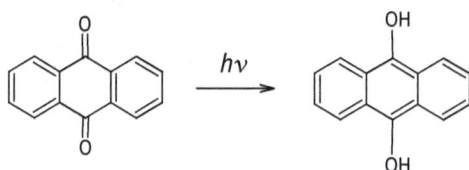

Figure 8.28: Photochemical reduction of an anthraquinone to the 9,10-dihydroxyanthraquinone.

As long as no irreversible follow-up reactions occur, reversible reoxidation of the 9,10-dihydroxyanthraquinone into the corresponding 9,10-anthraquinone is possible.

The reversible reoxidation of the dihydroxy form of a chemically reduced anthraquinone dye in alkaline solution is a central step in the technical application of vat dyes. In alkaline aqueous solution, the redox potential of oxygen is sufficient to permit rapid and complete reoxidation to 9,10-anthraquinone form.

At conditions of neutral solution pH, anthraquinone dyes, such as C.I. Acid Green 25, undergo rather stable photoreduction to the corresponding dihydroxy form, which re-oxidises only slowly and incompletely, even in the presence of air oxygen [9]. This behaviour is similar to the behaviour of vat dyes at low pH, where the rather stable leuco acid (the dihydroxy form of the vat dye) is formed, which then re-oxidises only slowly and incompletely.

As a result of their photochemical stability, anthraquinoid dyes also act as photosensitising dyes, which then induce photodegradation of surrounding polymer material and other dyes. The photodegradation can proceed via formation of singlet oxygen (eq. (8.2)). Examples for such dyes are C.I Vat Green 1 and C.I. Vat Orange 9.

Indigoid dyes: The photofading of indigo is of interest as this dye is of enormous relevance for the production of denim products. Indigo is an important example of the substrate-dependent light fastness. In dyeing on cotton, the light fastness according the blue standard scale rating (scale 1–8, with 1 being the lowest fastness and 8 the highest fastness) reaches a level of 3, while on wool substantially higher fastness is observed.

The fading of indigo follows a self-sensitised reaction pathway. As a first step, singlet oxygen is formed, which then reacts with indigo to form a dioxetane intermediate and decomposes to the yellow isatin [10] (Figure 8.29).

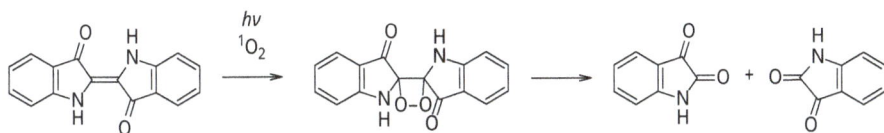

Figure 8.29: Photodegradation of indigo to isatin via a dioxetan as intermediate.

The high stability of the indigo dye in wool is explained by quenching of singlet oxygen by the amino acids in keratin. Thus, the formation of the dioxetan does not occur to a similar extent as it is the case for cellulose fibres.

Catalytic fading: In practical dyeing, only very rarely application of a single dye is sufficient to reach a certain colour; thus, the use of dyestuff combinations is required. A serious problem for commercial textile dyehouses appears from possible negative influence of certain dyes on the light fastness of others. Quite often the presence of yellow and orange dyes reduces the fastness of red, blue or violet dyes. A yellow dye will absorb visible light with short wavelength (violet) with high efficiency and thus will achieve an energetically higher level of excitation compared to a blue dye that absorbs in the yellow/orange range of the visible light.

The more rapid fading can be explained with the transfer of absorbed energy from the exited yellow/orange dyes to other dyes present in the mixture, which then degrade. Another reaction model bases on the formation of singlet oxygen by the yellow dye, which then chemically attacks the other dyes.

Photodegradation of fibres: The photodegradation of a textile fibre is a material-dependent property. In the case of undyed cellulosic fibres, rather low sensitivity to light is observed. Intense irradiation in the UV region can lead to yellowing of the fibre. Dye-sensitised photodegradation can follow different mechanisms, which are dependent on the type of dye present. Reaction mechanisms include the hydrogen and electron abstraction. In addition, dye-induced formation of singlet oxygen and hydrogen peroxide can lead to oxidative fibre degradation. The formation of hydrogen peroxide is a moisture-dependent process; thus, the presence of moisture also plays a significant role in the phototendering of cellulosic fibres.

In the photodegradation of polyamide fibres, the direct photolysis of the amide linkage can occur during irradiation at wavelengths shorter than 300 nm. The mechanism of dye-sensitised photodegradation depends on the type of dyestuff, reductive and oxidative mechanisms have been observed.

The photodegradation of polyester fibres poly(ethylene terephthalate) is caused by the absorption of UV light (290–320 nm) by the ester carbonyl group, which then leads to cleavage of the ester bond. Similar to the degradation of polyamide fibres, the incorporation of titanium oxide reduces the photostability.

Improvement of light fastness of dyes and pigments can be achieved through additives, stabilisers such as nickel complexes, sterically hindered phenols and amines and benzotriazoles (Figure 8.30).

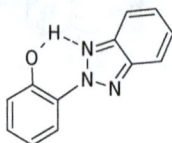

Figure 8.30: Structure of a benzotriazole-based photostabiliser.

References

[1]   Heinrich, Zollinger. Color Chemistry, Wiley-VCH, Weinheim, Germany, 2003.
[2]   Hunger, K. Industrial Dyes, Chemistry, Properties, Applications. Wiley-VCH, Weinheim, Germany, 2003.
[3]   Bechtold, T. Chapter 10, Natural colorants-Quinoid, Naphthoquinoid and Anthraquinoid dyes. In: Handbook of Natural Colorants, Ed. T. Bechtold & Rita Mussak. Wiley & Sons, 2009, 151–182. ISBN-10: 0470511990.
[4]   Komboonchoo, S., Bechtold, T. Sorption characteristics of indigo carmine as blue colorant for use in one-bath natural dyeing. Text. Res. J. 2010, 80, 734–743 doi:10.1177/0040517509342319.
[5]   Komboonchoo, S., Bechtold, T. Natural dyeing of wool and hair with indigo carmine (C.I. Natural Blue 2), a renewable resource based blue dye. J. Clean. Prod. 2009, 17/16, 1487–1493.
[6]   Bechtold, T., Berktold, F., Turcanu, A. The redox behaviour of CI Sulphur Black 1 – A basis for improved understanding of sulfur dyeing. J. Soc. Dyers Colour. 2000, 116, 215–221.
[7]   Manian, AP., Paul, R., Bechtold, T. Metal Mordanting in Dyeing with Natural Colourants: A Review. Color. Technol. 2016, 132, 107–113.
[8]   Bechtold, T., Turcanu, A., Ganglberger, E., Geissler, S. Natural dyes in modern textile dyehouses – How to combine experiences of two centuries to meet the demands of the future?. J. Cleaner Prod. 2003, 11, 499–509.
[9]   Peters, AT., Freeman, HS. Advances in color chemistry series – Volume 4, Physico-Chemical Principles of Color Chemistry. Blackie Academic & Professional, London, 1996.
[10] Kuramoto, N., Kitao, T. Contribution of singlet oxygen to the photofading of indigo. J. Soc. Dyers Colour. 1979, 95, 257.

Take home messages

The classification of dyes can be done by a number of characteristics:
- Application (acid dye for wool, reactive dye vat dye, ink-jet printing)
- Basic chromophore (anthraquinoid, azo dyes, phthalocyanine dyes, metal complexes)
- The material to be dyed (wool, cellulose)
- The solubility pigments that require either binder systems or are embedded into the polymer mass and dyes (which have to show an affinity to a fibre material)

Irrespective of type of chromophore, dyes can be arranged to a group of dyes with regard to common conditions of application. Photodegradation of dyes bases on chemical degradation reactions following to the adsorption of light. Fading often includes presence of singlet oxygen as highly reactive species.

Quiz

1. Explain the molecular characteristics of C.I. Vat Yellow 1 (Figure 8.14) and indigo (Figure 8.15). Why reduction of these dyes is an essential step in dyeing?
2. Which optical brightener works better: the one with a quantum yield of 56% or the other with a yield of 62%? What happens to the rest?
3. In dyeing with a mixture of two dyes, the observed photofading is accelerated substantially. Can you give a possible explanation and propose a technical solution?

Exercises

1. Search for the following dye structures and classify them with regard to the chromophore: Acid Red 3; Disperse Yellow 2; Vat Blue 4.

Solutions

1. (Question 1): Both molecules do not contain any functional groups that could permit water solubility. Thus, reduction is required to transfer the molecules into their alkali soluble form.
2. (Question 2): The higher the quantum yield, the more of the absorbed light is emitted as fluorescent light. The rest of the light is absorbed and is either emitted via phosphorescence or transferred into heat.
3. (Question 3): Most probably one of the dyes stabilises its excited energy on costs of the other one, through transfer of its energy to the other, which then decomposes. Solution: Replace at least one of the dyes.
4. (Exercise 1): Acid Red 3: azo chromophore; Disperse Yellow 2: acridinone (9(10H)-acridone) chromophore and Vat Blue 4: anthraquinoid chromophore.

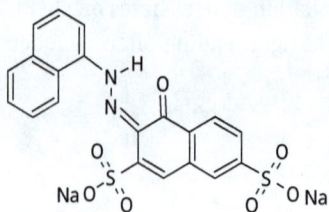

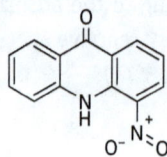

C.I. Acid Red 3   C.I. Disperse Yellow 2   C.I. Vat Blue 4

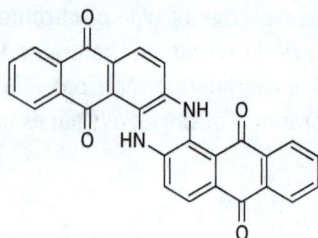

List of abbreviations/symbols

Φ quantum yield
I_e intensity of emitted light
I_a intensity of absorbed light
S_0 singlet ground state
T_1 first excited triplet state

9 Colour measurement

9.1 The perception of colour

In a physical sense the definition of colour could be described by the reflectance curves of a material. However, this does not include the fact that colour is a perceived signal which includes a substantially higher number of factors than the simple reflectance curve of a non-transparent sample.

Colour measurement is an experimental approach to transfer the perception of colour into an objective and measurable system [1]. In this chapter we will focus on the use of the CIELAB coordinates as descriptive three-dimensional colour space in which every colour is defined by its characteristic coordinates.

The CIELAB system is a technical tool with the attempt to standardise perceived colours in terms of colour definition and colour communication.

A high number of physiological and psychological factors influence the individual perception of colour and colour differences between two samples, thus a generalisation of the technical tool will always lack behind the individual colour perception. However, the advantage of objective description and comparability of colour perception provide substantial help and support in the process of colour definition, such as in textile dyeing, and thus is a widely used tool in textile and garment production.

The transformation of electromagnetic radiation into a physiological signal is performed by the eye. A general scheme of the structure of the human eye is given in Figure 9.1.

The optical system of the eye (cornea, pupil, lens) projects an inverted picture of the outside observation onto the retina. The light sensitive part of the human eye is the retina, where electromagnetic irradiation is transferred into a nerve signal, which then is transferred via the optic nerve to the brain.

On the retina the cones and rods are present, which are the actual receptors for light.

Cones are responsible for vision of colour and require a higher light intensity for proper signal formation. The rods are able to transfer lower amounts of light into nerve signals and mainly transmit information about lightness. Thus, the sensation of colour fades away at night and at low intensity of illumination.

The perceived colour will always depend on the intensity of illumination. As an example, a critical level of light intensity has to be kept as illumination in a museum to allow visitors a realistic impression of a coloured piece of artwork, while colour fading and photodegradation of a material will be more rapid at higher light intensity. Thus, intensity of light should be minimised to avoid any damage of the artwork.

Another difficulty arises from the non-uniform distribution of cones and rods on the retina. Near to the fovea (*macula retina*, the yellow spot of the retina) the distribution of cones is substantially higher than in the peripheral parts of the retina.

https://doi.org/10.1515/9783110795738-009

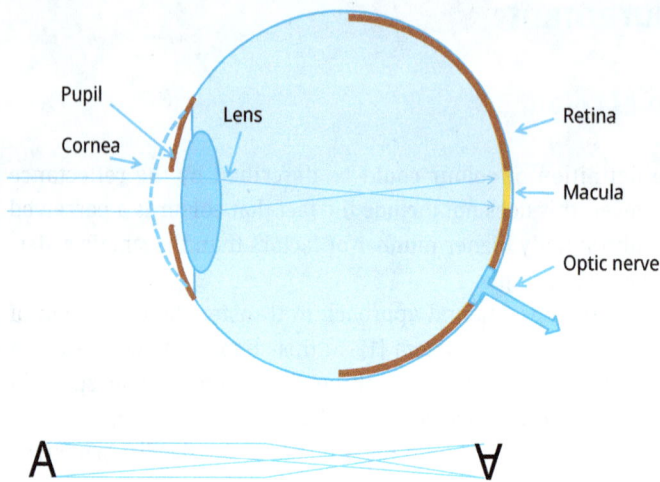

Figure 9.1: Scheme of the human eye and simplified drawing of the optical beam inside the eye.

The projected inverted image of a coloured textile sample is projected onto a certain area of the retina. As the distribution of colour-sensitive cells on the retina is not uniform, the size of the projected image on the retina will determine the signal formation and hence the colour perception. The size of the projected image on retina depends on the viewing angle under which a sample is observed. The geometry of the overall optical system including sample size, distance to the eye and projection to the retina will also influence the sensation of colour. While reflectance curves of a sample may be identical, the perceived colour will be dependent on the viewing angle (e.g. 2° or 10° observation). Often we would like to call such a phenomenon an optical illusion; however, the physiological background behind the observed colour difference is a fact and must be considered in the assessment of colour matching and when samples of substantially differing size should be compared.

As an example, a sample size of 7.7 cm diameter which is observed at a viewing distance of 25 cm will be seen under a viewing angle of 10°, while a sample with diameter of 1.6 cm at the same distance will be observed under a viewing angle 2°.

The signal formation on the retina occurs through a combination of photoreceptors, the retinal ganglion cells.

Three different types of ganglion cells are present on the human retina:

- Magnocellular cells: These are the large rods which mainly signal lightness. Speed of reaction is high so these cells are responsible for detection of movement. These cells represent a non-opposing mechanism forming an achromatic channel (A).
- Parvocellular cells: These cells are the small cones responsible for the red-green sensation, and thus for the perception of colour and form. Speed of reaction is moderate. These cells form an opposing mechanism for sensation of red-green colour (T).

- Koniocellular cells: These are very small cones which are responsible for short wavelength perception (blue-yellow). Again, these cells form an opposing mechanism for sensation of yellow-blue colour (*D*).

The physiological response to colour is related to the design of the final structure of the CIELAB coordinates system in terms of the L^*, a^* and b^* coordinates.

The achromatic channel A is related mainly to the L^* coordinate for lightness of a colour. The opposing chromatic channel T relates to the a^* coordinate (red-green axis) and the chromatic channel D relates to the b^* coordinate (yellow-blue axis).

Thus, in a simplified approach we can understand the perception of colour as a combination of three sensor responses, a non-opposing information for lightness of a sample and two opposing stimuli for position with regard to the red-green and yellow-blue quality.

The sensation of colour results from a combination of stimuli. Examples for contribution of the three responses to form the perception of a colour are listed in (Table 9.1).

Table 9.1: Examples for combination of perceived colour and stimuli responses of the *A, T* and *D* cells.

Colour	Intensity Achromatic A Lightness	Position Chromatic T Red-green	Position Chromatic D Yellow-blue
White	High	Middle	Middle
Orange	Middle	Red	Yellow
Blue-green	Middle	Green	Blue
Violet	Middle	Red	Blue
Black	Low	Middle	Middle

The descriptive characteristics of perceived colour can also be related to the physical characteristics of a measured reflectance curve. In Table 9.2 the measured characteristics of a colour are related to the psychophysical descriptors of colour. Hue characterises the general allocation of a tone, for example, red, green, brown. The term brightness is a characteristic for the 'purity' of a colour, the brightest colours would be defined by the colour of a certain spectral line of the visible light.

Table 9.2: Relationship between psychophysical parameters and descriptors for perceived colour (from Perez et al. [2] and Pridmore [3]).

Psychophysical parameter	Colour descriptor
Dominant wavelength	Hue
Luminance	Lightness, brightness
Purity	Chroma, colourfulness

From a physical point the definition of colour is clear and unambiguous. The perceived impression of colour, however, depends on a series of side effects which do not influence the reflectance curves and calculation of colour coordinates,and results from the complex processing of stimuli in the brain.

Often the change in appearance of a colour is recognised as an optical illusion; however, there is a physiological background behind the perception, which thus makes the observations objective and repeatable.

Important cases for chromatic effects are [2]:

- A change in luminance also changes the tone of a colour. Luminance and perceived colour are not independent.
- With increasing chroma the brightness of a colour of the same lightness apparently increases.
- Simultaneous contrast: The background surrounding a colour influences the perceived colour. Perceived brightness of a colour increases when the background is darker.
- Crispening effect: The differences between two stimuli are perceived to be larger when the background is similar to the two stimuli. This is of substantial relevance for colour assessment in a viewing chamber. The more different the colour of the surrounding background is, the lower the ability to differentiate between two similar colours.

9.2 The standard observer

In colour measurement, the ability of the human eye to differentiate between colours has been analysed by a large number of colour-matching experiments. In case of normal colour vision, the colour sensitivity of the human eye is described by the so-called standard observer in the tristimulus curves. Figure 9.2 shows the sensitivity curves for standard observers for 2° and 10° viewing angles. The differences in the curves reflect the different area covered on the retina by the projected picture of the sample. The differences in the distribution of receptors on the retina leads to differences in the sensitivity curves shown in Figure 9.2.

The different tristimulus curves $\overline{x}_\lambda, \overline{y}_\lambda, \overline{z}_\lambda$ indicate the sensitivity of the colour receptors in the human eye in a standardised form. For a given viewing angle, the curve $\overline{x}_\lambda$ indicates a wavelength-dependent sensitivity for light with a maximum near 600 nm and a second maximum near 440 nm. This corresponds to sensitivity for colours with red tone. The $\overline{y}_\lambda$ curve exhibits a maximum near 560 nm and represents sensitivity for green light. The $\overline{z}_\lambda$ curve with a sharp maximum near 450 nm is attributed to sensitivity in the blue range.

For our understanding we can assume that the three different sensitivity curves $\overline{x}_\lambda, \overline{y}_\lambda, \overline{z}_\lambda$ of the standard observer are related to the three different receptor cell systems

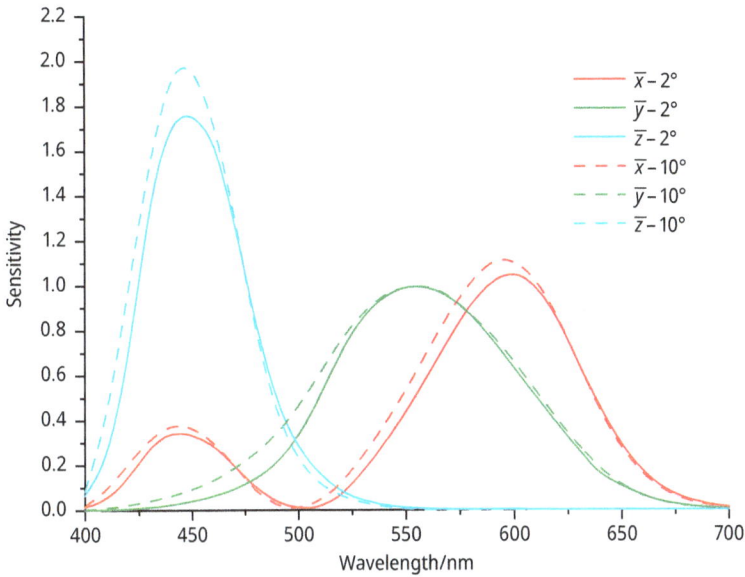

Figure 9.2: Tristimulus curves – sensitivity curves for the standard observer viewing angles 2° and 10°.

on the retina, the magnocellular cells (lightness, achromatic channel), the parvocellular cells (red-green colour, chromatic) and the koniocellular cells (yellow-blue, chromatic).

All curves almost come to zero below 400 nm and above 700 nm, thus colour measurement usually concentrates on the interval between 400 and 700 nm. Only light in this wavelength interval will lead to stimulation of the photoreceptors and thus will lead to perception of colour.

A white surface reflects the spectrum of a light source without specific changes in the respective intensity of the different wavelengths. Colour leads to a wavelength modulation in the light intensity of the reflected light. As indicated by the three tristimulus curves, a colour will lead to three independent signals, corresponding to the composition of the colour in terms of red (R), green (G) and blue (B).

A similar system is used in the RGB colour palette in the adjustment of every computer screen. The colour of a rectangular field or a cell drawn in almost any program is adjusted in the RGB scale. The contribution of a colour can be adjusted between arbitrary values set by the program (e.g. 0 for no intensity and 255 for maximum contribution). A colour with $R = B = G = 0$ will be black, and a colour with $R = B = G = 255$ will be white. Colours in between will be adjusted by the contribution of the different colours.

9.3 Colour specification through the CIELAB system

The CIELAB system bases on the standardisation of basic parameters which were required as condition to develop a uniform and generalised description of colour in a three-dimensional colour space.

The activity follows an initiative taken by the CIE, the 'Commission Internationale de L'Eclairage', the international commission on illumination.

Besides the standardisation of an average sensitivity of human perception in form of the tristimulus curves, the CIE also formulated standardised energy distributions for important light sources (Figure 9.3).

The important light sources are:

- Light source A – corresponds to light emitted from the surface of a black body with a temperature of 2,856 K, often simulated by the use of a tungsten filament lamp.
- Light source C – corresponds to a temperature of 6,774 K, representing daylight from the north on an overcast day.
- Light source D_{65} – corresponds to a theoretical energy distribution emitted by a 6,500 K black surface. While being a theoretical light source in many cases, D_{65} is used as light source for calculation of colour coordinates.
- Light source F-11 – a representative of the energy distribution of a fluorescent lamp (TL84) with characteristic narrow bands in the wavelength regions for blue, green and orange-red light.

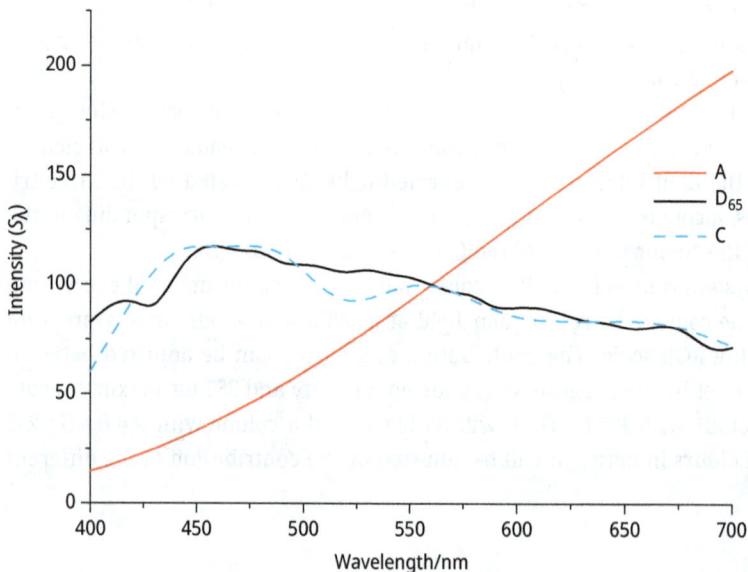

Figure 9.3: Energy distribution of important light sources A, D_{65}, C,.

9.4 Metamerism

Colours of two samples can be perceived as identical under a certain illumination such as daylight (light source C); however, they do not match under other light sources such as light sources A or F-11. This observation is an example of metamerism.

The origin of metamerism is found in the reflectance curves of the two dyeings which are not absolutely identical. Figure 9.4 shows an example between two nearly identical dyeings which are subject to metamerism. The dyeing curves are almost identical, however, a substantial difference is observed between 600 and 700 nm. Possibly the colour of the samples is perceived identical under light source C; however, if the intensity of a light source is substantially different in the region where the differences in the reflectance curves are more substantial, then the different curves will lead to a visible difference. This, for example, can be the case during observation under light source A.

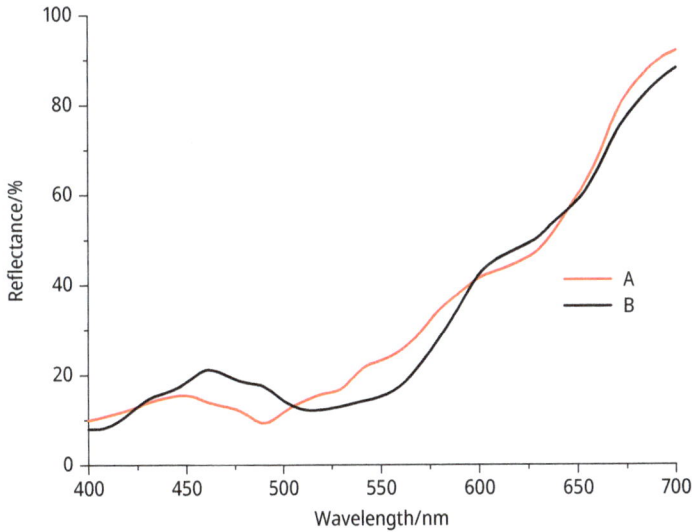

Figure 9.4: Reflectance curves of two dyeings which will exhibit metamerism.

In practice, metamerism is a critical issue as a number of reasons limit the degrees of freedom to minimise these light source–dependent colour differences.

In practical dyeing, often a colour sample is given by the designer or customer, and the dyer has to reproduce this colour almost exactly with the dyes available in his dyehouse. Often the given sample colours are water colours or on paper, while the desired shade is to be realised on a distinct textile material. This requires the use of different dyes and combinations of dyes. In an ideal case, the reflectance curves are very near to the reflectance curves of the colour sample; however, identity will be obtained only in case the same dyes are used as were present in the colour of the sample.

Another problem results from the combination of different materials in one garment, which should match in colour or harmonise in a combination of colours. As different dyes (e.g. metal complex dyes for polyamide, reactive dyes for cotton) will be used for different materials, the phenomenon of metamerism can lead to an undesirable colour difference between two parts of the garment, which however match under another light source.

Only in the case of identical material and identical dyes, identical reflectance curves will be achieved for sample colour and dyeing.

Modern methods for calculation of colour recipes used in dyehouses already permit prediction of possible metamerism. (Please compare later part in this section about colour recipe prediction.)

Metamerism must not be mixed with colour stability, which describes the change in perception of a single colour under changing light. As an example, the author's motorcycle at daylight looked blue; however, under street lighting (which were in authors younger days high-pressure mercury lamps) the colour was perceived as violet. This change in colour is not due to the differences in reflectance curves, but due to the energy distribution of the illumination. In case of usual light sources, the differences in perceived colour of a sample are not high enough to lead to the sensation of different colours. However, in case of a highly monochromatic light source, the human colour memory is sufficient in capacity to realise the change in perceived colour.

9.5 Tristimulus values

A general requirement on the mathematics of colour measurement is to design a three-dimensional system in which every colour is characterised through its unique coordinates.

To formulate a mathematical approach describing colour, we have to combine three basic informations:

1. The reflectance curve of the sample has to be measured as a characteristic property describing the spectral absorption of light. This is done by photometric methods and in an ideal case, should be independent of the measurement technique (in practice this is not the case).
2. The sensitivity of receptors in the human eye for certain wavelength has to be included in such a model. This is done by means of the tristimulus curves $\overline{x_\lambda}, \overline{y_\lambda}, \overline{z_\lambda}$ for a chosen viewing angle (2°, 10°).

3. The light source has to be included by introduction of the spectral energy distribution of a standardised light source (e.g. C, A, D$_{65}$, F-11).

A general formula for calculation of the tristimulus values X, Y, Z is given in eq. (9.1).

Tristimulus value (X, Y, Z) = reflectance × standard observer response × light source
(9.1)

The exact formulation is given in eqs. (9.2)–(9.4).

$$X = \int_{400}^{700} S_\lambda \, \beta_\lambda \, \overline{x_\lambda} \, d\lambda \qquad (9.2)$$

$$Y = \int_{400}^{700} S_\lambda \, \beta_\lambda \, \overline{y_\lambda} \, d\lambda \qquad (9.3)$$

$$Z = \int_{400}^{700} S_\lambda \, \beta_\lambda \, \overline{z_\lambda} \, d\lambda \qquad (9.4)$$

with S_λ being the intensity of the respective light source, β_λ being the reflectance of the sample at wavelength λ (value between 0 and 1) and $\overline{x_\lambda}$, $\overline{y_\lambda}$, $\overline{z_\lambda}$ being the respective values of the tristimulus curves.

Theoretically, the calculation of the three stimulus values X, Y and Z would correspond to an integral of the products $R_\lambda * S_\lambda * (\overline{x_\lambda}, \overline{y_\lambda}, \overline{z_\lambda})$ in the wavelength region between 400 and 700 nm. In practice, the calculation is done by the summation of the respective products with an interval of 10 nm (or 5 nm). The values for tristimulus curves and the characteristic energy distribution of a light source are available in standardised form in tables. The reflectance curves will remain as the only experimental values which are characteristic for the colour of the sample.

In case if a 10 nm interval is chosen for calculation, the summation of each tristimulus value will consist of 31 values. Due to the limited access to calculation capacity of computers in the initial phase of colour measurement, the intervals were set with 20 nm, thus leading to 16 data points.

It is important to consider that the wavelength interval used for the calculation of the tristimulus values can be extended to 360 nm as the lower limit and 780 nm as the upper limit. Thus, the interval used for a measurement and calculation of coordinates always have to be considered.

Today the majority of standard methods to calculate tristimulus values uses a 10 nm interval, which is sufficient to obtain results with an appropriate accuracy.

The tristimulus values are also standardised to obtain a value of $Y = 100$ for a perfect white and $Y = 0$ for a perfect black.

9.6 Colour coordinates

As indicated by the symbols X, Y, Z, the tristimulus values represent a first set of colour coordinates in a three-dimensional colour space.

However, the values are not directly related to the characteristics of the colour, for example, in terms of hue, chroma. Only the standardisation of white ($Y = 100$) and black ($Y = 0$) has already delivered a coordinate which is related to the perceived information of colour.

As a next step, the tristimulus values are converted into the chromaticity coordinates x, y, z which allow the formation of a three-dimensional colour space. These coordinates were used to generate the CIE colour space.

$$x = \frac{X}{X+Y+Z} \tag{9.5}$$

$$y = \frac{Y}{X+Y+Z} \tag{9.6}$$

$$z = \frac{Z}{X+Y+Z} \tag{9.7}$$

with $x + y + z = 1$.

The calculation of the chromaticity coordinates follows eqs. (9.5)–(9.7).

One horizontal layer of the three-dimensional CIE colour space is shown in Figure 9.5, which is called chromaticity diagram. For the chromaticity diagram only the x and y values are used. Instead of the z value, which is already dependent on the values x and y, the tristimulus value Y is used as the third coordinate and as a measure for lightness of the colour.

The chromaticity diagram exhibits a number of characteristics. Colours are present only inside a defined area. All colours on the plane are characterised by the same lightness Y. The borderline is formed by the most brilliant colours, which are represented by the respective wavelength of monochromatic light. For example, the right corner is formed by a red colour which would reflect only monochromatic light of 700 nm, the left corner is represented by a wavelength of 400 nm. The straight line between the two corners is called the line of purples. The colours on this line are obtained by mixing light of 400 and 700 nm.

All real colours are within the CIE colour space. Grey as an achromatic colour is found near the middle of the field.

Both size and shape of the field depend on the height of the value Y, as the field will be reduced to a single dot at $Y = 0$ (black) as well as at $Y = 100$ (white).

The CIE colour space could solve the problem to transfer reflectance values and perceived colour of a body into a three-dimensional system. Unfortunately, a study of MacAdam indicated a substantial weakness in the representation of colour in terms of x, y and Y coordinates. Geometric distances between coordinates of colours which

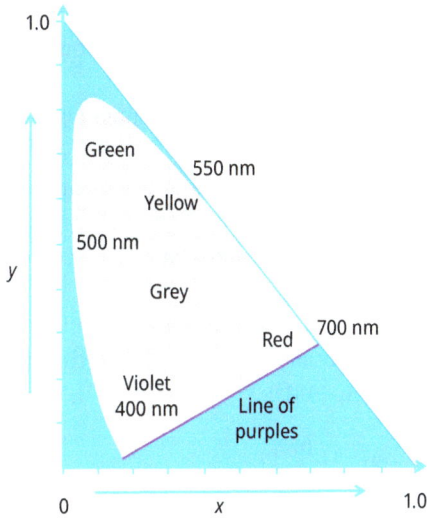

Figure 9.5: The CIE chromaticity diagram for a given value of Y.

exhibited a similar difference in tone were not the same at different places in the space and thus were dependent on the colour. While very small distances in coordinates already led to a perceived difference for violet tones, much higher distances were required to perceive a difference for green tones.

For the design of a technically useful colour space it is of high importance that mathematically calculated distances between two colours also correspond to a similar perception, independent of the colour tone. This non-uniformity of the CIE colour space required further transformation of the coordinates and led to the development of the widely used CIELAB system with the coordinates L^*, a^* and b^* (eqs. (9.8)–(9.10)).

$$L^* = 116 \left(\frac{Y}{Y_n}\right)^{\frac{1}{3}} - 16 \tag{9.8}$$

$$a^* = 500 \left[\left(\frac{X}{X_n}\right)^{\frac{1}{3}} - \left(\frac{Y}{Y_n}\right)^{\frac{1}{3}}\right] \tag{9.9}$$

$$b^* = 200 \left[\left(\frac{Y}{Y_n}\right)^{\frac{1}{3}} - \left(\frac{Z}{Z_n}\right)^{\frac{1}{3}}\right] \tag{9.10}$$

with X_n, Y_n and Z_n being the coordinates of the standard white body.

It is important to keep in mind that this transformation of the coordinates does not follow a pure mathematical approach, but bases on an empirical assessment of perceived colour differences between similar shades. The aim of the procedure is to expand and compress the colour space at the right places and in the right dimension to achieve better matching between perceived and calculated colour differences. In an ideal case a geometric distance between two colours corresponds to a similar observation of difference, independent if there are two light yellows, two pink shades or dark blues.

The value L^* stands for the lightness of a colour, white being represented by $L^* = 100$ and black by $L^* = 0$.

The a^* coordinate represents the position on the red-green axis. A positive a^* indicates a red tone and a negative value of a^* indicates green colour. The b^* coordinate describes the position on the yellow-blue axis, a positive b^* stands for yellow colour and a negative value of b^* indicates blue colour.

Besides the Cartesian coordinates L^*, a^* and b^* polar coordinates C^*_{ab} and h_{ab} are also in use. A colour then is described by L^* for lightness, the chromacity (chroma or saturation) C^*_{ab} and the hue or colour angle h_{ab} (eq. (9.11)).

$$C^* = \sqrt{a^{*2} + b^{*2}} \qquad h_{ab} = \tan^{-1}\left(\frac{b^*}{a^*}\right) \tag{9.11}$$

The length of the vector C^*_{ab} indicates the saturation of a colour. Similar to the borderlines in the CIE chromaticity diagram, the most brilliant colours (spectral lines of visible light) would define the maximum chromacity possible for a given lightness L^*. The quality of colour is defined by the colour angle h_{ab}. A value of $h_{ab} = 0°$ indicates a colour on the positive a^* axis, a value of $h_{ab} = 45°$ indicates a colour with positive a^* and positive b^*. Depending on C^* and L^* this can be light yellow, orange or dark brown colour, or even a slightly coloured grey (e.g. $L^* = 60$, $h_{ab} = 45°$, $C^* = 0.5$).

A representation of the CIELAB colour space is given in Figure 9.6.

The three-dimensional representation of the CIELAB colour space demonstrates some characteristics:

- The form of the space is not cylindrical or regular. There is an extension for colours in the range of light yellow, and there are more space for dark colours in the brown-blue shades.
- There is only one perfect white and one perfect black. However, there is already a high number of different 'blacks' with a slight coloured tone, which means that for a textile chemist real 'blacks' can have a high number of variations. Thus, from the point of view of a dyer black belongs to the group colours which is difficult to dye.
- There are no colours outside the space. The CIELAB space has been constructed artificially with input of a number of empirical factors, thus should not be understood as a purely mathematical construct. The borders of the colour space are defined indirectly by the range of wavelengths (400–700 nm) which can lead to a stimulation of the photoreceptors of the human eye.

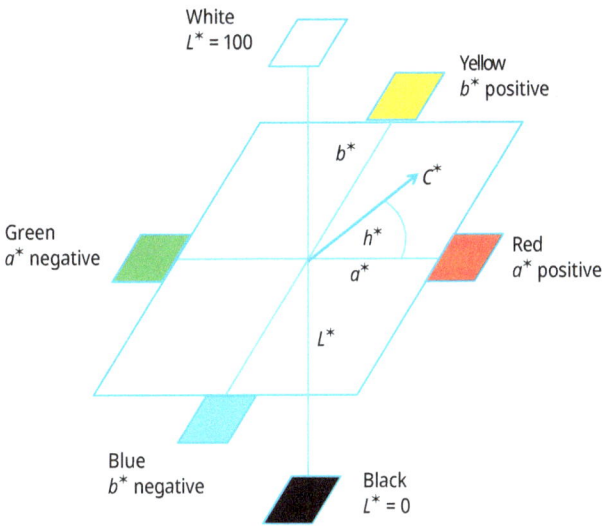

Figure 9.6: The CIELAB coordinates and a three-dimensional representation of the colour space.

An important consequence: The colour space explains the development of colour in a textile dyeing operation when using mixtures of dyes in terms of colour coordinates and indicates an important consequence for textile dyeing processes.

A dyeing with a single dye corresponds to a certain point in the CIELAB colour space. Dyeings with different colour depth of a single dye then form a line from light shades to darker shades.

A combination of two dyes yields a point which is in between the two individual lines; however, mixing two dyes does not only change the tone in terms of a^* and b^*, but also lowers L^* as textile dyeing follows the subtractive colour formation.

The dyeing of a targeted shade with use of dyestuff mixtures thus can be compared with target parachute jumping: There is no way to move upwards (in L^*) by adding dyes and there is no way to increase chromacity C^* without loss in L^*. Thus, the choice of the initial dyes to develop the targeted tone is essential, as these dyes permit a 'more easy landing' in the targeted area. Brilliant dyes with high C^* (a^*, b^*) are preferable as these dyes will permit higher flexibility in application than duller dyes (which for example may be cheaper or exhibit higher fastness).

Colour distances: In the CIELAB colour space the geometric distance between two colour coordinates is a measure of the visually perceived colour difference.

The distance is calculated from the coordinates L^*, a^* and b^* according to eqs. (9.12) and (9.13) and is called the total colour difference ΔE.

$$\Delta E = [(\Delta L^*)^2 + (\Delta a^*)^2 + (\Delta b^*)^2]^{1/2} \tag{9.12}$$

$$\Delta E = [(\Delta L^*)^2 + (\Delta C^*)^2 + (\Delta H^*)^2]^{1/2} \tag{9.13}$$

with ΔH^* being the metric hue difference $\Delta H^* = C^* \Delta h^* (\pi/180)$

The calculation of the total colour difference ΔE permits an assessment of colour differences by means of a measureable quantity. Usually a total colour difference ΔE of 1 is taken as the threshold value for colour differences to match with a given standard. The use of ΔE thus allows an automated decision for colour matching and is a central parameter for automated calculation of dyeing recipes (see the following).

However, certain limitations have to be kept in mind when ΔE is taken as the single value for a pass/fail decision.

- The uniformity of the CIELAB colour space is rather good; however, there are still deviations to perception for certain colours. Thus, large enterprises for garment processing which purchase materials from the international markets have developed more specific colour difference formulas for their important shades to achieve more reliable decisions.
- For an assessment of the absolute deviation in terms of ΔE, it is important to consider the contribution of the different coordinates, L^*, a^*, b^*. Differences of ΔE near 1, which result from a major contribution of a single coordinate such as a^*, will be visible even at a value of $\Delta E = 1$.

9.7 Measurement of reflectance curves

In measurement of transmission of a clear solution the attenuation factor of the transmitting light is determined. The measurement of the reflectance curves is also performed by spectrophotometry; however, a different approach is required to measure the reflectance of non-transparent bodies.

The general configuration of a device to measure reflectance curves contains the following elements:

- A light source which is able to provide light of sufficient intensity in the range of at least 400–700 nm.
- An illuminating and viewing device which is characterised by its respective geometric parameters.
- A monochromator which separates the light according to the respective wavelengths.
- A detector system to measure the light intensity.

The measurement of a reflectance curve is substantially more complex than the measurement of transmittance. This results from a number of factors which influence the intensity of the reflected light:

- The samples, in particular textiles, exhibit a structured surface, which is the combination of a series of structural levels: scattering pigments in fibres (e.g. TiO_2), rough fibre surface (e.g. cotton, wool) or highly lustrous filament fibres (e.g.

polyester fibres), yarn structure (e.g. filament, spun yarn), construction of the textile fabric (plain weave, twill, satin) and possible finishing (e.g. calendaring, coating). Besides scattering effects and shadow formation, reflective surfaces also alter the signal in colour measurement.

– Transparency of thin fabric leads to disturbing contributions from the background material placed behind the sample (in such cases, sample thickness has to be increased by the use of a sufficient number of layers).

– Depending on the light source, fluorescent dyes and optical brighteners contribute to the energy distribution of the reflected light.

The high number of sample characteristics which can influence the reflectance curves explains why the geometry of the measurement chamber is of decisive relevance for the results. Usually the reflected light is levelled out through the use of an integrating sphere (e.g. an Ulbricht sphere) which is a hollow white chamber which reflects light coming from the sample. Different geometric systems are in use, the major differences are with regard to the type and direction of incident light and positioning of the device to collect and measure the reflected light. Two symbols characterise the geometry used in a device: examples are 8°/d (8° angle of incident light and measurement of diffuse light), d/8° (illumination with diffuse light, measured at an angle of 8°), 0°/45° (vertical illumination at 0° angle, measured at an angle of 45°) (Figure 9.7).

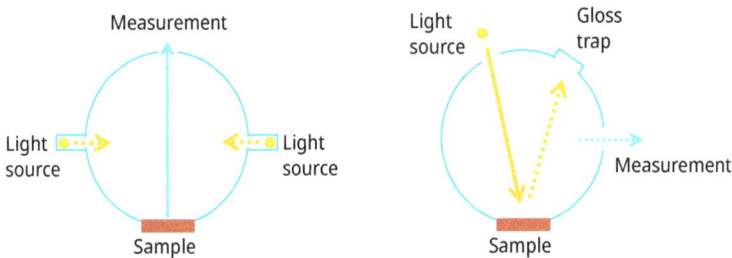

Figure 9.7: Measurement geometries d/0° and 8°/d.

In some configurations the beam obtained by direct reflection of the incident light is removed by absorption in a gloss trap. The disturbing effects of fluorescence can be minimised by the position of the monochromator. In case the monochromator is placed behind the sample chamber, the sample will be illuminated with the full spectrum of light; thus fluorescence effects will also influence the intensity of detected light. When the sample is irradiated only with monochromatic light, the effects of fluorescence through optical brightener on colour measurement can be lowered.

It is important to consider that a direct comparison of results such as colour coordinates in the CIELAB colour space which were calculated from results obtained with different measurement geometries has to be made with great care.

Less critical is a comparative assessment of smaller colour differences measured between two samples; however, for example, effects due to direct reflectance, fluorescence and sample structure can still lead to substantial disparity between measurement geometries.

9.8 Determination of whiteness, yellowness

A substantial part of textiles and also other materials (e.g. paper, paint) are produced in white colour. Due to high relevance, sensitivity of eye for small deviations in 'whiteness' and the effect of frequently used fluorescent whitening agents, the measurement of whiteness uses more specific approaches.

Standard samples for white bodies can be $BaSO_4$, MgO and special polymers (sintered PTFE, spectralon®).

In particular, when fluorescent whitening agents are used, the light source (e.g. UV content) and the position of the monochromator are of decisive influence for the measured result.

In case the sample is irradiated with monochromatic light, fluorescence will not play a substantial role and the reflectance of a sample for a certain wavelength will be measured. In case the sample is illuminated with polychromatic light and the monochromator is placed behind the sample, fluorescence due to UV light will contribute to the reflectance values measured in the visible region of light.

Due to the minimal colour contribution, only a single value is calculated as measure for whiteness. Different whiteness indices are used. The formula for calculation of a certain whiteness index includes colour coordinates with specific weight factors, thus the contribution of colour and shade is emphasised in different way. The choice of a whiteness index is also dependent on the product (textile, paper, paint) and on regional preferences.

Representative approaches used at present to describe the degree of whiteness are:

Y: The tristimulus value Y represents the luminance factor and in some cases can be used for description of the degree of whiteness. A weakness of the use of Y as single factor is that all contributions from colour are neglected.

Paper brightness: The reflectance of a sample is measured at filtered light (maximum 457 nm, width half of height 44 nm) [4].

Whiteness according to Berger: The formula which has been developed by Berger [5] considers contributions of colour components through introduction of the tristimulus values X and Z (eq. (9.14)).

$$WI_{Berger} = Y/3 + k_1 Z - k_2 X \qquad (9.14)$$

with k_1 and k_2 being constants which depend on the illumination and the measurement geometry used. In case of an ASTM method, k_1 takes the value of 3.108 and k_2 the value of 3.831 [6].

Whiteness according to Hunter and Stensby: The Hunter formula uses a direct combination of the L^* coordinate for lightness and the b^* value for the position on the yellow-blue axis (eq. (9.15)), which was then modified further by Stensby (eq. (9.16)) [7, 8].

$$WI_{\text{Hunter}} = L^* - 3b^* \tag{9.15}$$

$$WI_{\text{Stensby}} = L^* - 3b^* + 3a^* \tag{9.16}$$

Whiteness according to Taube: Another example of whiteness has been developed by Taube who took the difference between the blue and green contributions in terms of tristimulus values (eq. (9.17)).

$$WI_{\text{Taube}} = 3.388\,Z - 3\,Y \tag{9.17}$$

The differences in the formula to calculate whiteness indices are the result of individual approaches to reduce a complex property into a single number. Depending on the formula used, the influence of minor contributions of colour components varies, and it is to be considered in the assessment of a given whiteness index that application of another formula could yield different results.

Yellowness: Ageing of white samples in many cases leads to the formation of coloured components, which usually cause the appearance of a slightly yellow shade. Yellowing, for example, can be a serious problem during storage of white goods and formation of coloured degradation products due to light, nitrous oxides, ozone and temperature.

Thus, yellowness is determined by the use of specialised formulas and calculated as yellowness index. As an example, the calculation of the yellowness index YI according ASTM is given in eq. (9.18) [6]. The coefficients C_x and C_z depend on light source and geometry used for the measurement of the sample reflectance (e.g. for light source D_{65}, the values are $C_x = 1.3013$, $C_z = 1.1498$).

$$YI = 100\,(C_x X - C_z Z)/Y \tag{9.18}$$

In a simplified approach, the CIELAB coordinate b^* can be used as a measure for yellowness.

9.9 Colour strength – the Kubelka–Munk function

For transparent samples, such as an aqueous solution of a molecular disperse dye, a linear relationship between absorbance and concentration is defined by the Lambert–Beer law.

For reflectance curves, the dependency between concentration of dye and the absorbance of light is formulated according to the Kubelka–Munk function (eq. (9.19)) (Figure 9.8)

$$K/S = (1-\beta)^2/2\beta \tag{9.19}$$

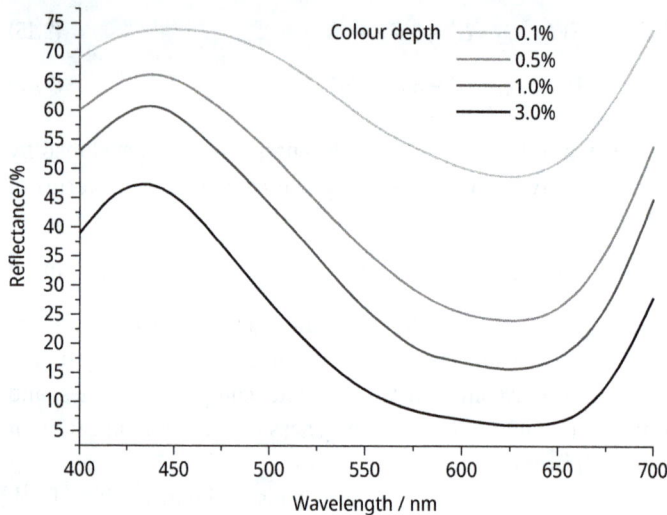

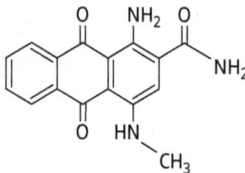

Figure 9.8: Chemical structure and reflectance curves of dyeings of a C.I. Disperse Blue 5 with different colour depth (0.1; 0.5; 1.0; 3.0 wt%) on cellulosetriacetate.

The reflection at a given wavelength is measured and given as β, where the reflection at a given wavelength λ has a value of β between 0–1.

The value K/S represents the K/S Kubelka–Munk value, which depends on the dyeing and the dyestuffs used. The Kubelka–Munk value consists of two contributions: K is an absorption constant which describes the wavelength-dependent light absorption of a dyeing and is mainly dyestuff dependent, S represents a factor for light scattering and is mainly dependent on the properties of the textile substrate. In many cases, a separation between K and S is not possible and the term K/S ('K over S') is used for further calculations.

$$\frac{K}{S} = (aK_a + bK_b + cK_c + K_w)/(aS_a + bS_b + cS_c + S_w) \tag{9.20}$$

$$\frac{K}{S} = (aK_a/S_w) + (bK_b/S_w) + \ldots(K_w/S_w) \tag{9.21}$$

In eq. (9.20), the scattering coefficients for every dye used (S_a, S_b, S_c) and the scattering coefficient for the textile material (S_w) are represented separately. The absorption coefficients K_a, K_b, K_c, K_w stand for the absorbances of each dye and the absorbance of the blank material. The coefficients a–c in eqs. (9.20) and (9.21) stand for the concentration of respective dye applied in the recipe to obtain the final colour.

When the scattering of the colourant is neglected, the formula simplifies to eq. (9.21).

For a dyeing with a single dye, eq. (9.21) represents a linear relation between the dye concentration applied and the measured K/S value, as the factor K_w/S_w is a constant value for a given material.

In Figure 9.9, an example of the linear relation between K/S value and dyestuff concentration is shown. Also in the graph, the dependency of L^* from the dyestuff concentration is shown, which is not linear.

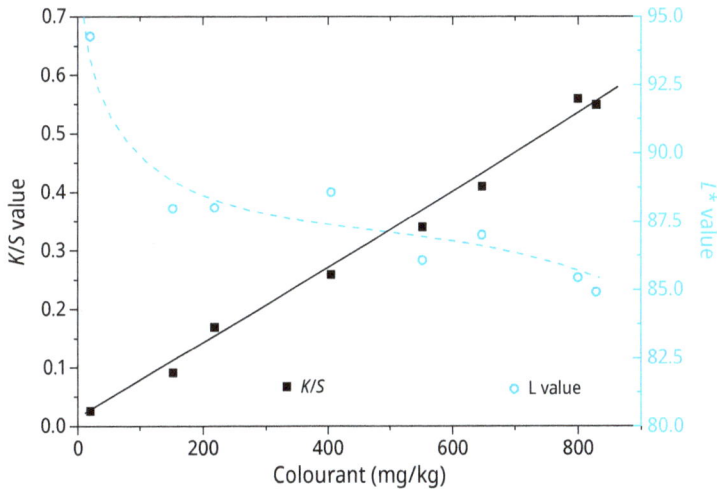

Figure 9.9: Dependency of K/S value and L^* value on the concentration of an iron-based pigment present in cotton fabric (from Kongdee and Bechtold) [9].

The CIELAB colour coordinates permit prediction of colour differences and an computerised assessment of colour matching. The K/S relation describes the dependency of the reflectance curve from the dyestuff concentration.

In case of dyestuff mixtures, additive effects can be assumed as long as dyestuff concentration is not too high and the dye molecules can be accepted to be independent.

9.10 Prediction of dyeing recipes

The Kubelka–Munk function defines a relationship between light absorption/reflectance curves and dyestuff concentration. Based on a set of calibration dyeings with individual dyes using a standardised recipe and different dyestuff concentrations, a calculation of the K/S values for given wavelength intervals between 400 nm and 700 nm can be achieved.

The respective absorbance curve can be calculated by the combination of the K/S values for each dye according to the concentrations of the individual dyes in a recipe. Starting from an initially proposed recipe, the optimum result can be obtained through iteration.

The CIELAB coordinates of the expected dyeing and the total colour difference ΔE between simulated dyeing and the coordinates of the targeted colour can be calculated from the absorbance curve. In case the colour difference ΔE is higher than 1, the analysis of the colour difference in terms of ΔL^*, Δa^*, Δb^* will indicate the direction of optimisation, for example ΔL^* too dark, too light (reduces/increases the concentration of all dyes); Δa^* reduces/increases the concentration of red/green components, Δb^* reduces/increases the concentration of yellow/blue components.

Through stepwise iteration the colour difference from the target colour can be minimised. In an ideal case the expected colour difference is smaller than unity.

This system is widely used in dyehouses for calculation of dyeing recipes, however, the results of the method have to be considered with care:

- The standard dyeings were performed with a standard recipe. The real dyeing will be performed with a recipe which is determined according to the total amount of dyestuff applied, thus recipes will often be different. A typical case is the dyeing of a dark shade with a minor dyestuff component to adjust the final tone. The chemicals will be applied in concentrations defined by the full dyestuff concentration, while calibration curves for low dye concentrations had been prepared with lower concentration of auxiliaries and chemicals.
- The liquor ratio in exhaust processes will not always be the same in particular when fibre blends are used, the liquor ratio of the individual type of fibre is different to the liquor ratio of the material as a whole.
- The standard dyeings were performed on a certain material (fibre properties, fabric structure, pretreatment, orientation of synthetic fibres) which usually differs more or less from the product to be dyed.
- The dyestuff gamut available might differ from the dyes used in the target sample, thus the overlap with the reflectance curve of the sample can be limited. Very brilliant shades can be obtained only by the use of dyes with very high chroma. Thus, if there are no dyes with high brilliance in the chosen gamut of dyes, a desired brilliant colour may not be achieved through dyestuff combination with these dyes.

Opacity: An important factor for the assessment of textiles particularly in base-layer products such as leggings and underwear is the opacity of the material. The term opacity describes the amount of light which passes through a material. While transparency of solutions can be determined through the measurement of transmission (intensity of light passing through the sample/initial intensity of light), the structure of a textile material prevents such an approach as sample positioning can also influence the outcome.

Thus, measurement of reflectance is used when a different backing material (black, white) is positioned behind the sample.

Opacity is then calculated by comparison of the lightness-dependent CIE tristimulus value Y with black and white backings (eq. (9.22)).

$$\text{Opacity} = Y_{\text{Black backing}} / Y_{\text{White backing}} \tag{9.22}$$

The value for opacity can vary between 1 (completely opaque) and 0 (completely transparent).

9.11 Gloss/lustre

Many textile samples exhibit lustre. Plain glossy surfaces can be obtained through calendaring and coating. High reflectance is also observed for filament yarns in the absence of any incorporated pigments, which lead to scattering. A distinct problem in colour measurement arises from the difficulty to differentiate between the reflectance of a white surface and the gloss.

As observers, we have learned to distinguish between lustre as directed reflectance of light following the law of reflection and the appearance of a white surface. We can distinguish between these two phenomena because our eyes receive different information with regard to reflected light and similar information for a white surface. The differentiation becomes even clearer with a slight movement of the head. The look of the white surface will be independent on the viewing position, while gloss will depend on our position with regard to the sample and light source. For example, in a printed advertisement of a new car, the glossy parts of the automobile are printed as white areas, however, we have learned that this indicates the glossy part of the surface. Hence, we know how to differentiate between gloss and white from our experiences.

Objective measurement of lustre is difficult for several reasons:

The reflectance is dependent on the angle between light source, sample and detector. Though reflectance curves can be measured at different angles, the question of interpretation still remains. Figure 9.10 exhibits reflected intensities on different plane surfaces.

In practical colour measurement, the use of an integrating Ulbricht sphere helps to overcome problems from angle dependency, and gloss traps are used to remove the directly reflected beam by a black surface. However, measurement of samples

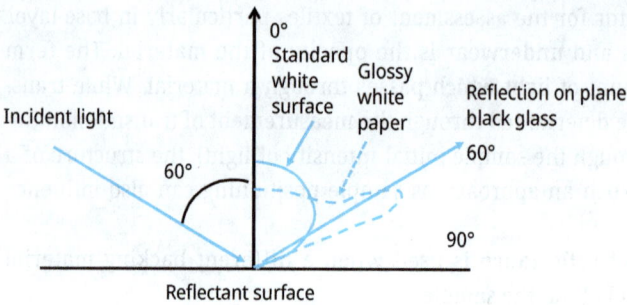

Figure 9.10: Reflectance curves on standard white, glossy white paper and a black glass plate.

with high lustre has to be considered with care, as the geometry of a measurement device used will be of high influence on the experimental result.

References

[1]   Gulrajani, M.L. (ed.), Colour measurement – Principles, advances and industrial applications, 2010, Woodhead Publishing, Philadelphia, USA.

[2]   Perez, V.V., De Fez Saiz, D., Martinez Verdü, F. Colour vision: theories and principles, Chapter 1 in M.L. Gulrajani (ed.), Colour Measurement – Principles, Advances and Industrial Applications, 2010, Woodhead Publishing, Philadelphia, USA.

[3]   Pridmore, RW. Effects of luminance, wavelength and purity on the color attributes: brief review with new data and perspectives. COLOR Res. Appl. 2007, 32/3, 208–222.

[4]   ASTM, E 985-97. Standard Test Method for Brightness of Pulp, Paper, and Paperboard (Directional Reflectance at 457 nm), American Society for Testing and Materials, West Conshohocken, PA, 2007.

[5]   ASTM, E 313-05. Standard Practice for Calculation Yellowness and Whiteness Indices from Instrumentally Measured Color Coordinates, American Society for Testing and Materials, West Conshohocken, PA, 2005.

[6]   Berger, A. Weissgradformeln und ihre praktische bedeutung. Die Farbe. 1959, 8, 187–202.

[7]   Hunter, RS. New reflectometer and its use for whiteness measurement. Op. Soc. Am. 1960, 50, 44–48.

[8]   Stensby, PS. Optical brighteners and their evaluation. Soap Chem. Spec. 1967, 43/7, 80.

[9]   Kongdee, A., Bechtold, T. The complexation of Fe(III) ions in cellulose fibres – A fundamental property. Carbohydr. Polym. 2004, 56, 39–44.

Take home messages

To describe colour, we have to distinguish between the physical description of light absorption, reflectance and the perceived colour.

Physical description is possible through measurement of reflectance curves and is thus independent of physiological responses or type of illumination.

The perceived colour is a combination of three factors: wavelength-dependent energy distribution of the light source, the reflectance curve of the sample and the wavelength-dependent sensitivity of the observer.

Perceived colour is described by the coordinates L^*, a^*, b^* in the three-dimensional CIELAB colour space. In this space, colour differences can be described by the geometric distance ΔE between two colours, which permits an automated assessment of colour matching.

The concentration dependent change in reflectance curve is described by the Kubelka–Munk relation, which represents a linear relationship between concentration of dye/pigment on the fabric and reflectance at the wavelength of maximum absorption.

The structure of a sample (thickness, opacity, gloss, roughness) as well as the geometry of the device used for measurement of reflectance, all together will influence the measured value of reflectance.

Quiz

1. Which of the two colours is lighter and more brilliant: 1. $L^* = 78$, $a^* = -18$ $b^* = -12$; 2. $L^* = 45$, $a^* = 15$, $b^* = 8$. What colour we will percieve?
2. Why whiteness can be described with a single coordinate? Which coordinate should be considered in the description of yellowing?

Exercises

1. Calculate the colour differences for the following pairs of dyeings, which dyeings you would accept: Sample $L^* = 65.2$, $a^* = 4.2$, $b^* = -6.1$, experimental result: $L^* = 64.6$, $a^* = 3.7$, $b^* = -6.3$; sample $L^* = 48.2$, $a^* = -6.2$, $b^* = 7.1$, experimental result: $L^* = 48.3$, $a^* = -5.5$, $b^* = 6.5$; sample $L^* = 72.2$, $a^* = 12.2$, $b^* = -3.1$, experimental result: $L^* = 71.1$, $a^* = 11.0$, $b^* = -2.7$.

Solutions

1. (Question 1): Number 1 is lighter and more brilliant, the shade is turquoise, as we have a negative value for a^* and b^*. Number 2 will be a darker orange thus it is near to brown.
2. (Question 2): The colour space terminates with a single point at $L^* = 100$, which is the perfect white. As the contribution of colour coordinates a^* and b^* is rather low for white coordinates, the description of the level of whiteness in one coordinate is

sufficient for the major part of applications. For the description of yellowing, b^* coordinate is important.

3. (Exercise 1): The values are inserted into the formula for the colour difference (1) $\Delta E = 0.81$ difference in colour < 1, ok; (2) $\Delta E = 0.93$ difference < 1, ok; (3) $\Delta E = 1.68$ >1, thus deviation is too high. Actually you should check the perceived difference at different light sources to confirm this, possibly the difference is a border case and still acceptable.

List of abbreviations/symbols

X,Y,Z	tristimulus values
L^*	CIELAB coordinate for lightness
a^*	CIELAB coordinate red-green
b^*	CIELAB coordinate yellow-blue
C^*	CIELAB coordinate for chroma
h^*	CIELAB coordinate for hue
YI	yellowness index
x, y, z	chromaticity coordinates
$\overline{x}_\lambda, \overline{y}_\lambda, \overline{z}_\lambda$	standard observer response
ΔE	total colour difference
WI	whiteness indices
K/S	Kubelka–Munk value

10 Dye chemistry

10.1 Overview

The application of pigments in the mass colouration of man-made fibres has been discussed briefly in Chapter 2 as this process comes under fibre production.

The application of dyes in textile dyeing operations can be performed at any stage of the textile production. Based on the material to be processed different equipment is required, which will be discussed in more detail in Chapter 13.

The general principles of dyestuff chemistry in application will be discussed in this chapter, as chemical reactions during application for example of a reactive dye or a vat dye in principle will remain similar and independent on the respective process chosen in the production line [1, 2].

Often intimately blended fibres have to be dyed or combinations of yarns made from different material have to be processed. In a few cases, a dyestuff system is able to dye different classes of fibres; however, in the majority of processes two dyeing processes have to be used in combination. In such a case, the properties of the most sensitive fibre component determine the conditions that can be used without damaging of the material (e.g. polyester/wool, cotton/viscose and polyamide/viscose).

In the following discussion of dyeing methods, the focus is placed on dyeing processes that are of substantial technical importance; thus, special processes or dyeing concepts with minor relevance are mentioned only very briefly or missed out.

10.2 Disperse dyes

10.2.1 Dyeing processes with disperse dyes

Disperse dyes are used for dyeing of polyester fibres and selected dyes can also be used for dyeing of polyamide fibres and acrylic fibres.

In advance to the dyeing process, the polyester fibres require some preparatory treatment:
- Spin finishes, anti-static preparations, grease and dirt have to be removed by appropriate washing procedures.
- In case of a woven fabric sizes have to be removed, which had been added for strengthening the warp yarn to withstand the abrasive stress during weaving.
- Non-relaxed synthetic polymer fibres still remain under internal tension. Thus, uncontrolled relaxation of yarn and fabric can occur during later washing procedures, with the result of low-dimensional stability.

https://doi.org/10.1515/9783110795738-010

Thus pretreatment steps before dyeing include desizing, washing and thermal relaxation/fixation under controlled dimensions of the textile goods. The fixation/relaxation can be achieved through hot water treatments, steaming processes or hot air treatment (e.g. on a tenter).

Usually the whiteness index of the fibres is high enough to make bleach processes unnecessary.

From the point of dyestuff chemistry, the dyeing process for polyester fibres mainly has to consider physical–chemical aspects. The dyeing process in the simplified form follows a general scheme:
- Addition of disperse dye
- Mobilisation of disperse dye (dissolution and sublimation)
- Sorption on the fibre surface
- Diffusion into the fibre matrix
- Reductive clearing of the dyed material

The dyeing process occurs via the monomolecular form of the dye that can be present in dissolved, molten or gaseous state. The process can be performed from an aqueous dyebath, as hot air-driven thermosol process.

The solubility of a disperse dye in water is in the range of 20–50 mg/L. For dyeing of a shade with 2 wt% colour depth, which means that the amount of dye used is 2 wt % of the dyed goods, 20 g dye is used per 1 kg of goods. When a liquor ratio of 1:10 is used for the dyeing, the concentration of dye in the dyebath is at 2 g/L, which is nearly 100 times above the maximum solubility of the dye.

As a chemical characteristic, disperse dyes do not contain ionic groups that would provide good solubility in water, but at least contain polar groups, for example, a hydroxyethyl group that increases the low solubility of the dye molecule.

A high number of polar groups could increase solubility in water; however, at the same time the partition coefficient according to the Nernst-type sorption isotherm is reduced, which means that more dyestuff remains in the dyebath as part of the sorption equilibrium.

Among other factors such as rate of dissolution, molecule size of the dye and the dyeing rate of a disperse dye will depend on the actual solubility under dyebath conditions. When the concentration in solution is very low, only a small concentration gradient can be built up between the dyebath and the fibre surface, which then slows down the diffusional transport processes (compare Chapter 6).

Representative examples for disperse dyes are given in Figure 10.1.

Polyester dyeing from aqueous solution is usually performed as exhaust process. In this batchwise dyeing technique, the material is immersed into the dyebath, which contains the dispersed dye.

In case the pretreatment had been performed on the same machine, the material is already wetted. Temperature of the dyebath is raised rather rapidly as long as the

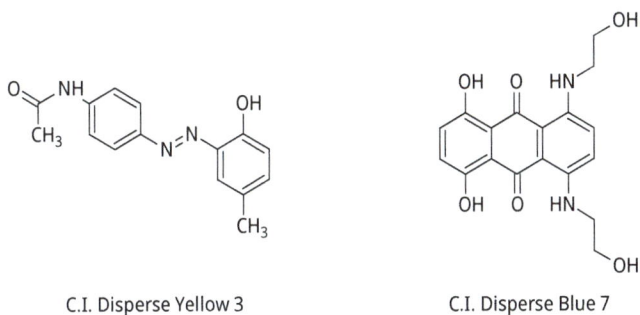

Figure 10.1: Representative examples for disperse dyes.

C.I. Disperse Yellow 3   C.I. Disperse Blue 7

bath temperature is substantially below the glass transition temperature T_g of the fibres, which in case of a standard polyester fibre is near to 65 °C.

Above this temperature, the disperse dyestuff commences to diffuse into the fibre matrix. Thus, from this temperature, the dyeing rate and dye exhaustion have to be controlled by accurate temperature control.

The dyeing rate is a dyestuff specific characteristic, thus to obtain uniform dyeing combinations of disperse dyes have to be selected with regard to similar dyeing rates. Even dyestuff exhaustion requires an appropriate dyebath circulation, respectively, movement of the textile goods to guarantee equal exhaustion at every site of the material. As both mechanical movement and dyebath circulation are subject to technical limitations, the control of the exhaustion rate by temperature control is of decisive importance.

As a guideline, the dyeing rate in a yarn dyeing apparatus should not exceed 2% of the total amount of dye within the time needed to circulate the full bath once through the goods. In yarn dyeing, a specific dyebath circulation of 35 L per 1 kg of goods and minute is a representative value. At a liquor ratio of 1:10, then 1 kg yarn is dyed in 10 L of dyebath, which circulates 3.5 times through the goods to be dyed. An exhaustion of 2% may be achieved within 1/3.5 min, respectively, 7% exhaustion of the dye initially present in the bath may be permitted within 1 min. Thus, the time for total dyestuff exhaustion (100%) should not be shorter than 14 min.

The dyebath temperature is then increased to a temperature of 115–120 °C, to obtain full dyebath exhaustion and to support diffusion of the dye in the fibre.

In many cases, the extraction equilibrium of the dyestuff in polyester fibres follows a Nernst-type sorption isotherm.

The high dyebath temperature requires the use of high-temperature (HT) dyeing apparatus that allows work in pressurized vessels.

Once the dyeing equilibrium has been established, the dyebath temperature is reduced to 70 °C and the dyebath is released.

A representative dyeing diagram for the dyeing process is shown in Figure 10.2.

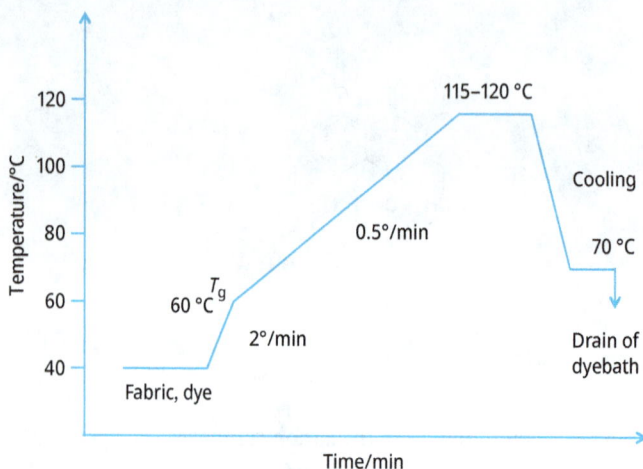

Figure 10.2: Representative dyeing curve for HT polyester fibre dyeing with disperse dyes.

Polyester fibre–wool blends require more gentle dyeing conditions due to the lower chemical stability of the wool fibres. In such cases, or when the dyeing process should be performed in non-HT equipment, for example, a open jigger, a jet or an overflow dyeing apparatus, lower dyeing temperatures have to be used. In such cases, the addition of carriers with the aim to reduce T_g of the polyester fibre is required. While chlorinated aromatics had been widely used as carriers today less toxic and less polluting systems are in use, which are esters of benzoic acid, for example, benzyl benzoate, methyl salicylate and dimethylphthalate.

Auxiliaries that are used in the dyebath for HT-dyeing are a dispersing agent (e.g. a non-ionic surfactant such as an alkyl-polyethyleneglycol), a levelling agent, a weak organic acid to adjust pH 4–6 and in case of a carrier dyeing the respective amount of carrier system.

A critical issue in HT-dyeing is the release of low-molecular-weight polyester oligomers from the polyester fibres into the dyebath. These oligomers are soluble in the dyebath at elevated temperature and precipitate at lower temperature, which for example can be sites with less bath circulation. Deposition of oligomers on valves of the dyeing apparatus can lead to technical problems due to malfunction of the equipment. Deposition of oligomers on the dyed goods can occur during cooling down of the dyebath and lead to visual grey-white depositions on the material. Use of appropriate dispersants helps to stabilise the oligomers in the dyebath in dispersed form.

Thermosol dyeing is another method to transfer disperse dyes from the solid phase into the fibre matrix. The thermosol dyeing is performed as a continuous process.

The disperse dye is padded on the fabric, which is then pre-dried rapidly in an infrared dryer. The pre-drying with infrared irradiation intends to minimise migration, which is the technical term for a redistribution of dispersed dye particles with the movement of the evaporating liquid phase. The sensitivity of a dye to exhibit migration is also dependent on the particle size of the dispersed dye. Based on drying conditions, the different dye particles move to the fabric surface which leads to uneven dye distribution and irregularities in final colour.

Water-soluble polymers can act as anti-migration additives through increase of the viscosity of the padding liquid, in particular, during the phase of drying and thus prevent redistribution of dispersed dye particles.

After pre-drying the partially dried material is dried completely with hot air and then heated up to 200 °C for 30–60 s. At 200 °C the dispersed dye melts or dissolves in the surface layer of the auxiliaries present on the fibre, selected low molecular weight dyes also sublimate into the fibre.

The thermosol process thus consists of two main steps: melting or dissolution of the dye, respectively, sublimation of the disperse dye and dissolution of the dye into the fibre matrix.

Instead of hot air also contact heat or superheated steam can be used to achieve the required temperature.

Main components in the padding liquor for thermosol dyeing are the dispersed dye, an anti-migration product and a dispersing agent. The pH of the bath should not be above pH 6–7.

10.2.2 Reductive cleaning

At the end of the dyeing process, the major part of the disperse dye has been dissolved in the polymer matrix of the fibre; however, a small amount of dye is still present on the surface of the fibres. In particular for medium and dark shades the presence of weakly bound dyestuff at the fibre surface can lead to inacceptable fastness properties, as this dye can be released during laundry and mechanical action.

The deposited dyestuff can be removed through a reductive cleaning process. In a representative example, the bath contains a dispersant and a strong reducing agent (e.g. $Na_2S_2O_4$/NaOH at 70 °C). The action of the reducing agent leads to two different principles for removal of the disperse dye (eqs. (10.1) and (10.2)):

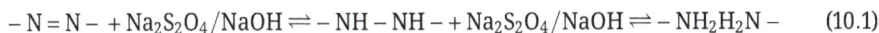

$$-N=N- + Na_2S_2O_4/NaOH \rightleftharpoons -NH-NH- + Na_2S_2O_4/NaOH \rightleftharpoons -NH_2H_2N- \quad (10.1)$$

$$AQ + Na_2S_2O_4/NaOH \rightleftharpoons AQ(OH)_2 + NaOH \rightleftharpoons AQ(O^-Na^+)_2 \quad (10.2)$$

A reductive cleavage of the azo group present, which is part of the chromophore in such disperse dyes, leads to destruction of the azo chromophore and formation of amino components that exhibit higher solubility in the treatment bath.

The anthraquinoid chromophore is stable against the action of reducing agents; however, this is reduced by chemicals into the corresponding dihydroxy form, which then dissociates in the alkaline bath as a leuco di-anion. Care has to be taken that a reoxidation of the hydrazo form or of the leuco di-anion does not occur on the fibre surface before release of the reductive rinse bath.

As the temperature of the treatment bath is below T_g of the polyester fibre, no release of dyestuff from the polymer matrix of the polyester fibre occurs; thus, the dye inside the fibres remains unchanged.

When the process is performed above T_g the exhaust equilibrium (Nernst-type isotherm) will be established with the treatment bath. The dyestuff diffuses from the fibre into the treatment bath to build up the sorption equilibrium. As the dye is destroyed continuously under the reductive conditions present in the treatment bath, a major part of the dye present in the fibre is removed. Thus, such conditions are applied when a colour discharge, respectively, lightening of a dyeing with colour deviation dyeing has to be achieved.

A special dyeing technique applicable for dispersed dyes utilises supercritical CO_2 as liquid phase. Representative conditions for application of supercritical CO_2 are 130 °C temperature and >20 MPa pressure (>200 bar). The batch process appears very straight-forward, as the solvent is removed easily by evaporation and the dyed fibre remains. The rather complex technical installation and the need for appropriate dyes are retarding drawbacks of this "water-free" dyeing technique. A possible transfer into a continuous dyeing process still appears difficult.

10.3 Direct dyes

Direct or substantive dyes exhibit a high substantivity towards cellulose fibres, which marks the difference to other apparently similar dyes (e.g. the class of acid dyes).

Chemical characteristics of direct dyes exhibit three functional elements (Figure 10.3):
-   The chromophore: This can be an azo chromophore, an anthraquinoid structure, a phthalocyanine structure, a metal complex and so on.
-   Solubilising groups, mainly sulphonate groups, are present to achieve sufficient solubility in the dyebath.
-   Polar groups, for example, hydroxyl groups and heteroelements containing functional groups provide functionalities to establish Van der Waals interactions and hydrogen binding to the substrate.

Direct dyes are usually applied from their dissolved state in aqueous dyebaths. The sorption process follows the Freundlich-type isotherm. Addition of salt increases dyestuff sorption substantially; thus, in exhaust dyeing addition of 5–20 g/L of Na_2SO_4 (Glauber Salt) or NaCl is typically made to the dyebath to increase dyestuff sorption and complete dyebath exhaustion. Also a smaller amount of sodium carbonate (Na_2CO_3, soda ash) of 1–5 g/L is used with the aim to improve levelling of the dyestuff.

The principle of the increased dyestuff through addition of salt is explained in Chapter 5 on the basis of the sorption equilibrium and the involvement of Na^+ ions.

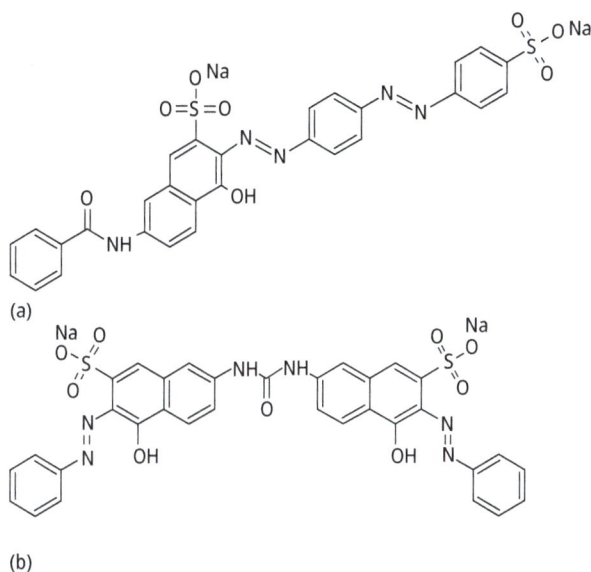

(a)

(b)

Figure 10.3: Representative structures of direct dyes: (a) C.I. Direct Red 81 and (b) C.I. Direct Orange 26.

The action of soda ash as levelling and retarding agent is due to a change in zeta potential of the surface of the cellulose fibres towards more negative values. As a result, a higher repulsion between negatively charged dyestuff and fibre surface occurs, thus reducing the affinity of the dyestuff to the fibre.

The general principle of the dyeing processes with direct dyes is based on an increase in dyebath temperature to a level where rapid adsorption and build-up of an exhaust equilibrium is achieved.

In exhaust dyeing, temperatures near to the boil are common, sometimes also HT conditions are used. In continuous processes, fixation can be achieved through steaming (e.g. 102 °C, 1–3 min) but also cold pad batch processes can be used. In cold pad batch dyeing, the dyestuff solution is padded on the fabric, which is then rested at room temperature for 8–24 h to complete dyestuff fixation.

The dyestuff binding is based only on attractive forces; thus, fastness to hot treatment in laundry is limited. At room temperature the build-up of dyeing equilibrium must be slow as otherwise poor fastness with regard to any water-based treatments will result.

In general, the dyestuff sorption is a reversible process. At the moment where conditions are present that permit mobility of the sorbed dye, an equilibrium state with the surrounding liquid phase begins to be established. For the removal of any unfixed dye, rinsing is sufficient to remove the non-bound dye.

Fastness properties towards hot treatment in aqueous phase (e.g. 90 °C washing) are limited and require special aftertreatments to reduce dyestuff mobility. Improvement of wet fastness can be achieved by different chemical principles:

- Use of polycationic substances that form an insoluble precipitate with the sulphonate groups of the adsorbed anionic dye.
- Cross-linking procedures with the use of reactive chemicals such as formaldehyde condensation products.
- Formation of metal complexes with the dye molecule. Here considerable shift in colour must be considered as part of the dyeing process. Actually this process could be better considered as an in situ formation of a metal complex dye with higher wet fastness.

Several advantages of direct dyes explain ether use despite their limited wet fastness:
- The dyeing technology is rather simple.
- Compared to other dyes the application uses little chemicals, and high exhaustion of the dyebath can be achieved. The process is thus rather clean with regard to water pollution and chemical consumption.
- Dyes with high fastness to light are available, which is of interest for applications, where this is an important requirement, for example, curtains and textiles for furniture.

10.4 Reactive dyes

10.4.1 Chemistry of reactive dyes

The characteristic property of a reactive dye is its ability to form a covalent bond between the dyestuff molecule and the fibre substrate (Figure 10.4).

Any substrate that exhibits a substantial number of functional groups and act as a nucleophilic group can be dyed with reactive dyes. Most important functional groups of fibres are hydroxyl groups ($-OH$) and amino groups ($-NH_2$).

In cellulose fibres (cotton, viscose, flax), the hydroxyl groups of the glucopyranose rings are accessible and are reactive sites; in protein fibres (wool, silk), the side groups of amino acids in the protein chain are the reactive nucleophiles (e.g. $-OH$ in serine and $-NH_2$ in lysine).

The structural characteristics of a reactive dye are:
- The chromophore system (azo, anthraquinone, phthalocyanine, etc.)
- Solubilising groups (e.g. sulphonate groups)
- Reactive anchor groups that establish the bridge between fibre and dyestuff.

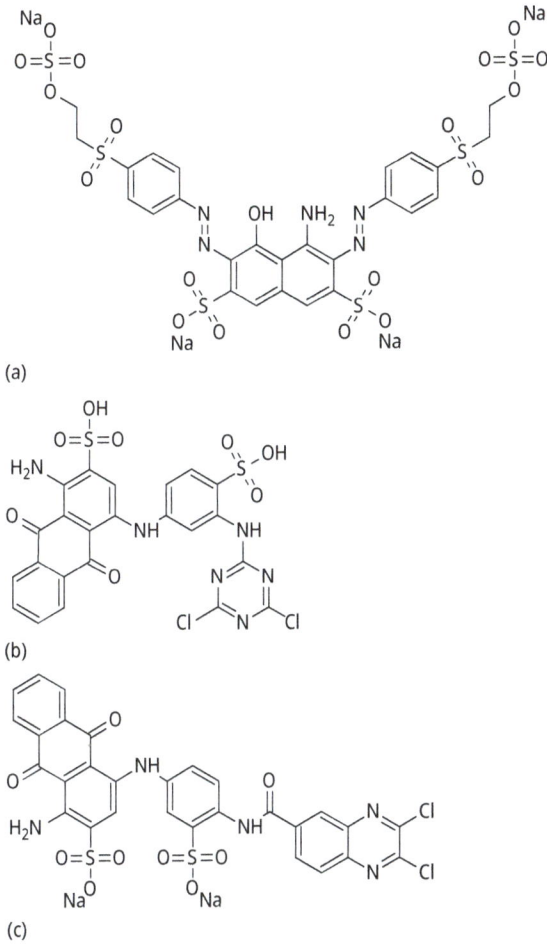

(a)

(b)

(c)

Figure 10.4: Representative structures of reactive dyes: (a) vinylsulphone type (C.I. Reactive Black 5), (b) dichloro-triazine type (C.I. Reactive Blue 4), (c) dichlorochinoxalin type (C.I. Reactive Blue 29), (d) trichloropyrimidine type, (e) chloroacetic acid and chloropropionic acid-type anchor and (f) bromoacrylamido reactive dye (C.I. Reactive Blue 69).

Two different types of reactions lead to the formation of the covalent bond between substrate and dyestuff (Figure 10.5):

In case of heterocycles with reduced π-electron density, a substitution reaction leads to the formation of an ester bridge between the heterocycle and the fibre polymer.

In case of vinylsulphone anchors (sulphato-ethyl-sulphone anchors), the reaction formally consists of two reaction steps: At first an elimination reaction of the sulphato group leads to formation of a vinyl sulphone group; as a second step, a nucleophilic addition (Michael addition) to the unsaturated bond leads to the formation of an ether bond between the dyestuff and the fibre molecule.

(d)

(e)

(f)

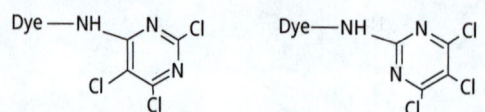

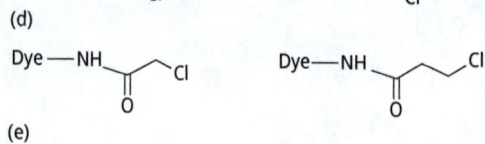

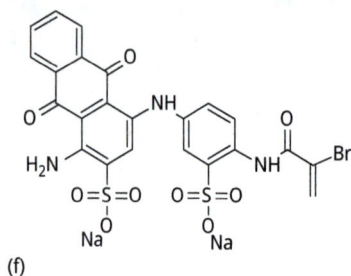

Figure 10.4 (continued)

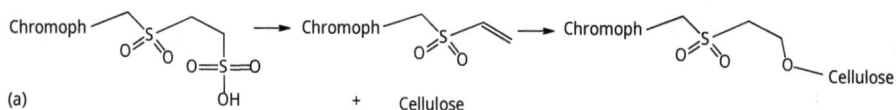

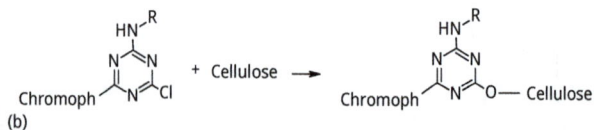

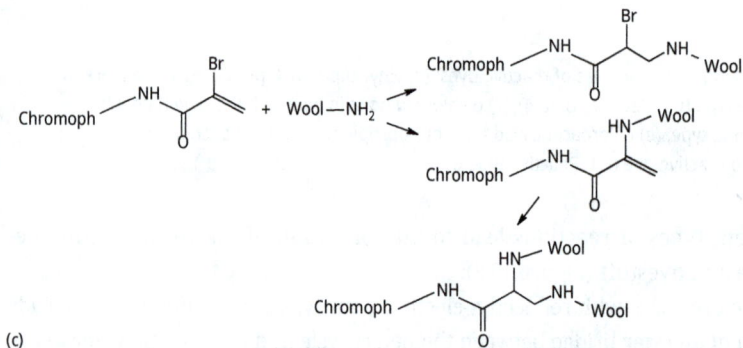

(a)

(b)

(c)

Figure 10.5: Formation of a covalent bond in reactive dyeing: (a) vinyl sulphone dye, (b) monochlorotriazine dye with cellulose and (c) a bromoacrylamido dye with wool.

The reactive dyes are water-soluble molecules that bind to the fibre in a chemical reaction. Following diffusion of the dye into the fibre structure (e.g. through voids and pores), the sorption of the dye to the fibre surface proceeds. Then the chemical reaction between dyestuff and fibre molecules occurs.

The conditions required to initiate the binding reaction are dependent on the reactivity of the anchor group, which thus also defines the field of application. Anchor groups with different reactivity are available (Figure 10.6).

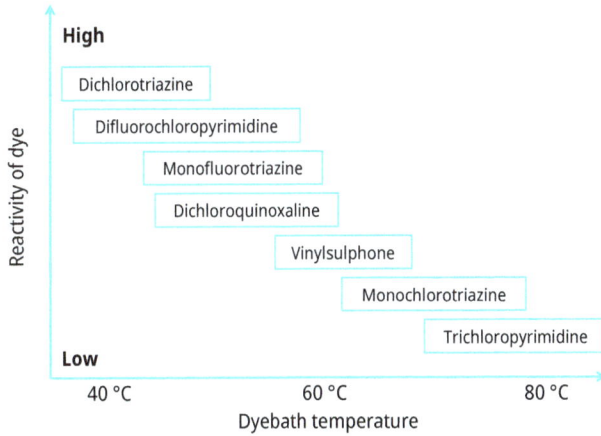

Figure 10.6: Order of reactivity of important representatives of anchor groups used in reactive dyes and temperature of application (from Leube et al. [2], Zollinger [3]).

The conditions to be applied for maximum dyestuff fixation depend on the characteristics of the fibre substrate, the substantivity of the dye and the reactivity of the anchor group.

- The reactivity and accessibility of the fibre substrate and also the chemical nature of the fibre determine the maximum amount of alkali used to initiate the binding reaction. Neutral conditions are used in dyeing of wool with a vinyl sulphone reactive dye (e.g. C.I. Reactive Black 5). Reactivity of such dyes towards the fibre keratin is sufficient to achieve dark shades in exhaustion processes. Reduced amounts of alkali are used for dyeing of viscose fibres with reactive dyes as this type of cellulose fibres exhibits higher dyestuff sorption and reaction rate, when compared to cotton cellulose.
- Temperature and pH values of a dyebath are adjusted to maximise the probability of reaction for the dye with the fibre substrate and to minimise hydrolysis. Dyebath temperature of highly reactive dyes thus must be kept lower than the reactive dyes with low reactivity of the anchor group. Such dyes permit dyeing at the boil.
- The substantivity of the reactive dyes is determined by a Freundlich-type sorption isotherm. Based on the individual substantivity of these dyes, substantial amounts

of salt are added to the dyebath to increase dyestuff sorption and thus maximise the dyestuff fixation. Concentrations of up to 50 g/L NaCl or Na_2SO_4 are used, preferably NaCl for reasons of higher solubility.

Hydrolysis: Hydrolysis of the reactive dye anchor occurs as a side reaction between the aqueous dyebath and the dyestuff. The loss of the reactive anchor lowers the efficiency of the dyeing step and reduces the colour depth obtained. In case a trichromy of dyes is used, the hydrolysis rates of the individual dyes should not differ much. Otherwise, a reduced colour depth and also a shift in shade of the colour (h*-coordinate) may occur.

Hydrolysis reactions of reactive anchors are formulated in Figure 10.7. The hydrolysis reaction cannot be avoided completely; however, for a reproducible dyeing the share of a dye which hydrolyses has to be kept constant. At the same time, for an efficient and ecological dyeing the release of high concentrations of the intensively coloured hydrolysates has to be minimised.

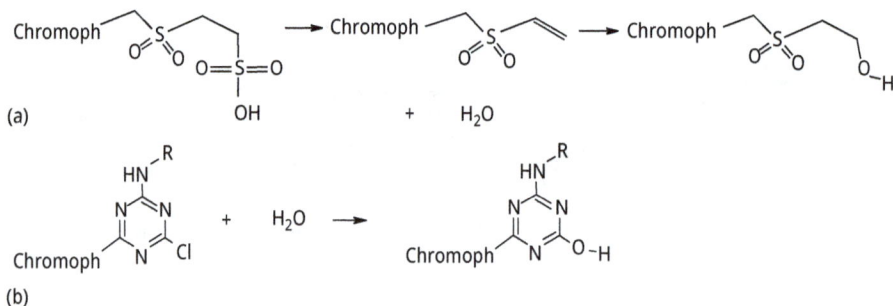

Figure 10.7: Hydrolysis reaction of a vinyl sulphone-type reactive anchor and of a monochlorotriazine anchor.

In comparison to dyes that only adsorb on the fibre substrate (e.g. direct dyes), the overall fixation reaction of a reactive dye is more complex. While the sorption equilibrium of the dyestuff itself could be described by a Freundlich-type isotherm, the overall reaction is determined by other factors:

- The fixation reaction represents an irreversible chemical reaction that removes adsorbed dye from the equilibrium between dyestuff in solution and adsorbed dye. At the moment, when the fixation reaction is initiated, for example, through dosage of alkali the sorption equilibrium is disturbed by a follow-up reaction.
- At the same time, when dyestuff fixation is initiated by the increase of pH, also the hydrolysis reaction in solution (and also in the fibre) gains importance. This reaction also influences the sorption equilibrium of the reactive dye, as dissolved reactive dye is removed from the sorption equilibrium by a competing irreversible chemical reaction.

– In practice another complication arises from the fact that dyeing processes must be finished within the limited time. Thus, full equilibrium is not reached and both the sorption equilibrium and the dyestuff fixation are influenced by dyeing kinetics.

As an estimate, the hydrolysis reaction increases by a factor of 10 for an increase in dyebath pH of single unit, which is the expected result for a pseudo-first-order reaction (assuming constant pH value and decreasing dyestuff concentration). Increase in temperature also increases reaction rates for fixation and hydrolysis by a factor of 2–3 per 10 °C.

A representative curve for dyestuff fixation is given in Figure 10.8. The dyeing process shown in Figure 10.8 is an exhaust process with addition of alkali to initiate dyestuff fixation. The required amount of salt already has been added at the start of dyeing. For the first 30 min of dyeing the unreacted dye only sorbs on the fibre. The exhaustion curve indicates that less than 50% of the dyestuff is adsorbed on the textile substrate during this phase. Compared to a direct dye or an acid dye for wool, dyestuff exhaustion is fairly low. More than 50% of the dye is still remaining in the dyebath and is thus available for both fixation on the fibre and hydrolysis. The exhaustion curve indicates that sorption equilibrium has been reached.

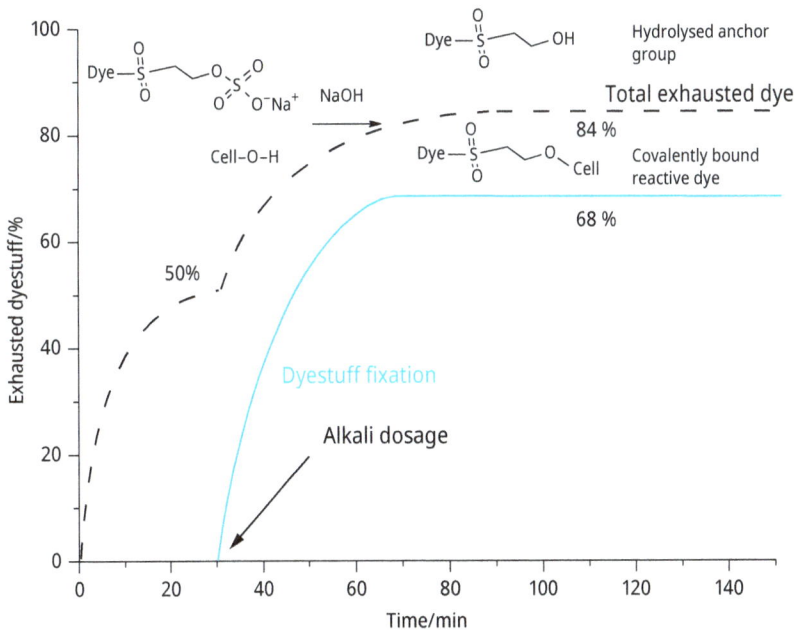

Figure 10.8: Dyestuff fixation in reactive dyeing. Exhaustion process with alkali addition through controlled dosage.

After 30 min of dyeing the dyestuff fixation is initiated by the addition of the al-kali. The binding of the dye on the substrate removes unreacted, sorbed dye from the exhaustion equilibrium, which thus stabilises through sorption of dye from the dye-bath. As a result, the fixation curve of the dye increases from initially zero to a level of 68% (in this example). In a parallel competitive reaction the hydrolysis of the anchor group leads to formation of non-reactive coloured molecules, the "hydrolysate", which also exhibits a certain substantivity to the fibres. A new second sorption equilibrium between dissolved hydrolysed reactive dyes and sorbed hydrolysed dyes is established. Thus, at the end of the dyeing process the observed degree of exhaustion (84%) does not reflect the amount of covalently bound reactive dye (68%) but also includes the sorbed hydrolysate (16%). In solution, only hydrolysate (16%) is present at the end of the dyestuff fixation process. In Figure 10.8, sorption equilibrium and dyestuff fixation are completed at 90 min of dyeing. A further prolongation of the process would not be required as no levelling phase is possible with covalently bound dyes. A scheme of the chemical reactions proceeding during reactive dye fixation is given in Figure 10.9.

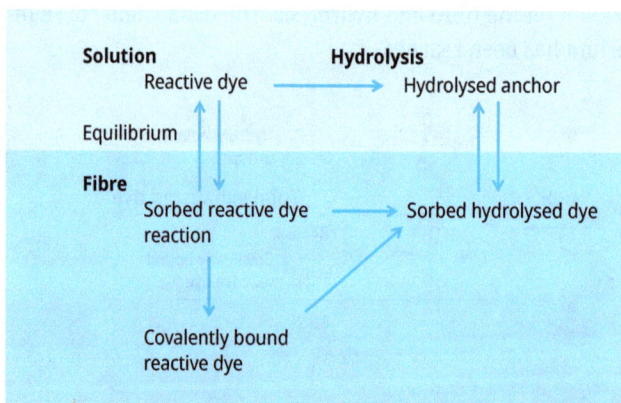

Figure 10.9: Reaction scheme for the proceeding reactions and established chemical equilibria in reactive dye fixation.

The adsorption of the hydrolysed reactive dye generates an important problem in reactive dyeing. Substantial amounts of hydrolysed reactive dye molecules are still released from the dyed goods into the rinsing baths. Thus considerable efforts are required to remove this sorbed dye from the dyed goods to avoid a negative influence on the fastness properties of the dyeing. In particular, for dark shades a higher number of rinsing baths and elevated washing temperatures are required to remove excess hydrolysate from the textile material.

In cases of certain anchors, for example, vinyl sulphone dyes, the hydrolysed molecules can exhibit lower solubility in water as a sulphate-mono-ester group has been replaced by a hydroxyl group, which increases the affinity of the hydrolysate molecules to the substrate.

To consider the appropriate dyeing conditions for a certain type of anchor group used in the dyes, different groups of reactive dyes have been formed.

Thus, for continuous cold pad-batch applications, reactive dyes with low substantivity and high reactivity are used (e.g. vinyl-sulphone-type reactive dyes). For application in exhaustion processes at higher temperatures, dyes with higher substantivity and reduced reactivity are preferably used.

Furthermore, also substrate-dependent (cellulose, wool) groups of reactive dyes are available, as the reaction conditions applicable for protein fibres require dyestuff fixation at neutral or slightly acidic pH.

Stability of the dyestuff fibre bond: As long as the dyestuff fibre bond is stable and the excess hydrolysate has been removed by proper rinsing of the dyeing, good fastness properties in aqueous systems can be achieved. Each class of anchor groups, however, exhibits specific sensitivity to hydrolysis in acidic or alkaline medium. The cellulose-ether bonding formed by vinyl sulphone dyes is very stable to acid hydrolysis, however, sensitive to hot alkaline solutions, which for example could be achieved in a hot laundry using a heavy-duty washing powder. The substitution anchors (halogen-substituted heterocycles, e.g. fluorochloropyrimidines) exhibit substantially higher resistivity against hydrolysis by alkaline solutions.

Bifunctional reactive dyes/heterobifunctional dyes: The limited degree of dyestuff fixation due to competing hydrolysis of the anchor group led to the development of bifunctional (trifunctional) reactive dyes. The use of two anchor group increases the probability of dyestuff fixation substantially. Under the assumption the dyestuff fixation follows the law of statistics, a dyestuff fixation of 60% would also correspond to the formation of 40% of hydrolysed dye through inactivation of the anchor group. In case of a bifunctional dye with identical anchor groups, the remaining 40% hydrolysate would get another 60% chance to bind to the fibre; thus, 24% of the 40% hydrolysate would form a covalent bond through a binding reaction with the second anchor group.

An important example for a bifunctional reactive dye is C.I. Reactive Black 5 (Figure 10.4.a), which contains 2 sulphato-ethyl-sulphone groups in the molecular structure.

Bifunctional dyes such as C.I. Reactive Black 5 also can form a reactive bond to the fibre structure (e.g. cellulose) with both anchor groups. In such a case, the dyestuff acts as a cross-linker, which stabilises the cellulose structure. As a result modified properties of the cross-linked cellulose fibre can be observed (e.g. with regard to fibrillation) [4].

Hetero-bifunctional reactive dyes contain two different anchor groups, which also should provide better fastness properties as the two anchor groups combine their individual chemical strength (Figure 10.10). The prerequisite for such an approach to be functional is, however, that both functional groups have been successfully linked to the fibre polymer.

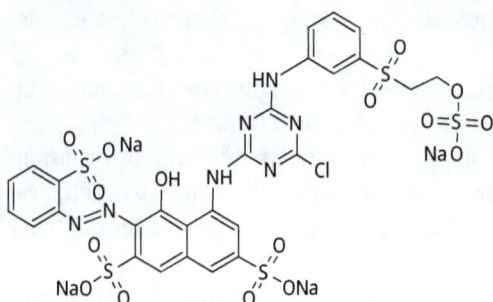

Figure 10.10: Representative structure of a heterobifunctional dye.

10.4.2 Important dyeing techniques

Through the wide range of reactivity and substantivity of reactive dyes available on the market, the application in all major dyeing processes is possible.

Exhaust dyeing: For exhaust dyeing, dyes with higher substantivity are preferable. Substantial amounts of salt are used in the dyebaths, which can range up to 50 g/L NaCl. "Short" liquor ratios of 1:10 and even shorter are in use to reduce the losses of unfixed and hydrolysed dye in the dyebath. As the hydrolysis reaction is dependent on the concentration of dye applied and the volume of dyebath used per kg of goods, a smaller dyebath volume is preferable. For a given colour depth a reduction in the dyebath volume from 1:15 to 1:7.5 will reduce the losses of dyestuff in the spent dyebath by 50%. The high concentrations of salt used (e.g. 50 g/L NaCl) lead to the release of substantial amounts of salt in the spent dyebath, for example, if 1 kg of goods is dyed in 10 L of dyebath (liquor ratio 1:10) a concentration of 50 g/L NaCl corresponds to a total consumption of 500 g (!) NaCl per 1 kg of the dyed material.

The concentration of dyestuff in the dyebath depends on the liquor ratio used (the mass of goods compared to the volume of dyebath, e.g. 1:10, 1:15) and on the colour depth to be dyed in wt% of dye per mass of goods (eqs. (10.3)–(10.5)):

$$\text{Colour depth (\%wt)} = 100 \times \text{mass of dyestuff (kg)}/\text{mass of goods (kg)} \qquad (10.3)$$

$$\text{Liquor ratio} = \text{mass of goods (kg)}/\text{volume (mass) of dyebath (L, kg)} \quad (10.4)$$

$$\text{Dye concentration (g/L)} = 10 \times \text{colour depth (wt\%)}$$
$$\times \text{mass of goods (kg)}/\text{volume of dyebath (L)} \qquad (10.5)$$

At the moment where dyestuff fixation begins, the dyeing process is not longer in an exhaustion equilibrium. Thus, fixation process has to be initiated following a levelling period and the process of fixation has to be controlled carefully, for example, by dosage of the alkali or by careful control of temperature. NaOH and Na_2CO_3 are used as alkali.

Cold pad batch: An important process in the application of reactive dyes is the semi-continuous cold pad batch dyeing. Reactive dyes with moderate reactivity (e.g. vinyl sulphone dyes) and low substantivity are applied on the fabric through padding.

The low substantivity supports the prevention of tailing, differences in colour due to dyestuff exhaustion and reduction of actual dyestuff concentration in the padding liquid.

Tailing is the result of a gradual dyestuff exhaustion by the goods that occur during the resident time of the material in the pad liquid. This is the time between dipping into the pad box and squeezing off the excess pad liquor. When dyestuff exhaustion exceeds a critical limit, the dyestuff concentration in the dyebath reduces gradually with the length of the padded material. As a result, the colour depth at the end of the batch is lighter than at the beginning. Such long-distance variations of effects during continuous treatments are called tailing.

In cold pad batch dyeing with reactive dyes, the dyestuff is padded together with the amount of alkali required for dyestuff fixation. To prevent hydrolysis of the dyestuff in the pad box the alkali solution is dosed permanently to an equivalent volume of dyestuff. The volume of the pad box is kept to a minimum, to keep dwell time in the pad box at maximum 5 min with the aim to avoid colour changes due to hydrolysis in the pad box. The maximum dwell time is dependent on the dyes used and the temperature of the pad liquid and the alkali concentration used.

The dyestuff concentrations in the pad liquor depend on the colour depth to be achieved. In exhaust dyeing, the colour depth defined in wt% dyestuff is used to dye a given mass of goods, while in continuous dyeing operations the colour depth is defined in mass concentration (g/L) of dye used in the dyebath. The colour used for dyeing a certain mass of goods thus depends on the pickup, which is the mass of dyebath padded on a given mass of goods (eq. (10.6)). In many cases, the density of the pad bath is assumed to be unity (ρ =1 g/mL), thus volume and mass of pad bath can be replaced directly. The definitions of pickup and colour depth are as follows (eqs. (10.6), (10.7)):

$$\text{Pickup} = 100 \times \text{mass (volume) of dyebath (kg, L)/mass of goods (kg)}$$
(10.6)

$$\text{Colour depth (g/kg)} = \text{concentration of dye in pad liquor (g/kg or g/L)} \times \text{pickup}$$
(10.7)

$$\text{Colour on fabric} = \text{colour depth (g/kg)} \times \text{degree of dyestuff fixation} \quad (10.8)$$

Similar to the exhaust dyeing, the amount of dyestuff used (colour depth in g/kg) is taken as a measure of colour depth, thus not considering the actual degree of fixation. The degree of dyestuff fixation, however, is dyestuff dependent, material dependent and difficult to determine in practice (eq. (10.8)). Thus all measures have to be taken to stabilise the degree of dyestuff fixation on highly reproducible level.

For a medium shade dyeing, for example, 6% in colour depth, a dyestuff amount of 60 g dyestuff is required to dye 1 kg of goods. When a pickup of 75% is achieved in the padding process the dyestuff concentration in the padder must be 80 g/L.

The high concentration of dyestuff in the padding liquid requires adaptation of the recipes.

1. For darker shades, dyes with high solubility have to be selected to avoid the presence of insoluble matter in the padding liquid.
2. Addition of urea in concentrations of up to 100 g/L supports the solubility of the dyes, however, also increases the costs for the dyeing process.
3. Instead of mixtures of Na_2CO_3/NaOH often sodium silicate (waterglass) is used in combination with NaOH to formulate a buffered alkaline dyebath.

After padding the material is rested for 4–24 h to complete dyestuff fixation. During this time, the wet material is packed into plastic film to avoid uptake of CO_2 from the ambient atmosphere, which would reduce pH and thus lead to lighter edges of the dyed fabric and also prevents evaporation and drying. During the period of dyestuff fixation, the fabric rolls are rotated to avoid settling or dripping off of dye liquor in the fabric roll, which would cause uneven dyeing and also leads to severe difficulties to roll off the fabric at the end of the fixation.

After fixation has been completed, the dyeings are washed and rinsed in continuous washing units, which depend on the material used and can be open-width or rope washing machines.

Pad dry and pad steam processes: The rate of fixation can be increased by the use of dry heat or steam. In such processes, fully continuous operation is possible as the duration of the fixation step (30 s to 2 min) fits into the design of continuous dyeing units.

In pad dry processing, the reactive dye is padded together with the alkali ($NaHCO_3$ and Na_2CO_3). Then, $NaHCO_3$ is thermally converted into Na_2CO_3 which initiates dyestuff fixation [5]. The dyestuff fixation proceeds during the heating and evaporation of the pad liquid; thus, fixation occurs from very short liquor ratio. Addition of urea is possible to increase dyestuff solubility as well as to prolong the phase of drying. As an alternative procedure also steaming instead of drying is possible. In this case, the heat transfer is achieved through steam, which prolongs the available duration for dye fixation.

In pad steam dyeing with reactive dyes, a two-step treatment is used. In the first step, the dye is padded on the fabric and dried at 110–140 °C. Then the fabric is padded in an alkali solution with high salt content (up to 250 g/L NaCl) and the required amount of alkali (dependent on the reactivity of the dyes and the colour depth). For solubility reasons, also urea can be added. The fixation of the dye is then achieved in saturated steam at 102–103 °C.

The dyed material is then rinsed as already described for cold pad batch dyeing.

An overview about the continuous dyeing processes is given in Figure 10.11.

Reactive dyes for protein fibres: In reactive dyeing of wool the stability conditions of the protein fibre define the pH range of application. Compared to the cellulose fibres, the adsorption of the dyestuff is rather slow and requires elevated temperatures. The pH value should be kept below pH 7, typical pH conditions for dyestuff fixation are between pH 4 and 6, to minimise fibre damage. Higher pH value (e.g. pH 8.5)

Cold pad-batch

Padding (dyestuff and alkali) ⟶ Fixation at ambient temp. ⟶ Rinsing

Pad-dry

Padding (dyestuff and alkali) ⟶ Fixation > 150 °C. ⟶ Rinsing

Pad-steam

Padding (dyestuff) drying ⟶ Padding (alkali) ⟶ Steam fixation ⟶ Rinsing

Figure 10.11: Main steps in continuous dyeing with reactive dyes.

would accelerate dyestuff fixation and reduce time for dyestuff fixation; however, protein hydrolysis can lead to substantial losses in fibre strength.

Silk exhibits higher stability to alkaline conditions; thus, in the exhaust dyeing of silk Na_2CO_3 can be applied at a temperature of 50 °C to initiate dyestuff fixation.

Stripping of dyeing: The stripping of reactive dyes can be achieved through both oxidative and reductive processes.

In oxidative processes, NaOCl or $NaClO_2$ can be used to destroy the chromophore. In reductive processes, for example, $Na_2S_2O_4$/NaOH leads to a reductive cleavage of the azo chromophore:

$$-N=N- \; + 2S_2O_4{}^{2-} + 4OH^- \rightleftharpoons -NH_2 + H_2N- \; + 4SO_3{}^{2-} \tag{10.9}$$

The reductive decolourisation is only possible in case of reactive dyes that contain an azo group. As important representatives contain anthraquinoid- and phthalocyanine-based chromophores, the applicability of the reductive discharge is limited.

The reductive cleavage of azo dyes can also occur during treatment in hot alkaline baths in the presence of reducing sugars (glucose). Such conditions for example can be found in washing at the boil with the addition of heavy-duty laundry detergents.

Both oxidative and reductive stripping leave parts of the dye molecule bound to the fibre structure.

Another strategy to light up failed reactive dyeings uses the hydrolysis sensitivity of the covalent bond between the anchor group and the fibre structure. Based on the anchor system of the dyeing treatment in acidic or alkaline baths at the boil can be used to remove parts of the dye from the substrate.

Applicatory profile: From the point of applicability and coloristic range, the reactive dyes offer an excellent combination of properties.

A major weakness arises from the limited fastness properties, in particular wet fastness, light fastness and combinations of wet treatment and exposure to light (e.g. weather fastness).

After treatment by addition of polycationic substances that reduce dyestuff solubility through formation of salt bridges is a method to improve fastness properties.

The limited bath exhaustion and the release of considerable amounts of hydro-lysed dye into the wastewater can cause problems with regard to heavily coloured wastewater as limits for light absorbance of wastewater from dyehouses exist [6].

10.5 Vat dyes

Vat dyes contain a quinoid structure as their characteristic sign. The dyestuff in its oxidised form is insoluble in water and most solvents, as no solubilising groups are present in the molecular structure. The reduction of the quinoid structure is an essential step in the application of vat dyes, as the reduced (leuco) form of the dye is soluble in alkaline solution. The reduced form of the vat dye is the fibre affine form that exhausts from the dyebath (Figure 10.12).

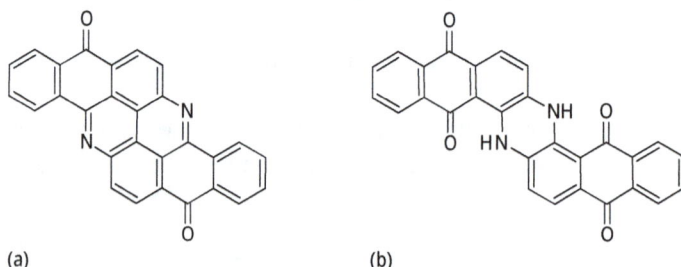

(a)   (b)

Figure 10.12: Characteristic structures of vat dyes: (a) C.I. Vat Yellow 1 and (b) C.I. Vat Blue 4.

The term quinoid structure must not be understood too strictly, as for example indigo (C.I. Vat Blue 1) also belongs to the group of vat dyes, however, has an indigoid structure. Due to its unique field of application, indigo will be discussed separately.

An anthraquinoid structure is also found in many chromophores present in direct dyes, reactive dyes and disperse dyes; however, in these classes of dyes the anthraquinoid structure only contributes to colour development and does not take a significant role in the dyeing process.

The basic steps in the application of vat dyes are summarised in Figure 10.13.

The application of a vat dye consists of the following steps:
- Application of the vat dye in insoluble, disperse form (pigmentation)
- Reduction of the dyestuff by appropriate reduction agents
- Dissolution of the reduced (leuco) form of the dye
- Exhaustion of the dye from the dyebath/adsorption of the reduced dye in continuous dyeing processes
- Levelling of the dye to achieve uniform dyestuff distribution

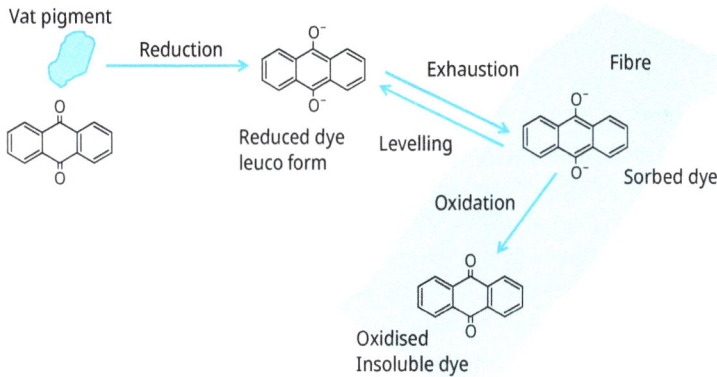

Figure 10.13: Basic steps in the application of vat dyes.

- Oxidation of the reduced dye into the insoluble pigment form
- Soaping to achieve the final particle distribution and tone
- Rinsing to remove chemicals and neutralise the material

The order of the first steps can vary, for example, in some cases the dyestuff is reduced first in the concentrated form, a stock vat is produced and then the reduced dye is added to the dyebath.

Application of the dispersed dye: As commercial form of a vat dye, usually the oxidised state is delivered either as finely milled powder or as liquid/paste form. The milling of a vat dye has to be performed to a particle size in the dimensions of approximately 1 μm; thus, special milling techniques are required to achieve a uniform particle size distribution. Bigger particles could be either filtrated on the outer layers of a yarn cone in case of yarn dyeing or lead to dark spots in continuous dyeing processes.

The dispersed form of the dye has no affinity to the material, which is favourable for application in continuous processes as the risk of tailing is low.

Stabilisation of the dispersed form in the dyebath has to be achieved by the addition of suitable dispersing agents (e.g. lignosulphonates, Setamol WS®; non-ionic surfactants).

Agglomeration of the dispersed dye would lead to non-uniform dyeing either through dyestuff filtration in yarn dyeing or deposition of larger agglomerates in continuous dyeing.

Dyestuff reduction: The reduction of the dispersed vat dye leads to the reduced form of the anthraquinone, which dissociates and dissolves in alkaline solution as leuco form of the vat dye. The reduction power of a reducing agent can be characterised by the redox potential that can be achieved in aqueous alkaline solution. In case of vat dyes, a redox potential of more than −1,000 mV (vs. Ag/AgCl, 3 M KCl reference electrode) is required to achieve stable conditions of dyestuff reduction.

The required amount of reducing agent is the sum of three factors:

Based on the amount of dyestuff used, a stoichiometric amount of reducing agent is consumed to reduce the dyestuff into the leuco form. A representative estimate of the required consumption of dithionite for vat dye reduction is approximately 200 g $Na_2S_2O_4$ (hydrosulphite, sodium dithionite) per 1 kg of commercial vat dye. Besides the dyestuff, also the oxygen initially present in the dyebath and dissolving in the dyebath at the surface of the dyebath and the oxidative load brought into the bath by the textile material have to be compensated by an excess of reducing agent.

A number of reducing systems have been developed, which produce sufficiently negative redox potential in the alkaline solution:

Two major groups of reducing agents exist:

1. Reducing agents that deliver sulphoxylic acid HSO_2H, HSO_2Na:

$$\text{Dithionite } Na_2S_2O_4 = NaOSO - SO - ONa + H_2O \rightarrow NaHSO_3 + HSO_2Na$$

$$\text{Formaldehyde sulphoxylate } HOCH_2SO_2Na + H_2O \rightarrow CH_2O + HSO_2Na$$

$$\text{Formamidine sulphinic acid } H_2NCNHSO_2Na + H_2O \rightarrow H_2NCONH_2 + HSO_2Na$$

For dithionite also the formation of a sulphoxylate radical anion $(SO_2)^-$ has been discussed as possible intermediate species with strong reducing properties.

Based on the reactivity of the reducing agent and the release of the reducing agent, sulphoxylic acid proceeds either at room temperature or requires elevated temperature.

2. Organic reducing agents, for example, hydroxyacetone and glucose:
    In general, organic carbohydrate-based reducing agents require higher concentration of alkali and an increase in temperature to achieve sufficient reduction power for application in vat dyeing processes.

Sodium dithionite: In alkaline solution sodium dithionite $(Na_2S_2O_4)$ is a powerful reducing agent which is able to achieve a reduction potential more negative than −1,000 mV (vs. Ag/AgCl, 3 M KCl) in alkaline solution.

The basic reactions of $Na_2S_2O_4$ and NaOH in vat dyeing are given in eq. (10.10) for the action of dithionite, eq. (10.11) for the reduction of the vat dye and in eq. (10.12) as the sum of both reactions:

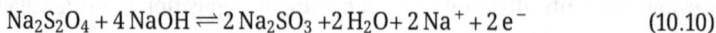

$$Na_2S_2O_4 + 4\,NaOH \rightleftharpoons 2\,Na_2SO_3 + 2\,H_2O + 2\,Na^+ + 2\,e^- \tag{10.10}$$

$$AQ(\text{vat dye}) + 2\,Na^+ + 2\,e^- \rightleftharpoons AQ^{2-} + 2\,Na^+ \tag{10.11}$$

Full reaction:

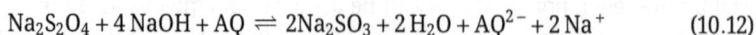

$$Na_2S_2O_4 + 4\,NaOH + AQ \rightleftharpoons 2\,Na_2SO_3 + 2\,H_2O + AQ^{2-} + 2\,Na^+ \tag{10.12}$$

The reaction of dithionite with dissolved oxygen is as follows:

$$Na_2S_2O_4 + 2\,NaOH + \frac{1}{2}\,O_2 \rightleftharpoons 2\,Na_2SO_3 + H_2O \qquad (10.13)$$

Reactions (10.12) and (10.13) indicate an important aspect to be considered in vat dyeing. To achieve stable dyestuff reduction an excess of reducing agent is used. The amount of reducing agent is dependent on the process, the equipment used and the material. Normally the non-dyestuff-dependent consumption of reducing agent for a given process is known. The reaction of the reducing agent consumes alkali; thus, the concentration of alkali decreases during the dyeing; however, the alkalinity of the dyebath must be kept sufficiently high to maintain the leuco form of the dyestuff in dissolved state during the exhaustion and levelling phase.

Sodium dithionite is widely used in vat dyeing because rapid build-up of the required redox potential and stable reduction conditions can be maintained. The chemical however is hazardous in storage and handling due to rapid decomposition and risk of self-ignition in the presence of water. Difficulties can result from dyestuff destruction through overreduction. The primary product sulphite rapidly oxidises to sulphate, which then leads to notable concentrations of sulphate in the wastewater [6].

Formaldehyde-sulphoxylate (HOCH₂SO₂Na): Sodium hydroxyl methane sulphinate (Rongalit C®) is prepared by the reaction of sodium dithionite with formaldehyde in alkaline solution (eq. (10.14)):

$$Na_2S_2O_4 + 2\,CH_2O + 2\,H_2O \rightleftharpoons HOCH_2SO_3Na + HOCH_2SO_2Na \qquad (10.14)$$

The reactivity of the formaldehyde sulphoxylate is much lower in comparison to $Na_2S_2O_4$. Thus, solutions and aqueous preparations (e.g. printing pastes) containing $HOCH_2SO_2Na$ are stable at room temperature and become activated at elevated temperature (e.g. in the hot dyebath, during steaming of prints).

When Zn-dithionite is used as a starting material, the zinc formaldehyde sulphinate is obtained $((HOCH_2SO_2)_2Zn)$. The use of a Zn-based chemical has been reduced substantially, as a substantial load of Zn in the wastewater results from these chemicals.

Formamidine sulphinic acid-Na-salt – H₂NCNHSO₂Na: The sodium salt of thiourea dioxide (formamidine sulphinic acid sodium salt) is more reactive at lower temperature when compared to the formaldehyde sulphinic acid and thus can be used to replace sodium dithionite in vat dyeing. The use of formamidine sulphinic acid instead of dithionite reduces the sulphate content in the wastewater from vat dyeing by 50%, however, increases nitrogen content.

Hydroxyacetone (HO-CH₂-CO-CH₃): Hydroxyacetone requires higher temperature and relative high alkali concentration to develop the reduction potential which is required for stable vat dye reduction. The build-up of reducing conditions is under kinetic control, thus breakdown of potential in the presence of higher amounts of oxygen occurs.

Hydroxyacetone, for example, is used for HT-dyeing of vat dyes in exhaust dyeing. As an advantage, the phenomenon of overreduction of sensitive vat dyes (e.g. C.I. Vat Blue 6) can be avoided with the use of hydroxyacetone. The strongly smelling chemical can be used to lower the sulphate concentration in the effluents, however contributes to the chemical oxygen demand (COD) load.

Glucose also belongs to the organic carbohydrate-based reducing agents and also requires higher alkalinity and elevated temperature to establish stable reducing conditions.

A common difficulty in the application of all reducing agents results from the problem to monitor the concentration of reducing agent during the dyeing process. Redox potential measurements can be used to measure the conditions in the dyebath; however, lack in reliability and weak correlation between concentration of reducing agent and measured signal height prevented a wider distribution of such a technique in practice.

As alternative reducing agents also electrochemical processes e.g. the use of alkali stable Fe(II)-complexes which can be cathodically regenerated have been proposed in the literature. In such a technique, a regenerable reducing agent is used to transfer reduction equivalents from a cathode to the dispersed dye present in the dyebath [7].

Three different groups of vat dyes such as IK, IW and IN have been formed, with regard to their behaviour in technical application.

The differences between these groups lie in the concentration of reducing agent and alkali used for dyestuff reduction, in the dyeing temperature and in the addition of salt to compensate for low affinity of the leuco form to the fibre.

IK (K = cold, *ger.: kalt*) dyes are dyed at low dyebath temperature (20–30 °C) and in the presence of salt to improve bath exhaustion. A prominent representative for IK dyes is indigo, which is dyed in completely different processes (see Section 10.6).

IW (W = warm, *ger.: warm*) dyes are applied in the temperature range of 45–50 °C. Addition of salt improves the bath exhaustion. In exhaust dyeing, levelling of such dyes can be achieved at dyebath temperatures above 80 °C; however, at the end of the dyeing process the bath should be cooled to 50–60 °C to achieve higher dye exhaustion.

IN (N = normal, *ger.: normal*). These dyes exhibit high affinity to the cellulose material; thus, no addition of salt is required. The dyebath temperature should be at least 60 °C; however, in practice, exhaust dyeing bath temperature is between 80 and 95 °C to achieve uniform dyeing. Compared to IK and IW dyeing, higher concentrations of alkali and dithionite are used, the absolute value being dependent on the dyeing process, liquor ratio and the colour depth, respectively, the concentration of dye used.

Dyestuff adsorption: Only the leuco form of the vat dye exhibits affinity to the fibre. At a bath temperature of 60 °C rapid exhaustion within 10–15 min is observed, which however is rather uneven. Thus, in batch processes a levelling phase has to follow the phase of dyestuff adsorption. The redistribution of the adsorbed dye can be accelerated at elevated temperature. In exhaust processes, the bath temperature of the levelling phase is above 80 °C.

As long as the reduction potential in the bath is sufficiently negative to maintain the dyestuff in its reduced state, a sorption equilibrium between dye in the dyebath and the dye sorbed on the fibre can be established, which is the chemical basis for a levelling of dyeing (Figure 10.14).

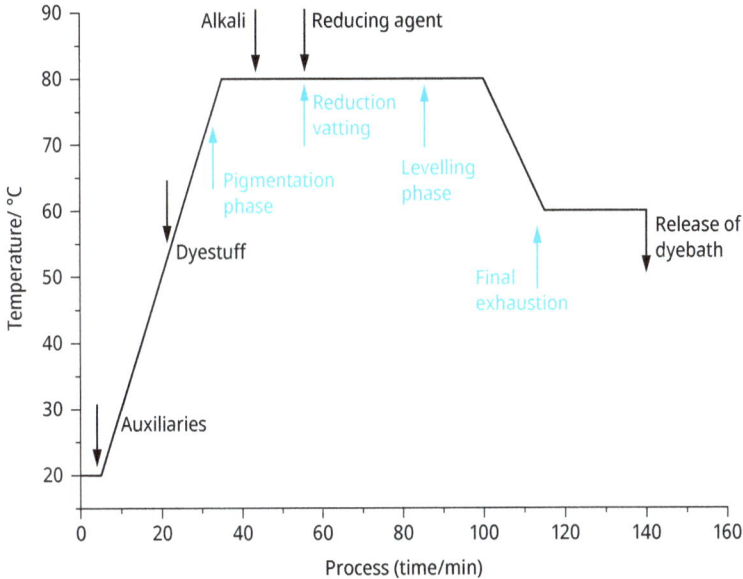

Figure 10.14: Example for a temperature diagram of an exhaust dyeing process with use of vat dyes. The first step is deposition of the oxidised pigment (pigmentation) followed by dye reduction and levelling phase.

The formation of a uniform dyeing can be supported by the use of levelling agents.

The function of a levelling agent can be through formation of an aggregate with dissolved dyestuff (dyestuff-affine levelling agent) or through competitive adsorption on the fibre (fibre-affine levelling agent). In the case of vat dyeing, auxiliaries that exhibit an affinity to the leuco form of the dye are used. The function of a dyestuff-affine levelling agent is to increase the total concentration of dissolved leuco dye present in solution. As the overall rate to achieve sorption equilibrium is dependent on the total dyestuff concentration in solution, an increase in dyestuff concentration in solution shortens the time to achieve equilibrium distribution of sorbed dye (compare Chapter 6). An important prerequisite for proper function of the levelling agent is the equilibrium constant for aggregate formation and the rate of binding and release of dye. In case the equilibrium constant is too high, substantially lighter dyeing is obtained. Thus this chemical would act as a stripping agent. In case the formation of the aggregate is irreversible, no levelling function is obtained (Figure 10.15).

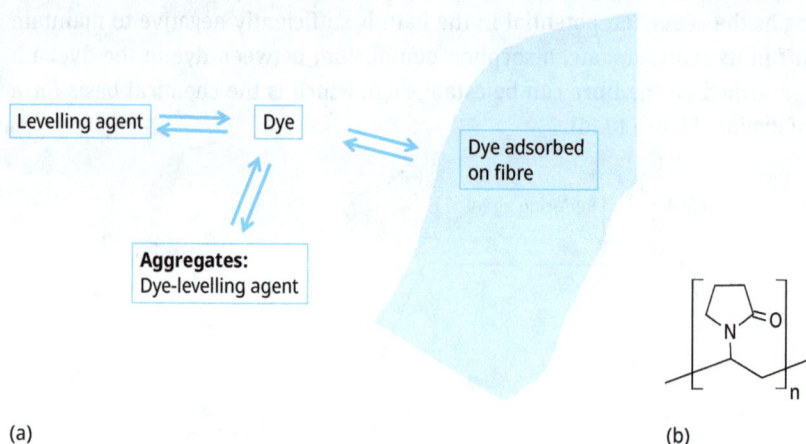

(a)   (b)

Figure 10.15: Function (a) of a dyestuff-affine levelling agent and (b) structure of polyvinylpyrrolidone (PVP).

Levelling agents can be surface-active substances (e.g. fatty alcohol ethoxylates) or polymeric compounds [e.g. polyalkylene amides, oligomeric aminoamides, polyvinylpyrrolidone (PVP)] [8].

Important examples in vat dyeing are based on oligomeric amides (Peregal P®), which are used to support dyestuff redistribution, and PVP (Albigen A®), which forms very stable aggregates with reduced vat dye and is thus used in stripping procedures.

No chemical bond exists between the sorbed dye and the fibre substrate; thus, the sorption equilibrium begins to establish at the moment, when the vat dye is reduced in alkaline solution. On the contrary to reactive dyes, thus levelling of already dyed samples is possible through redistribution of reduced dye.

10.5.1 Oxidation/soaping

At the end of the dyeing process, the dyebath is removed and following a rinsing process the oxidation of the reduced vat dye is performed.

Reproducible and complete oxidation can be achieved by means of a number of oxidants:

Oxygen (O_2) dissolved in water leads to a rapid oxidation of vat dyes, however, is only applicable in special cases (e.g. indigo dyeing) as in many cases uniformity of the oxidation is not sufficient.

Hydrogen peroxide (H_2O_2) permits rapid oxidation at low costs and without secondary problems in the wastewater. Overoxidation can lead to unwanted change in the colour of certain dyes.

m-Nitro-benzene-sulphonate (Ludigol®) is used as a mild oxidant for vat dyeing, as it permits controlled reoxidation without the risk of overoxidation, however, at higher costs compared to hydrogen peroxide. In addition, the product contributes to the COD load in the wastewater (Figure 10.16).

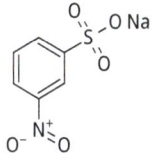

Figure 10.16: Structure of *m*-nitrobenzene sulphonic acid.

In selected applications, also the reoxidation by use of halogen-based oxidants is possible (e.g. NaOCl, NaClO₂).

Through the oxidation step the reduced vat dye is brought back into the oxidised form, which is insoluble in water. The oxidised dye desorbs, precipitates and agglomerates to the pigment form (eq. (10.15)):

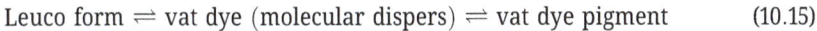

$$\text{Leuco form} \rightleftharpoons \text{vat dye (molecular dispers)} \rightleftharpoons \text{vat dye pigment} \qquad (10.15)$$

The pigment form is finely divided and agglomeration to the final particle size is achieved through a hot treatment in water (soaping). Through the change in particle size, the colour of the dyeing is changed to the final state and the excellent fastness properties are obtained as less fixed pigments become removed.

Leuco acid formation: Reduced vat dyes are soluble in alkaline solution, as long as the dissociation of the phenolic groups of the reduced quinoid system is complete. When pH is lowered, for example, through reaction with dithionite to a value below 12, the leuco acid of certain vat dyes begins to precipitate, also chemical rearrangements of dyes can occur (eq. (10.16)).

The leuco acid exhibits no affinity to the fibre, thus in selected cases the formation of the leuco acid was also used to steer the affinity of the dyestuff in exhaust processes:

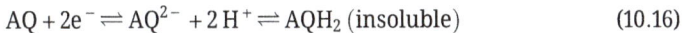

$$\text{AQ} + 2e^- \rightleftharpoons \text{AQ}^{2-} + 2\,\text{H}^+ \rightleftharpoons \text{AQH}_2 \text{ (insoluble)} \qquad (10.16)$$

Overreduction: Overreduction occurs with certain blue vat dyes (e.g. C.I. Vat Blue 4, C.I. Vat Blue 6, C.I. Vat Blue 14) in the presence of high concentrations of $Na_2S_2O_4$. A representative example for the reactions occurring in overreduction is shown in Figure 10.17. At first, the dyestuff molecule is reduced into two stages, and four phenolic groups are formed in the molecule before the irreversible molecular rearrangement begins which leads to unwanted changes in colour. Overreduction can be avoided by the use of alternative reducing agents, for example, hydroxyacetone or by addition of redox buffers (e.g. glucose, Na-nitrite).

Overoxidation: Overoxidation appears through the step of reoxidation. In particular, at high pH value the dyestuff molecule formed as a result of the oxidation step is still sensitive to further oxidation. The process is reversible and thus can be corrected by use of mild reducing agents in the soaping bath, for example, glucose (Figure 10.18).

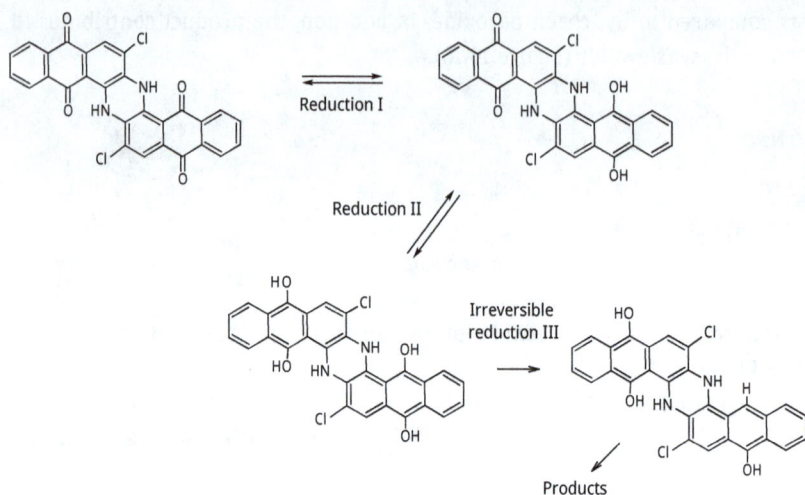

Figure 10.17: Overreduction of C.I. Vat Blue 6 (Indanthrene Blue BC). The reaction scheme has been formulated on the basis of leuco-acids instead of the dissociated leuco-anions.

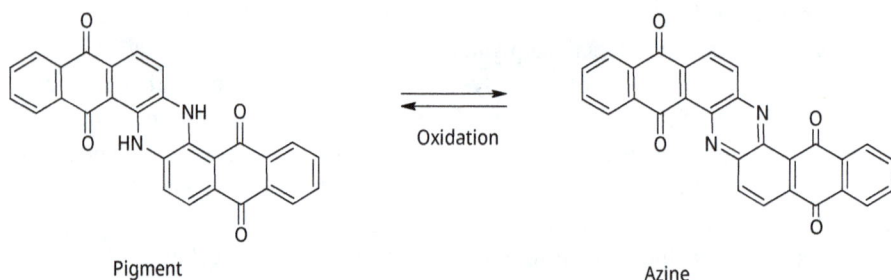

Figure 10.18: Overoxidation of C.I. Vat Blue 4 (Indanthrene Blue RS) to the azine form.

Application of vat dyes: The characteristic steps of vat dyeing can be realised with almost every dyeing equipment of both discontinuous and continuous plants. The usual order is application of oxidised dye (pigmentation), reduction and dyestuff fixation, levelling, rinsing, oxidation and soaping, and rinsing.

In batch dyeing, processes with the use of dyeing apparatus that exhibit low surface are favourable. Jet dyeing or exhaust processes in overflow or winch-type dyeing machinery requires huge amounts of reducing agent to maintain proper dyestuff reduction or use of a nitrogen atmosphere to reduce oxidative load through air oxygen.

Continuous processes mainly use pad steam techniques:

In the first step, the non-fibre-affine pigment is padded on the material in the presence of a wetting agent and a thickening agent (e.g. polyacrylate, alginate) to avoid migration of the dispersed dye during intermediate drying. Then the dried fabric is padded with an alkaline solution of reducing agent and dispersing agent. In the following steaming process (45–60 s), the dyestuff is reduced and adsorbs on the fibre. The steaming must be performed in an oxygen-free steamer, or at least low oxygen content in the steaming atmosphere has to be achieved to avoid rapid oxidation of the reducing chemicals.

At the end of the steaming process, the concentration of alkali on the fabric and the residual concentration of unfixed dye on the fabric both are reduced through a short rinse in the water seal of the steamer. The material is then rinsed, oxidised, soaped and rinsed in the washers.

A direct application of the dye in reduced form is not possible as the maximum solubility of the reduced dye in the dyebath is insufficient for the most purposes and the high affinity of the leuco-dye to the fibre material leads to considerable risk of tailing.

The only relevant exception where continuous vat dyeing is performed with a fully reduced dyebath is observed in indigo dyeing for denim production.

The fastness properties of vat dyes are superior; however, the more comparable complex application and the lack of highly brilliant red shades explain their preferred use in applications where highest fastness properties are required (e.g. green hospital clothing, colour woven fabric, working clothes and terry towels).

A particular colouristic strength of the vat dye gamut lies in the range of high fastness blue or black dyeing.

Vat dyes offer interesting approaches to bath recycling. In reactive dyeing, the non-fixed dye in the spent dyebath bears a hydrolysed anchor group and thus is not regenerable. The non-exhausted vat dye which is remaining in the spent dyebath could be collected in its oxidised pigment form through filtration. The dye could be used again for dyeing procedures; however, in many cases, use of mixtures of dyes makes dyestuff regeneration difficult.

The high bath exhaustion and presence of unfixed dye in form of the water-insoluble pigment reduces problems of coloured wastewater, which are of minor relevance.

Light-induced fibre degradation: A remarkable property of a number of yellow and orange vat dyes is their ability to increase light-induced degradation of cellulose. The light absorption of the dyestuff leads to an excited state of the dye. Different models have been proposed to explain the degradation of the fibre polymer:

- The dyestuff oxidises the substrate under hydrogen abstraction and forms a radical (eq. (10.17)).
- The photoreaction leads to formation of the leuco form of the vat dye, which then re-oxidises under production of hydrogen peroxide, which oxidises the fibre polymer (eq. (10.18)):

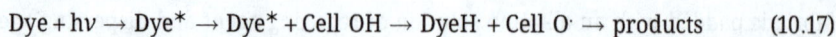

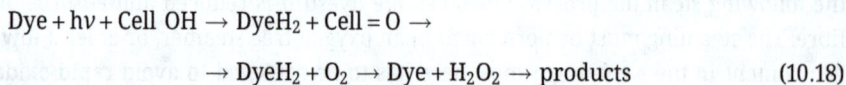

$$\mathrm{Dye} + h\nu \rightarrow \mathrm{Dye}^* \rightarrow \mathrm{Dye}^* + \mathrm{Cell\ OH} \rightarrow \mathrm{DyeH^{\cdot}} + \mathrm{Cell\ O^{\cdot}} \rightarrow \mathrm{products} \qquad (10.17)$$

$$\mathrm{Dye} + h\nu + \mathrm{Cell\ OH} \rightarrow \mathrm{DyeH_2} + \mathrm{Cell} = \mathrm{O} \rightarrow$$

$$\rightarrow \mathrm{DyeH_2} + \mathrm{O_2} \rightarrow \mathrm{Dye} + \mathrm{H_2O_2} \rightarrow \mathrm{products} \qquad (10.18)$$

10.6 Indigo

10.6.1 Synthesis of indigo

Indigo (C.I. Vat Blue 1) is one of the few examples where a natural dye has been re-placed by a chemically identical synthetic molecule. The reason for this arises from the comparable simple structure of the dye and the fact that efficient routes for the synthesis of this molecule have been developed at the beginning of the twentieth century.

Originally, the dyes were used to dye woollen clothes (e.g. for uniforms), where high fastness of the dark blue dyeing was requested. The light fastness on cellulose-based cloth, however, was substantially lower. A mark of 3 on the scale between 1 and 8 represents a typical value for indigo (with 1 being the mark for the lowest fastness and 8 for the highest level of light fastness). Thus indigo on cotton and cellulose tex-tiles was replaced by other vat dyes, which exhibited a higher level of fastness. The production of indigo was on the way to be closed down, up to the point where the phenomenon denim for jeans appeared (approximately 1970) (Figure 10.19).

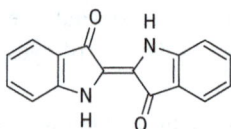

Figure 10.19: Structure of indigo, C.I. Vat Blue 1; (2-(1,3-dihydro-3-oxo-2H-indazol-2-ylidene)-1,2-dihydro-3H-indol-3-one).

Today the production of denim fabric for jeans products is the only relevant application for indigo and has reached an enormous production volume. The annual production of synthetic indigo can be estimated with 50,000 metric tonnes [9]. The annual production of indigo-dyed fabric can be estimated to be in the dimension of 5 billion units [10], which corresponds to 7–10 billion m^2 and an estimated consumption of 20% of the global cotton production (total production of cotton 25–30 Mio metric tonnes) [11].

The situation of indigo is thus unique among the vat dyes, as this dye is synthes-ised almost completely for a unique application.

A representative route for indigo synthesis is given in Figure 10.20 (developed by Heumann-Pfleger).

The characteristic sign of synthesis is the use of rather simple and thus cheap chemicals. The removal of by-products and the purification of the product are achieved

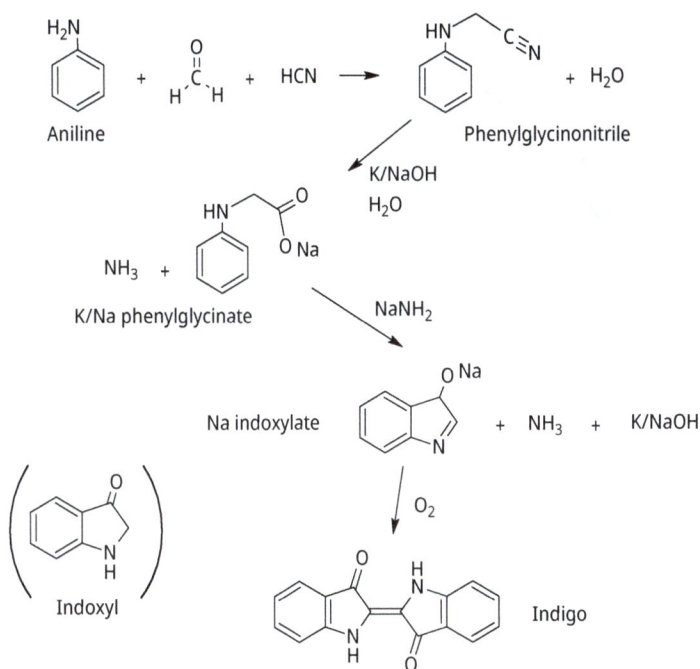

Figure 10.20: Indigo synthesis according to the Heumann–Pfleger process.

via the oxidation of the primarily formed leuco indigo to the insoluble indigo and collection of the precipitated pigment.

The product is then delivered to the dyehouses in different form. Important representatives for products are solid indigo granules (dyestuff content 90–95wt%), pastes containing indigo in dispersed form (20 – 40 wt%) and indigo solution, which contains the catalytically hydrogenated indigo dissolved in an alkaline solution (up to 40 wt%).

10.6.2 Application of indigo

Indigo dyeing for denim production is completely different to standard vat dyeing procedures.

Usually only the warp yarn is dyed in a continuous dyeing process. The fabric is then woven using an undyed weft, which gives the denim its characteristic appearance. To achieve an appropriate wash-down of the indigo dye during the later garment wash, a ring dyeing is required, where only the outer surface of the yarn should be dyed with the indigo dye (Figure 10.21).

Usually in a dyeing process, the intention is to achieve complete penetration of the material with dyestuff and distribute the dyestuff uniformly in the material.

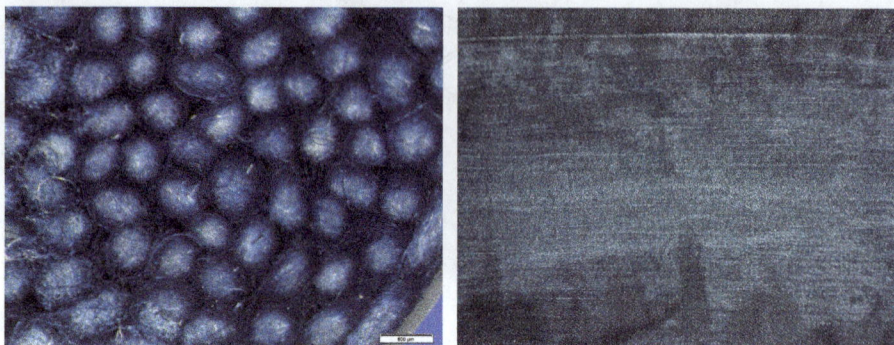

Figure 10.21: Ring-dyed yarn and wash down of fabric.

The ring dyeing requires a concentration of the dyeing process on the surface of the yarn; thus, in the dyeing every strategy which is not recommended in conventional dyeing is utilised.

– The indigo dyeing is performed on pre-wetted material; thus, the core of the yarn is already filled with water. Instead of a prewetting also an alkalisation step can be placed before the indigo dyeing. Penetration of the dyebath into the yarn thus requires replacement of the liquid which is already there.

– The indigo is present in the dyebath in the reduced state and exhibits a certain affinity to the cellulose material. Thus, time of immersion is kept short (10–15 s) to avoid penetration, and then excess dyebath is squeezed off.

– The dyebath is kept at room temperature (20–25 °C) to increase dyestuff affinity, lower bath exchange between material and dyeing box and to slow down diffusion of the dissolved leuco indigo into the core of the yarn.

– pH of the dyebath is adjusted near to pH 11.5. In this pH region, indigo is present in its mono-ionic form which exhibits substantially higher affinity to the cellulose than the di-ionic form, which is prevailing at higher pH (above pH 12.5) (Figure 10.22). At low pH (e.g. pH 10), the precipitation of leuco indigo (indigo white) occurs. This substance is almost insoluble in the dyebath, thus leuco indigo exhibits only low dyeing rate and affinity to the fibre [12].

The affinity of reduced indigo (including the mono-ionic form) is not sufficient to dye the required dark shades within one dip into a dyebath containing 1–2 g/L of reduced indigo dye. This would require much higher concentrations of indigo in the bath, which then would lead to substantial problems of rub-fastness through deposition of less fixed pigment particles and an undesired redder tone of the indigo dyeing. Thus, the indigo dyeing is performed in repetitive steps. After the first dipping into the dyebath, the excess liquid is squeezed off and the dyeing is oxidised for approximately 120 s in air (skying), then the material is dipped again into the indigo bath, squeezed and air oxidised again.

(a)

Indigo (leuco acid)
Low solubility
Prevailing at pH < 9.5

Mono-ionic form
High affinity
9.5–12.5

Di-ionic form
lower affinity
12.5 <

$pK_1 = 9.5$ $pK_2 = 12.7$

(b)

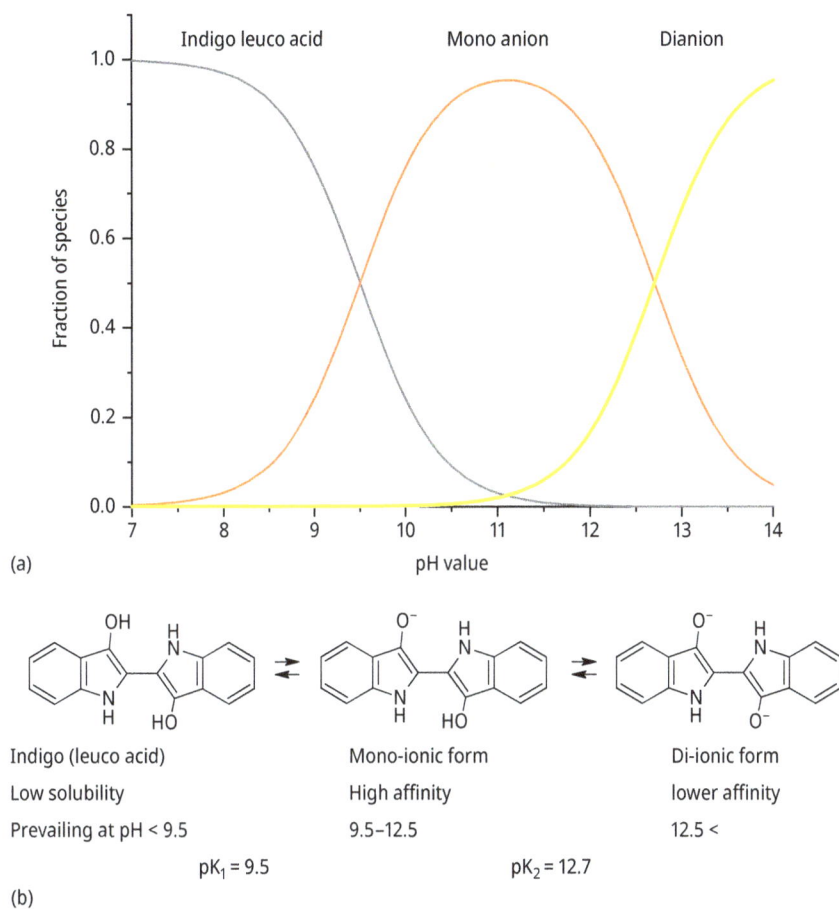

Figure 10.22: Species distribution of reduced indigo as function of dyebath pH [12].

The time of immersion into the dyebath and the conditions of the bath (low excess of reducing agent, pH 11.5, ambient temperature) prevents a reduction of already fixed pigment and thus a fresh dyeing equilibrium is established. The oxidised indigo dye deposited during the first dipping is not reduced and thus does not participate in the exhaust equilibrium. Based on the machinery and the colour depth to be dyed, this procedure is repeated up to eight times and more.

The intake of oxygen into the pad box leads to a constant oxidative load which is compensated by constant addition of reducing agent and caustic soda. The reaction of dithionite also consumes alkali, thus monitoring of the dyebath conditions (redox potential and pH) is required to maintain long-term stable dyebath conditions and to avoid tailing. The redox potential in solution must be lower than -700 mV (vs. a Ag, AgCl, 3 M KCl reference electrode); otherwise, substantial amounts of pigment will be present in the dyebath and lead to a redder tone of the blue dyeing.

Representative conditions of a commercial indigo dyeing plant are shown in Table 10.1.

Table 10.1: Representative dimensions of an indigo dyeing plant (data dependent on dimension of plant and quality of yarn to be dyed).

Production (cotton yarn)	15,000 kg/day
	11.9 kg/min
Hours of operation	21 h/day
Warp speed	35 m/min
Depth of shade	2 wt% indigo
Consumption of $Na_2S_2O_4$	50–126 kg/day
	40–100 g/min
Water consumption	3–5 L/kg of goods
	45–74 m^3/day
Composition of dyebath	
Wetting agent	0.5 g/L
Indigo	1–2 g/L
NaOH	pH 11.5–12
Temperature	20–25 °C
Redox potential	<−700 mV (vs. Ag, AgCl, 3 M KCl)

The dyebath is replenished constantly by the addition of indigo vat, alkali and sodium dithionite. As the dyeing process is a continuous wet in wet treatment the pickup of the material entering the dye box is put to lower values, for example, 65 wt% compared to the pickup of the dyeing padder, which is set to 80 wt%. The volume difference can then be used to dose indigo vat and alkali into the dyebath without bath overflow. Sodium dithionite has to be added in the solid form.

10.6.3 Chemistry of indigo reduction

Indigo has to be dosed into the dye containing box in the reduced form. A concentrated stock vat has to be prepared to handle the volume balance in the dyebox and to prevent a dyebath volume overflow, which would cost substantial amounts of chemicals and indigo dye.

Different methods for dyestuff reduction have been studied in the past [13].

Bacterial reduction

Bacterial reduction of indigo represents the oldest method to prepare an indigo vat and is still applied in traditional indigo dyeing, in particular when natural indigo is used. The reduction is executed by a bacterial consortium (e.g. *Clostridium isatidis*). The mechanism of the electron transfer from the bacteria to the dispersed indigo is still under investigation, however, may involve electron shuttles (e.g. soluble anthraquinones) [14]. Bacterial reduction requires larger volume and proceeds rather slowly. The capacity of such a bath to compensate oxygen intake through handling of the bath is rather low, however, acceptable for traditional dyeing procedures. The pH of the bath should be kept above pH 11.0 to achieve proper dissolution of indigo.

Sodium dithionite, (hydrosulphite) $Na_2S_2O_4$

Sodium dithionite is the major reducing agent used to prepare a stock vat. A typical stock vat contains 80 g/L indigo, 1 g/L dispersing agent, 70 mL/L NaOH (50 wt%) and 65 g/L sodium dithionite ($Na_2S_2O_4$). Higher indigo concentrations are critical to prepare as the solubility of the leuco indigo in the salt containing stock vat comes to a limit.

Besides safety aspects during handling of dithionite, concerns arise from the high sulphite (SO_3^{2-}) and sulphate (SO_4^{2-}) load in the wastewater.

Catalytically hydrogenated indigo

Indigo can be reduced catalytically by hydrogen on Raney nickel as catalyst (Raney nickel is prepared from a 50:50 Ni:Al alloy by dissolution of aluminium in alkaline solution) [15]. The alkaline solutions of leuco indigo are stable; however, presence of oxygen has to be avoided and the solutions have to be handled under nitrogen atmosphere. Today, 40 wt% indigo solutions mark the ecological standard in indigo dyeing, as the chemicals used in stock vat preparation have become unnecessary. The salt load (sulphite, sulphate) from indigo dyeing thus can be reduced to half the value, when compared to dithionite stock vat.

The remarkably high concentration in the pre-reduced indigo solutions is due to low salt concentrations in the product and use of Na^+/K^+ mixtures as counterions of the leuco indigo.

Hydroxyacetone ($HO\text{-}CH_2\text{-}CO\text{-}CH_3$)

Among the organic reducing agents, hydroxyacetone has been used to reduce the dithionite consumption in indigo dyeing. The limited reducing power at low alkalinity and the temperature dependence of the process required special techniques to increase reduction rate [16, 17]. As disadvantages, the product exhibited an intensive smell and also contributes to the COD load in the effluents. Reducing sugars (glucose and fructose (!)) can be used for preparation of the indigo stock vat as higher concentration of alkali and elevated temperature can be applied in this step. Such a process leads to a substantial reduction in the overall consumption of sodium dithionite in the overall dyeing process [18].

Other reducing agents

In principle, substances that can form a redox potential above -700 mV at a pH of 11.5 could be used to reduce indigo. Thus, also zinc powder, Fe(II)sulphate or $NaBH_4$ could be used to prepare a stock vat.

Electrochemical reduction

In disperse form, the indigo particles can be reduced either by direct contact to a cathode through electrochemically generated hydrogen (electrocatalytic reduction) or by means of an electrochemically regenerable redox system (a mediator) which is able to form a sufficient negative reduction potential in alkaline solution. Suitable chemical systems are based on iron complexes and anthraquinoid systems (Figure 10.23) [19].

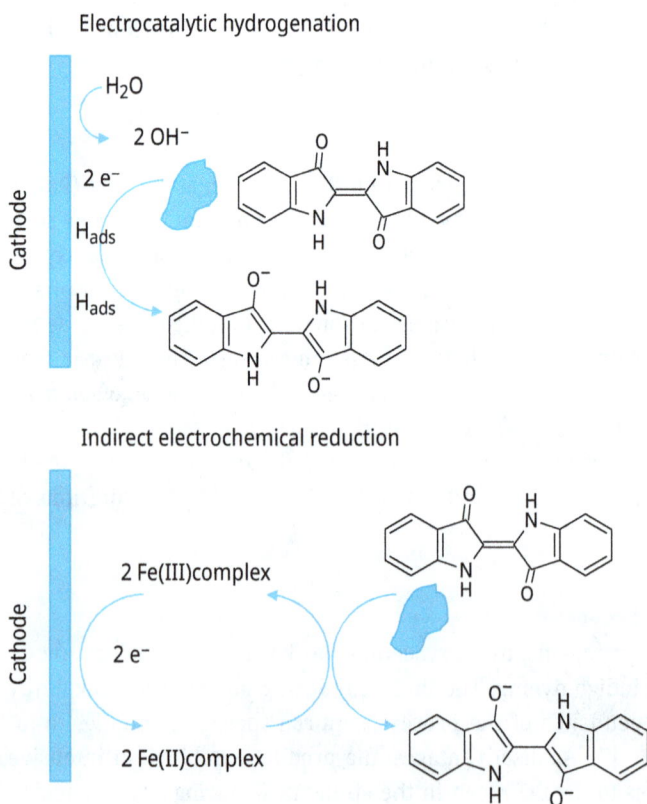

Figure 10.23: Electrocatalytic and indirect electrochemical reduction.

The direct reduction of indigo leads to a reduced product, which still requires the addition of a reducing agent in the dyeing process. In case of indirect reduction, the mediator serves as a regenerable reducing system which can also be applied in the dye box. In this case, the mediator system has to be recovered from the effluents (e.g. by ultra/nanofiltration).

Technical aspects of indigo dyeing: In technical scale, yarn dyeing with indigo is performed on plants that are designed for denim production only.

The continuous operating machinery can process the material either open width in form of the warp beam (slasher dyeing) or in form of cables in rope dyeing. Such cables, for example, have been assembled by winding together 400 yarns (400 ends). Many variations are in use; a very representative arrangement of treatments is shown in Figure 10.24.

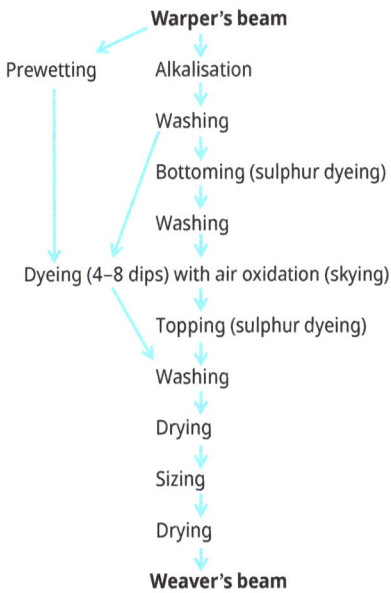

Warper's beam
↓
Prewetting   Alkalisation
↓
Washing
↓
Bottoming (sulphur dyeing)
↓
Washing
↓
Dyeing (4–8 dips) with air oxidation (skying)
↓
Topping (sulphur dyeing)
↓
Washing
↓
Drying
↓
Sizing
↓
Drying
↓
Weaver's beam

Figure 10.24: A representative example for indigo dyeing on a slasher-type dyeing range.

In slasher-type dyeing units, the sizing is performed directly after the dyeing step and the sized full warp beam is delivered at the end of the machine.

A representative scheme for the mass balances of a slasher-type indigo dyeing range using a simple prewetting and 4 dips is shown in Figure 10.25.

As shown in Figure 10.25, a water volume of 45,000 L water is released per day of operation for processing of 15,000 kg of yarn. A recovery of oxidised indigo from the wastewater is possible through ultrafiltration; however, costs of such an installation make an amortisation based on possible savings in indigo dye difficult.

The dyed indigo warp is then woven to fabric using an undyed weft to obtain the characteristic blue denim. The garment is then produced from this material.

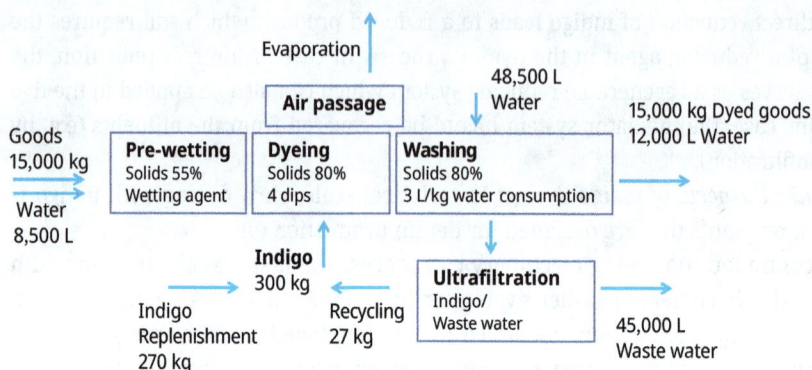

Figure 10.25: A representative example for mass and volume flow in a slasher-type dyeing range (prewetting only and 4 dips into the indigo bath).

Darker more dull shades are produced by bottoming and topping. Here an additional dyeing with sulphur dye (usually sulphur black) is added before (bottoming) or after (topping) the indigo dyeing. The disadvantage of bottoming is the constant release of sulphur black from the dyed yarn into the indigo box, which thus enriches in sulphur black.

In case of rope dyeing, 12/24/36 ropes are dyed in parallel and wound up separately at the end of the dyeing. Before sizing the rope has to be separated into single threads and then combined to a warp (rebeaming); thus, rope dyeing requires more manual work force than slasher dyeing.

10.6.4 Garment wash – fading

The raw untreated denim garment is processed further to remove indigo from the surface of the products and thus to develop the characteristic used-look appearance of the products.

Many processes are in use to light up the fabric and to remove indigo dye locally (Table 10.2).

A huge number of variations exist, as the combination of these treatments will lead to the final appearance of the product (compare chapter 12).

Backstaining: During wet treatment, redistribution of removed blue indigo dye to the undyed white weft yarn can occur, thus reducing contrast in the product and lowering whiteness of the weft yarn.

Yellowing: Through the limited light fastness of indigo and in particular in the presence of photochemically produced smog components (e.g. NO_x and O_3), a degradation of indigo to the yellow isatin occurs, which causes local degradation of the blue dye and leads to appearance of yellowing (eq. (10.19)):

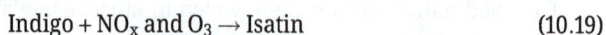

$$\text{Indigo} + NO_x \text{ and } O_3 \rightarrow \text{Isatin} \tag{10.19}$$

Table 10.2: Processes to develop the used look of denim-type garment.

Mechanical processes	Principle	Treatment
Water-based processes	Abrasion	Brushes, sand paper, (sand blasting = forbidden)
		Stonewash (pumice stone)
	Oxidants	NaOCl
		KMnO$_4$
		Ozone O$_3$
	Enzymes	Cellulases
		Oxidoreductases
Dry heat	Dyeing	Direct dyes, pigment dyes
	Reduction	Glucose, NaOH
	Light absorption	Laser treatment

10.7 Sulphur dyes

Despite the low number of individual dyes available, the class of sulphur dyes represents an important group of colourants for cellulose fibres. These dyes offer high colour strength in particular for dark blue and black shades at an acceptable level of fastness and comparable low costs. With an annual production of approximately 70,000 tonnes, the C.I. Sulphur Black 1 is most probably the dye with highest mass of production volume in terms of mass. A share of 10 wt% of the dyestuff market is estimated to be covered by C.I. Sulphur Black 1 [20].

The synthesis of sulphur dyes is based on the reaction of sulphur, polysulphides and sulphides with organic precursors in melt (180–350 °C, baking process) or solution (solvent reflux process).

The non-stoichiometric reaction leads to the formation of polymer dyestuff products, which are linked together through disulphide and polysulphide bridges. The structure of many sulphur dyes is thus proposed on the basis of the chemical structure of the organic precursor substances (Figure 10.26):

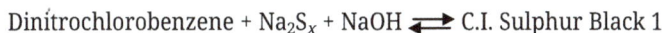

$$\text{Dinitrochlorobenzene} + \text{Na}_2\text{S}_x + \text{NaOH} \rightleftharpoons \text{C.I. Sulphur Black 1}$$

Similar to the vat dyes, the transfer from an insoluble oxidised form into a reduced soluble form is achieved through a reduction step, which opens the disulphide bridges (eq. (10.20)). In quinoneimine-based sulphur dyes also a reduction of the quinoneimine groups is possible (compare Figure 10.25). In alkaline solution, the corresponding thiolate groups are formed and the dyestuff fragments dissolve in water:

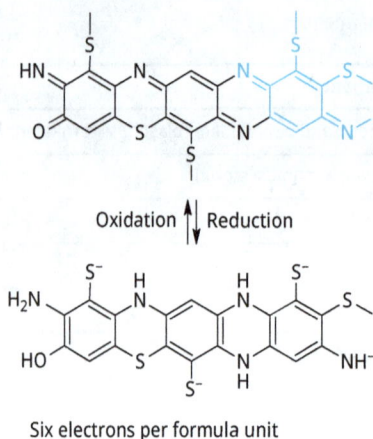

Oxidation ↕ Reduction

Six electrons per formula unit

Figure 10.26: The proposed structure of C.I. Sulphur Black 1. From Bechtold et al. [21].

$$R - S - S - R' \rightleftharpoons R - S - H + H - S - R' + 2\,NaOH \rightleftharpoons R - S^- Na^+ + R' - S^- Na^+ \qquad (10.20)$$

The redox potential required to reduce sulphur dyes with −500 mV is substantially lower than for indigo (−700 mV) or for vat dyes (−950 to −1,000 mV, all potentials vs. Ag/AgCl. 3 M KCl reference electrode). Thus also other reducing agents that establish a redox potential of −400 to 500 mV already are sufficient to achieve stable reducing conditions in the dyebath:

- Na$_2$S (sodium sulphide), NaHS (sodium hydrogensulphide, sodium sulphhydrate), NaHS$_x$ (sodium polysulphide) in NaOH at pH >10
- Glucose, hydroxyacetone or other reducing sugars in combination with alkaline solutions (NaOH)
- Combinations of glucose and dithionite also can be used

Despite the ecological concerns with regard to the use of sulphide-based reducing system, these dyeing results are still favourable in terms of colour yield, shade and rub-fastness that can be achieved.

An alternate route to reduce sulphur dyes is the direct cathodic reduction. As sulphur dyes can transport more than one reducing equivalent per formula unit, these dyes can serve as their own mediator systems to achieve cathodic dyestuff reduction and to build up the required redox potential in the dyebath [22].

Higher concentrations of alkali used in glucose-based systems increase the risk of bronzing, which also goes ahead with lowered rub fastness (bronzing = formation of finely divided oxidised dye which leads to a redder appearance of the dyed surface).

In sulphur dyeing, a gradual reduction of the polymeric dye molecule into lower molecular weight fragments occurs. The size of the dye fragments formed determines the affinity to the cellulose fibre. Above a certain redox potential the dyestuff is not dissolved completely (e.g. >−500 mV) however to intensive reduction (e.g. <−700 mV) leads to the formation of low-molecular-weight fragments, which then show reduced affinity to the fibre and lead to lighter dyeings in this case [21].

Powerful reducing agents lead to reduction in colour depth, a phenomenon called overreduction.

The usual dyeing process for sulphur dyeing is either performed at temperatures between 80 and 110 °C for 30–60 min as exhaust processes or for 30–60 s in pad steam process at a temperature of 102–105 °C. On the contrary to vat dyeing processes, the solubility of sulphur dyes permits the continuous padding of dyestuff and reducing chemicals in the same bath.

A particular application is the sulphur dyeing on indigo dyeing ranges for production of black denim (bottoming or topping). In this case, the dyeing is performed as a continuous process by immersion of the yarn/rope in one to three dye boxes containing the reduced sulphur dye at 80–90 °C bath temperature. Total time of immersion into the dyebath is around 60 s.

At the end of the dyeing process the dyebath is washed off to lower pH value and the oxidation of the dye is initiated.

In case of vat dyes, a reoxidation occurs at single dye molecules. In sulphur dyeing bridges between thiolate groups of different dye fragments have to be formed to regain the full size of the chromophore. Thus in case of sulphur dyes controlled oxidation is even more important.

A number of oxidants have been used in the past, some of them are not longer acceptable with regard to environmental issues. Relevant processes include:

– Oxidation with H_2O_2 and acetic acid (pH 4–4.5) or in slightly alkaline solution.
– Oxidation with bromate ($NaBrO_3$) at pH 4–4.5 (critical with regard to handling and release of BrO_3^-). The limit for bromate in drinking water has been set in Germany with 0.01 mg/L.
– Chlorite- and hypochlorite-based processes require careful process control. This process is critical with regard to AOX in the wastewater. The wastewater limit for AOX content has been set in Germany with 0.5 mg/L.

The application of $K_2Cr_2O_7$ potassium (or sodium) bichromate in diluted acetic acid (pH 4–4.5) is mentioned here for completeness only, however, should not be used any more for ecological reasons. The limit for Cr(III) in the wastewater in Germany has been set as 0.5 mg/L. No Cr(VI) is permitted to occur in wastewater from sulphur and vat dyeing.

Problem arises when the oxidation had not been terminated at the stage of the disulphide groups, but had continued to the formation of sulphonate groups (eq. (10.21)). Then the presence of such groups led to problems with reduced wet fastness of the dyeing and acid-catalysed damage of the fibre polymer:

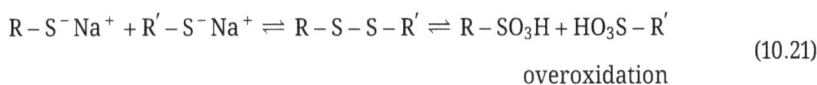

$$R - S^- Na^+ + R' - S^- Na^+ \rightleftharpoons R - S - S - R' \rightleftharpoons R - SO_3H + HO_3S - R'$$

$$\text{(10.21)}$$

$$\text{overoxidation}$$

The oxidised dyeing is then rinsed and washed at elevated temperatures of 60–95 °C to remove any weakly bound dye.

In case of sulphur black dyeing, a buffer is added to the last rinse bath (Na-acetate, soda) to buffer and neutralise any sulphuric acid which is formed during storage of the dyeing at elevated temperature and high humidity.

Groups of sulphur dyes

Based on the initial form of the dyestuff, three different types of sulphur dyes can be distinguished:

Sulphur dyes (C.I. Sulphur dye) are present as amorphous solid dyestuff or aqueous dispersion which has to be transferred into solution state by a reducing step at elevated temperature.

Pre-reduced sulphur dyes (C.I. Leuco sulphur dye) contain already a certain amount of reducing agent and are present in the concentrated reduced form. The amount of reducing agent in the formulation is rather high, thus based on the application, no further addition of the reducing agent is required.

Water-soluble sulphur dyes (C.I. Solubilised sulphur dye) have been chemically protected in reduced form through formation of a Bunte salt (eq. (10.22)):

$$\text{Dye} - \text{S}^- + \text{NaHSO}_3{}^- \rightleftharpoons \text{Dye} - \text{S} - \text{SO}_3\text{Na (Buntesalt)} + \text{H}^+ + 2\,\text{e}^- \tag{10.22}$$

$$\rightarrow\ +\ \text{Na}_2\text{S}_8 \rightarrow \text{C.I. Sulphur Black 1} \tag{10.23}$$

Through formation of a Bunte salt, a water-soluble dye is obtained which is insensitive against oxidation by air. During dyeing, the protective group is removed by the addition of a reducing agent (e.g. sulphide) and the free sulphur dye is released (eq. (10.23)).

A special group of sulphur dyes is called sulphur vat dyes. An important representative is the Hydron Blue R® (C.I. Vat Blue 43), which is synthesised in a similar process such as C.I. Sulphur Black 1. Instead of 2,4-dinitrophenol in this case, 4-[(9*H*-carbazol-3-yl)amino]phenol is heated with Na_2S_8 in butanol (eq. (10.24)):

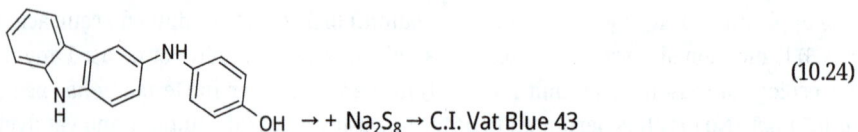

$$\tag{10.24}$$

OH → + Na$_2$S$_8$ → C.I. Vat Blue 43

With regard to its constitution, the dye clearly belongs to the sulphur dyes. However, a more negative reduction potential is required in the dyebath to achieve proper dyestuff reduction. $\text{Na}_2\text{S}_2\text{O}_4$ is required as a reducing agent; thus, such dyes are called sulphur vat dyes.

Relevant applications

The most important representative for sulphur dyes is C.I. Sulphur Black 1, which is used in a limited number of applications:

- Black denim: Only the warp yarn is ring dyed on indigo dyeing ranges and combined with undyed weft. Typical concentrations of commercial sulphur dye in the dyebath then range from 100 to 200 g/L.
- Garment dyeing: The full piece of clothing (e.g. a trouser) is manufactured first and then dyed in special dyeing machines (e.g. drum dyeing machines).
- Exhaust dyeing on jet: The conventional exhaust dyeing process requires either an inert nitrogen atmosphere in the dyeing machine to reduce air oxidation on the surface of the dyebath or substantial amounts of reducing agent have to be added to remove the oxygen inside the machine (1 m^3 of air consumes 1.4 kg $Na_2S_2O_4$ and 1.4 L of 32 wt% NaOH).
- Pad steam continue dyeing: Conventional continuous pad steam processing is applicable.

Further important applications for sulphur dyes are:

- Dyeing of velvet in brown, blue and black shades
- Dyeing of working clothes (typically blue- and grey-coloured working clothes are dyed with sulphur vat dyes)

10.8 Acid dyes and metal complex dyes

A common property of these dyes is their mechanism of interaction with the fibre polymers, which leads to strong binding of the dye to the material.

The binding of the dye at least partly can be understood to follow an ion-exchange mechanism. Ionic bonds are formed between charged groups present in the fibre polymer and the charged dyestuff molecule. The dyes are important for colouration of protein fibres (wool, silk) and polyamides, which contain ammonium groups at low pH. The amount of ammonium groups in wool can be assumed with 850 mmol/kg, in silk with 250 mmol/kg and in polyamide with 30–50 mmol/kg.

These cationic groups serve as binding anchor to build an ionic bonding with negatively charged groups, for example, sulphonate groups in the acid dyes and metal complex dyes.

The anionic groups in such dyes thus have two functions:

- To achieve water solubility of the dye
- To build an ionic bond linkage to the fibre

Besides the ionic bond, also secondary attractive forces (Van der Waals forces) contribute to strengthening of the dyestuff–fibre interaction (Figure 10.27).

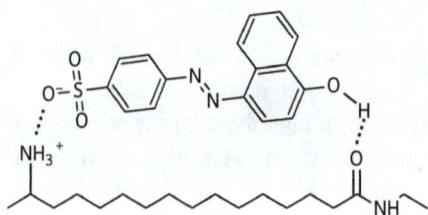

Figure 10.27: Binding of an acid dye.

In case of metal complex dyes, also the coordination of the centre ion can contribute to the dyestuff binding strength.

Dyestuff sorption follows a Langmuir-type isotherm, thus also saturation effects can be observed. The dyestuff sorption bases only on reversible equilibrium reactions; thus, inappropriate conditions (e.g. high pH and high alkalinity) will weaken the dyestuff binding and lead to an unacceptably high release of dye. Thus low washing temperatures are required to keep the rate of desorption low and only neutral to weakly acid pH is allowed in washing.

The protonation of amino groups leads to formation of ammonium groups. Acidic pH also reduces the dissociation of the carboxylic groups that are present as carboxylic acids. Thus, the charge of the protein fibre depends on the pH of dyebath. Thus, for a given temperature, sorption of an anionic dyestuff will be higher with lower pH value.

Based on their affinity to protein fibres, dyes have been assorted into different classes (Table 10.3).

Table 10.3: Colouristic groups of acid dyes.

Affinity	Bath pH	Molecular size of dyestuff	Levelling
Low	Acidic	Small	Good
Average	Weakly acidic	Average	Moderately
High	Neutral	Large	Poor

In practice, a combination of temperature and pH control is applied to control bath exhaustion of the acid dyes.

The affinity of the dyes can be controlled by addition of salt (sodium sulphate) (for detailed explanation of the mechanism, compare Chapter 4). Also levelling agents are added in particular in case of dyes with high affinity. Fibre-affine levelling agents such as anionic surfactants (alkyl polyglycol ether sulphates, polycarboxylates) reversibly block the binding sites on the fibre. Dyestuff-affine cationic levelling agents form soluble aggregates with the dye (polyglycol ethers, fatty amines, fatty alkyl polyamines). Also amphoteric compounds that combine both effects are used.

In metal complex dyes, the chromophore is built from an organic ligand and a metal centre ion.

Depending on the structure of the complex we distinguish between 1:1 metal complex dyes and 1:2 metal complex dyes (Figure 10.28). The numbers describe the ratio between the centre ion and ligand. Cr or Co are used as most important centre ions. A special case of metal complex dyes are chrome dyes. These are acid dyes that are transferred into a chromium complex during the dyeing procedure.

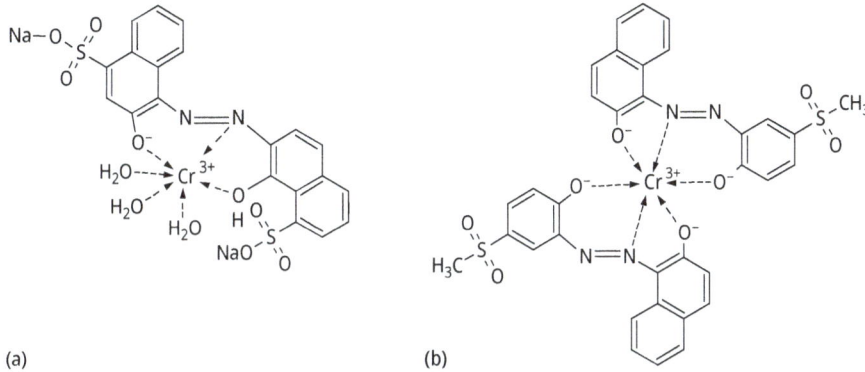

(a)   (b)

Figure 10.28: Representative structures of metal complex dyes: (a) C.I. Acid Blue 158 and (b) C.I. Acid Violet 78.

Besides the presence of sulphonate groups ($-SO_3Na$) water solubility of 1:2 metal complex dyes is achieved through incorporation of sulphonamide groups ($-SO_2NH_2$) or methylsulphone groups ($-SO_2CH_3$).

The different molecular size leads to differences in dyeing conditions. The smaller 1:1 complexes are dyed at lower pH (2.5–4) while 1:2 complex dyes are dyed in the range of pH 4.5–5.5.

In continuous processing padding of the acidic dyestuff solution followed by steaming and rinsing can be applied. In case high dyestuff concentrations are used, for example, e.g. for dyeing black, the maximum solubility of the dye may be exceeded in the padder box. Under such circumstances, part of the dye is present in dispersed form. During the steaming process, the dyestuff concentration in the padded liquid reduces through dyestuff fixation in the fibre, which allows the dispersed metal complex dye to dissolve and adsorb on the fibre.

The metal content of the metal complex dyes contributes to the chromium (cobalt) concentrations determined in the wastewater. Thus attention has to be paid to release of unfixed metal complex dyes. A representative value for the limit of the chromium(III) concentration in the textile wastewater to be released to a communal wastewater treatment plant is 0.5 mg/L (chromium (VI) 0.1 mg/L).

10.9 Naphthol dyes

A special group of dyes is formed by the naphthol dyes (Naphthol AS®), where the azo dye is synthesised directly on the fabric. The name of this group of dyes origins from β-naphtholate as one of the first naphthol components used for the azo coupling.

The dyeing process consists of two main steps. In a first step the activated aromatic component (naphthol-component) is deposited on the fabric (= impregnation). The affinity of the naphthol component is rather low, which favours the application in continuous processes and requires attention in exhaust processes. A general scheme is shown in Figure 10.29.

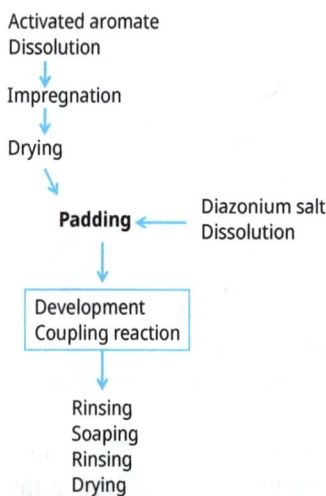

Activated aromate
Dissolution
↓
Impregnation
↓
Drying
↘
Padding ← Diazonium salt
Dissolution
↓
Development
Coupling reaction
↓
Rinsing
Soaping
Rinsing
Drying

Figure 10.29: General reaction scheme for the application of naphthol dyes.

Through the absence of solubilising groups, special conditions are required for dissolution of the naphthol component. In the first step, the naphthol component is dissolved in alcohol with the addition of alkali and in presence of protective colloids. For protection against air oxidation, some naphthols require addition of formaldehyde, which protects the naphthol through formation of a more stable intermediate compound.

The solution is then applied on the fabric either through padding or exhaust processes. In continuous processes the material is dried; in exhaust processes a controlled intermediate rinse in the presence of salt is performed to remove excess naphthol present in the dyebath. The colour depth is defined by the concentration of naphthol used. The diazonium salt is used in over-stoichiometric excess as this component is thermally sensitive and the active concentration of this component is not absolutely defined.

The azo coupling is initiated through addition of the diazonium salt (= developing of the dyeing). The developing solution is stored at room temperature to avoid thermal decomposition of the diazonium salt. Dispersing agents are added to stabilise released dyestuff lake in finely dispersed state. The aqueous solution must be buffered carefully. The buffer system is chosen according to the optimum coupling pH of the components used. The buffer capacity is calculated for a chosen colour depth used and is designed to neutralise alkali present in the impregnated textile material.

The coupling reaction usually proceeds rapidly and can be observed to complete within few seconds to minutes. Excess pigment (colour lake) is removed through rinsing and hot washing (soaping). The hot treatment also modifies particle size of the insoluble azo dye and thus develops the final shade of the dyeing.

The solubility of the components used for the azo coupling results from the dissociation of the phenolic hydroxyl group in the naphthole and from the dissociated diazonium group initially present in the diazonium salt.

Thus, after neutralisation the final product does not carry functional groups that contribute to water solubility and the dye is practically insoluble in water.

Selected combinations of naphthol dyes thus exhibit very high fastness, which reaches the level of vat dyes.

A particular strength of naphthol dyes is in the colour range of yellow, red, blue and black dyeings.

Selected combinations of different naphthols can be dyed in combination, where the same diazonium salt is used in surplus. Lack of diazonium salt would lead to kinetically determined selection of the more reactive naphthol. In this case, then, variations in shade would appear, dependent on the local conditions prevailing during coupling.

10.10 Cationic dyes/basic dyes

The general characteristic of a basic dye is the cationic group, which leads to formation of an ionic bonding to anionic groups present in the fibre polymer.

Representative examples for basic dyes are given in Figure 10.30.

The dyes contain polar groups for solubilising (e.g. amino groups); however, no sulphonate groups are present. A wide variety of chromophores is used in basic dyes.

The major application of basic dyes is in dyeing of polyacrylic fibres that have been copolymerised in combination with anionic copolymers (compare Chapter 2).

The binding of the dyestuff primarily occurs through formation of an ionic bond. In the dyebath T_g of the fibre polymer has to be exceeded to permit accessibility of the dyestuff into the fibre. Thus, temperature control is of high relevance, in particular during the initial phase of rapid dyestuff sorption.

The dyestuff binding in the fibre follows a Langmuir-type isotherm. In case the dyestuff concentration offered in the dyebath exceeds the limit for fibre saturation S_F (fibre saturation value), no further dyestuff sorption will be possible.

Depending on the molar coefficient of extinction and the concentration of dye used only, a certain share of S_F will be used in practice. In case combinations of dyes

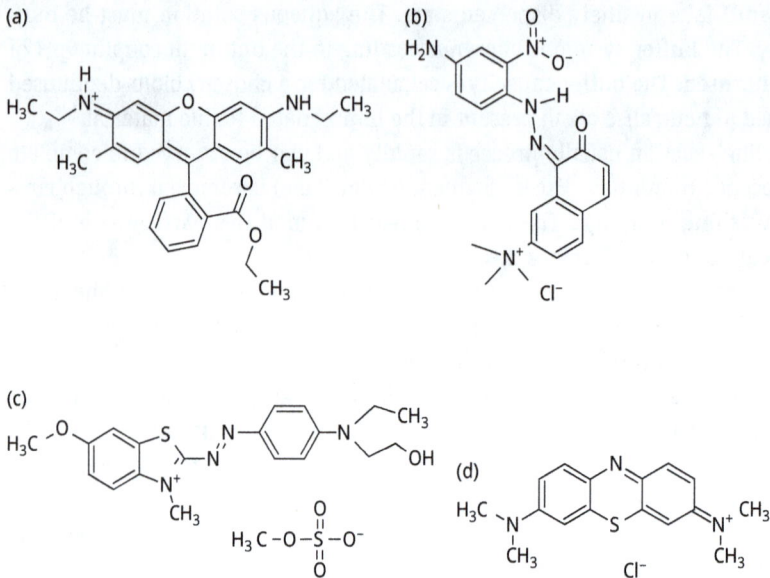

Figure 10.30: Representatives for basic dyes: (a) C.I. Basic Red 1, (b) C.I. Basic Red 118, (c) C.I. Basic Blue 41 and (d) C.I. Basic Blue 9 (Methylene blue).

are used, the sum of these shares must not exceed S_F, because the sorption capacity of the fibre will be insufficient and an uncontrolled amount of each dye will remain in solution. As a result, colour variations and low levelness of the dyeing will result.

A scheme of a polyacrylic dyeing is shown in Figure 10.31.

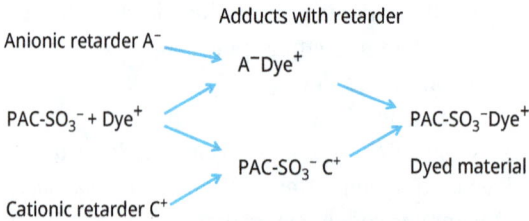

Figure 10.31: Functional scheme of a dyeing with cationic dyes and retarders.

Retarders can be added to the dyebath to support levelling of a dyeing. Anionic retarders reversibly form soluble adducts with the dye in solution. Thus, the effective concentration of dyestuff c_s in the concentration gradient for dyestuff adsorption is lowered (compare Chapter 6) and the rate of dyestuff sorption is reduced. The strength of the adduct must be lower than the binding strength between fibre and dyestuff; otherwise, the dye will not be released completely from the adduct and substantially lighter dyeings will be obtained. Condensation products between naphthalene sulphonic acid and formaldehyde are chemical representatives for anionic retarders.

Cationic retarders (e.g. quaternary ammonium compounds) compete for the adsorption and ionic binding at the fibre. Apparently, the number of binding sites on the fibre S_F is reduced and the rate of dyestuff sorption is lowered. The cationic retarder thus acts like a colourless dye, which also blocks a share of the available binding sites. In the calculations of fibre saturation through dyestuff binding, the concentration of cationic retarders used has to be considered. In case the amount of retarder has been chosen to high, the dyestuff sorption will remain incomplete.

Dyeings on polyacrylic fibres can be obtained with high colour depth, high brilliance and good fastness.

The cationic dyes can be applied in exhaust processes as well as in continuous dyeing processes.

In exhaust dyeing careful temperature control in the range of high dyestuff absorption rate is required.

As an example, the dyeing can be started at 60 °C and heated up to 70–75 °C rapidly. The actual interval of temperature is dependent on the fibre type. Then very slow and well-controlled increase in temperature with a gradient of 0.25–0.5 °C/min is required to maintain rate of dyestuff sorption low. At the boil, the dyestuff sorption is completed within 20–60 min, based on concentration of dye and exact bath temperature.

A special dyeing method with cationic dyes is performed at the site of fibre formation during the wet spinning process. Directly after removal of the spinning solvent by water, the fibre is still present in the gel state. In this state, the fibre is highly accessible and dyestuff sorption and binding occurs extremely rapid (seconds to minutes). Thus, the dyeing process can be performed during the fibre spinning before the formed fibres are drawn and dried. During drying the gel stat collapses and irreversibly incorporates the dyestuff.

10.11 Natural colourants

10.11.1 General aspects

The major principle of colour development in nature results from absorption of electromagnetic irradiation in the visible region of 400–700 nm. However, in nature colour also results from other principles, such as refraction of light (rainbows) or interference (butterfly wings) (compare Chapter 7).

In the group of natural colourants, we can summarise all kinds of coloured material which can be obtained from natural sources.

Based on the source of the natural colourant we can distinguish between minerals (red ochre, α-Fe_2O_3), plant-derived materials (e.g. flavonoids, Canadian golden rod) and colourants sourced from animals (e.g. indigoid colourants from molluscs).

In the group of natural colourants, both coloured pigments and dyes are found. While a huge number of plant extracts are heavily coloured only a restricted amount of these compounds can be used as dyes, as a specific affinity to the textile substrate is a requirement.

Natural colourants can be grouped according to their chemical structure into a few main classes [23]:

- Flavonoid dyes (including anthocyanins)
- Naphthoquinoid and anthraquinoid dyes
- Indigoid dyes
- Tannins
- Carotenoids
- Chlorophylls.

Usually the natural colourant is extracted from the plant source by an aqueous extraction step. Solvent extraction usually leads to substantial losses of organic solvent, which have to be considered in the assessment calculations of the ecological profile.

While the dyestuff content in a synthetic dye range is high and ranges from 40 to 95 wt%, the dyestuff content in natural sources is relatively low. Usually only a few wt % of the dry plant material is suitable as colourant and considerable amounts of raw material have to be extracted and then disposed.

The non-selective extraction process leads to the formation of an extract with a complex composition, containing the desired natural colourants and a high number of other components. In the classification of the natural colourant, the chemical constitution of the major component that serves as dyestuff is considered. In practice, a natural colourant will always contain a number of extracted dyes, with varying composition, dependent on the plant source, the variety of the plant, growing conditions as well as the time of harvesting and extraction.

For many natural dyes, the aqueous extract can be applied without further additives as a direct dye; however, in many cases also mordants are used to increase the sorptive binding of the dye (mordanting with tannin) or to form metal complexes with the natural dye (formation of dye lakes through mordanting with metal salt).

10.11.2 Major classes of natural colourants

Natural colourants usually represent a complex mixture of dyes; thus, the classification of a natural dye is based on the most relevant species present. For a number of dyes that have to be isolated as a chemical compound also a colour index (C.I.) number has been assigned to the respective chemical structure.

Indigoid dyes: The two important representatives for natural dyes with indigoid structure are indigo (C.I. Natural Blue 1) from plant sources and the historically relevant Tyrian purple (6,6′-dibromo-indigo, C.I. Natural Violet 1, C.I. 75800) from sea snails of the Muricidae family, for example, *Bolinus brandaris* or *Hexaplex trunculus* (Figure 10.32).

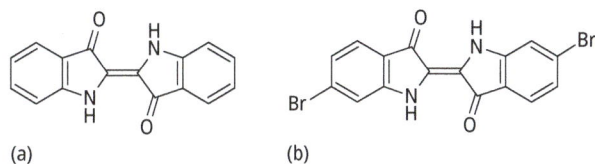

(a)   (b)

Figure 10.32: Chemical structure of indigo (C.I. Natural Blue 1) and Tyrian purple (6,6′-dibromo-indigo, C.I. Natural Violet 1).

The Tyrian purple was collected from the hypobranchial glands of the gastropod mollusc. While colour and fastness of the purple dye was acceptable, the low dyestuff content, the laborious and disputable way of production stopped the use of Tyrian purple with the appearance of synthetic dyes.

The most important blue natural dye is indigo. In almost every culture, plants have been cultivated to produce indigo. In general, the plant material does not contain the indigo dye in oxidised blue form, but contain indoxyl-glycosides as precursors for indigo. Based on the region and climate, different plants have been cultivated (Table. 10.4).

Table 10.4: Important indigo plants: botanical names, climate and region of growth, chemical nature of the precursor (from Bechtold [24]).

Plant	Botanical name	Climate/region	Precursor
Indigo-plant	*Indigofera tinctoria* L.	Tropical/India, Africa, North, Central and South America	Indican Indigo-β-D-glucoside
Woad	*Isatis tinctoria Isatis indigotica*	Temperate/ Europe, China	Isatan B Indoxyl-5-ketogluconate
Dyer's Knotweed, Ai	*Polygonum tinctorium* AIT.	Subtropical and temperate/ Europe, Japan	Indican Indigo-β-D-glucoside

In the first step, the water-soluble glycosides have to be extracted from the plant material. Once the plant material has been removed, in the second step, the enzymatic hydrolysis occurs to release the intermediate indoxyl. Indoxyl is sensitive to oxygen and in an ideal case dimerises in the presence of oxygen via leuco indigo to indigo. The insoluble blue indigo dye settles and is collected by filtration.

Indigo losses occur in case the insoluble indigo is formed in the plant material before aqueous extraction is completed, or in case the oxidation of indoxyl proceeds without dimerisation to the yellow isatin. Formation of isatin represents an irreversible loss in indigo yield. Figure 10.33 shows a general scheme for the reactions in natural indigo formation.

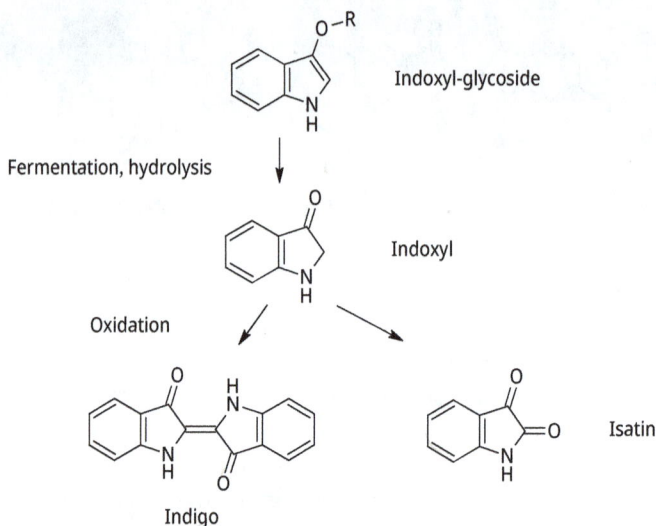

Figure 10.33: General reaction scheme for natural indigo production: Hydrolysis of indoxyl-glycosides, oxidation to indigo and side reaction to isatin (R = glycosidic-bound carbohydrate).

The application of indigo chemically corresponds to the processes in vat dyeing, respectively, the dyeing with synthetic indigo.

As a difference, the major application of natural indigo is in traditional handicraft techniques in combination with bacterial reduction in fermentation vats (which by the way explains the name vat dyeing). In traditional dyeing, a fermentation indigo dyebath is used for a long time and replenished constantly with the addition of fresh indigo. The fixation on the substrates (wool, silk, cotton, flax, etc.) occurs through air oxidation. A high number of dips are required to obtain a deep colour. The bacterial reduction leads to a rather low stability of the vat against air oxidation, as no surplus reducing agent is present in the dyebath.

The use of natural indigo in modern denim ranges is possible, as the principle chemistry of synthetic and natural indigo is identical. However, consideration of the rather low indigo content of the plants (1–2 wt% of dry plant material), eco-balances and resources consumption, land use and competition for farming of food favours the use of synthetic indigo in denim production [25].

Flavonoids and anthocyanins: Flavonoids represent an important class of yellow–purple dyes. The colour of many food products and fruits is based on the presence of flavonoids.

Flavonoids (e.g. flavonols, chalcones, aurons) are mainly yellow dyes while anthocyanins exhibit orange–red–purple colour. Similar to indigo precursors, the flavonoids and anthocyanins in plant material are present in the form of glycosides. The aglycones of anthocyanins are named anthocyanidins (Figure 10.34).

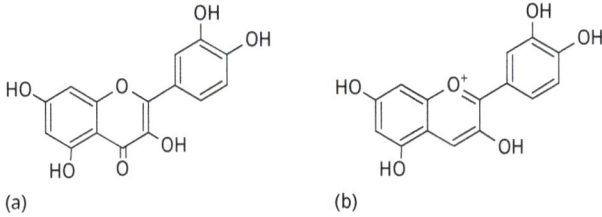

(a)   (b)

Figure 10.34: Chemical structures of (a) quercetin (flavonoid) and (b) cyanidin (anthocyanidin).

Many plant sources can be used to extract the yellow flavonoid dyes, and examples are given in Table 10.5 [26].

Table. 10.5: Representative plant sources for extraction of flavonoids (from Bechtold [24]).

Plant	Botanical name	Main colourants	Colour Index	Part of plant
Weld	Reseda luteola	Luteolin Apigenin	C.I. Natural Yellow 2	Plant except roots
Roman Chamomile	Chamaemelum nobile	Apigenin	C.I. Natural Yellow 1	Flower
Onion	Allium cepa	Quercetin		Outer shell of fruit
Black oak	Quercus velutina	Quercetin	C.I. Natural Yellow 10	Bark

Flavonoids can be used as direct dyes, in this case, the extract is used directly, or with the addition of mordants. In mordanting, either metal salt or tannins are added to strengthen the adsorptive dyestuff binding. Tannins support the binding of dyes through hydrogen bond interaction and Van der Waals forces. In case of metal salt-based mordants (e.g. Fe- or Al-salts), a metal complex is formed, which similar to metal complex dyes exhibits a higher affinity to the fibre material. Metal salt mordanting, in particular iron and copper, leads to a distinct change in colour, often more dull shades are formed (e.g. olive, brown). Based on the order, pre-mordanting (before the dyeing), after-mordanting (after dyestuff application) or meta-mordanting (direct addition to the dyebath) are used.

Often natural colourants are used in the context of environmental-friendly and "green" processes. The use of heavy metal mordants such as chromium, tin or copper salt-based mordants thus should not be considered any longer [27].

Anthocyanin dyes offer a highly interesting range of colours; however, the limited stability of the dyes limits their application considerably. The dyes can undergo pH-dependent irreversible rearrangement to colourless or less coloured products

(Figure 10.35). Improvement of stability has been achieved by metal complexation and co-pigmentation.

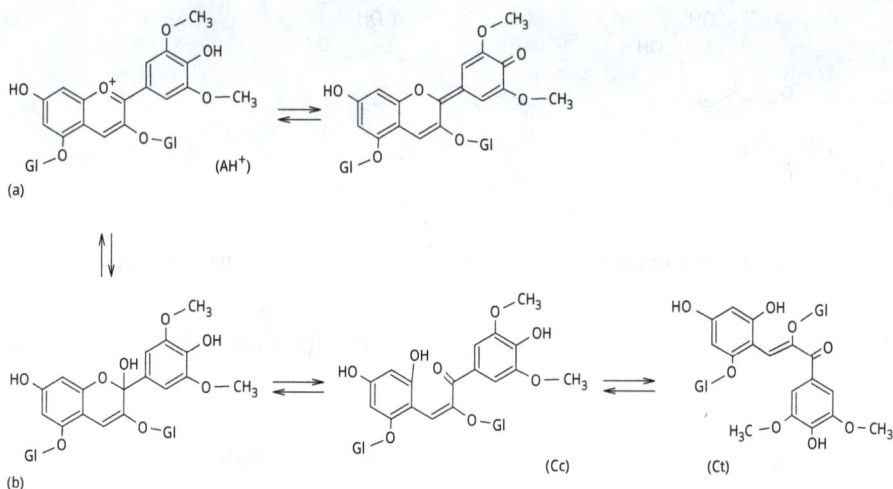

(a)

(b)

Figure 10.35: Chemical transformation of anthocyanins (e.g.: malvidin 3,5,-diglycoside; Gl = glycosidic group) as a function of pH.

The rather stable orange-red/violet oxonium/flavylium ion AH^+ is present at pH = 1.0. At pH 4–6 the corresponding quinoid base A appears, which is still coloured. Also the colourless hemiketal B is formed, which then tautomerises with ring opening, to yield the colourless *cis*- and *trans*-chalcone forms Cc and Ct [28].

Copigmentation stabilises the dye molecules through formation of inter- and intra-molecular complexes. In a similar approach, metal complexes can be used to stabilise the anthocyanin dyes. Representative sources for anthocyanins are given in Table 10.6.

Table 10.6: Representative plant sources for isolation of anthocyanins (from Bechtold [24]).

Plant	Botanical name	Main colourants	Colour Index	Part of plant
Elder	*Sambucus nigra*	Cyanidin-glycosides		Fruit
Vine	*Vitis vinifera*	Malvidin-glycosides		Fruit
Privet	*Ligustrum vulgare*	Malvidin-, Cyanidin- Delphinidin-glycosides	C.I. Natural Black 5	Berries
Hollyhock	*Alcea rosea*	Malvidin- Delphinidin- glycosides		Flowers

Quinoid, naphthoquinoid and anthraquinoid dyes
Many relevant yellow, orange and red natural dyes contain a quinoid group as a characteristic structural element. An example for a quinoid dye is the red-orange carthamin (C.I. Natural Red 26), which is extracted from blooms of safflower (*Carthamus tinctorius*) (Figure 10.36).

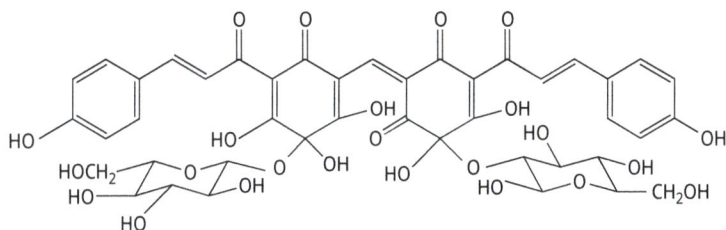

Figure 10.36: Chemical structure of carthamin (C.I. Natural Red 26).

Two important representatives of naphthoquinone dyes are lawson (2-hydroxy-1,4-naphthoquinone) and juglon (5-hydroxy-1,4-naphthoquinone) (Figure 10.37).

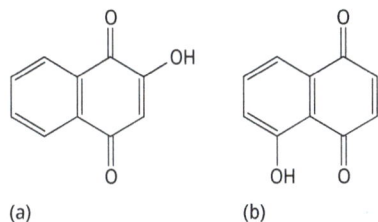

(a) (b)

Figure 10.37: Chemical structures of (a) lawson (C.I. Natural Orange 6) and (b) juglon (C.I. Natural Brown 7).

Lawson (C.I. Natural Orange 6) is the major dyeing component, which is extracted from henna (*Lawsonia inermis*). In the plant lawson is present in its glycoside form, which then hydrolyses after extraction. The aglycon is of importance in hair dyeing. In combination with natural indigo, the so-called black henna is obtained. Both dyes are used to dye hair in brown shades. Due to the low molecular weight of the dyes, diffusion and sorption are achieved at moderate temperature, which makes them suitable for dyeing of hair.

Juglon (C.I. Natural Brown 7) is extracted from green walnut shells.

Many plant sources contain natural colourants with anthraquinoid structure. Often the highest content of dye is observed in the roots of the plant. Important representatives are madder (*Rubia tinctorium*, C.I. Natural Red 8) and hedge bedstraw (*Gallium mollugo*, Natural Red 14) [29]. Also fungi and lichen or insects (Kermes, *Kermes vermilio*, C.I. Natural Red 3; cochineal, *Dactylopius coccus*, C.I. Natural Red 4) serve as sources for anthraquinoid dyes [30] (Table 10.7).

Table 10.7: Representative examples for insect dyes (from Bechtold [24]).

Insect	Name	Main colourants	Colour Index	Host plant
Kermes	*Kermes vermilio*	Kermesic acid	C.I. Natural Red 3	Kermes oak
Cochenille	*Dactylopius coccus*	Carminic acid	C.I. Natural Red 4	Nopal cactus
Lac-insect	*Kerria lacca*	Laccaic acids	C.I. Natural Red 25	Not specific

Similar to flavonoids and indigo, a glycosidic form is present as precursor of the colourants, from which the free anthraquinoid dye is released through hydrolysis.

Important representatives are alizarin (1,2-dihydroxyanthraquinone) and pseudopurpurin (3-carboxy-1,2,4-trihydroxyanthraquinone) (Figure 10.38). The presence of phenolic groups enables these molecules to form stable metal complexes, which then are called lakes. Important metal ions are Al^{3+} and Ca^{2+}.

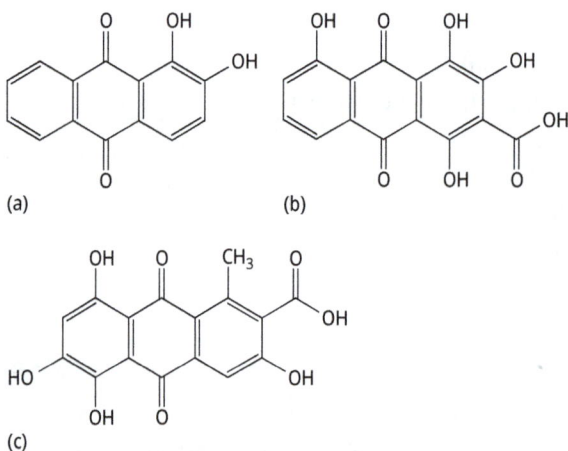

Figure 10.38: Chemical structure of (a) alizarin (C.I. Pigment Red 83), (b) pseudopurpurin (C.I. Natural Red 14) and (c) kermesic acid (C.I. Natural Red 3).

Through formation of a dye lake, dyestuff fixation and fastness properties are improved substantially. Brilliant shades and good fastness properties can be obtained with madder extracts, thus there is still considerable interest in such plant dyes. Attention has to be paid to remove harmful by-products, for example, lucidin (1,3-dihydroxy-2-hydroxymethyl-anthraquinone), by oxidation to less critical products.

An important group of red dyes is obtained from insects: kermes, cochenille and lac. Kermesic acid is shown in Figure 10.38 as a representative example.

In traditional dyeing, such dyes were widely used because of their brilliant shade and acceptable fastness properties. Today, these dyes find wider application as colourants in cosmetic product formulations (e.g. lipstick).

10.11.3 Tannin-based dyes

Tannins are of interest both as mordants and as natural colourant. Chemically, these substances consist of a complex mixture of condensed polyphenolic compounds. Gallic acid is the basic structural unit of gallotannins. In *hydrolysable tannins*, gallic acid is esterified with sugar molecules; in *condensed tannins* complex polyphenolic structures are present (Figure 10.39).

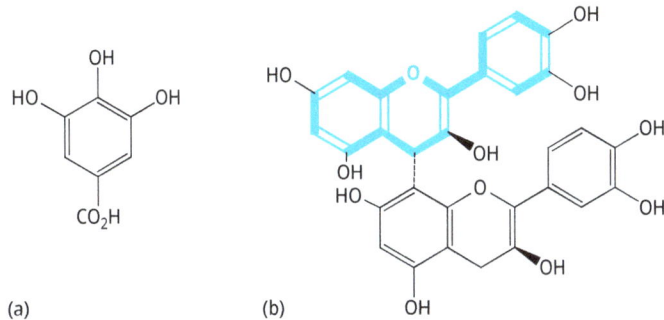

Figure 10.39: Representative structures of (a) gallic acid and (b) a condensed tannin ((+)-Catechin-(+)-Catechin) (flavan structure is highlighted in blue).

Tannins are present in almost every part of a plant. High contents are found in bark, wood and gallnut. Important sources are shown in Table 10.8. In contrary to other

Table 10.8: Representative plant sources for extraction of tannins (from Bechtold [24]).

Plant	Name		Main colourants	Content (wt%)	Colour index
Aleppo gall on oak tree	*Quercus infectoria*	Gallnut	Turkey tannin	50 – 70	
Sicilian Sumach	*Rhus coriaria*	Leaves, twigs	Gallotannin	23–35	C.I. Natural Brown 6
Sticky alder tree	*Alnus glutinosa*	Bark	Gallotannin	20	
Pomegranate	*Punica granatum*	Pomegranate fruit bark	Gallotannin	28	C.I. Natural Yellow 7
Scits pine	*Pinus silvestris*	Bark	Tannin	17	
Cutch/catechu	–	Bark, leaves	Tannin	–	C.I. Natural Brown 3
Tea	*Camellia sinensis*	Leaves	Tannin	25	

natural colourants, the tannin content in plant material is much higher and dependent on source and can reach levels of 50% of the plant weight (Table 10.8).

When tannins are applied as mordants in advance to the dyeing process (e.g. premordanting), an increase in dyestuff sorption leads to higher colour depth and increased fastness of the dyeing.

The polyphenolic structure enables tannins to form intensively coloured and stable complexes with many metal ions (e.g. iron, copper).

Carotenoid dyes

Carotenoid dyes are based on a polyisoprenoid structure. The yellow, orange and red dyes are found in plants and animals. Carotenes are hydrocarbons, and xanthophylls also contain oxygen atoms. Due to the lack in polar and solubilising groups, these dyes often exhibit low water solubility. As an exception, aqueous extracts from annato seeds and saffron flowers can be used for dyeing of wool, silk and cellulose fibres.

Important representatives for carotenoid dyes are α-carotene (carrots, *Daucus carota*; red palm oil), lutein (green leafy vegetables), bixin (annatto seeds, *Bixa orellana, C.I. Natural Orange 4*) and crocetin (saffron, *Crocus sativus*, C.I. Natural Yellow 6) (Figure 10.40).

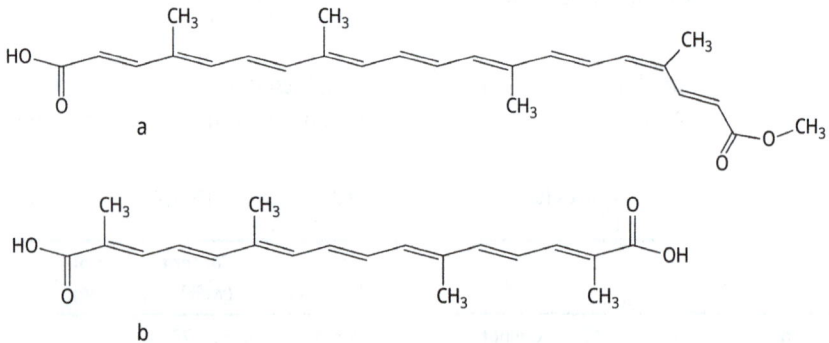

Figure 10.40: Chemical structure of (a) bixin (C.I. Natural Orange 4) and (b) crocetin (C.I. Natural Yellow 6).

Chlorophylls

Chlorophylls would be of high interest for textile dyeing, as the pigment is the most abundant in nature; however, lack in chemical stability and high production costs still prevent wider use. Replacement of the chelated magnesium by other metal ions (e.g. copper) leads to the semisynthetic chlorophyllins, which often are water soluble and exhibit higher stability. Up to present their application for colouration of paper, textiles and leather is still limited.

Applicatory aspects of natural colourants

Often natural colourants are estimated as green, sustainable and environmental-friendly products. However, to achieve this expectation a number of requirements have to be fulfilled [31, 32]:

- The plant sources should be collected from agricultural by-products and food wastes (e.g. barks, onion peel) to avoid the use of farmland for growing of dye plants.
- Aqueous extraction must be favoured, as solvent-based processes will have to be handled with care to avoid substantial losses of solvent with the extracted material.
- In an ideal case, the hot extract is directly used as dyebath to save thermal energy.
- Use of mordanting should be restricted to uncritical metal ions (iron, aluminium, calcium) and tannins
- The number of baths used in exhaustion processes should be low to keep water consumption reasonable.
- The dyeing processes must allow mixing of dyes from different sources to widen the range of colours available.

10.12 Pigment dyes

In the classical sense of dyeing, colouration with the use of pigment is based on a different process. The insoluble pigment that does not exhibit affinity to the fibre must be fixed on the substrate by means of a binder system.

Pigments can be insoluble organic or inorganic substances. Organic pigment dyes can be obtained with almost all types of chromophores (azo-dyes, metal complex dyes and phthalocyanine dyes, anthraquinoid pigments). Due to the absence of any solubilising groups, the colourant is insoluble in water. The most important pigment in textile dyeing is carbon black. Important white pigments are TiO_2 and Al_2O_3. Other inorganic pigments are metal oxides such as iron oxides (Fe_2O_3 ochre, mixed Fe-oxides like Indian red).

An important group of pigments also results from doping of crystal structures, for example, doping of the spinel $MgAl_2O_4$ with metal ions such as iron, chromium, zinc, manganese and cobalt. According to the ligand field theory, the energy splitting of electron orbitals in the dope metal ion leads to light absorption in the visible range of light and thus to the formation of intensively coloured inorganic pigments.

A small particle size (≤ 0.5 μm) is required to achieve even distribution and fixation on the material and to avoid settling in the dyebath.

In the application of pigments in textile dyeing, a binder system is required to attach the pigment particles on the surface of the textile. Representative cases for binder systems are based on polyacrylics and respective copolymers, polyurethanes, latex and styrol–butadiene copolymers.

The addition of a binder polymer to the flexible textile structure leads to a reduction in material flexibility. The increase in stiffness and a reduced softness results from a lower mobility of fibres in the yarn, respectively, the fabric.

In continuous dyeing, the dyebath is padded on the fabric, dried and the binder is thermally cured (e.g. at 160 °C for 60 s). Besides dispersing agents also anti-migration additives have to be used to prevent dyestuff redistribution during drying, which is called migration.

Migration: After padding to a pickup of 75 wt%, a considerable amount of capillary water is present in the fabric. During the first phase of drying, evaporation mainly occurs at the surface of the fabric; thus, water moves from inside the yarn to the surface of the fabric. This movement of water transports significant amounts of dispersed pigment to the site of evaporation. As a result, colour differences appear, depending on the rate of evaporation. Antimigration auxiliaries help to prevent the dye re-distribution of dyes through an increase in viscosity of the dye liquor, thus act as thickeners during evaporation of the padded dyebath. Chemical representatives for migration inhibitors are polysaccharides and acrylic copolymers.

10.13 Textile printing

10.13.1 General aspects of printing

The localised dyeing of textiles (e.g. to develop a coloured pattern) is called printing.

The general chemistry of application utilises the same concepts as described in dyeing; however, the possible separation between dye application and fixation permits a much higher number of variations and specialised methods. Thus in this part representative examples for techniques, and chemical concepts are explained.

In a most simple approach, the dye containing paste is printed on the textile fabric and the dye is fixed at elevated temperature (e.g. through steaming, drying). Then the excess paste containing dye and thickener is removed by thorough washing.

The physical chemistry of printing is related to the dyeing processes. The printing paste represents the dyebath. Thus, a very short liquor ratio is achieved. The dye transfer corresponds to the exhaustion process; however, viscosity of the dyebath is substantially higher. Thus processes will require both heat and time. The washing steps after the dye fixation have to be considered very carefully as in contrary to exhaust processes, substantial amounts of dye are still present in the paste and care has to be taken to remove the thickened paste completely. Staining of unprinted areas has to be avoided.

In traditional printing *block printing* was used, a method where the pattern was cut into wooden blocks, or was built through an assembly of metal stripes, nails and so on. The pattern then was the elevated part of the block, similar to a stamp. These elevated parts were then covered with printing paste and the block was pressed on the fabric. The transferred dye was then fixed on the fabric through appropriate methods.

In *resist printing*, a paste or material is printed on the fabric which prevents access of the dye to the fibres. Wax resist represents a physical approach for such a concept. The hydrophobic wax covers parts of the fabric which remains undyed during the following dyeing step. The wax is removed after the dyeing has been fixed and the pattern is formed by combination of dyed and undyed parts. A chemical form of a resist printing, for example, prevents dyestuff fixation or coupling through pH variation and formation of a chemical barrier.

In *discharge printing*, chemicals are applied, which locally destroy the dyestuff. Printing of reducing agents, for example, can be used to achieve reductive destruction of an azo-group-containing dye, thus leading to a "white pattern". In case a suitable dye is added to the printing paste, which is able to dye under conditions of discharge, a *colour discharge* printing is obtained. As an example a fabric that has been dyed with suited reactive dyes can be discharge printed with vat dyes, which are fixed on the fabric under the reductive conditions applied for discharge.

The composition of a printing paste is similar to composition of a dyebath

Dependent on the printing technique further chemicals are added. Representative additives found printing pastes are:

Humectants: Often the printed fabric is dried for intermediate storage before the fixation step. If steam fixation is used, the humectants support the water uptake into the printing paste and formation of a liquid phase, which is required to achieve dyestuff mobility and fixation. Examples are urea, glycerine and glycols.

Wetting agents support access to the fabric (ionic and non-ionic surfactants).

Reducing agents for discharge printing: The reducing agents must be stable to air oxygen at room temperature and must build up a sufficiently negative reduction potential to reduce both the azo-groups for the discharge process and the vat dyes for the colour fixation. Examples are sodium hydroxyl-methane-sulphinate (Rongalit C®; $HOCH_2\text{-}SO_2Na$) (Figure 10.41).

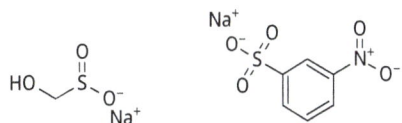

Figure 10.41: Chemical formula of sodium hydroxyl-methane-sulphinate and *m*-nitrobenzensulphonate (Ludigol®).

In case of discharge printing *mild oxidising agents* are padded on the fabric to protect reduction-sensitive dyeings from unwanted reductive decolourisation (e.g. during steaming). The oxidant stabilises the azo-dyes at sites where no discharge chemicals were printed on the fabric and also contributes to development of a sharp contour between discharge printed and non-printed areas. A representative is *m*-nitrobenzenesulphonate, which is also used in vat dyeing as a mild oxidant for the reoxidation of vat dyes.

Control of dye fixation can be achieved through addition of *carriers* (e.g. polyester fibres) or retarders (for basic dyes).

The chemical nature of a *resist agent* depends on the class of dyestuff to be printed (e.g. acid or alkali to shift pH).

Antimicrobial substances are added to avoid microbial growth in the paste and thus prevent reduction of the chain length of the polysaccharides during storage. Microbial degradation of thickener molecules leads to reduction in viscosity and resulting problems during printing.

Thickeners: The viscosity and shear properties of a paste are adjusted by use of appropriate thickeners. Typical solid content in a printing paste is between 2.5 and 16 wt%. Native and modified polysaccharides, as well as synthetic polymers (e.g. polyacrylate and polyvinyl alcohol) are in use.

In the majority of applications, the printing pastes exhibit structural viscous behaviour. The viscosity reduces under shear stress. In addition often a flow limit is desirable, which means that flow of the paste is achieved only above a certain shear stress.

Important technical requirements for thickeners are:
- Technically defined rheological properties
- Simple preparation of paste, rapid and complete swelling/dissolution properties, reasonable level of costs
- Stability under conditions of storage, no chemical degradation, hydrolysis
- Free from undissolved matter which could block the printing screen
- Compatibility with chemicals and dyes used (no competition for dye fixation (e.g. in reactive dyeing)
- High degree of dyestuff fixation, minimum retain of dyestuff
- Easy and complete wash off after dyestuff fixation
- Biodegradable, low COD load of the paste

Recycling of printing pastes from preparation in surplus and screen cleaning has been studied extensively. The complex composition of the pastes, the removal of the dyestuff and the standardisation of the regenerate are still challenges to be solved.

Important representatives for thickeners in printing pastes are:

Native polysaccharides: Non-ionic thickeners can be based on starch (corn, rice), gum arabic, gum tragacanth. These native products require longer time for preparation due to lack in cold solubility.

Modified polysaccharides: Through heat treatment of starch non-ionic hot-water-soluble types are obtained (British gum). Also etherified or esterified starch derivatives are used as thickeners. Other examples are hydrothermally degraded gum resins, galactomannans (e.g. carob flour) and cellulose derivatives (e.g. methylcellulose, hydroxyethylcellulose, hydroxypropylcellulose).

Anionic thickeners include alginate, which is obtained from brown algae and is of particular importance as thickener in reactive dyeing. Carboxymethylcellulose, carboxymethyl starch and xanthan belong to the group of anionic polysaccharide-based thickeners.

Representative structures are shown in Figure 10.42.

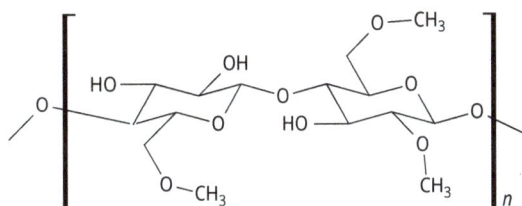

(a)

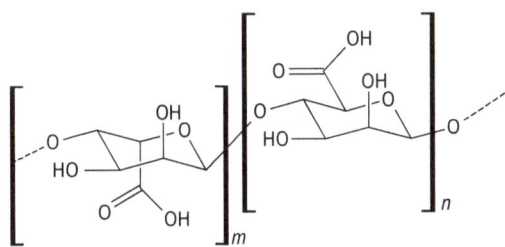

(b)

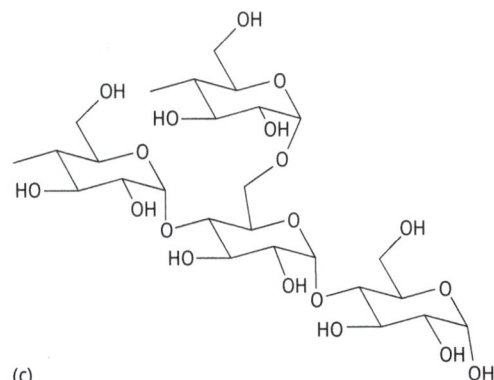

(c)

Figure 10.42: Chemical formula of (a) methylcellulose and (b) alginate (m = glucuronate units, n = mannuronate units), starch.

Synthetic polymer thickeners: A wide range of water-soluble polymers is in use as thickeners: Polyacrylic acid, copolymers between maleic acid anhydride and ethylene, styrene and polyfunctional monomers (e.g. divinylbenzene). These thickeners exhibit particularly high structural viscosity, which is of interest in printing through fine screen apertures.

Emulsion thickeners: Thickening of paste through formation of emulsions containing naphtha has been widely used in textile printing. The increase in viscosity was a result of the combination of a thickened aqueous phase with an emulsified organic liquid phase (naphtha). During drying of the print, both the aqueous phase and the organic solvent phase were evaporated; thus, considerable emissions of VOC (volatile organic hydrocarbons) were released. The use of emulsion thickeners thus has been stopped for environmental reasons by legislation through definition of low limits for VOC. Thus, in Europe naphtha-containing emulsion thickeners have disappeared completely.

An enormous variety of printing techniques and machinery are technically available. Thus, this chapter concentrates on a few major techniques: screen printing, transfer and ink-jet printing.

10.13.2 Screen printing

In screen printing the permeability of a screen defines where paste is printed on the fabric. The paste is pressed through the screen by means of a squeegee (wipes, rollers). The viscosity of the paste is adjusted to such a level that sufficient paste is transferred to the fabric; however, the contour (stand) of the print is high enough to obtain sharp borders between printed and non-printed areas.

Major aspects that define the amount of printed paste per area are:
- Viscosity of the paste, which should be shear thinning (pseudoplastic). When the paste is pressed through the screen, the viscosity reduces considerably as a result of high shear stress. On the fabric, a rapid recovery of the viscosity should appear to maintain the printed paste in position.
- The size of the apertures of the screen gauze and the open area define the permeability of a paste. The gauze is made of synthetic material (mainly monofilament).
- The pressure applied by the squeegee to press the paste through the screen. In flat screen printing, a squeegee operation can be repeated, which is not possible in rotary screen printing.

Pattern repeat: In flat screen printing, a plane stencil is used and the process of printing is repeated stepwise. The paste is pressed top-down to the fabric. After printing, the screen is lifted and the fabric moves forward the exact length of one pattern repeat. In rotary screen printing, hollow cylindrical screens are used and the paste is pressed from inside-out to the fabric. Flat screen printing is an intermittent process, while rotary screen printing is a continuous process and thus substantially more productive than flat screen printing. The dimensions of the stencil define the maximum pattern repeat. In any case the length of a stencil, respectively, the circumference defines the maximum length of a pattern repeat.

Registration: In colour, printing several printing pastes can be placed sequentially on neighbouring areas on the fabric. The correct placement of every colour without gaps or overlapping is called *registration.* The correct adjustment is achieved by slight movement of the position of screens relative to each other.

To obtain a stable positioning of the flexible fabric, the material is glued on an endless rubber belt, which transports the material under the screens.

After the printing step the fresh print is dried.

The stencil is prepared by transfer of the created pattern on a screen, which had been coated with a photosensitive resist. The screen is covered with a film, where open areas had been printed with an opaque layer; areas to be closed are illuminated to harden the resist. The photosensitive resist then can be washed off at sites, where no light had been received. The impermeable part then can be strengthened by application of polymer layers or electrolytic deposition of nickel layers. Another method for preparation of the stencil uses a laser to produce open areas on an initially completely closed screen (laser engraving).

Transfer printing: In this technique, first, the pattern is printed on a disposable transfer material (e.g. paper). In thermotransfer printing, the paper is placed in contact to the fabric and then the pattern is transferred to the textile surface with application of pressure and heat.

Selected disperse dyes can be transferred to polyester fibres through sublimation (e.g. 60 s, 200 °C). Thermoplastic binder systems can be applied to transfer the dye physically on almost every substrate. By this technique also complete pictures can be transferred to the textile in form of a thermoplastic film (e.g. in T-shirt printing).

10.13.3 Ink-jet printing

In ink-jet printing the dye containing ink is placed on the textile substrate in form of tiny drops and without any screen. The pattern is formed through a combination of ink deposition and movement of the printer heads. The requirement of a drop formation technology defines the specifications for an ink and the applicability of a dye system.

The use of jet printing was introduced first in carpet printing, as the dimensions of the pattern and the amount of dye to be applied were high compared to other applications. In carpet dyeing, the resolution of the dyeing is comparably low, thus mechanical control of the dye jet is sufficient (e.g. through valves or control of the direction of the dye jet).

With regard to drop formation different ink-jet printing technologies have been developed [33, 34].

In general, two general principles have to be distinguished:
- Continuous ink-jet technology (synchronous droplet ejection technology): a continuous beam of tiny droplets is generated by the print head (representative values for drop formation are in the dimension of 100,000 drops/s). The droplets are charged electrostatically and can be deflected by an electrical field. The height of

the electrostatic deflection field thus can be used to steer deposition of colour droplets on the fabric.

– Drop on demand technology: In this technology droplets are formed by the printer head only in case colour deposition is required by the pattern to be printed.

Different techniques can be used to generate the tiny droplets in the printer head:

– Piezoelectric elements change their physical dimensions to push liquid through the nozzle (piezo ink-jet).

– Thermal energy leads to vapour formation (bubble-jet).

– Electrostatic or acoustic effects also can be used to eject ink from the head.

A special technical concept for application of high volumes of ink uses a mechanical system (flatjet technology) to generate drops (Chromotex printers). As a result of the mechanical approach, the selection of a printing ink is more straightforward.

The requirements for a suitable ink depend on the principle of the droplet formation used in the printing head. Thus usually only use of very well-defined printing inks is possible in ink-jet printers. The use of in-house modified inks is rare, as the risk of an irreversible damage of the expensive printer heads is considerable. The cost of a printing head thus hampers simple replacement of inks by more experimental formulations. This makes a big difference to conventional screen printing, where a wide range of freedom in composition and design of recipes for a printing paste exists.

General requirements for ink-jet inks are [33]:

– low viscosity (to permit drop ejection and formation);

– surface tension <25–40 mN/m required for drop separation and formation;

– low content of undissolved matter (particle size <1 µm);

– appropriate conductivity (>100 mS for continuous ink-jet, 0 mS for piezo ink-jet); and

– non-corrosive for the material used in the printer head (pH, redox-active chemicals, ingredients with abrasive properties).

A representative composition of an ink-jet ink is summarised in Table 10.9.

An important difficulty of ink-jet printing arises from the fact that for each colour a separate (group of) printer head(s) is required. Colours are formed by subtractive mixing of dyes during printing. Thus, primary dyes of maximum brilliance and high colour fastness are required, independent on the colours to be printed.

The specific requirements to the composition of the ink lead to a concentration on high value markets.

Table. 10.9: Representative compositions of ink-jet inks. From Tawiah et al. [33].

Function	Chemical basis	Concentration (wt%)
Liquid solvent	Deionised water	60–90
Humectants, viscosity controller	Water-soluble solvents, glycols, alcohols, PEGs	5–30
Colourant	Dye, pigment	1–10
Reduction of surface tension (wetting, penetrating, foam control)	Surfactant, defoamer (silicones, tributylphosphate)	0.1–10
Prevention of microbial growth	Biocide	0.1–0.5
Masking of multivalent ions	Complexing agent	>1
In case of a pigment print		
Pigment binder	Polymer emulsion	10–15

Important examples for technical applications of ink-jet technology are:

– pigment printing, using a paste which contains dye and printer, followed by thermal fixation;
– disperse dyes on polyester, followed by thermal fixation;
– reactive dyes on cellulose textiles that have been previously impregnated with alkali/urea/alginate thickener to achieve fixation during following steaming; and
– acid dyes on wool, silk, nylon followed by steam fixation.

The chemistry behind the dye fixation and the following wash off of surplus dye and printing auxiliaries follows the usual methods to finish a print.

10.13.4 Special printing techniques

Non-fabric printing: Printing techniques can be applied also on pre-stages in textile production, e.g. sliver yarn, warp beams, thus leading to special irregular patterns.

Flock printing: Flock printing is a representative for techniques where an adhesive is printed in the first stage. In the second step, an effect material is fixed to the adhesive.

In case of flock print, short staple fibres (flocks, 0.3–3 mm) are fixed on the adhesive layer by means of an electrostatic field (20–60 kV) which also orients the fibres in direction perpendicular to the fabric surface, thereby a velvet-like surface structure is obtained.

Many other effect materials can be fixed on the adhesive layer:
– Finely chopped thin plastic particles lead to a glitter effect
– Metal film can be fixed as conductive surface layer or to achieve a metal effect

Batik printing: In the classical processes, a wax was printed on the fabric to form a resist. After dyeing with usual dyeing procedures, the resist was removed. In a special case indigo was used for dyeing. During dyeing of the fabric the waxy layer cracks and the cracks become marked with dye, which leads to the characteristic appearance of this article.

Two-phase printing: In two-phase printing, the pattern printing and dyestuff fixation processes are separated into two stages. As an example after printing of a reactive dye and intermediate drying, the print is fixed in a continuous process through padding in rather concentrated alkali solution. Similarly, vat dyes can be printed and fixed in the reducing agent.

References

[1]   BASF, Ratgeber. Färben und Ausrüsten von Polyesterfasern, B 363 d, 8.74 (1974) BASF Aktiengesellschaft, Ludwigshafen, Germany, 1974.

[2]   Leube, H., Rüttiger, W., Kühnel, G., Wolff, J., Ruppert, G., Schmidt, M., Heid, C., Hückel, M., Flath, H-J., Beckmann, W., Brossmann, R., Söll, M., Sewekow, U. Textile dyeing. in ULLMANN's Fibres, Volume 2, Textile and Dyeing Technologies, High Performance and Optical Fibres, Wiley VCH Verlag, Weinheim, Germany, 2008.

[3]   Zollinger, H. Color Chemistry, Wiley-VCH, Weinheim, Germany, 2003.

[4]   Bui, HM., Ehrhardt, A., Bechtold, T. CI Reactive Black 5 dye as a visible crosslinker to improve physical properties of lyocell fabrics. Cellulose 2009,16/1,27–35.

[5]   Trotman, ER. Dyeing and Chemical Technology of Textile Fibres, 6th ed. Charles Griffin & Company Ltd. Bucks, England, 1984.

[6]   Bechtold, T., Burtscher, E., Hung, Y. Treatment of textile wastes. in Handbook of Industrial and Hazardous Wastes Treatment, 2nd edition, ed. Wang, LK., Hung, Y-T., Lo, HH., Yapijakis, C. Marcel Dekker, Inc. New York, 2004, ISBN: 0-8247-411-5.

[7]   Bechtold, T., Turcanu, A. Electrochemical reduction in vat dyeing – Greener chemistry replaces traditional processes. J. Cleaner Production. 2009,17,1669–1679, DOI 10.1016/j.jclepro.2009.08.004.

[8]   Fischer, K., Marquart, K., Schlüter, K., Gebert, K., Borschel, E-M., Heimann, S., Kromm, E., Giesen, V., Schneider, R., Wayland, RL. Textile auxiliaries. in ULLMANN's Fibres, Volume 2, Textile and Dyeing Technologies, High Performance and Optical Fibres, Wiley VCH Verlag, Weinheim, Germany, 2008.

[9]   Schnitzer, G. https://pubs.acs.org/cen/science/89/8943sci3.html accessed 01.01.2018.

[10] Garcia, B. Reduced water washing of denim garments. Chapter 13 in Denim. Manufacture, Finishing and Application, ed. Paul R. Woodhead Publishing, Cambridge, UK, 2015.

[11] Schrott, W., Paul, R. Environmental impacts of denim production. Chapter 10 in Denim. Manufacture, Finishing and Application, ed. Roshan Paul, Woodhead Publishing, Cambridge, UK, 2015.

[12] Etters, NJ. Effect of dyebath pH on color yield in indigo dyeing of cotton denim yarn. Text. Chem. Colorist. 1989, 21/12, 25–31.

[13] Blackburn, RS., Bechtold, T., John, P. Indigo reduction methods and pre-reduced indigo products: A historical review and recent developments. Color. Technol. 2009, 125, 193–207.

[14] Nicholson, SK., John, P. The mechanism of bacterial indigo reduction. Appl. Microbiol. Biotechnol. 2005, 68, 117–123.

[15] Schnitzer, G., Suetsch, F., Schmit, M., Kromm, E., Schlueter, H., Krueger, R., Weiper,-Idelmann, A. Dyeing cellulose-containing textile material with hydrogenated indigo. US 5586992 1996.

[16] Marte, W. Biokompatible Denim Färbetechnologe, International Textile Bulletin 1995, 1, 33–37.

[17] Poulakis, K., Bach, E., Knittel, D., Schollmeyer, E. Einfluss von Ultraschall auf die Verküpungsgeschwindigkeit von Indigofarbstoffen mit α-Hydroxyaceton als Reduktionsmittel. Textilveredlung. 1996, 31/5–6, 110–113.

[18] Laksanawadee, S., Setthayanond, J., Karpkird, T., Bechtold, T., Suwanruji, P. Green reducing agents for indigo dyeing on cotton fabrics. J. Cleaner Production. 2018, 197, 106–113.

[19] Bechtold, T., Burtscher, E., Kühnel, G., Bobleter, O. Electrochemical processes in indigo dyeing. J. Soc. Dyers Colour. 1997, 113, 135–144.

[20] Blackburn, RS., Harvey, A. Green chemistry methods in sulfur dyeing: Application of various reducing D—sugars and analysis of the importance of optimum redox potential. Environ. Sci. Technol. 2004,38,4034–4039.

[21] Bechtold, T., Berktold, F., Turcanu, A. The redox behaviour of C.I. Sulphur Black 1 – a basis for improved understanding of sulphur dyeing. J. Soc. Dyers Colour. 2000,116,215–221.

[22] Bechtold, T., Burtscher, E., Turcanu, A. Direct cathodic reduction of leuco sulfur black 1 and sulfur black 1. J. Appl. Electrochem. 1998,28,1243–1250.

[23] Schweppe, H. Handbuch der Naturfarbstoffe, Ecomed Verlagsges, Landsberg/Lech, 1993.

[24] Bechtold, T. Colorant, natural. In: Luo R. (Ed.) Encyclopedia of Color Science and Technology, Springer-Verlag Berlin Heidelberg, ISBN 978-1-4419-8071-7.

[25] Bechtold, T., Turcanu, A., Geissler, S., Ganglberger, E. Process balance and product quality in the production of natural indigo form *Polygonum tinctorium Ait*. applying Low-Technology Methods. Bioresource Technol. 2002,81,171–177.

[26] Bechtold, T., Mahmud-Ali, A., Mussak, R. Chapter 31. Natural dyes from food processing wastes – Useage for textile dyeing. in *"Waste management and co-product in food processing"*, ed. Waldron KW. Woodhead Publishing Ltd, Cambridge, England, ISBN 1 84569 025 7, March 2007, 502–533.

[27] Manian, AP., Paul, R., Bechtold, T. Metal mordanting in dyeing with natural colourants: A review. Color. Technol. 2016, 132, 107–113.

[28] McClelland, RA., McGall, GH. Hydration of the flavylium ion. 2. The 4'hydroxyflavylium ion. J. Org. Chem. 1982, 47, 3730–3736.

[29] Derksen, GCH. Red, Redder Madder – Analysis and isolation of anthraquinones from madder roots (*Rubia tinctorium*), Dissertation Wageningen University, Wageningen, The Netherlands, ISBN 90-5808-462-0, 2001.

[30] Räisänen, R. Anthraquinones from the fungus *Dermocybe sanguinea* as textile dyes. Dissertation, Department of Home Economics and Craft Science, University of Helsinky, Helsinky, ISBN 952-10-0537-9, 2002.

[31] Bechtold, T., Turcanu, A., Ganglberger, E., Geissler, S. Natural dyes in modern textile dyehouses – How to combine experiences of two centuries to meet the demands of the future? J. Cleaner Prod. 2003, 11, 499–509.

[32] Bechtold, T., Mussak, R. ed. Handbook of Natural Colorants, John Wiley & Sons, Chichester, 2009.

[33] Tawiah, B., Howard, EK., Asinyo, BK. The chemistry of inkjet inks for digital textile printing. BEST: Int. J. Manage., Inf. Technol. Eng. 2016, 4/5, 61–78.

[34] Malik, SK., Kadian, S., Kumar, S. Advances in ink-jet printing technology of textiles. Indian J. Fibre Text. Res. 2005, 30, 99–113.

Take home messages

Based on the technique of application and the materials to be dyed, different classes of dyes can be distinguished (e.g. reactive dyes, direct dyes, acid dyes, vat dyes).

The major characteristics found in a class of dyes, for example, presence of solubilising groups, reactive anchors are the similar; however, the type of chromophore present in a certain dye may differ.

Dyes with similar applicatory properties can be collected in a group of dyes that can then be used in combination to achieve a certain shade.

The recipe used in a dyeing process with a selected class of dyes depends on the process conditions and process design (e.g. continuous, discontinuous, pad-dry, pad-batch) and on the textile material (fibres, yarn, fabric).

Printing processes apply in many cases the same techniques for dyestuff fixation; however, the recipes differ as a localised application and fixation has to be achieved.

Quiz

1. Why reactive dyes should exhibit a perfect fastness to washing? Why there are limitations in reality?
2. Why we cannot use a reactive dye for polyester (PET) fibres?
3. What is the reason for poor rub fastness in vat dyeing, indigo and sulphur dyeing? What to improve?
4. Which dyes theoretically will permit dyeing of the following intimate blends in one bath: Wool/cotton, wool/polyamide, polyamide/polyester, wool/silk, viscose/silk.
5. Why a levelling phase in reactive dyeing does not make sense at the end of the dyeing process?

Exercises

1. Transfer (as an estimate) a pad-batch recipe with 20 g/L and pickup of 75% into % dye used in exhaust dyeing.
2. Calculate the average dwell time in a padder box with 25 L/12 L content when fabric (150 g/m^2 and 1.8 m width) is processed at a pickup of 70% with a production speed of 20 m/min.
3. Develop a graphic representation that demonstrates the consumption of NaCl in g NaCl per 1 kg of goods in exhaust dyeing as a function of liquor ratio used (between 1:5 and 1:50) for a bath concentration of 30, 40 and 50 g/L NaCl.
4. The dyestuff fixation of a 2% reactive dyeing has been determined with 70% for a liquor ratio of 1:10. When we assume similar concentrations of hydrolysed dye at

different liquor ratio, what will be the amount of dye fixed at liquor ratio of 1:6, 1:15 and 1:20.

5. Calculate the theoretical amount of $Na_2S_2O_4$ and NaOH to achieve reduction of 50 g C.I. Vat Yellow 1; 50 g C.I. Vat Blue 4, and 50 g indigo (C.I. Vat Blue 1).

6. Which volume of oxygen at normal conditions is required to oxidise 1 kg of dithionite? Which amount of NaOH as 50% NaOH solution is consumed by this process?

7. Calculate the sulphate load in mg/L from a vat dyeing process whose 6 g/L sodium dithionite in the dyeing. The effluent of dyeing is mixed with five times the volume from other processing baths.

8. A continuous indigo dyeing range operates at the following conditions: 15 kg yarn per min. Pickup for wet material entering the dye box: 65 wt%, pickup after dyeing 80%. The dyeing intends to fix 2 wt% of indigo on the fabric. What is the minimum dyestuff concentration in the stock vat to prevent a volume overflow in the dyebox?

9. A metal complex dye exhibits a total chromium content of 1.3%. During continuous dyeing, a fixation of 95% of the used dye is achieved. The dyestuff concentration in the padder is 120 g/L, pickup of the dye liquor is 75% (75 mL of liquid per 100 g of goods). In the following rinsing, a total amount of 8 L/kg water is used. What concentration of chromium is expected in the effluents leaving the dyeing range.

Solutions

1. (Question 1): Because the covalent linkage between the substrate and the dye binds the dye molecule. Limitations in fastness arise from hydrolysis of the covalent bonding and from the presence of sorbed hydrolysed reactive dye.

2. (Question 2): Because there are only few reactive groups available in the polyester fibre and the accessibility of these groups is very limited.

3. (Question 3): In case substantial amounts of pigment have been deposited on the surface of the material, these pigments will contribute to a low fastness to rubbing. Improvements: Better dispersion of particles through dispersing agent, higher stability of the reduced state, rapid wash off of the dyebath without oxidation of remaining dyebath on the material.

4. (Question 4): Wool/cotton: reactive dyes; wool/polyamide: acid dyes, metal complex dyes; polyamide/polyester: disperse dyes; wool/silk: acid dyes, metal complex dyes, reactive dyes; viscose/silk: reactive dyes.

5. (Question 5): As the dye should be fixed by a covalent bonding, there is no equilibration possible, thus a levelling phase is not applicable.

6. (Exercise 1): A concentration of 20 g/L applied with a pickup of 75% correspond to m(dye) = 0.75 × 20 = 15 g/kg; thus, in an equivalent exhaust recipe, we start with the use of 1.5% dye.
7. (Exercise 2): At first, we calculate the removal of dyebath from the pad-box: Assume a density of 1 kg/L. m = 0.150 kg/m^2 × 1.8 m × 20 m/min × 0.7 = 3.78 kg/min or 3.78 L/min. The average resident time is then for the 25 L box: t = 25 L/3.78 L/min = 6.6 min; and for the 12 L box t = 12 L/3.78 L/min = 3.17 min.
8. (Exercise 3): For 1 kg of goods we require 5–50 L of dyebath, this corresponds to 150 g (30 g/L, 1:5)–2500 g (50 g/L, 1:50). If the formula $M = LR \times c(\text{NaCl})$ is transferred to a calculation programme, a graphic representation is obtained.

	c(NaCl)		
LR 1:	30	40	50
5	150	200	250
10	300	400	500
15	450	600	750
20	600	800	1,000
25	750	1,000	1,250
30	900	1,200	1,500
35	1,050	1,400	1,750
40	1,200	1,600	2,000
45	1,350	1,800	2,250
50	1,500	2,000	2,500

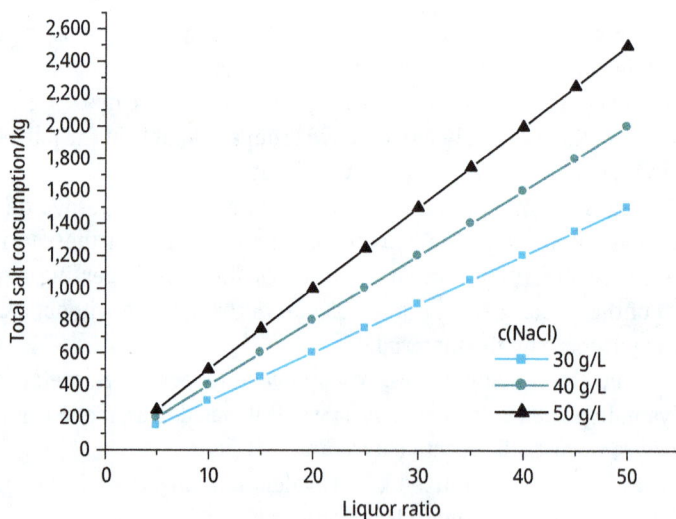

9.   (Exercise 4): At first, we calculate the equilibrium concentration in the bath: At a liquor ratio of 1:10 the final concentration in the bath will be:
$c(bath)$ = 20 g/kg * 0.3/10 L/kg = 0.6 g/L dyestuff, 14 g will be fixed on the fabric. The equilibrium concentration in the bath is almost independent from the liquor ratio; thus, we can assume: at LR 1:6 the total amount of dye in the bath will be: m = 6 * 0.6 g/L = 3.6 g, at LR 1:15 m = 15 * 0.6 g = 9 g; at LR 1:20 m = 20 * 0.6 g = 12 g; The dyestuff fixed on the fabric thus will be the rest:
LR 1:6, 16.4 g (!); LR 1:15, 11 g; LR 1:20, 8 g(!).

10.   (Exercise 5): Molar mass of 50 g C.I. Vat Yellow 1 = 408,4 g/mol, $M(Na_2S_2O_4)$ = 174 g/mol, $M(NaOH)$ = 40 g/mol; For 1 mol of dyestuff, we need 1 mol of dithionite (two electrons) and 4 mol of NaOH
$m(Na_2S_2O_4)$ = 50 × 174/408.4 = 21.3 g
$m(NaOH)$ = 50 × 40 x 4/408.4 = 19.6 g
M(C.I. Vat Blue 4) = 442.4 g/mol; M(indigo, C.I. Vat Blue 1) = 262.3 g/mol analogous: 19.6 g $Na_2S_2O_4$, 18.1 g NaOH; 33.2 g $Na_2S_2O_4$, 30.5 g NaOH.

11.   (Exercise 6): 1 mol of $Na_2S_2O_4$ consumes ½ mol of O_2 and requires 2 mol of NaOH.
$n(Na_2S_2O_4)$ = 1,000 g/174 g/mol = 5.7 mol, this corresponds to a consumption of m (O_2) = 5.7 × 16 = 92 g O_2 and approximately 64.4 L oxygen, approx. 306 L of air. m (NaOH) = 5.7 × 40 × 2 = 456 g NaOH. This corresponds to 912 g NaOH 50 wt%.

12.   (Exercise 7): One mole of dithionite will lead to 2 mol of sulphate. Thus $m(SO_4^{2-})$ = 96 g/mol × 2 × 6 g/L/174 g/mol = 6.6 g/L sulphate initially, which then are diluted 6 times to 1.1 g/L sulphate (1,100 mg/L).

13.   (Exercise 8): The full amount of dyestuff has to be added in the difference of the two pickups; thus, the 15% difference in pickup defines the maximum volume: 0.15 L of dyebath per 1 kg of goods. In this amount, we have to transport 20 g/kg of indigo dye. Thus, c(indigo) = 20 g/0.15 L = 133 g/L indigo in the stock vat.

14.   (Exercise 9): At first we calculate the amount of dye released from the fabric:
m(dye) = 120 g/L * 0.75 L/kg * (1–0.95) = 4.5 g/kg released into 8 L of wash water. The concentration of dye in the wash water thus will be
c(dye) = 4.5 g/kg/8 L/kg = 0.56 g/L. The chromium content thus will be
c(Cr) = 0.56 g/L × 0.013 g(Cr)/g(dye) = 0.0073 g(Cr)/L= 7.3 mg(Cr)/L.

List of abbreviations/symbols

HT	high temperature
COD	chemical oxygen demand
PVP	polyvinylpyrrolidone
PAC	polyacrylic fibre
AOX	adsorbable halogenated organic compounds
S_F	fibre saturation value (for cationic dyes)

c_s	concentration of dyestuff in solution
C.I.	colour index
PEG	polyethyleneglycol
T_g	glass transition temperature
VOC	volatile organic compounds

11 Pre-treatment

Depending on the type of fibre, fabric and product type, a number of processing steps are performed in advance to dyeing and printing.

In this chapter, only the most important steps are summarised to demonstrate the different principles of pretreatment.

11.1 Sizing

In fabric production, the warp yarn is subjected to substantial abrasive forces during weaving. Thus, the yarn has to be reinforced with a polymer coating, the so-called size. Due to the high number of cycles of a weaving loom (up to 700 turns/min), the productivity of such a unit is high. As a result, with increasing speed the costs of a single-warp yarn breakage rise. A high number of yarn breakages will lower the product quality; however, it will also lead to a dramatic reduction in productivity of such a loom. Thus, the appropriate sizing and climate control in the weaving hall are of high economic significance.

The chosen type of size used depends on the fabric, fibre type and yarn construction, as well as on the weaving loom used.

Native polymers (starch, galactomannan), modified natural products (CMC) and synthetic polymers (polyvinylacohol, poly(meth)acrylates) are in use. The most important group of sizes are *starch type* products.

Starch consists of two groups of polysaccharides (Figure 11.1): the unbranched amylose (15–25 wt%) with a degree of polymerisation (DP) of 100–1,000 and the branched amylopectin (75–85 wt%) with a DP of up to 10^6.

To dissolve unmodified starch in water, the starch grains are heated under elevated pressure and high shear rate to achieve gelatinisation (the formation of a gel state). Application has to be performed near to the boil to avoid retrogradation, which is the formation of hydrogen bond interlinkages between neighbouring chains, leading to reduced sizing properties, deposits on rollers and skin formation.

With *modified starches* gelatinisation can be achieved at reduced temperature. These products exhibit lower viscosity in solution and reduced tendency for retrogradation. Representative examples are starches which were modified by acid hydrolysis, oxidative degradation or chemical modification such as starch esters (phosphate starch, acetyl starch) and starch ethers (hydroxyethyl-starch, hydroxypropyl-starch, carboxymethyl-starch CMS).

Viscosity of starch sizes can be increased by addition of borate, which acts through formation of a complex with starch. Addition of urea leads to a reduction in viscosity. Starch degrading agents also reduce viscosity, either through hydrolytic action of acid or oxidative degradation (peroxo(di)sulphates).

https://doi.org/10.1515/9783110795738-011

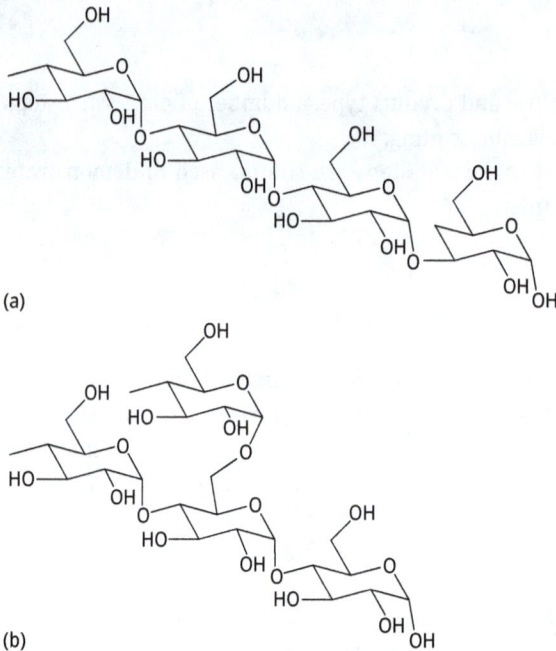

(a)

(b)

Figure 11.1: Chemical formula of (a) amylose and (b) amylopectin.

In waste water, starch and starch derivatives do not cause problems with regard to toxicity or accumulation. However, the high chemical oxygen demand (COD) of starch-based sizes lead to a considerable COD load in the waste water (900–1,000 mg/g).

Modified celluloses: The most important representative of cellulose-based sizes is carboxymethylcellulose (CMC). The biological degradation of CMC is rather slow, the COD value is near to 900 mg O_2/g.

Synthetic sizing agents base on water soluble synthetic polymers. Major products are polyvinylalcohols (PVA), poly(meth)acrylates, polyester condensates (Figure 11.2).

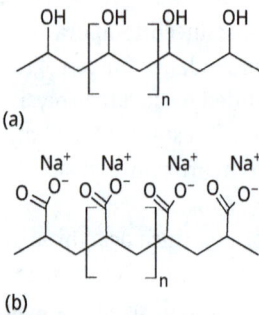

(a)

(b)

Figure 11.2: Representative chemical formula of synthetic sizes (PVA, acrylate).

Polyvinylalcohols offer high variability in properties, high film strength and flexibility. The polymer is produced through polymerisation of vinylacetate, which is then hydrolysed in a second step. The water solubility can be adjusted with the degree of hydrolysis of the acetate ester. Maximum solubility is observed after hydrolysis of approximately 88% of all ester groups initially present. Copolymerisates with methylmethacylate are also in use.

PVA sizing can be removed by dissolution in water, which makes the product well suited for recovery through reconcentration by ultrafiltration. Several difficulties have to be solved to introduce a size recovery on industrial scale:

-   The recovery site of used size should be near to the sizing plant to permit reuse without delay and to avoid high volumes of tanks required for intermediate storage of the recuperated size.
-   Microbial growth in the regenerate has to be controlled (e.g. by temperature).
-   The quality of the regenerate has to be standardised with regard to concentration as well as for the actual molecular weight distribution and purity.

After adaptation of a microbial consortium, biodegradation of PVA is possible. The COD value is approximately 1,700 mg O_2/g product.

Poly(meth)acrylates are available as copolymerisates from a number of different building blocks: acrylic acid and salts, acrylic acid esters and amides, acrylonitrile, methacrylic acid. The polymers exhibit good water solubility, which permits removal through swelling and dissolution in water. These sizes are not biodegradable; however, these hydrophobic copolymers are adsorbed at the activated sludge and thus can be removed from the effluents in a communal wastewater treatment plant. A typical value for the COD ranges between 1,300–1,700 mg O_2/g.

Polyester condensates are synthesised similarly to polyethylene terephthalate (PETP) through polycondensation of aromatic dicarboxylic acids with diols; however, in addition, sulphonated aromatic dicarboxylic acids are added to get a water soluble copolymer as the product. This size is mainly used for the sizing of polyester filament yarns. In presence of polyvalent cations, precipitation of the size can occur.

Further additives are added to the size to improve performance. Important additives are:

-   Viscosity regulators for starch-based sizes.
-   Sizing fats are added to reduce dry splitting, smoothen the size and lower antistatic charge. Representative products are emulsified fats (sulphated fats and oils), fatty acid esters in combination with emulsifiers and polyethylene glycols.
-   Wetting agents are added to guarantee stable pick-up of sizing liquor by the yarn, which often exhibits limited wettability such as in sizing of raw cotton yarn.
-   Defoamers are added to sizes which tend to the formation of foam during padding. Representative agents base on paraffin oil, phosphoric esters, fatty acid esters. Silicone oil is very efficient in defoaming; however, it must be considered critically with regard to complete removal before a dyeing process.

- Preservatives (fungicides, bactericides) have to be added to prevent microbial growth in the size during application, storage and possible regeneration. Representative examples are formaldehyde and formaldehyde precursors, isothiazolin-type compounds.

Both surfactants and silicone oil can reduce the adhesion of the size film on the yarn and thus reduce the overall performance of the size.

11.2 Desizing

Sizing belongs to the preparatory steps of the material for the weaving process and is thus usually counted as part of the fabric production.

The first wet treatment of a woven fabric in a textile dyehouse is the removal of the size to prepare the material for the textile chemical processes such as scouring, bleaching, dyeing and so on. The technique of removal depends on the size used.

Starch sizes are degraded to achieve reduction of chain length and better solubility in water. The degradation of starch can be achieved by different methods:

- Enzymatic degradation by action of amylases (bacterial amylases, pancreas amylases, malt diastase). The stereospecific action of the enzymes hydrolyses only the α-glycosidic bond of starch and leaves the β-glycosidic bond of cellulose unchanged. Conditions apply depending on the type of enzyme, where the representative conditions are: pH 5–7 and 50–60 °C temperature. (Figure 11.3)
- Oxidative degradation of starch also leads to the formation of fragments with reduced molecular weight and improved solubility. Peroxo compounds (hydrogen peroxide) are mainly used. The cold pad-batch bleaching process uses a combination of oxidative starch degradation and bleach effects to obtain an efficient desizing in combination with the first bleach step.

Enzymatic degradation

Starch ⟶ Oligosaccharides, dextrins, glucose

Figure 11.3: Enzymatic degradation of starch size.

Starch degradation is an irreversible process, independent of the process chosen, thus regeneration of starch sizes from waste water is not possible.

Water soluble sizes (PVA, acrylates) can be removed by direct washing processes. When the size on the raw fabric first comes in contact with water, rapid uptake with swelling of the size coating appears. After swelling is completed, dissolution of the size proceeds. The high viscosity of the boundary layer between swollen size and wash liquid makes the process substantially slower compared to a wash off of low molecular weight soluble substances. Desizing can be performed as a continuous wash process, as long as sufficient time for swelling and dissolution of the polymer is permitted.

Modern continuous washing machines for removal of water soluble sizes collect rather concentrated streams of sizes, which could be regenerated and recycled. In practice, such concepts are very rarely found due to a number of difficulties:

- The distance between the site of sizing and fabric production, and the desizing in the dyehouse should be short enough to permit economically reasonable transport.
- Viscosity of the regenerated size increases with concentration, which limits the maximum concentration in the collected wash solution. Further reconcentration through membrane filtration is required to achieve the minimum necessary concentration for reuse.
- A considerable content of additives and impurities accumulates in the regenerate. These components must be considered critically as a reduction in the efficiency of the weaving process can occur (yarn breakage, dust formation).
- Standardisation of the composition of a size is required to avoid a high number of variations collected with the regenerate, which would also prevent a wider use of the regenerate.

11.3 Alkaline extraction

Cotton fibres in raw state contain a considerable amount of non-cellulosic materials, such as pectins, waxes, proteins and inorganic salts (see Chapter 2).

Cellulose fibres are acid sensitive and quite alkali resistant, thus a hot alkaline prewash (scouring) in presence of detergents and complexing agents is used to remove the non-cellulosic material. In particular, the complete removal of waxes is of importance to achieve uniform and rapid wettability and high absorbency in later stages of dyeing.

Also, non-fibrous impurities, for example, seed husk residues, are partly hydrolysed and removed. A total weight loss of 3–7 wt% of the material occurs through this stage of processing.

In cotton farming, approximately 8–10% of the total worldwide production of insecticides, herbicides and fungicides are used to achieve the expected high crop per area of farmland [1]. The first wet stage in textile chemical processing of cotton textiles is the desizing or the prewash/scouring of the fabric. Thus, it is not a surprise that substantial concentrations of such bioactive agents can be found in the waste water of textile mills situated in countries where no cotton farming takes place, but where the first wet stage in cotton processing is executed.

Besides a substantial concentration of alkali (20–40 g/L, 0.5–1 M NaOH), the treated solution contains wetting agents (to achieve rapid and uniform access), dispersants/emulsifier (to stabilise removed fatty components in the solution) and complexing agents (to bind earth alkali metal ions and heavy metal ions).

11.4 Prewashing of textiles from synthetic fibres

Knitted products do not require a desizing. For cotton fibres an alkaline scouring is performed. For fabric made from synthetic fibres a thorough prewash is sufficient to remove spin finishes, preparations and silicone oil before a dyeing process is possible.

The complete removal of the preparation is a prerequisite for a uniform dyeing.

In particular, the removal of silicone-based preparations used in knitting of polyurethane filaments requires attention, as these substances are more difficult to be emulsified and removed. In addition, these additives are used in comparable high amount such as up to 10 wt%. Often non-ionic detergents are recommended. Stable emulsification of the silicone oil has to be warranted to avoid coalescence and formation of oil droplets which leave mark during the following dyeing processes.

11.5 Setting of synthetic fibres

In the fibre spinning process, synthetic fibres are produced in a longitudinal form (see Chapter 2). A number of further deformation steps occur during the following stages of yarn/fabric formation (e.g. yarn twist, knitting loops, weaving). As a result, considerable tension had been packed inside a fabric on fibre scale. These local forces lead to undesirable changes in dimensions of a fabric during thermal processes, hot washing, ironing and so on.

The process of setting intends to remove these internal stresses through thermal relaxation. The temperature for setting is adjusted above the glass transition temperature T_g to achieve mobility of the polymer chain segments present in the amorphous parts of the fibre, however, the temperature must be kept clearly below the melting temperature T_m to maintain the orientation and order of the crystalline regions. Appropriate fixation conditions are of high relevance to achieve a number of technological requirements:

- Dyestuff uptake is determined by the state of order in the amorphous regions, thus differences in the thermofixation will change dyestuff uptake and also the final colour obtained in dyeing.
- Stabilisation of form and desirable reduction in shrinkage also depend on conditions applied during fixation.
- To achieve permanent fixation, temperature of this step has to be chosen approximately 30 °C above the highest temperature expected to occur in the following processes.

Thermosetting can be achieved by dry heat, steaming, hot water.

The position of the thermosetting process in the textile chemical processing depends on the material and product. Thermosetting of raw material bears the risk of burning in of spin finishes and preparations. Thermosetting in a later stage of processing permits the formation of permanent creases during the preceding treatment steps. Fixation of fully fashioned tights is also possible following the exhaust dyeing.

11.6 Alkalisation (causticising, mercerisation)

For cellulose textiles, treatment in concentrated alkali is a highly relevant process to modify the fibre properties. Processing conditions are dependent on the type of fibre:

Cotton: In cotton treatment, two major groups of processes are in use: causticising and mercerisation, the difference mainly being the alkali concentration applied during the treatment.

In mercerisation, the material is treated typically with alkali concentrations of 20–24 wt% NaOH (240–300 g/L, 6–7.5 mol/L) usually at room temperature for 25–40 s.

The process leads to a substantial reorganisation of the cellulose structure. Native cellulose in cotton is present in the allomorph crystal structure of Cellulose I. During the alkali treatment, NaOH enters the crystalline regions of the cellulose and a pseudo-stochiometric product is formed, the so-called alkali cellulose. During the immersed state in the NaOH solution the fibres are present in highly swollen state. Upon dilution of the NaOH solution with water, the cellulose structure reorganises; however, the unit cell of the crystalline lattice ends up with the allomorph crystal structure of Cellulose II (Figure 11.4).

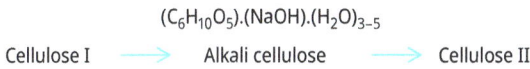

$$(C_6H_{10}O_5).(NaOH).(H_2O)_{3-5}$$

Cellulose I $\longrightarrow$ Alkali cellulose $\longrightarrow$ Cellulose II

Figure 11.4: Formation of alkali cellulose and Cellulose II during alkalisation.

The high swelling leads to considerable shrinkage forces and the fabric must be kept under tension to maintain the dimensions. Temperature control of the process is important as the formation of alkali cellulose is an exothermal reaction which releases considerable amounts of heat. Mercerisation at 20 °C thus requires constant cooling of the NaOH process bath, which is expensive due to substantial energy costs for cooling. The alkali treatment with application of tension is an important characteristic of mercerisation, when compared to alkalisation where no tension is applied to the fabric. Mercerisation leads to a number of positive effects:

– A considerable increase in lustre is obtained.
– Tensile strength of the fibres increases.
– Dyestuff accessibility and fixation increases.
– The kidney-like cross-section of the cotton fibres changes to a round-shaped oval form of mercerised fibres.

- Dyestuff uptake by unripe and dead cotton fibres improves, thus unwanted appearance of light fibres in dyed fabric is reduced.
- The dimensional stability of the material increases.

In causticising, the major intention of the treatment is to achieve higher dyestuff uptake, improve uniformity in dyeing of dead and unripe fibres and achieve higher dimensional stability.

In practical application, the treatment in alkali lasts only for 30–60 s, thus only the outer part of the cotton fibres in a yarn is transformed into Cellulose II form. Complete transformation would require several hours of treatment in alkali. This, however, is not possible due to increased fibre degradation through oxidative effects and cellulose reactions at high alkali concentrations.

At the end of the alkali treatment, the concentration of alkali is first reduced by rinsing in diluted alkali solutions. Thereby, the highly swollen state of the fibre collapses and the cellulose reaches its final state. First rinse steps are performed with diluted alkali solution to achieve a minimum concentration of NaOH in the effluent, which is sufficiently high to permit a recovery through reboiling [2, 3]. A hot rinse then follows and finally a cold rinse with water is performed. In case residual amounts of alkali are critical for the following steps (e.g. reactive dyeing in exhaust processes), a neutralisation of any core alkali with the use of a weak organic acid, such as acetic acid, is performed.

Regenerated cellulose fibres must be handled with great care as cellulose with low DP exhibits a solubility maximum near to 10 wt% NaOH solutions. Normal viscose fibres thus are not mercerised. Modal fibres and lyocell-type fibres which exhibit higher molecular weight can be treated in concentrated alkali under certain precautions.

While cotton fabric can be entered into the alkali solution both in dry or in wet state, regenerated cellulose fabric should enter into the concentrated alkali solution in dry state and must be washed out rapidly with the use of hot water. This permits a fast dilution of the concentrated alkali in the washing compartment and thus minimises the time in which critical concentrations are present in the fibres.

For stabilisation of the fabric dimensions and reduction of fibrillation also, alkalisation at lower concentrations such as 120–160 g/L NaOH are in use.

Other alkali solutions like KOH or LiOH or organic amines can be used to replace NaOH. The type of base used also changes the effect of the treatment as the type of cation (Li^+, Na^+, K^+) also influences the swelling of cellulose [4, 5].

The influence of the type of cation results from the major differences in accessibility into the cellulose structure [6]. Depending on the type of alkali used, different sectors of the porous cellulose fibres are reorganised, thus leading to substantial differences in the effects of the later finishing steps.

Technically for alkalisation of regenerated cellulose fibres also KOH solutions are in use. Representative concentrations are near to 250 g/L (4.4 M) KOH. This treatment is executed as a cold pad batch treatment.

A special technique used for alkalisation of cotton fabric is the treatment in liquid NH_3 at a temperature of −33 °C for 6–12 s. This process is often followed by a finishing process where cross-linking chemicals are applied under the conditions of moist curing. By the action of NH_3, Cellulose I is transferred to Cellulose III. The NH_3 is either evaporated at the end of the process or washed off in water. The technical justification for the choice of this rather expensive combination of processes comes from the high level of crease-recovery properties and abrasion resistance that can be realised with such techniques.

When an intensive alkalisation in aqueous solution follows the NH_3 treatment, the Cellulose III structure is transformed back to the Cellulose II form.

11.7 Alkalisation of polyester fibres

Fabric made from 100% polyester fibres can also be treated in concentrated alkali solutions. Representative concentrations are 10–20 g/L NaOH (0.25–0.5 M), which is substantially lower compared to alkalisation of cellulose fibres. The temperature is in the range of 95–110 °C and the duration of the process is up to 1 h. The chemical background is completely different. The main reaction with aqueous alkali solutions is a hydrolytic cleavage of the ester bond at accessible sites near to the fibre surface (Figure 11.5). Alkali soluble oligomers are formed, which dissolve in the hot treatment bath. The polyester group of PETP can be hydrolysed with caustic soda solutions to form soluble degradation products, in an ideal case glycol and the disodium salt of terephthalic acid.

The reaction is accompanied with substantial weight loss (10–20 wt%) and considerable reduction in tensile strength of up to 35% of the initial value. The process is mainly located onto the surface, as the access of the aqueous alkaline solution into the apolar fibre is limited.

When the polyester hydrolysis is performed under well-controlled conditions, a reduction in fibre diameter can be achieved, thus leading to thinner fibres with lower linear density.

The treated fibres exhibit silk-like lustre and handle, reduced lustre (as a result of the more uneven fibre surface), higher hydrophilicity (water uptake and wettability), reduced development of static charge and apparently higher dyestuff uptake (as a result of the reduced lustre).

The process is usually performed batchwise, for example, in jet or beam dyeing apparatus.

As a disadvantage of such a technique, substantial amounts of organic matter are released into the waste water together with the used caustic soda treatment bath.

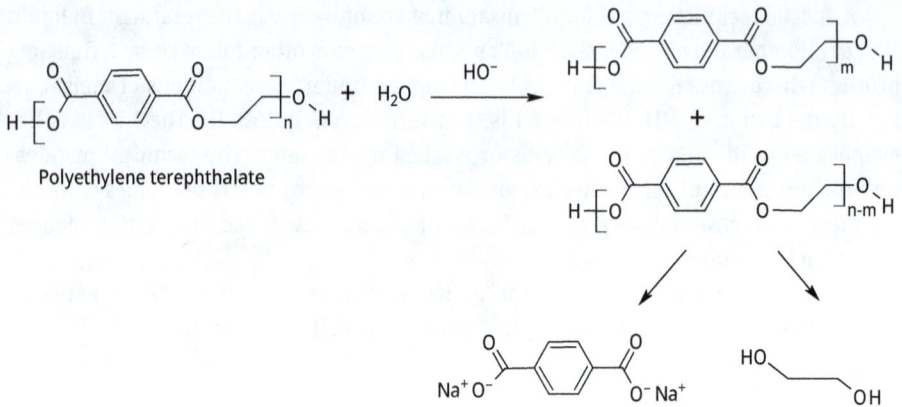

Figure 11.5: Hydrolysis/peeling of PETP to terephthalic acid and glycol.

11.8 Bleaching

11.8.1 General aspects

Natural fibres like cotton, flax, wool contain coloured substances which must be removed before dyeing of brilliant shades can be considered. In case of cotton and flax, coloured components result from lignin and related polyphenolic components. In case of wool, oxidative or reductive decolourisation of melanin pigments is necessary.

Synthetic fibres usually require only a careful washing to remove spin finishes and preparations, such as silicones; however, bleaching is not required and is performed only in special cases such as for white products.

The choice of the bleach process and the intensity of the treatment both depend on the material to be treated and the colour depth applied in the following dyeing process.

11.8.2 Peroxide compounds

Hydrogen peroxide H_2O_2: Hydrogen peroxide represents the most widely used oxidant in textile bleaching.

Usually aqueous solutions with a content of up to 35 wt% are in use. At higher concentrations of peroxide, the risk of uncontrolled decomposition increases.

During storage, hydrogen peroxide tends to decompose slowly in presence of catalytically active impurities such as dust (eq. (11.1)).

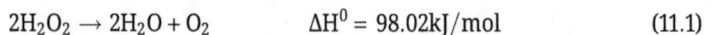

$$2H_2O_2 \rightarrow 2H_2O + O_2 \qquad \Delta H^0 = 98.02kJ/mol \qquad (11.1)$$

The reaction is exothermal, thus heat is released during the decomposition reaction. In case of slow decomposition, small amounts of heat are dissipated to the environment; however, at higher concentration or in the presence of high amounts of activating catalysts exothermal decomposition can develop into an explosion.

Cotton is the most important textile material to be bleached with peroxide. Thus, this fibre will be used as a representative to discuss the chemistry of hydrogen peroxide in bleach processes.

Important catalysts for decomposition of peroxide are heavy metal ions such as Cu^{2+}, Fe^{3+}, Ag^+, $Mn^{2+/4+}$ and corresponding metals. Also, precious metals such as Pt and Rh catalyse the decomposition. Thus, any heavy metal ions in solution must be either complexed or masked in the presence of appropriate stabilisers (complexing agents, adsorbents like magnesium or calcium water glass, phosphonates).

The apparatus for peroxide bleach should be manufactured from stainless steel.

The presence of iron or other heavy metal residues (e.g. from rust, soil) on the fabric will lead to localised uncontrolled decomposition of the peroxide, which instead of a desired bleach reaction will rapidly degrade the cellulose fibre polymer (= catalytic damage) [7].

Hydrogen peroxide is a weak acid with $pK_s = 2.4 \ 10^{-12}$. Thus, dissociation of hydrogen peroxide occurs in the alkaline solution (eq. (11.2)).

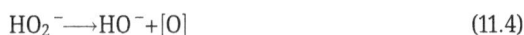

$$H_2O_2 + OH^- \longrightarrow HO_2^- + H_2O \tag{11.2}$$

$$H_2O_2 \longrightarrow H_2O + [O] \tag{11.3}$$

$$HO_2^- \longrightarrow HO^- + [O] \tag{11.4}$$

Peroxide bleach is performed in the alkaline solution, thus the oxidising agent is the hydroperoxide anion. The redox potential (vs. standard hydrogen electrode) of hydrogen peroxide depends on the pH value and ranges from +470 mV (pH 8–8.5) to +250–+400 mV (pH 11–12.5) [8]. It is important to consider that measurements of redox potential indicate the level of oxidation power, however, they are not indicated if possible chemical reaction pathways exist, which lead to a bleach effect.

In presence of catalytically active substances or at high alkalinity (e.g. above pH 12), the decomposition first leads to the formation of monomolecular oxygen (O), which leads to uncontrolled and unselective oxidation reactions and fibre damage (eq. (11.3)).

The stability of the hydroperoxide ion is lower than that of peroxide, thus alkalinity increases the tendency of decomposition to elemental oxygen (eq. (11.4)). The level of fibre damage can be determined by the analysis of the DP, however, local damage (holes) will not be reflected in the DP value.

A typical composition of a peroxide bleach contains H_2O_2, NaOH, stabilising and complexing auxiliaries, detergents.

At neutral or acidic pH the oxidative power of hydrogen peroxide is low, thus such conditions are usually not applied.

The concentration and conditions chosen depend on the machinery used, temperature and duration of the process. For full whiteness, often two peroxide bleach stages are combined in a consecutive series.

For a long time, a combination of hypochlorite bleach and peroxide bleach had been used to obtain a high degree of whiteness. The use of hypochlorite in textile bleaching, however, disappeared mainly because of the formation of halogenated products (AOX = adsorbable halogenated compounds) which are released with the waste water.

From the view of environmental aspects, the bleach processes based on hydrogen peroxide are recommended because under normal conditions formation of critical by-products is negligible.

Peracetic acid (peroxoacetic acid): An alternative peroxide-based bleaching agent is peroxoacetic acid (peracetic acid).

The synthesis usually bases on the chemical equilibrium between water, hydrogen peroxide, acetic acid and peracetic acid (eq. (11.5)). In the presence of an acid catalyst (e.g. sulphuric acid), the equilibrium establishes within few hours. The theoretically possible preparation through the reaction of hydrogen peroxide with the anhydride of acetic acid is not recommended as the process can yield diacetyl peroxide. This product then separates due to its low solubility in water in the form of a concentrated explosive liquid.

$$CH_3 - CO_2H + H_2O_2 \longrightarrow CH_3 - CO_3H + H_2O \tag{11.5}$$

The reactivity of peracetic acid is substantially higher compared to hydrogen peroxide, thus bleach operations at low temperature are possible. The redox potential of peracetic acid is in the range of +580–+680 mV at pH 5–6. The bleach reactions follow a different chemical pathway. Through homolytic cleavage of the O–O bond two radicals are formed (eq. (11.6)). The formed hydroxyl radical .OH is very reactive and acts as a bleaching agent.

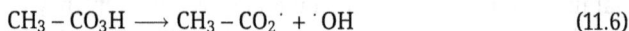

$$CH_3 - CO_3H \longrightarrow CH_3 - CO_2^{\cdot} + {}^{\cdot}OH \tag{11.6}$$

The difference in reaction chemistry also explains the high antimicrobial activity of peracetic acid. Thus, besides its use for textile bleach processes another application is found in laundry, where an in situ disinfection can be achieved through an addition of small amounts of peracetic acid to the last rinse of a wash process.

Peracetic acid is also formed as a product from a reaction between hydrogen peroxide and a chemical activator TAED (tetraacetylethylenediamine) [9] (Figure 11.6).

The use of an activating system in textile bleach operations has been proposed and tested; however, the costs of the system TAED + peroxide are rather high compared to the use of a hydrogen peroxide–based bleach.

Caroate, Peroxymonosulfuric acid (2 KHSO$_5$ · KHSO$_4$ · K$_2$SO$_4$): Peroxomonosulfuric acid exhibits a high oxidation potential (+1,440 mV at acidic pH) and could be used as bleaching agent and oxidant in pretreatment processes (eq. (11.7)). The product is

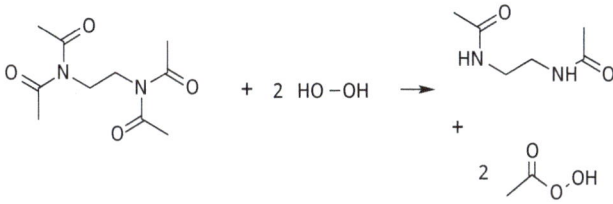

Figure 11.6: Formation of peracetic acid from hydrogen peroxide and TAED.

handled as a triple salt between persulfate, hydrogensulfate and sulphate. As a result, considerable amounts of the product are required to deliver the same reducing equivalents compare to the 35 wt% hydrogen peroxide solution.

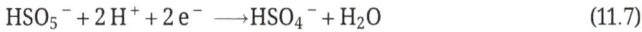

$$HSO_5^- + 2\,H^+ + 2\,e^- \longrightarrow HSO_4^- + H_2O \qquad (11.7)$$

11.8.3 Halogen-based oxidants

The use of halogen-based oxidants must be considered with care, as the formation of halogenated products in the waste water (AOX) is limited strictly in Europe both with regard to concentration and load of AOX.

Sodium hypochlorite NaOCl/Cl₂: Sodium hypochlorite is still in use as a bleaching agent in special applications (e.g. garment wash of jeans), the anti-felt treatment of wool (chlorine Hercosett® process), discharging of mismatching dyeings and industrial laundry processes.

The major chemical in use is an aqueous solution of NaOCl, which has been prepared by the introduction of chlorine into caustic soda solution. In the first step, hypochlorous acid and hydrochlorous acid are formed, which are then neutralised by caustic soda (eqs. (11.8) and (11.9)).

$$Cl_2 + H_2O \rightarrow HOCl + HCl \qquad (11.8)$$

$$Cl_2 + 2\,NaOH \rightarrow NaOCl + NaCl \qquad (11.9)$$

Above pH 10 the major species present is the hypochlorite anion (OCl⁻), which exhibits a redox potential of +850 mV and is a strong bleaching agent. At pH 4–6 the major species is the undissociated hypchlorous acid HCl. At pH 2 chlorine is mainly present.

Upon acidification, the equilibrium (eq. (11.8)) is shifted backwards and chlorine is released. At low concentrations of hypochlorite, an aqueous solution of Cl₂ (pH = 2) is formed which then tends to form chlorinated products (redox potential +1,250–+1,400 mV).

At high concentrations, hypochlorite acidification of the bath leads to the formation of such amounts of elemental chlorine that the solubility in water is exceeded and a dangerous deliberation of gaseous chlorine will occur.

Besides the aqueous solution of hypochlorite, a number of chlorine-liberating compounds can be used to generate hypochlorite or chlorine in situ, examples are

N-chloro tosylamide sodium salt (p-Toluenesulphonic acid chloroamide sodium salt, chloramine T) and sodium dichloroisocyanurate (NaDCC) (Figure 11.7).

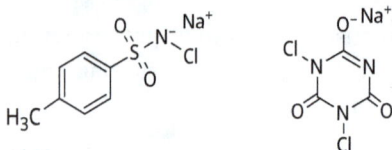

Figure 11.7: Chemical formula of N-chloro tosylamide sodium salt (chloramine T) and sodium dichloroisocyanurate (NaDCC).

Similar to paper and pulp production, the use of chlorine/hypochlorite-based bleach processes has almost disappeared in Europe as the formation of AOX is no longer tolerated (wastewater limit for AOX calculated as Cl 0.5 mg/).

At the end of a hypochlorite bleach, a dechlorination has to follow to remove any residual oxidant. In case a peroxide bleach follows the hypochlorite bleach, the peroxide stage will neutralise any chlorine left (eq. (11.10)). An active neutralisation of the residual chlorine can also be achieved with reducing agents such as sodium thiosulphate $Na_2S_2O_3$ or sodium sulphite Na_2SO_3 (eq. (11.11)).

$$NaOCl + H_2O_2 \longrightarrow NaCl + H_2O + O_2 \tag{11.10}$$

$$2\,Na_2S_2O_3 + NaOCl + 2\,HCl \longrightarrow Na_2S_4O_6 + 3NaCl + H_2O \tag{11.11}$$

Sodium chlorite NaClO$_2$: Sodium chlorite is a strong bleach agent for weakly acidic baths (pH 3–4.5). At this range mainly chloric acid ($HClO_2$) is present (redox potential +1,040–+1,200 mV). The product is either handled as an aqueous solution (30 wt%) or as a mixture in combination with NaCl. In acidic conditions at pH values near pH 2, the concentration of the toxic ClO_2 increases (eq. (11.12)).

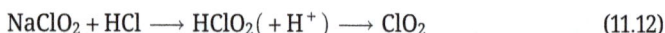

$$NaClO_2 + HCl \longrightarrow HClO_2(+H^+) \longrightarrow ClO_2 \tag{11.12}$$

A possible application is for bleaching of cellulose fibres (cotton, flax) in combination with PETP and PAC. The risk of fibre damage is lower compared to a hypochlorite bleach; however, the corrosivity of the chemical, the toxicity of volatile ClO_2 and the AOX content in the waste water prevent a wider application.

The chemical is not applicable for PA- or PUR-containing material as fibre damage and yellowing will occur.

11.9 Singing

Singing is usually the first process performed in the pretreatment with the fabric produced from staple-fibre yarns. As a high number of fibre ends are present in the fabric, in particular the weakly bound fibre ends stick out from the surface and form an irregular surface. A clean surface, free of long fibres, is a particular requirement for fabric which should be printed on in a later stage. The long fibres would deteriorate the sharp contour of the print, cause reservations and lead to a fuzzy surface. These fibre ends are removed thermally in the singing process.

During singing the fibre ends are burned off through direct action of gas flames, or IR radiation. The material passes through the flames with such a speed that only tiny fibre ends are burnt off. In case of PETP fibres, the temperature of the flames has to be sufficiently high to permit burning of the fibres and to avoid melting and formation of spherules from molten fibres.

In case of a machine stop, the burners have to be shielded off immediately to avoid burning of the fabric. The ignition of the singed material is prevented by rapid cooling on cooling drums or by direct immersion in the desizing bath, which extinguishes any glowing dust.

11.10 Carbonisation

Raw wool contains a considerable amount of impurities and non-fibrous components (see Chapter 5). Thus, in the first step the raw wool is washed. Besides soil, dirt and sweat, wool fat (lanolin) is also removed during the wash. The first wash of raw wool is usually performed near the site where the wool has been sheared.

Depending on the type of the product, the order of the next processing steps can vary.

Carbonisation is a cleaning step with the intention to remove tightly bound cellulosic impurities which were not removed mechanically. The process bases on different sensitivities of wool and cellulose towards the hydrolytic action of acids (H_2SO_4, HCl, lewis type acids such as $AlCl_3$, $MgCl_2$, $ZnCl_2$).

The material is impregnated with diluted acid, excess acid is removed and the wetted material is dried/ heated slowly. As an example, wool is impregnated with 2–6 wt% H_2SO_4 and then heated up to a maximum temperature of 130 °C. The cellulosic parts are hydrolysed and partly carbonised. The brittle impurities are then removed mechanically or through the following washing and neutralisation steps.

The action of sulphuric acid also leads to the chemical modification of wool keratin.

Besides hydrolytic degradation of the keratin, sulphatation (eq. (11.13)) and formation of sulphamic acid (eq. (11.14)) also occur.

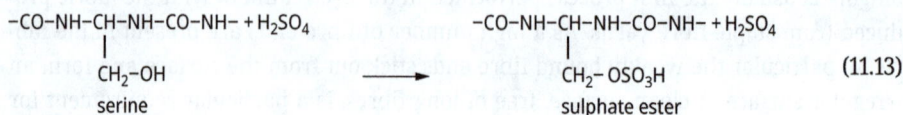

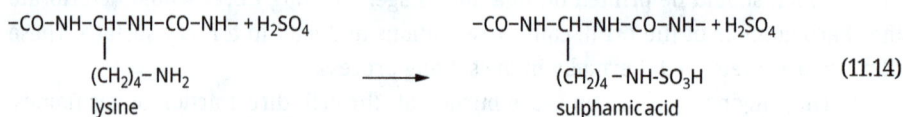

$-CO-NH-CH-NH-CO-NH-\ +H_2SO_4$ $-CO-NH-CH-NH-CO-NH-\ +H_2SO_4$

     CH_2-OH ⟶ CH_2-OSO_3H (11.13)

     serine   sulphate ester

$-CO-NH-CH-NH-CO-NH-\ +H_2SO_4$ $-CO-NH-CH-NH-CO-NH-\ +H_2SO_4$

     $(CH_2)_4-NH_2$ ⟶ $(CH_2)_4-NH-SO_3H$ (11.14)

     lysine   sulphamic acid

The formation of sulphate esters and sulphamic acid modifies dyestuff uptake and alkali solubility of wool due to a higher number of negative charges and solubilising groups.

11.11 Reductive bleach

Wool is often bleached in a two-stage process, one being an oxidative step followed by reductive bleach. Usually hydrogen peroxide is used to perform the oxidiative stage, at pH 8–10, followed by a treatment with reducing agents such as sodium bisulphite, thiourea dioxide, hydrosulphite. Also, $NaBH_4$ (sodium boranate) in combination with sodium bisulphite has been proposed for reductive bleaching of wool [10].

While raw white wool exhibits photobleaching under irradiation, bleached wool exhibits rapid photoyellowing both in dry and in wet state.

Similar to other polymers, the photooxidation of wool is supposed to occur via a radical chain reaction. Coloured chromophores found in oxidised wool are supposed to exhibit either a high content of proline, tyrosine or tryptophan.

Fluorescent whitening agents (FWA) are known to speed up photoyellowing considerably.

The first step of the reaction is supposed to be a photoinduced electron transfer from electron-rich amino acids to the irradiated FWA, which then leads to the formation of H_2O_2 and O_2. radicals. Catalytic metal ions can increase the rate of decomposition of H_2O_2 and accelerate photoyellowing.

The follow-up reactions between H_2O_2 or O_2. with wool lead to the formation of coloured products.

Strategies to improve photoyellowing of bleached wool include:
- careful removal of any catalytic metal ions with appropriate complexing agents (e.g. ethylenediaminetetraacetic acid, EDTA)
- addition of UV absorbers, which, however, inactivate FWA
- careful removal of any FWA present in wool, for example, in case of wool–polyester blends, where a disperse FWA is used for the polyester.

11.12 Wool anti-felt treatment

The anti-felt treatment of wool can be allocated as a pre-treatment process both before dyeing as well as a finishing process.

The traditional process utilises a combination of oxidative and reductive treatments in combination with the coating of the fibre surface by polymers, with the aim to prevent shrinkage of woollen fabric during repetitive washing/tumbling cycles (superwash finished wool).

Felting is the result of the unidirectional friction which comes from the scales forming the cuticle. As a general approach, shrinkage in washing processes is reduced when the mechanical stability of the edges of the scales is lowered. In addition, the scales are covered with a polymer film.

Different chemical concepts are in use. The general principle follows the sequence:

1. Oxidative treatment – 2. Reductive treatment – 3. Polymer coating

Stages 1 and 2 are subtractive processes, which remove materials from the wool fibres, while stage 3 is an additive process.

The Chlorine–Hercosett Process: In the first step, wool is treated with a cold acidic solution of chlorine. Typical concentrations are pH 1.5–2, 10–15 s dwell time, temperature below room temperature and chlorine consumption in the dimension of 1.5 wt% of the wool.

In the chlorination step covalently bound lipids are removed from the surface of the wool hair and the cystein disulphide bonds are oxidised to the corresponding sulphonic acids (cysteic acid) (eq. (11.15)).

$$R - S - S - R' + 5Cl_2 + 6H_2O \rightarrow R - SO_3H + HO_3S - R' + 10\,HCl \qquad (11.15)$$

The reactivity of chlorine is very high, thus the oxidative attack is concentrated on the scales forming the wool cuticle. The mechanical strength of the edges of the scales is reduced and felting will be lowered as the unidirectional friction effect reduces.

The treatment solution is then washed off to reduce the acid content and a treatment with sulphite/hydrogensulphite near neutral pH follows. After rinsing with water, a polymer is deposited on the wool surface to cover the scales with a polymer layer and reduce friction further. Typical polymers can base on polyamide-epichlorhydrine polymers (Hercosett), polyurethanes as well as polyisocyanates.

The chlorine-based antifelt treatment of wool delivers high-quality products, however, it is no longer acceptable with regard to environmental aspects. The chlorination process leads to the formation of substantial amounts of AOX products which are released into the waste water and can exceed legal wastewater limits by far.

The search for alternative peroxide-based processes leads to the development of less-polluting techniques.

Caroate based wool treatment (EXP-Process): In the first step of the process the wool is treated with an acidic mixture of potassium persulphate (Peroxymonosulphuric acid 2 $KHSO_5 \cdot KHSO_4 \cdot K_2SO_4$) [11]. Due to the lower reactivity of the persulphate, prolonged reaction time for the oxidation is required. The oxidative reaction leads to the formation of oxidised cystein bondings (eq. (11.16)).

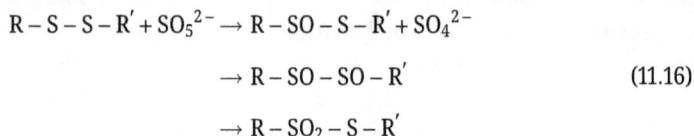

$$R - S - S - R' + SO_5^{2-} \rightarrow R - SO - S - R' + SO_4^{2-}$$
$$\rightarrow R - SO - SO - R' \tag{11.16}$$
$$\rightarrow R - SO_2 - S - R'$$

Similar to the chlorine process, a rinse follows to reduce the acid concentration, and a treatment with a sulphite/bisulphite solution is performed. The treatment with sulphite via a number of redox reactions leads to the cleavage of the oxidised cystein bonds and formation of S-thiosulphate esters (Bunte salts). (eq. (11.17))

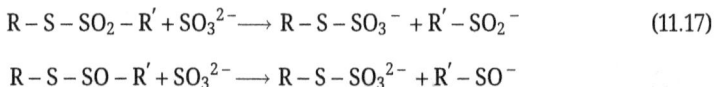

$$R - S - SO_2 - R' + SO_3^{2-} \longrightarrow R - S - SO_3^- + R' - SO_2^- \tag{11.17}$$
$$R - S - SO - R' + SO_3^{2-} \longrightarrow R - S - SO_3^{2-} + R' - SO^-$$

The sulphenic acid (R–SOH) and the sulphinic acid (R–SO$_2$H) will undergo rapid disproportionation to the corresponding thiol (RSH), disulphide (R–S–S–R) and sulphonic acid (R–SO$_3$H).

After careful rinsing, the wool is treated with a polymer to coat the surface and reduce unidirectional friction. As no chlorine is involved in the reaction, the formation of AOX is not relevant. The only concentrations of solutes in the waste which are worth to be mentioned are the concentrations of released sulphate and sulphite. Sulphite will oxidise rapidly to sulphate. A representative value for the wastewater limit for the sulphate concentration is 200 mg/L SO_4^{2-} for release into a communal wastewater treatment plant. This limit has been set to prevent corrosion of concrete tubes which can accelerate at higher sulphate concentration.

Different alternative methods to reduce the mechanical strength of the cuticle scales have been proposed for anti-felt treatment of wool. Important processes include enzymatic processes and plasma treatment.

The treatment of wool with proteases leads to hydrolysis of the surface proteins and thus to a reduced friction. Similar atmospheric plasma (corona discharge) or low pressure plasma (glow discharge) treatments are applied to degrade the outer layers of the cuticle by the action of free radicals, electrons and ions generated in the plasma.

Such processes are often used in combination with addition of a polymer coating. Also, the addition of silicone-based polymers is in use; however, the use of silicon-based products has to be considered with care when a dyeing process is planned to follow.

11.13 Cationisation

Through modification of cellulose fibres with reactive compounds which bear a cationic group, the negative charge of the cellulose material can be reverted to a positive charge. Representative systems are shown in Figure 11.8 [12].

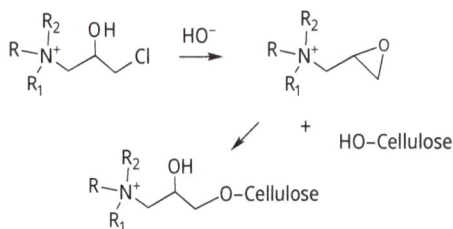

Figure 11.8: Cationisation of cellulose by reactive components.

While the binding of a reactive dye usually occurs through the generation of a covalent bonding in the cellulose fibre, a second mechanism of binding is possible in case the cellulose fibre is cationised. In this case also, the ionic bonding of the negatively charged dye with the cationic groups in the cellulose fibre permits a strong binding of the dye.

During a reactive dyeing, dyestuff hydrolysate is generated as a by-product of the hydrolysis reaction of the anchor groups. These molecules still bear a negative group and thus will be attached to the cationised cellulose. As a result, high degree of bath exhaustion and high rates of wet fastness can be achieved on cationised fibres.

In the dyeing of cationised cellulose, the ionic binding to the cellulose will depend on the distribution of the cationic groups in the material, thus the dyeing process will highlight the uneven distribution of cationic reagent in the material. The process of cationisation needs to be performed with the same accuracy as a dyeing process. Correction of an imperfect dyeing will be difficult.

11.14 Degumming of silk

Raw silk consists of two components, the two silk filaments which consist of fibroin (70–80 wt%) and the surrounding coating of sericin (30–20 wt%), a gummy substance which binds together the two filaments.

During silk processing the sericin is removed through degumming. Usually hot, slightly alkaline extraction in presence of detergents is used to dissolve the sericin.

Sericin is recovered from the degumming bath and is used in cosmetic formulations.

Degummend silk exhibits a substantial affinity to bind metal ions, which is used in the production of weighted silk. To compensate for the weight loss from degumming, silk can be treated with $SnCl_4$, Na_2HPO_4 and water glass. The increase in weight can reach values of more than 50 wt% of fibroin. At higher level of weighting, the durability of fibroin decreases substantially.

11.15 Production of microfibres

Microfibres usually exhibit a linear density of less than 1 dtex. The production through classical spinning processes is difficult for several reasons:
- The mechanical stability of the very thin fibres leaving the spinneret is very low. In case of melt spinning and solution spinning, process stability is limited.
- For a given number of bores, the productivity of the spinneret reduces with linear density of the fibres.

As a result of the technical difficulties to produce very fine diameters of synthetic fibres, alternative technologies were developed to produce such fibres in technical scale. Besides the hydrolytic removal of fibre polymer to reduce the fibre diameter of polyester fibres also bi-component fibres (polyester/polyamide) are used to prepare microfibres through splitting. In this process a bi-component fibre is produced which consists of two non-compatible polymers. By action of heat, chemicals (e.g. alkali) and intensive mechanical treatment the two components of the fibre then are split into microfibres of low diameter.

References

[1] https://www.proplanta.de/Baumwolle/Pflanzenschutz-Pflanzenbauliche-Basisinformationen-Baumwolle_Pflanze1171633468.html (accessed 26.01.2018)

[2] Bechtold, T., Burtscher, E., Sejkora, G., Boblerter, O. Moderne Verfahren zur Laugenrückgewinnung. Internationales Textilbulletin – Veredlung. 1985, 31, 5–26.

[3] Bechtold, T., Gmeiner, D., Burtscher, E., Bösch, I., Bobleter, O. Flotation of particles suspended in lye by the decomposition of hydrogen peroxide. Sep. Sci. Technol. 1989, 24, 441–451.

[4] Zhang, W., Okubayashi, S., Bechtold, T. Fibrillation tendency of cellulosic fibres –part 3. Effects of alkali pretreatment of lyocell fibre. Carbohydr. Polym. 2005, 59, 173–179.

[5] Zhang, W., Okubayashi, S., Bechtold, T. Fibrillation tendency of cellulosic fibres – Part 4, Effects of alkali pretreatment of vairous cellulosic fibres. Carbohydr. Polym. 2005, 61, 427–433.

[6] Jaturapiree, A., Manian, AP., Bechtold, T. Sorption Studies on Regenerated Cellulosic Fibres in Salt-Alkali Mixtures. Cellulose. 2006, 13, 647–654.

[7] Kongdee, A., Bechtold, T. Fe(III) Complex catalyzed damage of cellulose fibers during bleaching. AATCC Rev. 2005, 5/4, 30–34.

[8] Döcke, W., et al. Appretur: Vorappretur, Bleichen, Trocknen, mechanische und chemische Appretur. VEB Fachbuchverlag, Leipzig, 1990.

[9] Buetzer, P. Peroxyessigsäure: Einfach aber wirksam. CLB. 2012, 64/3, 96–115.

[10]   Johnson, NAG., Russel, I. Advances in Wool Technology, Woodhead Publishing Ltd, Cambridge, England, 2009.
[11]   Bechtold, T., Mahmud-Ali, A., Siroky, J., Riehl, M., Krüger, M. Superwash ohne Chlor – Ein neuer Standard wird technische Realität. Melliand Textilberichte. 2012, 3, 148–150.
[12]   Schindler, WD., Hauser, PJ. Chemical Finishing of Textiles, Woodhead Publishing, Cambridge, England, 2004.

Take home messages

Before dyeing or finishing operations can be begun, an appropriate pre-treatment is required.

To remove impurities, waxes and preparations, an intensive prewash is a general standard. The actual conditions depend on the chemical stability of the fibre and on the substances to be removed. Comparable harsh conditions are used in the pre-treatment of cotton fibres.

Natural fibres usually require a bleach operation to achieve a fully white colour.

For synthetic fibres in most cases, a wash in combination with a fluorescent whitening agent is sufficient.

Stabilisation of fabric dimensions in case of synthetic fibres is achieved by thermal relaxation, in case of cellulose fibres an alkali treatment is used, for wool a steam setting can be applied.

Depending on the character of the fabric and the product to be manufactured, a wide range of special pre-treatments is used. Examples are singing, carbonisation, cationisation, mercerisation, anti-felt treatment of wool.

Quiz

1.   What is the difference between alkalisation and mercerisation?
2.   What happens during thermal fixation of synthetic fibres?
3.   What is the difference between desizing and degumming?
4.   Which bleach agents are favourable and why?
5.   What is the principle of anti-felt treatment of wool?

Exercises

1.   Estimate the COD value of starch, CMC, PVA (90% hydrolysed) based on their average formula of a monomeric unit. Calculate the values in mg O_2/g of sizing polymer.

2. How much of caroate (2 $KHSO_5 \cdot KHSO_4 \cdot K_2SO_4$) is required to deliver the same amount of oxidation equivalents as available in a solution containing 9 g/L 35% hydrogen peroxide (density 1.13 g/ml)?
3. Compare the active chlorine content (calculated as elemental chlorine) liberated from 1 L of a NaOCl solution (150 g/L NaOCl), 1 kg of solid Chloramine T and solid NaDCC (consider both products as 100% content).
4. A raw fabric made of PA 6,6 (90 wt%) and elasthane(10 wt%) is washed in a jet using a liquor ratio of 1:7 followed by two rinsing baths. The elasthane contains 10 wt% silicone oil. What is the expected concentration of silicone oil in the entire stream of waste water?

Solutions

1. (Question 1): Mercerisation is performed at higher alkali concentration and under tension, which is not necessarily the case in alkalisation.
2. (Question 2): The fibre polymer is heated above T_g and thus chain flexibility in the amorphous regions is increased to such an extent that internal stress from drawing of fibres, spinning and fabric formation are decreased.
3. (Question 3): Desizing is the process of removal of size from warp yarn in woven fabric. Degumming is the process of removal of sericin from raw silk.
4. (Question 4): Peroxide-based bleaching agents are favourable as chlorine-based agents tend to form AOX as by-products. Reductive bleaching delivers products that can exhibit increased photoyellowing during later use.
5. (Question 5): Shrinkage of wool during wash processes can be reduced by chemical weakening of the scales of the cuticle. This is achieved by a combination of oxidative and reductive processes, followed by the addition of a polymer as protective coating.
6. (Exercise 1): The unit formula of
CMC: $C_8H_{12}O_7$ M = 220 g/mol; 1 mol CMC delivers $8CO_2$ and $6H_2O$, thus 22–7 oxygen atoms are required, which corresponds to 7.5 O_2; 1 g of polymer will require 7.5 × 32/220 g O_2 = 1.09 g = 1,090 mg oxygen.
PVA: C_2H_4O 90% M = 44 g/mol; 1 mol delivers 2 CO_2 and 2 H_2O, thus 6–1 oxygen atoms are required, which corresponds to 2.5 O_2; 1 g of polymer will require 2.5 × 32/44 g O_2 = 1,818 mg oxygen.
Non-deacetylated PVA: $C_4H_6O_2$ 10% M = 86 g/mol; 1 mol delivers 4 CO_2 and 3 H_2O, thus 11–2 oxygen atoms are required, which corresponds to 4.5 O_2; 1 g of polymer will require 4.5 × 32 / 86 g O_2 = 1,674 mg oxygen. With a distribution of 90:10, the final value will be 0.9 × 1,818 + 0.1 × 1,674 = 1,804 mg oxygen per 1 g of substance.
7. (Exercise 2): The molar H_2O_2 concentration in the solution is $C(H_2O_2)$ = 9 × 0.35/34 = 0.0926 mol/L. An equivalent amount of peroxide is present in a solution which contains 0.0926 × 307.38 g/L = 28.4 g/L.

8. (Exercise 3): M(NaOCl) = 74.44 g/mol; 1 mol of NaOCl is equivalent to 1 mol of Cl_2 M(Cl_2) = 70.9 g/mol. A solution of 150 g/L NaOCl is thus equivalent to 150 × 70.9/ 74.44 = 143 g/L Cl_2. Chloramine T: M = 281.69 g/mol; 1 kg of solid chloramine T corresponds to 1,000 × 70.9/281.69 = 251 g/kg Cl_2. NaDCC: M = 219.95 (255.9 g/mol): 2 g/ mol, delivers two Cl_2. Thus, 1 kg of solid NaDCC corresponds to 1,000 × 70.9 × 2/ 219.95 = 644 7 g/kg (dihydrate: 554.1 g/kg) Cl_2.

9. (Exercise 4): The total wastewater stream for 1 kg will be 3 × 7 L. The amount of silicone oil released per 1 kg of fabric will be m = 1 kg × 0.1 × 0.1 = 10 g. Thus, the concentration in the wastewater stream will be 10 g/21 L = 0.48 g/L.

List of abbreviations/symbols

DP	Degree of polymerisation
PETP	Polyethylene terephthalate
CMS	carboxymethylstarch
CMC	carboxymethylcellulose
COD	chemical oxygen demand
PVA	polyvinylalcohol
T_g	glass transition temperature
T_m	melting temperature
AOX	Adsorbable halogenated products
TAED	Tetraacetyl ethylenediamine
PA	polyamide
PUR	polyurethane
FWA	fluorescent whitening agent
EDTA	ethylenediaminetetraacetic acid

12 Finishing

12.1 General aspects

During the preparation of a textile product, pretreatment and textile dyeing/printing operations can be understood as general textile chemical concepts to produce a fabric with a certain desired colour and pattern. During the finishing processes, a certain portfolio of properties is developed further and new properties can be added to the product. Based on the applied process conditions, we can distinguish between physical and chemical finishing. Physical finishing includes sanding, brushing, calendering and shearing. Chemical finishing includes the addition of softeners, crosslinking, coating, chemical treatment to achieve hydrophobic properties and so on.

On the basis of the desired effect, chemicals can be added one sided (e.g., kiss roll applicator, spraying and coating) or double sided (padding). In addition, application of foam is possible to minimise the amount of liquid added on the fabric and to reduce energy costs for drying.

In this chapter, the chemistry of the finishing processes will be highlighted, and the major focus will be on chemical processing.

12.2 Easy-care/durable press finishing

Cellulose-based textiles (cotton, flax and regenerated cellulose fibres) exhibit a substantial tendency to form creases, in particular, when a combination of pressure and moisture acts on the cellulose. Because of the presence of water in the cellulose structure, polymer chain mobility in the amorphous parts of the cellulose fibre increases and plastic deformation of the cellulose occurs.

Typical conditions that lead to crease development can be observed during laundry and drying operations as well as during wear at conditions of high humidity.

The chemical strategy to reduce crease formation intends to reduce chain mobility in the amorphous regions of the fibre. This will lead to a more rapid and complete recovery of the plain fabric structure after there are no more the deformation forces. Because of the high number of reactive hydroxyl groups in cellulose, chemical crosslinking of cellulose chains is comparatively easy to achieve.

The formation of chemical joints between the cellulose chains leads to a number of parallel effects [1]:
–   reduction of crease formation during handling in dry or wet state;
–   increased dimensional stability and
–   modified physical properties, for example, tensile strength and abrasion resistance.

https://doi.org/10.1515/9783110795738-012

The formation of crosslinks also reduces the accessibility and swelling of the cellulose structure in the presence of water. The swelling of the cellulose fibres leads to the development of substantial shrinkage forces and the dimensions of a fabric will change. The extent of shrinkage can directly be related to the swelling, namely, the water retention value. Crosslinking reduces the uptake of water, thus resulting in higher dimensional stability.

The modification of the fibre structure and the reduced chain mobility reduce the ability of the fibres to distribute tensional load; thus, a reduction in tensile strength appears particularly when high amounts of crosslinker have been used.

The effect is reversible; thus after hydrolytic removal of the crosslinks, the strength of the fibre recovers.

Usually the crosslinking is performed by using acidic catalysts; thus, hydrolytic damage occurs in parallel to the desired formation of crosslinks. Substantial losses both in strength and in abrasion resistance of finished fabric can appear when conditions for crosslinking were not appropriate or concentrations of applied chemicals and catalysts were too high.

By far the vast majority of finishing chemicals for crosslinking of cellulose is based on formaldehyde urea resins. Typical crosslinking agents are N,N′-dimethylol urea (DMU), N,N′-dimethoxymethyl urea, N,N′-dimethylol-dihydroxyethylene urea (DMDHEU), N,N′-dimethyl-dihydroxyethylene urea (DMeDHEU). The synthesis and the general crosslinking reaction are shown in Figure 12.1 and Figure 12.2.

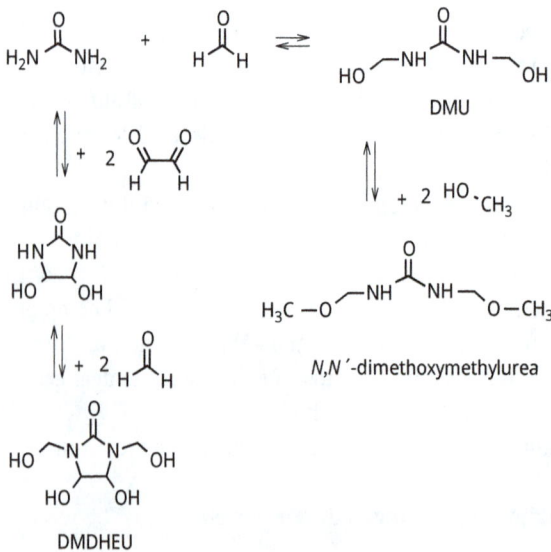

Figure 12.1: Synthesis of crosslinkers (DMU and DMDHEU).

Another group of crosslinkers is based on melamine structures (2,4,6 triamino-1,3,5 triazine), which is then modified with formaldehyde and possibly capped with methanol, analogous to DMU that is modified to yield *N,N'*-dimethoxymethylurea (Figure 12.1).

The crosslinking with cellulose is catalysed by Lewis acids ($MgCl_2$, $AlCl_3$ and $NaBF_4$), organic acids (citric acid and maleic acid), NH_4Cl and other ammonium salts. Chemically seen the reactive functional groups in DMDHEU are half aminals ($N-CH_2-OH$) or half acetals ($-NH-CH-OH$). During the crosslinking in both cases, a mixed aminoacetal is formed. While half acetals can be formed with catalysis of both bases or acids, the reaction to form an acetal requires acid catalysis.

Technically in the first step, the aqueous crosslinker solution is dried on the fabric, and then the condensation reaction is initiated at an elevated temperature, for example, 150–170 °C for 30–180 s. Under such conditions, a concentrated molten phase consisting of crosslinker, catalyst and residual water is assumed to be present in the fibre as a reactive crosslinker solution.

The crosslinking reaction of DMDHEU and DMeDHEU with cellulose is formulated in Figure 12.2.

In the case of DMDHEU, the reaction can proceed on the methylol groups or on the hydroxyl ethyl groups, which, in the case of DMeDHEU, are the only reactive sites available for crosslinking.

With exception of DMeDHEU, the other crosslinkers shown in Figure 12.1 are derived from formaldehyde as the precursor and thus exhibit a tendency to release

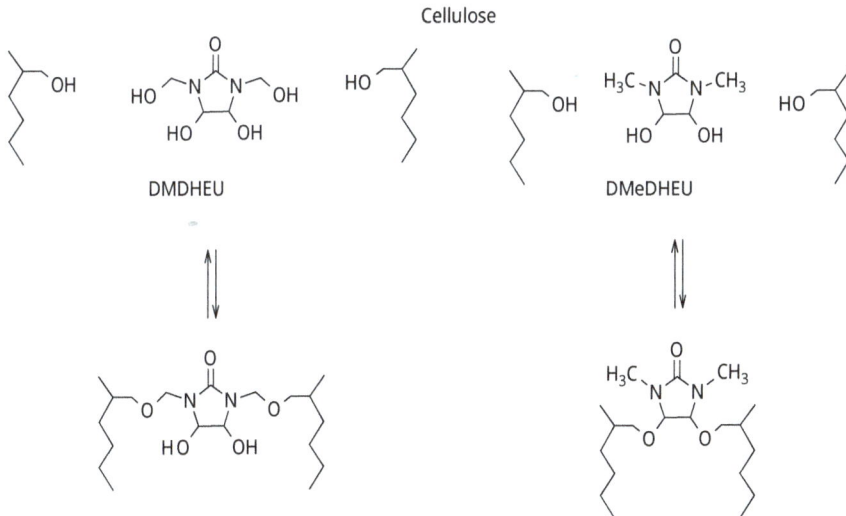

Figure 12.2: Crosslinking reaction between DMDHEU and C6-hydroxyl groups of cellulose and crosslinking with DMeDHEU.

formaldehyde in the finished textile. To minimise the release of free formaldehyde, formaldehyde scavengers, for example, methanol and diethyleneglycol, are added to the technical formulation. From the formation of acetals with formaldehyde scavengers, the free formaldehyde content, which is analytically determined, reduces substantially; however, mobile precursors are still present in crosslinked products [2].

Today, the formaldehyde content is usually determined by the acetylacetone method based on the Japanese Law Method 112 [3]. For baby clothes, a formaldehyde level of less than 20 ppm (milligram formaldehyde per kilogram of goods) is a recommended limit. Thus, analytical results for formaldehyde content have to stay below the detection limit of the method, which is at this level. For normal clothes, a value of up to 75 ppm has been accepted as a general limit for the maximum content of formaldehyde.

A special technology to achieve the highest level of easy care uses a modified process of crosslinking, the so-called moist crosslinking. In the first step, the fabric is impregnated with a solution containing the crosslinker and an acid catalyst (mainly organic acids). The material is then dried with careful control of conditions to a residual moisture content of 14–16 wt%, which leaves sufficient residual water to permit acid catalysed crosslinking of the cellulose without substantial hydrolysis. The conditions of drying have to be controlled very accurately as high content of residual moisture will lead to incomplete crosslinking, while overdrying to low moisture content will increase acid-catalysed hydrolysis and strength loss of the material. After several hours of resting at room temperature, the material is washed to remove the catalyst and unfixed crosslinker and is dried.

By moist crosslinking, a very favourable combination of high crease recovery angles both in dry and wet state can be achieved. A crosslinking operation always fixes a certain state of the fabric, which is determined by the specific conditions prevailing when the chemical links are formed.

Thus in pad-dry-cure crosslinking, the dry state will be the one that had been fixed. As a result, high crease recovery angles are obtained with the dry fabric, while the prevention of crease formation in wet state is weaker.

In the case of moist crosslinking, the fixation of the polymer structure in the amorphous zones has been made in the presence of substantial amounts of moisture; thus, crease recovery of wet fabric samples is also on an excellent level.

For cotton fabric, a combination of liquid ammonia pretreatment with moist crosslinking is used, which delivers the highest level of crease recovery in combination with soft handle and high abrasion resistance of the finished fabric. 1,2,3,4-Butanetetracarboxylic acid (BTCA) has been proposed as a possible formaldehyde-free crosslinker. The reaction is catalysed by sodium hypophosphite ($NaPO_2H_2$) as a catalyst. A reaction scheme is shown in Figure 12.3 [4, 5].

The higher costs for the chemicals and catalyst, as well as the reducing properties of the sodium hypophosphite that can destroy azo groups in reactive dyes, limit the

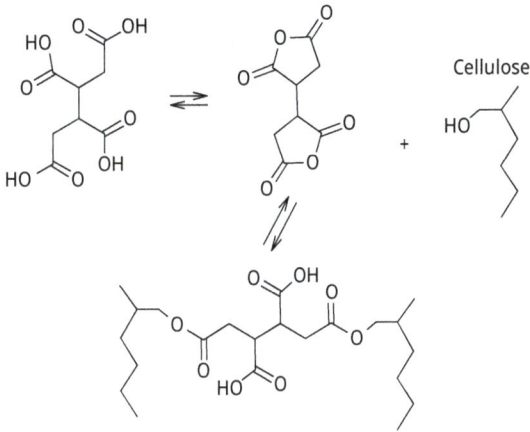

Figure 12.3: Reaction scheme of cellulose crosslinking with BTCA.

applicability of this system. The crosslinking is based on the formation of ester groups; thus, hydrolysis under alkaline washing operations also reduces the durability of the finish.

Wet crosslinking: Wet crosslinking of cellulose fabric is found in the production of lyocell fibres to prevent fibrillation; however, the chemistry behind these processes is completely different from that of dry crosslinking. Wet crosslinkers utilise the chemical reactions found in anchor groups of reactive dyes (compare Chapter 2). The chemicals are applied from aqueous solution; thus, such a step stabilises the fibre structure in a highly swollen state. The water retention value of a crosslinked lyocell fibre is thus higher than that of a non-crosslinked fibre. As examples the crosslinked lyocell fibres exhibit a water retention value of 70–80 wt%, while the non-crosslinked fibre exhibits a water retention value of 60–70 wt% [6].

12.3 Softening

Softness is a complex sum of properties that are influenced by a number of technical parameters, for example, flexibility, bendability, elasticity and compressibility and surface characteristics (roughness and structure), and is thus difficult to measure. Often softness is assessed by hand that depends on the individual's perception, thus resulting in many complaints and reclamations.

Softeners intend to increase the flexibility of the textile structure by reducing the friction between fibres. In addition, plasticisation of fibres can be a part of the softening effect. Softeners can be applied at different stages of the textile chemical treatment: Finishing with crosslinkers usually leads to a reduction in softness. Thus in the case of a pad-dry-cure process, suitable softening agents are added directly to the finishing recipe. In exhaust dyeing of yarn, softeners are added to improve the processability of the yarn for following winding, weaving and knitting. In household laundry, softeners can be added to the last rinse. On the basis of the application, the softener system has to be chosen with regard to material, process of application, bath exhaustion, fixation and permanence and compatibility with other chemicals.

Different principles of softening can be distinguished as follows:

– Use of surface-active softening chemicals that usually consist of a nonpolar fatty hydrocarbon chain (e.g., fatty alcohol and fatty amine) and a polar site. Polar groups can bear an anionic or a cationic charge, or a non-ionic element, for example, a polyethylene glycol, which actually can be understood to be slightly cationic due to partial protonation of the ether oxygen.

Examples for softeners with anionic groups in softeners are alkyl sulphates $R-O-SO_3Na$ and sulphonates $R-SO_3Na$. Examples for softeners with cationic groups are quaternary ammonium salts, for example, fatty alkyl$-N^+(CH_3)_2R$.

Nonionic softeners, for example, can be based on ethoxylated fatty amines $R-NH-(CH_2-CH_2-O)_n-H$, fatty alcohols $R-O-(CH_2-CH_2-O)_n-H$ or fatty amides $R-CONH-(CH_2-CH_2-O)_n-H$. Amine oxides $R-N^+(CH_3)_2-O^-$ can also be used.

These softeners adsorb on the fibre and form a fatty layer, with a regular orientation of the polar groups, where the direction depends on the surface charge of the fibre. For example on a negatively charged cellulose fibre, the adsorption of a cationic softener will be with the positively charged ammonium group oriented towards the surface of the fibre.

– An important class of softeners is based on paraffin and polyethylene structures $(-CH_2-CH_2)_n$. Polyethylene dispersions are often used as additives to crosslinking recipes as these products are indifferent in chemical behaviour and thus do not interfere with the crosslinking reaction. As the products are not linked to the surface of the fibres, the permanence of the softening is limited and dependent on the washing conditions applied.

– Silicones: By using silicone-based softeners, a high level of technological properties is obtained, for example, soft hand, sewability, crease recovery, abrasion resistance and so on.

The polydimethyl siloxane structure has often been modified by the introduction of polar groups, for example, amino, hydroxyl or cationic groups. The polarity of these groups increases the binding to the polar fibrous substrate (e.g., cellulose, wool and polyamide). Because of the hydrophobic nature of the mainly nonpolar polydimethyl siloxane $-(O-Si(CH_3)_2)_n-$, the finishing with silicones of low polarity results in hydrophobic properties. By introducing higher amount of hydrophilic groups, more hydrophilic silicone softeners are obtained. Representative structures of silicones are shown in Figure 12.4.

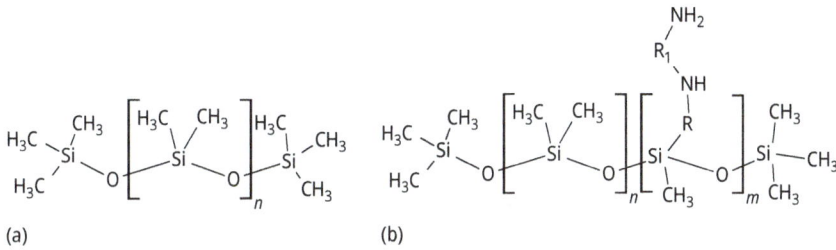

Figure 12.4: Representative structures of silicone softeners (n = 50–200; m = < 20) [1].

Hydrocarbon-based softeners and siloxane-based softeners are applied as an emulsion; thus, the stability of the emulsion has to be considered with care. Breakdown of the softener emulsion leads to uneven deposition, which can lead to visible stains. The oily character of the silicone softeners can also lead to changes in perceived colour, for example, deepening of shade through reduced reflection of light.

Amino functional softeners can also lead to yellowing of finished material either through oxidation of the amino group or through fluorescence quenching of the fluorescent whitening agent. The presence of residual amounts of hydrocarbons as well as of silicone softeners in the waste water has to be examined with care as limits for the release of such apolar substances may exist.

12.4 Hand building finishes

Hand building finishes intend to modify the handle of a product to achieve the perception of fullness (increase perception of bulkiness and weight) or stiffness (mainly bending resistance). The effects can be permanent or non-durable.

Typical examples for non-permanent hand building finishes are starch-based products and polyvinyl alcohol. In fact, partially hydrolysed polyvinyl acetate is used, which thus is a copolymer between vinyl alcohol and vinyl acetate.

Durable hand builders can consist of emulsions of vinyl acetate and acrylic copolymers (e.g., polyacrylates and polymethacrylates). Because of their insolubility in water, the wash permanence is good and can be improved further by incorporating crosslinking monomers, for example, N-methylol-acrylamide. In addition, melamine-based dispersions or solutions of precondensates can be used in a similar chemical approach as applied in easy-care finishing. Another group of hand building systems uses polyurethane-based additives.

12.5 Water-repellent finishes

Historically the production of textile material that is water repellent belongs to the oldest finishing operations. The major function of a water-repellent finished textile is to protect against permeation by water, mainly rain.

Besides the reduction in surface energy, the fabric itself should be constructed in an appropriate manner. The material has to be dense enough to avoid permeation of tiny droplets formed by the kinetic energy of a falling drop. This, for example, can be achieved by the introduction of mechanical barriers, for example, in the form of a membrane, which is laminated behind the outermost layer of an outdoor jacket.

A very basic approach is to use a completely impermeable coating, for example, polyvinyl chloride or polyurethane, which however leads to substantial disadvantages in wear comfort because of a very low water vapour permeability of the coating.

If the open structure of a fabric need to be maintained highly porous, the wettability for water has to be reduced. The surface energy of water is rather high, with a value of 72.8 mN/m. Spontaneous spreading of a drop of water on a surface occurs in case the surface energy of the material is comparable in height or is higher than the surface energy of water.

Typical values for the surface energy of olive oil are at 32 mN/m. This value is similar to the surface energy of polypropylene fibres or silicon oil. Fluorocarbons exhibit a very low surface energy of 10–20 mN/m, which is even substantially lower than the surface energy of oil.

Thus, plant oil, mineral oil, silicone and fluorocarbon-based finishes will be able to repel water; however, oily substances will be repelled only by fluorocarbon finishes.

Paraffin-based repellents: A traditional method to produce a hydrophobic surface on textiles utilises the deposition of Zr^{4+} or Al^{3+} salts of fatty acids. The cation serves as an ionic anchor that links the fatty acids to the surface of the fabric.

The finish is rather cost efficient; however, permanence is limited as removal of the anchor also releases the hydrophobic fatty acid. In addition, the high content in combustible material increases flammability of the material. The fatty acids can also be combined with melamine-based crosslinking groups or bound into dendrimer structures.

Silicone-based water repellents utilise the hydrophobic character of methyl- and alkyl siloxanes $(-(R_2Si-O)_n-$. Permanence can be improved through reaction between silanoles, silanes and a catalyst (e.g., tin octanoate $Sn(CH_3(CH_2)_6 -COO)_2$); eq. (12.1)). The catalysed reactions between a vinyl silane and a silane can also lead to bridge formation and crosslinking (eq. (12.2)).

$$- O - Si(CH_3)_2 - OH + H - Si(CH_3)_2 - O - \qquad - O - Si(CH_3)_2 - O - Si(CH_3)_2 - O - + H_2$$

Silanol   Silane   Formation of a crosslink

$$(12.1)$$

$$(12.2)$$

vinylsilane   silane   coupling product

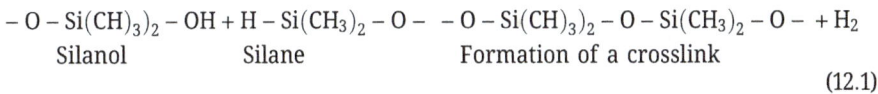

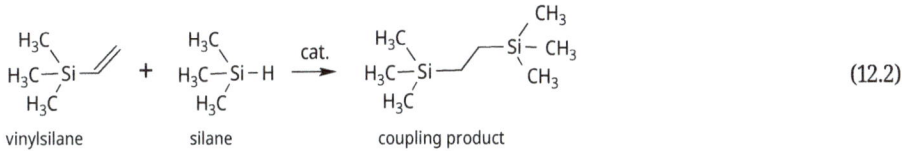

The application of silicones is straightforward and leads to favourable water repellency. However, durability is limited and also costs are comparatively high.

 Fluorocarbon-based water repellents: In the past for the vast majority of durable water-repellent finishes, fluorocarbon containing chemicals were used. The chemical concept of these finishes is based on the low surface tension and high chemical stability of highly fluorinated alkyl chains. To achieve a crosslinking of the repellent chemicals, the perfluorinated alkyl group was linked to a reactive anchor system, for example, an acrylic compound (Figure 12.5). The chain length of the fluorinated branch usually was 8–10 carbon atoms; thus, these finishes were summarised with the synonym "C8"-chemistry. A representative example for a crosslinked structure of a C8 fluorocarbon telomere in combination with an acrylic crosslinker is given in Figure 12.5.

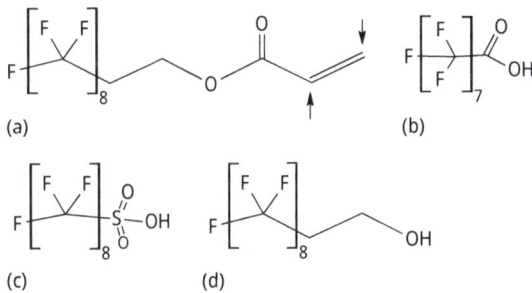

Figure 12.5: Structures of (a) crosslinking fluorocarbon repellent, (b) perfluorooctanoate, (c) perfluorooctane sulphonate and (d) fluorotelomer alcohol.

Instead of acrylic crosslinkers, the formation of polyurethane structures can also be used to support the formation of a permanent hydrophobic layer.

 The widespread use of the "C8"-based chemistry is evident, when the excellent performance of the finished products is considered. In addition to the water repellency, high rates of oil repellency could also be obtained with these chemicals, as the surface energy of the finished material was substantially lower compared to oily substances.

Today, the extensive use of fluorocarbon-based repellency agents is considered critically. In particular, the formation of perfluoro octane carbonic acid (PFOA) and perfluoro octane sulphonic acid (PFOS) has to be considered with care (Figure 12.5) because these substances are persistent, bioaccumulate and have toxic effects [7].

At present, manufacturers thus prefer the use of shorter hydrocarbon chain lengths ("C6"-chemistry) as these products cannot release PFOA or PFOS under normal circumstances. However with shorter chain length of the fluorocarbon chain, the repellency for oil decreases, while still good water repellency is obtained.

After cleaning with surfactants, for example, in household laundry, the coverage with surfactants reduces the water repellency, which can be reactivated through ironing, which is supposed to remove the surfactants from the fluorocarbon layer.

Extenders are chemicals that are based on non-fluorinated chemicals for water-repellent finishes, for example, quaternary ammonium compounds. These substances are added to the fluoro-chemicals with the aim to substitute part of the fluoro-compound without losses in performance and functionality.

Besides the use of wet chemical processes to introduce hydrophobic or hydrophilic groups on a textile surface, plasma processes can also be applied to produce a coating in molecular dimensions. Reactive fluorocarbon precursors (e.g., tetrafluoromethane, CF_4) or silicone precursors (e.g., hexamethyldisiloxane $(CH_3)_3Si-O-Si(CH_3)_3$ or trimethyl-vinyl-silane $(CH_3)_3Si-CH=CH_2$)) are injected into the plasma atmosphere. Through the action of radicals and the high energy in the plasma, reactive species are formed, which then build a coating on the surface of the textile substrate.

The need to replace the existing C8-fluorocarbon chemistry by alternatives led to several strategies:

– Structured surfaces: Through micro- and nanostructuring of the plain surface the effective contact angle between a water drop and the surface is increased, as the structure of the material requires bending of the water drop.
– The use of highly branched dendrimers with ordered star-like structure allows generation of highly hydrophobic surfaces; however, the surface tension still is close to hydrocarbons. Thus, no oleophobic properties will be achieved.
– Sol–gel materials: In this technique, a precursor substance (e.g., tetraethoxysilane) is hydrolysed with water. Silica nanoparticles are formed, which then build the silica-based coating (Figure 12.6). Based on the precursor substances used, a wide range of surface coatings can be realised.

12.6 Flame retardant finishes

From many applications, textiles with flame-retardant properties are requested; the major reason being safety of the persons or installations.

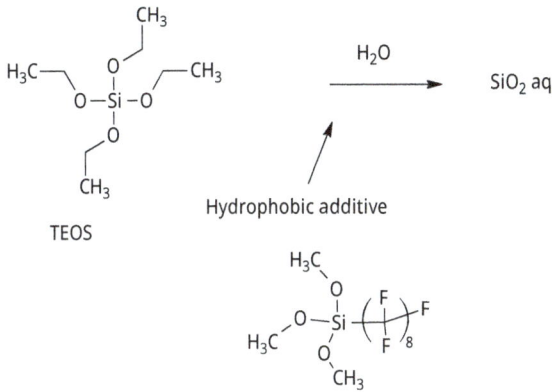

Figure 12.6: Formation of a sol–gel coating based on the hydrolysis of tetraethoxysilane.

The type of finish depends on several factors:
- Planned use: Finishing of clothes for garment or bedsheets requires consideration of mobility of chemicals as well as wear properties (handle and softness).
- Durability: Depending on the planned cleaning procedures, the finish must be permanent, for example, against soaking, washing and dry cleaning.
- Fibre type used: The chemistry of a textile fibre also determines the chemical concept that can be applied to reduce flammability.

The principle of a flame-retardant finish is to provide the material with an additional functionality that permits the interruption of the combustion cycle (Figure 12.7).

When a fibre material is heated above its decomposition temperature, pyrolysis begins and volatile products are released. Such products can be water, carbon dioxide, carbon monoxide and hydrocarbons; thus flammable products are also released. The composition of the volatiles depends on the temperature and the fibre type.

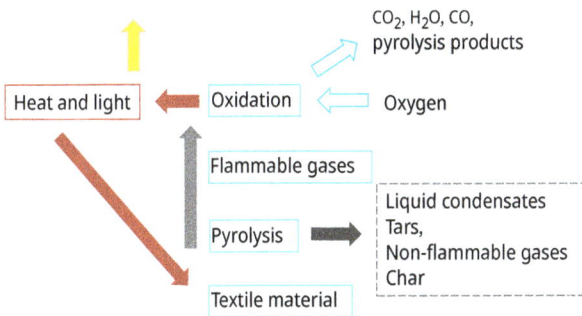

Figure 12.7: The combustion cycle [8].

If oxygen is available, the flammable part of the volatiles can ignite and lead to generation of heat, which increases the temperature of the fibre material further. Inside the flame, a chain reaction leads to flame propagation and continuous generation of energy. The heat of combustion of a fibre polymer can range from 17.0 kJ/kg for cellulose to 46.0 kJ/kg for PP fibres. This indicates that the hydrocarbon-based fibre PP releases more than twice the energy per mass by combustion compared to the already partially oxidised carbohydrate cellulose. Typical values for the decomposition temperature and auto-ignition temperature range from 300 to 400 °C.

As a result of the decomposition reaction, char, coal and ashes are also formed, which can build a barrier between the site of heat generation and the fibre material. In addition, the access of oxygen to the burning fibre can be retarded by the formation of char.

The need of oxygen for combustion also is illustrated by a method to determine the LOI (limiting oxygen index). The LOI is the minimum concentration of oxygen (vol%), which is required to maintain flaming combustion. A material with an LOI below 21 vol% oxygen will burn in air, while a material with an LOI above the oxygen content in air will not continue to burn. However, the LOI is not a suited method to assess flammability and burning behaviour of textiles. Specific test methods exist for the classification of a material for a certain application.

General textile chemical strategies to reduce flammability of fibres and textiles can be noticed at different steps of production:
– Material with low flammability as an inherent property or non-flammable fibres (e.g., glass fibres and asbestos) can be used.
– In case of man-made fibres, flame-retardant chemicals can be incorporated into the fibre matrix.
– Flame-retardant chemicals can be applied during textile finishing operations.

Different chemical principles can be chosen to achieve flame retardancy; often combinations are used:
– spreading and absorption of released heat and cooling can be achieved by endo-thermal processes running in parallel (e.g., melting, degradation and decomposition of inorganic and organic substances such as aluminium hydroxides, inorganic or organic phosphor compounds);
– reduction of release of flammable gases through barrier formation, char formation, coal formation (usually phosphor and nitrogen compounds are added to cellulosics to support char formation; metal complexes are used for wool and boron-based compounds form barriers);
– intumescent formation (formation of a coating/barrier with low flammability);
– separation of flame and access of oxygen through coating, for example, metal foil coating;

- release of non-flammable gases (water vapour from hydrate water or combustion);
- formulations that release chlorine or bromine, which interfere with chain reactions inside the flame (e.g., halogen compounds in combination with Sb-compounds).

Representative examples: Borates belong to the oldest chemicals used to produce flame-retardant cellulose fibres. The cellulose material is impregnated with boric acid or the corresponding salts (e.g., zinc borates). The flame-retardant properties are due to endothermal release of water and formation of a glassy residue (B_2O_3), which also acts as an intumescent (eq. (12.3)).

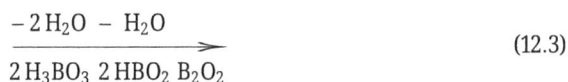

$$\underset{2\,H_3BO_3 \quad 2\,HBO_2 \quad B_2O_2}{\xrightarrow{-2\,H_2O \;-\; H_2O}} \tag{12.3}$$

In a similar manner, alumina tri-hydrate ($Al_2O_3 3H_2O$), magnesium hydroxide ($Mg(OH)_2$) or $CaCO_3$ can be used as flame retardants. Besides the endothermal release of water or CO_2, these chemicals also form a barrier for heat transmission to the fibres and thus reduce the decomposition and formation of flammable volatiles (eq. (12.4)–(12.6)).

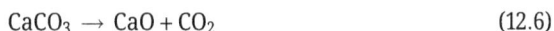

$$Al_2O_3 3H_2O \rightarrow Al_2O_3 + 3H_2O \tag{12.4}$$

$$Mg(OH)_2 \rightarrow MgO + H_2O \tag{12.5}$$

$$CaCO_3 \rightarrow CaO + CO_2 \tag{12.6}$$

These finishes are applied through the deposition of the respective *inorganic compound*. As there is no distinct affinity of the inorganic salts to the fibre polymer, the durability in wet processes and washing is low. In addition, substantial stiffness and reduced softness result from the deposition of high amounts of inorganic substances on the fibres.

    Char formation through decomposition of the fibre polymer in combination with a flame retardant finish is a major strategy in textile finishing operations. A certain content of phosphorous and nitrogen must be deposited on the textile material by using a combination of reactive chemicals. The components then decompose during combustion and reduce flame propagation by three effects:

- Heat is absorbed through endothermal processes.
- Non-flammable gases or gases with low flammability are formed as a result of the decomposition reaction (CO_2, H_2O and NH_3).
- Char is formed that then builds a barrier between flame and material; thereby, heat transfer to the textile material is reduced and access of oxygen to the residual fibres is hampered.

When crosslinking of the chemicals with the fibre polymer is achieved, a certain dura-bility for washing can be achieved. A representative example for the flame-retardant finishing of cellulosics is given in Figure 12.8 for *N*-methylol dimethylphosphonopro-pionamide (Pyrovatex CP®).

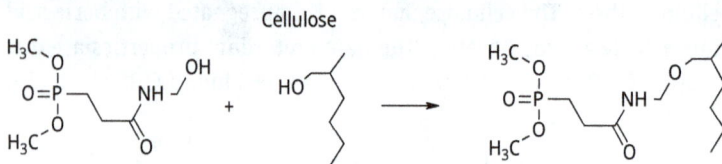

Figure 12.8: Reaction scheme for finishing of cellulose with *N*-methylol dimethylphosphonopropionamide.

As a more general rule, a phosphorous content of 1.5–4 wt% and a *P:N* molar ratio of 1:1 to 1:2 will lead to an acceptable level of flame retardancy. Less permanent systems could be based on ammonium polyphosphate $HO(PO_2NH_4-O)_n-H$ or diammonium phosphate $(NH_4)_2HPO_4$.

Chemical systems that are applicable for cellulosics as well as for synthetic fibres are based on a combination of antimony trioxide Sb_2O_3 with halogen compounds. The action of these combinations is observed both in the condensed phase, for example, by lowering of charring temperature and increasing the char yield, and in the vapour phase. The first step is the formation of volatile antimony halides SbX_3, which evapo-rate and effect the flame propagation through a radical chain reaction. For example, decabromodiphenyl oxide (DBDPO) is used as a halogen-delivering chemical (Fig-ure 12.9). A representative molar ratio between Sb:Br is 1:3 when a total amount of 5–8 wt% Br is applied on the material.

A strategy to provide fibres with flame-retardant properties is the incorporation of appropriate systems into the fibre matrix during the fibre production.

The chemicals have to be chosen based on the conditions applied during produc-tion, for example, temperature in melt spinning of thermoplastic fibres, alkaline/ acidic and reducing conditions during viscose production.

For viscose the incorporation of 2,2'-oxybis (5,5-dimethyl-1,3,2-dioxaphosphorinane)-2,2'-disulphide (Sandoflam 5060®) in an amount of 10–15 wt% leads to viscose fibres with flame-retardant properties (Viscose FR®). Another approach uses the incorpo-ration of polysilicic acid (e.g., 20 wt%), which forms a non-flammable silicate–carbon

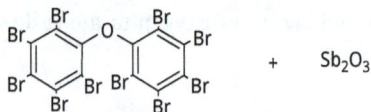

Figure 12.9: Antimony-halogen-based combination for fibre blends.

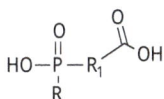

R = H or alkyl
R$_1$ = alkylene

Figure 12.10: Phosphinic acids as co-monomers for inherently flame-retardant polyester [8].

layer, which is even stable at temperatures above 500 °C, where carbon chars will be oxidised.

Inherently flame-retardant polyester fibres (e.g., Trevira CS®) are obtained through co-polycondensation of phosphinic acids as co-monomers. A general formula is given in Figure 12.10.

Incorporation of high amounts of additives during fibre production can lead to substantial reduction in fibre strength. Thus, a compromise between amount of additive and final strength of a fibre has to be found. In addition, the high amount of additive required substantially contributes to production costs.

The use of flame-retardant materials or high-performance fibres with reduced flammability (e.g., Nomex and Kevlar) increases safety substantially, particularly in areas where considerable risks exist, for example, petrol refineries and storage, firemen any rescue forces.

In any case the choice of a concept has to be considered carefully as the chemicals used (e.g., Sb_2O_3) may exhibit health hazards during use and in particular emitting toxic gases during decomposition and combustion.

12.7 Antistatic finishes

Static electricity appears at many stages of textile production and during wear, namely, use of textile products. The problem of static electricity is observed in fibres that contain very low humidity; thus, it is often observed in hydrophobic synthetic fibres, for example, polyester, polyamide and polyacrylic fibres. Natural fibres can develop static charging in overdried state, for example, at the outlet of a tenter, and in an environment with low humidity, for example, in winter.

During yarn production, static electricity on filament yarns leads to repulsion of filaments and ballooning of the yarn. Thus, appropriate preparation steps are added during fibre production to reduce the static electricity.

Static charges on fabrics can make handling difficult and can also lead to clinging of garment, which results in reduced wear comfort and aesthetic appearance. Even small electric shocks can result from the development of static charge. At the end of continuous textile drying machines where the fabric is transported over a longer distance without direct contact to any metal part e.g. tenter driers, the dried material exhibits high static charging at the outlet.

Static charge also contributes to an enhanced adsorption of dust and soil from air.

A serious problem of static electricity arises from clothing worn in workshops for electronic equipment. Through ongoing miniaturisation, the electrical power required to damage an integrated circuit or a device has decreased. In medical devices static charge can lead to malfunction and related risk of failure.

The origin of the static charge: When two materials come into contact with each other, positive or negative excess charges appear as a result of the transition of mobile charges at the contact area. During contact, a charged contact layer, the Helmholtz-electrical double layer, is formed. Electrons with low work of emission are transferred into the acceptor material. When the materials are separated, recombination is incomplete and residual charges remain [9]. Friction between materials is not necessary; however, this can intensify the formation of static charge (Figure 12.11).

The development of static charge is dependent on the combination of materials that are brought into contact. In the triboelectrical series, different materials have been ordered with regard to the sign (+/−) and the height of static charge, which is developed during contact with another material. As the development of the static electricity is influenced by many parameters, different triboelectrical series have been developed, where their order depends on the experimental procedures used and the physical state of material tested. In Figure 12.12 an example for a triboelectrial series is shown.

As a general behaviour, the positive end of the series is formed by the protein fibres and polyamide fibres, followed by cellulose-based fibres.

Slightly negative charge is developed by acrylics and cotton-based material. The negative end is formed by highly hydrophobic material (hydrocarbons, rubber and PTFE). The larger the distance in the triboelectrical row, the higher will be the charge that can be developed.

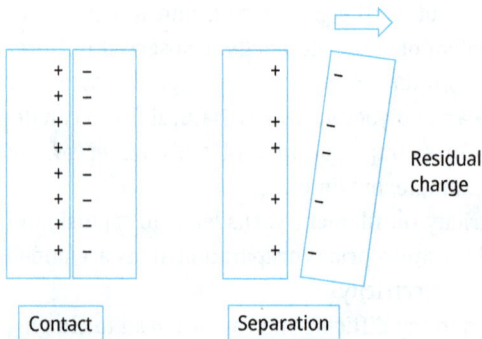

Figure 12.11: Formation of a Helmholtz-double layer and static charge as a result of charge separation between two materials.

```
+     Wool Wo
          Polyamide 6  PA6
             Silk  Ms
                Polyethylene terephthalate PET
                   Paper
Relative                Glass
charge
                           Polyamide 6,6   PA6,6
                              Cotton Co
                                 Polyacrylic PAC
                                    Polyvinylchloride PVC
–
                                       Rubber
```

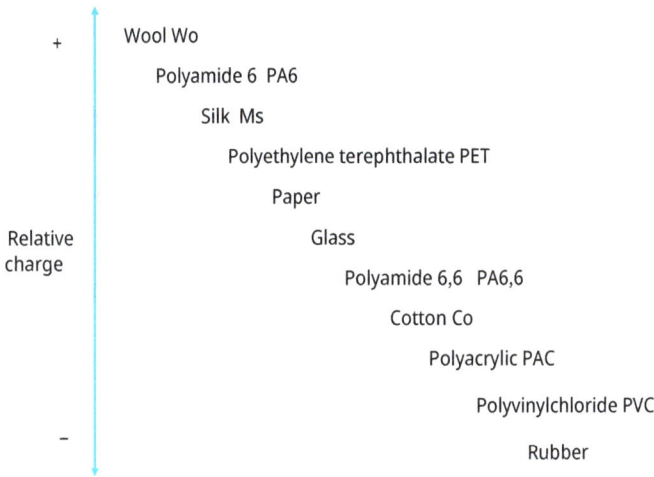

Figure 12.12: Triboelectric series of materials [10].

Charging processes: The charging process depends on the actual conditions; for example:
– combination of materials
– situation at the contact layer: impurities, deposits, oil moisture
– contact pressure
– roughness
– intensity of movement.

The first strategy to reduce static charge would be an appropriate selection of materials, which however often is restricted by technical requirements.

In addition to aspects with regard to the development of static charge, the rate of neutralisation is also of distinct relevance. Finishing operations usually intend to facilitate rapid neutralisation of static charge formed.

A textile surface can be compared to an electrical capacitor with capacity C. The discharge of such a capacitor will be characterised by eq. (12.7).

$$Q(t) = Q_0 e^{-t/T} \tag{12.7}$$

$$T = R\,C \tag{12.8}$$

The decrease of the initial charge Q_0 (in Coulomb) as a function of time $Q(t)$ depends on the relaxation constant T (in seconds), which is a product of the resistance for discharge R (Ohm) and the capacity C (Farad) of the charged textile (eq. (12.8)).

The majority of textile techniques to reduce static charging utilises the concept of a decrease of the electrical resistivity of the textile structure R to achieve a rapid charge neutralisation.

Techniques to modify electrical conductivity of fibrous material include the following:

1. *Increase of surface conductivity*: Polar centres in the fibre and at the fibre surface increase adsorption of moisture. This can be achieved by an antistatic finishing by using modified fibres, plasma treatment and grafting.
2. *Incorporation of conductive material* C, metal powder and carbon nanotubes into the fibre structure, with the aim to increase the intrinsic conductivity of the fibre material.
3. Use of *polymers with increased conductivity*, for example, polypyrrole fibres and carbon fibres.
4. *Deposition of conductive layers*, for example, through electroless deposition of metal layers, chemical vapour deposition (CVD) and sputtering.
5. *Combination with conductive materials* (metal wires and carbon).

The applicability of the different techniques depends on the characteristics of the textile product.

–   For technical products, less restrictions with regard to appearance, handle and design exist. The incorporation of a carbon into a fibre leads to black fibres, which is a restriction with regard to colour design for garments.
–   The incorporation of conductive fibres, for example, metal-based material, where coating with metal layers changes handle and appearance and also defines limits with regard to durability and care.
–   The use of conductive polymer fibres changes the properties of the fibre.

In garment antistatic properties often are developed by using appropriate finishing techniques. By the addition of chemicals, a layer of material is deposited on the electrically insulating fibre, which then exhibits substantial electrical conductivity to permit rapid neutralisation of static electricity.

The major principle is based on the deposition of hygroscopic substances that adsorb sufficient amounts of water to form a conductive layer. As a result these types of antistatic finishes will depend on the sorption of water and thus on the climatic conditions near the sample.

Non-permanent finishes: Typical chemicals are glycols, polyethylene glycols, quaternary ammonium salts with long alkyl chain, alkyl phosphonium salts and phosphoric acid ester salts. Representative examples are ditallowdimethylammonium chloride and an ethoxylated fatty acid ester (Figure 12.13).

$(R)_2N^+(CH_3)_2Cl^-$ $\qquad\qquad\qquad$ $R-CO-O-(CH_2-CH_2O)_x-H$

$R = C_nH_{2n-1}$ (n = 11–17)

Figure 12.13: Representative structures of non-permanent antistatic finishing. Chemicals: ditallowdimethylammonium chloride and ethoxylated fatty acid ester.

Durable antistatic finishes are formed by crosslinking of polyamines with polygly-cols. Another approach of a finish with higher durability utilises the deposition of a polyammonium ion on the surface of a synthetic fibre by the formation of an organic salt with low solubility. For example, such a conductive layer can be formed on poly-amide by the deposition of polydiallyl-dimethylammonium chloride with laurylether-sulphate as negative counter ion (Figure 12.14) [11].

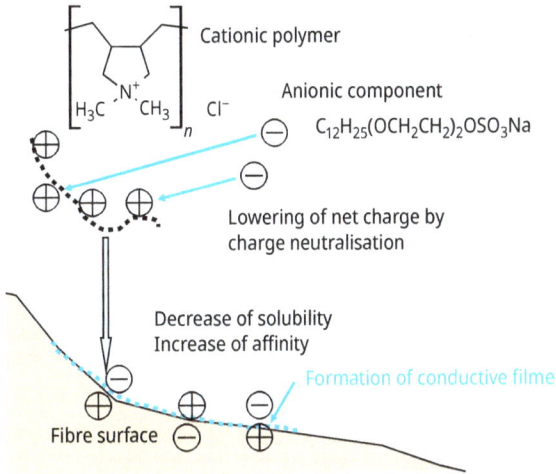

Figure 12.14: Formation of a conductive organic polymer layer as antistatic finishing.

Antistatic finishing must not alter other properties of a product, such as elasticity, handle and colour. Different methods to measure electrostatic charging are available. For testing charge development in carpets the static charging in walking tests with measurement of the charge development is used (e.g., AATCC Test Method 134 [3]). Dissipation test to monitor discharge by using an electrometer describes the neutrali-sation of static electricity.

In many cases, the electrical resistance of a fabric is measured under standar-dised conditions (DIN 54345, AATCC Test Method 76 [3, 9]). The resistance of the fabric is measured in conditioned state (20 °C, 65% r.h.) with a specially designed ring elec-trode. For example, for unfinished polyamide fabric, values of 10^{12}–10^{13} Ohms are measured, which should decrease to 10^9–10^{10} Ohms for a finished material.

The results demonstrate that a material that is understood as a conductive mate-rial in view of static charge is still considered as an insulator for many applications of technical textiles where higher currents, for example, mA–A should be transported by fibres.

12.8 Improvement of colour fastness

Under the term colour fastness, a complex pattern of properties is summarised, including fastness to water, sweat, washing, crocking and rubbing, light, peroxide, chlorine and combinations, for example, weather fastness. In fact, many fastness properties are defined by the chemical principle of the dyestuff system used and by the molecular structure and stability of the respective dye.

Thus, the improvement of colour fastness often concentrates on aspects where the chemical principle of the dyeing allows the dye to achieve mobility and re-equilibration. Establishment of a new exhaust equilibration leads to desorption of dyestuff molecules during conditions of use and care.

Polycationic compounds: Water fastness of direct dyeings and reactive dyeings can be improved by deposition of polycationic compounds. Polydadmac polydiallyl-dimethylammonium chloride (Figure 12.14) is a representative compound. The polycation forms insoluble ionic compounds with the anionic dyes and thus contributes to improved dyestuff fixation.

Direct dyes adhere to the fibre only through Van der Waals forces and hydrogen bonds; thus, establishing a new dissolution equilibrium under wet conditions is prevented.

In the case of reactive dyes, the dyes are bound covalently to the fibre; thus, the use of such a treatment does not seem necessary. However, both the presence of substantial amounts of adsorbed hydrolysate and slow hydrolysis of the anchor bonding lead to mobile dyestuff molecules, which then have to be kept attached to the fibre surface. Such polycationic additives also are used in the washing of darker reactive dyeings to shorten the process, as the full release of absorbed hydrolysate can require considerable time during technical rinsing operations.

The presence of such chemicals also influences other properties of the dyed product and can lead to the reduction of light fastness, colour change, greying, increased soil uptake and reduced uptake of softeners.

Syntans – condensation products of aromatic sulphonates: Syntans (synthetic tannins) are used in dyeings of polyamide fibres. Acid dyes and metal-complex dyes used for polyamide dyeing exhibit a negative charge. Thus, the mechanism of these anionic polymers is different. Representative formulas for such compounds are shown in Figure 12.15.

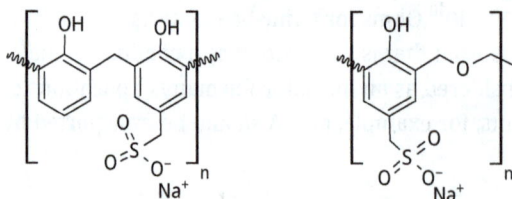

Figure 12.15: Condensation products of aromatic sulphonic acids (syntans) to improve wet fastness.

The function of these products is explained by different mechanisms. The negatively charged polymer forms a film on the surface of the fibre that prevents diffusion of negatively charged dye from inside of the fibre into the wash solution. Another explanation is based on aromatic interactions between the polymer and the chromophore of the dyestuff, thus strengthening the binding of the dye to the fibre.

12.9 Improving the light fastness

Absorption of light by a dyestuff leads to the formation of excited states, which then result in dissipation of energy and follow-up reactions that can lead to destruction of the dyestuff molecule.

The phenomenon of light fastness is complex and depends on a number of factors:

– *Fibre substrate*: The same dyestuff molecule exhibits different light fastness, depending on the chemical environment formed by the substrate.
– *Dyestuff molecule used and combination of dyes*: A substantial decrease in light fastness can occur through combination of incompatible dyes.
– *Colour depth of a dyeing*: In many cases, light fastness of light dyeings is lower compared to fuller shades.
– *Type of irradiation* (daylight, Xenon lamp, light emitting diodes, fluorescent lamp).
– *Climatic conditions*: Light fastness depends on the water content of a sample; thus, fastness in wet state, weather fastness (a combination of irradiation in wet and dry state) and in dry state will differ. The temperature also has a significant influence on light-induced degradation of dyes.
– *Auxiliaries*, for example, softeners and heavy metals, can influence light fastness of dyeings.

An important additive is the delustering agent TiO_2, which in particular in its anatase modification forms substantial amounts of peroxides when irradiated in the presence of moisture.

Improvement of light fastness can be achieved by two different strategies: Reduction of UV irradiation on the sample can be achieved by means of colourless UV absorbers. Chemically these components contain a larger aromatic system that selectively adsorbs in the UV region of the visible light (<400 nm) and thus does not contribute to the perceived colour. These components absorb the UV light and transfer it into vibrational energy (heat). Examples for such UV absorbers are shown in Figure 12.16.

Another approach involves additives that trap radicals that had been formed during UV absorption and consecutive chemical reactions. In many cases, these chemicals are sterically hindered molecules, for example, phenols or amines, which easily form rather stable radicals that then are trapped and stabilised (compare Chapter 7).

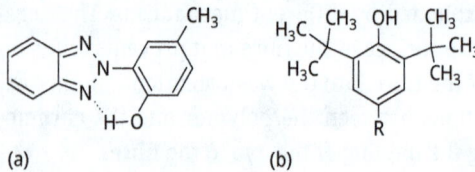

Figure 12.16: Additives to improve light fastness of dyeings: (a) UV absorbers (2(2'-hydroxy-5'-methylphenyl)benzotriazole and (b) radical scavengers (sterically hindered phenol light stabiliser).

12.10 UV protection

On contrary to the aspect of colour fastness, the term of UV protection addresses the reduction of UV irradiation which reaches the skin of a person, particularly, during intensive outdoor sports activities. More attention has been directed to the exposure of skin to sunlight. Long and intensive UV radiation is known as an important factor that influences the probability of sunburn, sun allergies and skin cancer. Thus, interest in textiles with the UV-protective function has increased substantially [12].

We have to distinguish different types of UV irradiation as the potential hazard depends on the wavelength, namely, the energy of the irradiation:

UVA 315–400 nm: penetrates the ozone layer to 98.8%, photon energy 315–350 kJ/mol, important for the production of vitamin D production and causes sunburn and cataracts.

UVB 280–315 nm: reaches the ground with an residual intensity of 1.1%, photon energy up to 400 kJ/mol mainly absorbed by the ozone layer, absorbed by DNA in human cells, with the risk of cancer.

UVC 100–280 nm: 100% absorbed in the stratosphere with the formation of ozone (max. $\lambda < 242$ nm), thus also increasing absorption for UVB.

The materials are characterised by a UVPF (ultra-violet protection factor). A mathematical representation for the UVPF is given in eq. (12.9).

$$\text{UVPF} = \frac{\sum_{280nm}^{400nm} E_\lambda S_\lambda \Delta_\lambda}{\sum_{280nm}^{400nm} E_\lambda S_\lambda T_\lambda \Delta_\lambda} \tag{12.9}$$

The UVPF considers the reduction in UV-light transmission by a combination of several factors:

T_λ the transmission of a sample for a certain wavelength; S_λ the solar spectral irradiation (W/m^{2}nm), which is a measure of the intensity of UV irradiation at a respective wavelength;

Δ_λ the wavelength interval considered in the calculation (e.g., 5 nm) and E_λ the relative erythemal spectral effectiveness. This is a factor that considers the potential risk for skin damage for a given wavelength.

The UVPF is a factor for the prolongation of time a person remains protected under UV irradiation in comparison to the time without clothing. A factor of 10 permits a person with given sensitivity for UV irradiation to stay 10 times longer under the same intensity of irradiation; a UVPF of 40 then indicates a factor of 40.

The experimental basis for the determination of the UVPF is a transmission measurement by spectrophotometry. A representative scheme for the measurement of UVPF is shown Figure 12.17.

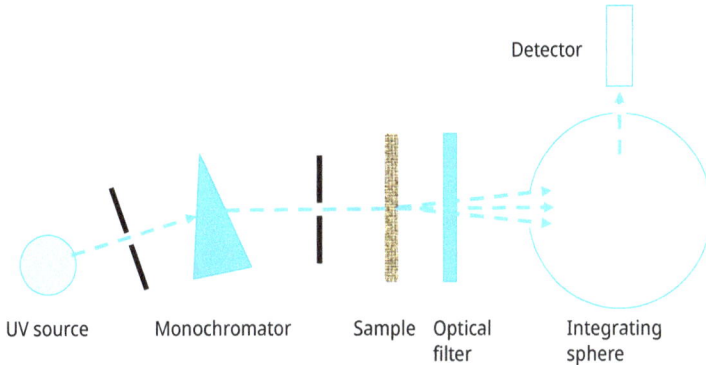

Figure 12.17: Scheme for the measurement of UVPF (optical filter removes VIS light).

Different geometries are used to determine UVPF. In the example given in Figure 12.17, the light of the UV lamp at first passes a monochromator to obtain irradiation of a certain wavelength interval. The light beam then impacts the sample, where different modes of interaction lead to absorption or weakening of the beam (Figure 12.18). Besides absorption, refraction and scattering lead to deflection of the beam, an effect that has to be considered in the measurement of the UVPF. Behind the textile sample, an optical filter removes any visible light, which is a prerequisite to remove fluorescent light, which could contribute to the signal in the detector. The problem of beam dispersion is solved by using an integrating sphere (Ulbricht sphere), which collects the light behind the sample.

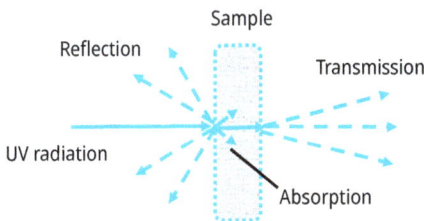

Figure 12.18: Major modes of interaction of an UV light beam with a textile sample.

A central problem in the measurement of UVPF is the low intensity of light that transmits a sample of high UVPF. Thus without using specialised equipment with high UV intensity, values of UVPF above 50 are often given as 50+. At a UVPF of 50+, the transmittance of the samples is less than 2.5%. At a UVPF of 12–24, the transmittance is between 6.7% and 4.2%.

The UVPF of a fabric depends on a number of factors:

- *Fibre material*: The UV absorption of a fibre polymer depends on the presence of functional groups (aromatic rings, ester and amide groups and nitrilo groups), which can absorb irradiation in the wavelength range of interest (250–400 nm). Natural fibres that contain residual lignin components (e.g., flax and hemp) exhibit higher UV absorption than 100% cellulose-based fibres.
- The *structure of the fabric* in terms of thickness of threads and fabric construction.
- *Cover factor*: Depending on the material construction, the skin is covered completely by the textile (cover factor = 100%) or open spaces in the fabric lead to incomplete covering (e.g., cover factor = 70%), which means that an area of 30% is irradiated without any cover. The cover factor can be increased by appropriate construction, for example, dense weaves or by physical processes such as calendering. In a simplified approach, the fibres can be assumed to absorb 100% of the UV irradiation. In such cases, the cover factor is directly related to the UVPF (eq. (12.10)) [1].

$$UVPF = 1/(1 - \text{cover factor}) \tag{12.10}$$

- *Colour of the fabric and additives*: The finishing of a fabric with UV-absorbing chemicals as well as the colour of a material both contribute to the overall UV absorption. However, there are limitations with regard to extensive use of chemical finishing as well as the choice of colour, as UVPF for the same garment should not depend strongly on the colour of the product.
- *The physical state (moisture content, wetness and stretch)*: This contributes to the permeability for UV. Stretch changes the cover factor of a product. The influence of stretch is of particular relevance for sports textiles that often are worn with significant stretch, for example, garment for triathlets and cyclists. Large amounts of water, namely, sweat, can reduce UVPF through reduced scattering of irradiation or increase UVPF in case of regenerated cellulosics as a result of fibre swelling and increase of cover factor.

Chemicals used for UVPF finishes are colourless substances that exhibit high absorbency in the wavelength interval between 300 and 320 nm. Representative examples for chemical structures are shown in Figure 12.19.

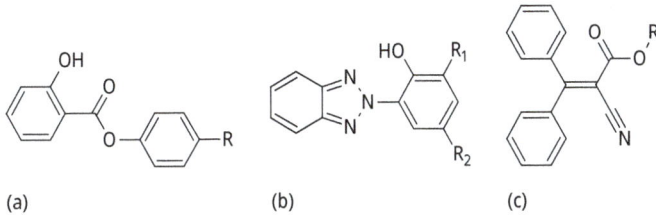

Figure 12.19: Representative structures for UV-absorbing chemicals for UVPF finishing:
(a) phenylsalicylate, (b) benzotriazole and (c) cyanoacrylate.

12.11 Antimicrobial finishing

Micro-organisms are part of our life and are almost everywhere in our environment. A complex consortium of micro-organisms lives on the human skin [13]. On clean skin, a population of 100–1000 micro-organisms per 1 cm^2 is observed. On normal skin, 10^2–10^7 aerobic bacteria/cm^2 and 10^2–10^6 anaerobic bacteria are observed.

A well-balanced distribution of individual species in the population is essential for a healthy condition of the skin. Besides bacteria, fungi are also present; the distribution of species depends on a number of conditions:

- Hygienic situation, gender, use of antiperspirants, site on body
- Physical activity, sweat production
- Clothing (material, chemicals) and climatic conditions near the skin.

In an analysis of the distribution of micro-organisms, the following species could be identified [14]: areobic coryneforms (mainly *C. xerosis*), propionibacteria (*P. acnes*, 47%; *P. avidum* 41%; *P. granulosum*, 3%), staphylococci (*Staphylococcus epidermidis*, 60%; *S. haemolyticus*, 12%; *S. hominis*, 6%; *S. capitis*, 3%; *S. saprophyticus*, 3%), micrococci (mainly *M. luteus*), gram negatives (mainly *acinetobacter*) and candida.

A share of approximately 30% of the micro-organisms could not be identified with the methods used. By DNA genotyping, an analysis of the micro-organisms in human sweat was performed [13]. Major species found in male sweat were *S. epidermis* (53%) and *proteus mirabilis* (24%), with a total number of 100 clones. In female sweat approximately 80 clones were found, major species being *S. epidermis* (10%), *pseudomonas sp.* (40%) and *enterobacter sp.* (40%). Differences in distribution between sweat from male and female subjects are mainly attributed to different habits in body cleaning (shower, detergents used and deodorants).

Microbial growth on textiles, in particular, bacteria and fungi, leads to a number of adverse effects:

- Hygienic difficulties can arise from uncontrolled growth of micro-organisms on textile structures. Examples for important products are medical textiles, wound dressings, hygiene wipes and products in food processing packaging. Infections and toxic by-products can lead to serious health concerns.
- In a less dramatic way, uncontrolled microbial growth leads to odour development (e.g., in sports cloths), discoloration of sensitive dyes, staining as well as fibre degradation, for example, in the case of fungal growth on cellulosic material.

The complex microbial situation on the human skin will be influenced by the use of antimicrobial agents in textiles. Thus when considering the use of antimicrobial concepts, we have to realise a distinct function of the product, which must not interfere negatively with the normal microbial situation of healthy skin.

Some natural fibres (e.g., wool) and man-made fibres (e.g., lyocell-type cellulosics) exhibit a reduced rate of bacterial growth, which leads to reduced odour development during wear. Such materials are bacteriostatic, as they do not support bacterial growth and in an ideal case do not interfere with the natural distribution of micro-organisms in the microbial consortium on healthy skin.

The use of antimicrobial products in textiles worn next to the skin has to be considered with care as these products will influence the natural flora of bacteria on the skin and indirectly then could cause irritation and allergic reactions.

Widespread use of a highly antimicrobial products is also in contradiction with the aspect of biodegradation of materials, as the sense of these products will be in contradiction to the request of an easy and rapid biodegradation of textile materials.

Different products and concepts can be used to provide antimicrobial properties to textile materials: Biostats control the growth and spread of micro-organisms (bacteriostats and fungistats). Biocides actively kill micro-organisms (bactericides and fungicides).

Chemicals can be deposited on a substrate and are active on the surface as well as in the near surrounding through controlled leaching of small amounts of the product.

During use and care operations (laundry and cleaning), the concentration of active product lowers below the threshold concentration required to provide full function of the antimicrobial property, which then begins to decrease.

Permanent products remain bound to the fibre surface and develop the functionality on the surface of the material. The substances lose their functionality when the surface is covered with a layer of killed bacteria, which form a barrier between the active biofilm and the antibacterial surface.

Leaching antibacterial substances:

- Formaldehyde is present in many finishes and provides antibacterial functionality when small amounts of free formaldehyde are released from the finishing.
- Triclosan: 2,4,4'-trichloro-2'-hydroxydiphenylether (Figure 12.20) is a widely used antibacterial substance with good activity against bacteria, but poor activity against fungi.

- Quaternary ammonium compounds also exhibit antibacterial properties; thus, some softeners are also non-permanent antimicrobial substances.
- Deposition of small silver particles in micro- or nano-scale has been widely used as a multifunctional antimicrobial agent. The functionality is based on the constant release of small concentrations of silver ions. These ions are supposed to act via different mechanisms, for example, deactivation of enzymes, disturbance of cell membrane function and DNA replication and generation of reactive oxygen species [15].
- Besides silver, copper-based products have also been proposed as antimicrobial textile finish. An example for a copper ion leaching product is Cu(I)-oxide (Cu_2O), which slowly releases small concentrations of copper ions, thus developing localised antimicrobial properties [16].

Bound antimicrobials include covalently bound quaternary ammonium compounds, polyhexamethylene biguanide (PHMB) and chitosan, which all are supposed to act by interacting with the anionic phospholipids present in the cell wall of the micro-organisms.

Another approach utilises the formation of chloroamines with hydantoin, which has been bound to the surface during textile finishing. Chlorohydantoin then serves as a rechargeable source of chlorine, which then is the active antimicrobial (Figure 12.20).

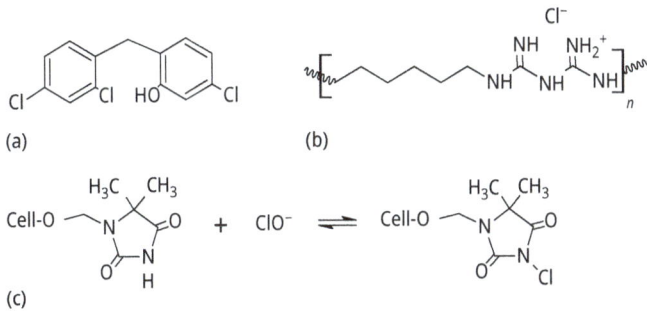

Figure 12.20: Antimicrobial compounds: (a) triclosan, (b) quaternary ammonium compound (polyhexamethylene biguanide) and (c) mechanism of chloramine formation and hypochlorite release.

A particular difficulty in testing of antimicrobial activity arises from the choice of the test method and the chosen stem of bacteria, fungi and so on. In a considerable number of tests, for example, AATCC Test method 147 (parallel streak method), leaching of the antimicrobial substance into the test medium (e.g., nutrient agar) is required to develop a zone of inhibition. In AATCC method 100, the textile sample is inoculated and incubated directly with micro-organisms, which permits the determination of immobilised antimicrobial agents [3]. In other methods, for example, the shaked flask test, according to ASTM E2149-01, the sample is shaken with a suspension of test micro-organisms and the reduction in viability of micro-organisms $R(\%)$ is determined according eq. (12.11)

[17]. In such methods, contact to a surface also leads to the antimicrobial effect as well as leaching of chemicals into the nutrient broth. Thus, it is important to consider the particular test method used to determine antimicrobial properties when test results and product applications are discussed.

$$R\% = ((A - B))/A \times 100 \qquad (12.11)$$

with $R\%$ being the percentage of reduction bacteria viability, A the number of viable bacteria in CFU/mL and B the number of viable bacteria after a given contact time (CFU, colony-forming units). Quantification of DNA on samples has also been used as a method to quantify bacterial growth on textiles [13].

12.12 Insect-resistant finishes – Mite protection

The major origin of insect-resistant finishes stems from approaches to protect wool against moths and beetles, which use the keratin as the nutrient source. The larvae digest wool keratin but cannot digest silk and other fibres. However, holes in non-keratin fibres can result from biting through such materials.

Two strategies for insect-resistant finishes exist: Contact poisons, usually nerve poisons interact in a less selective manner with insects on the surface of the finished fibres. Thus, such finishes can also be applied as finishes for anti-malaria textiles and anti-tick textiles.

The function of poisons through digestion of wool fibres is more specific. Examples for digestive poisons and nerve poisons as insect-resist finishes are shown in Figure 12.21.

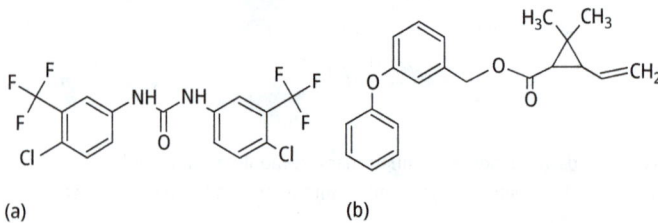

(a)   (b)

Figure 12.21: (a) Digestive poisons (Flucofenuron) and (b) nerve poisons (Permethrin) as insect-resistant finishes.

Mite protection: Dust mites belong to the Arachnida (a group of spiders). The major concern with regard to mites arises from their interrelation with allergies, asthma and other health problems.

Protective measures concentrate on reduced growth, through reduced temperature and moisture, washing, vacuum cleaning and use of tightly woven material, which forms a mechanical barrier between nutrient and mites.

Biocides and biostats are applied to stop mite growth, thereby also including a reduction of fungal and bacterial growth, which serves as element in the food chain of mites. In addition, a number of natural products exhibit insect-repellent properties. Representative examples are citronella (an essential oil, containing among others citronellal, citronellol and geraniol), neem (containing limonoids) and lemon eucalyptus (containing p-menthane-3,8,-diol) [18].

12.13 Enzymatic finishing – biofinishing

On the contrary to other finishing procedures that can be organised with regard to a distinct functionality that is intended, the term biofinishing includes a number of different processes. The common characteristic of these processes is the use of enzymes as an active system to develop the desired property.

Different enzymes are in use, and their name indicates the catalytical function of the protein (Table 12.1).

Table 12.1: Important types of enzymes used in biofinishing operations.

Enzyme	Function	Application
Amylase	Hydrolysis of starch	Desizing
Pectinase	Hydrolysis of pectins	Pretreatment of natural cellulose fibres (cotton, flax and hemp)
Catalase	Decomposes hydrogen peroxide	Removal of residual peroxide in advance to dyeing/printing operations
Cellulase	Hydrolyse cellulose	Anti-pilling, anti-fibrillation of regenerated cellulosics, biofinishing of denim, in detergents (laundry)
Protease	Hydrolysis of proteins	Anti-felt treatment of wool, household detergents (stain removal)
Lipases	Hydrolyse lipids	Removal of lipids (household detergents)
Peroxidase	Oxidative splitting of reactive dyes	Rinsing of reactive dyeing for higher fastness

The applications of amylases and pectinases are discussed in Chapter 11. Similarly, the use of proteases is discussed, together with other processes for anti-felt treatment of wool, in Chapter 11. Applications of enzymes in household detergents and washing agents will be discussed in Chapter 14.

The major focus in this chapter will be on the treatment of cellulose-based materials with cellulases. Important strains of fungi that produce cellulases are *Trichoderma*

and *Humicola* fungi. The so-called total crude cellulases are grown from fungal popu-
lations and contain up to five major activities of enzymes:
- β-Glucanases hydrolyse the cellulose chain from the side of the non-reducing
  endgroup, with the formation of glucose.
- Cellobiohydrolases hydrolyse cellulose with the release of cellobiose (glucose dimer).
- Endo-glucanases hydrolyse the cellulose at sites along the polymer chain.
- β-Glucosidases hydrolyse cellobiose to glucose, which is the final product of the
  hydrolysis reaction.

Representative conditions for applications of cellulases are in the pH range of 4–7 and
at temperatures up to 50–60 °C.

At the end of a process, an enzyme stop has to be set, for example, by an increase
in temperature or pH shift to alkaline conditions (>10) with the aim to denature the
proteins and thus prevent uncontrolled hydrolysis reaction during follow-up processes and storage.

The average pore dimension of cellulose fibres with the dimensions of 3–5 nm
permits an access of the larger enzymes into the pores. Because of their molecular
size in the dimension of 9.5–12.5 nm, the action of cellulases is concentrated on the
surface of the cellulose fibres and pores with larger diameter [19–21].

Higher activity of the cellulases is observed in regions where intensive agitation
occurs. Thus in particular, surface modifications can be achieved by a cellulose treatment. For example, the fibrous fuzz on the surface of a fabric can be removed by a
combination of cellulose hydrolysis and mechanical agitation. The hydrolysis of the
cellulose also leads to a reduction in the tensile strength of the fibres.

A typical application of cellulases is in the bio-polishing of cellulose textiles. Removal of small fibre ends from the fabric surface makes the colour appearance of the
dyed fabric brighter, leads to a softer handle and reduces pilling. However, the process is also accompanied by a strength loss, which has to be controlled carefully.

12.14 Denim finishing

A characteristic of denim products (jeans) is their used-look appearance. After the
production of the garment, a wide range of procedures is applied to widen the colouristic potential of a product, which mainly is available in blue or black colour. In this
chapter, mainly processes that lead to the development of the so-called wash-down
are summarised [22].

Many other processes, for example, desizing, overdyeing, softening and resin finishing, follow usual chemical concepts discussed in other chapters. When considering
the environmental impact of the garment wash, it is important to keep the figures of
the annual production of denim in mind, which at present is in the dimension of

several billions of pieces. Thus, minor problems can still lead to a considerable environmental burden.

The ring dyeing of the yarn is basic prerequisite to achieve a rapid removal of the (mainly) indigo dye from the yarn surface and to develop an intensive contrast between the white core of the yarn and the dyed surface.

The different mechanical and chemical treatments in combination intend to remove the dye from the surface without intensive damage of the yarn structure.

Abrasive processes: Abrasive treatments intend to imitate the use of the jeans; examples are sanding, brushing and sandblasting. Sandblasting has been banned with regard to health hazards due to aspiration and inhalation of silica dust, thus leading to severe silicosis. Dry ice (solid CO_2) has been proposed as an alternative, which does not form finely dispersed dust particles.

Stone washing: With the use of pumice stones, intensive abrasion of the garment can be achieved in big drum washers. The garment is washed with 0.5–3 times the weight pumice stones. Substantial removal of indigo dye into the wash liquor can lead to back staining which names a blue staining of the undyed weft and re-deposition of indigo on the garment. Substantial losses of abrasive material lead to significant release of sand into the waste water. The abrasive washing conditions also lead to rapid damage of the drum of the washer.

Often the pumice stones are also soaked in chemical solutions prior to their use to intensify the colour reduction by additional oxidative attack.

Oxidative treatments: The weft yarn in denim usually is an unbleached cotton yarn; thus besides the removal of the indigo on the blue warp yarn, a bleach of the weft yarn is achieved during the oxidative garment wash.

Important oxidative bleaching agents are hypochlorite (NaOCl), potassium permanganate ($KMnO_4$) and ozone (O_3). Peroxides (hydrogen peroxide H_2O_2 and persulphate K/$Na_2S_2O_8$) are also in use; however, reactivity towards indigo degradation is limited.

The major function of the reactive oxidants is based on the chemical attack of the chromophore and oxidation of indigo to isatin (Figure 12.22).

The oxidation with hypochlorite (pH 9–11.5) is a cheap and well-controllable process; however, environmental concerns arise from the release of halogenated organic compounds (AOX) into the wastewater. Following to the hypochlorite bleach, an antichlor treatment with appropriate reducing agents (e.g., bisulphite, thiosulphate and hydrogen peroxide) is required.

Potassium permanganate oxidation leads to the formation of brown insoluble MnO_2, which has to be removed with following treatments, for example, oxalic acid and sulphites. The chemical reaction behind the dissolution of MnO_2 is the reduction of the Mn^{4+} ions of MnO_2 to the readily soluble Mn^{2+}. The process leads to considerable load of manganese in the wastewater. The presence of residual traces of manganese ions also increases the tendency of yellowing of treated products through catalytic oxidation of indigo to the yellow isatin [21].

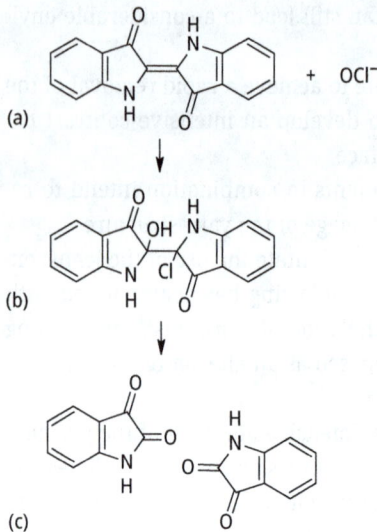

(a)

(b)

(c)

Figure 12.22: Oxidative degradation of (a) indigo via the reactive intermediate compound (b) to (c) isatin.

A combination of oxidising agent and stone wash can be achieved by soaking the pumice stones in KMnO$_4$ solution, which thus concentrates the oxidative attach near the abrasive effect of the pumice stones.

Oxidation of indigo with ozone (O$_3$) is an efficient technique as the formation of pollutants is reduced substantially. However, the high reactivity of the generated ozone makes the process difficult to control and requires direct installation of the ozone generator in short distance to the garment washer.

Reductive treatments: Reducing agents can also be applied to remove indigo from the dyed garment. Relevant chemicals are alkaline solutions of reducing sugars, for example, glucose or sulphinic acid derivate (R–SO$_2$Na). These agents remove the indigo through reduction of the dyestuff, which however leads to considerable back staining of the reduced indigo onto the undyed weft yarn and white pockets. Thus, appropriate auxiliaries have to be used to stabilise the indigo dye in the solution.

Enzymatic processes: An important method to achieve a less polluting garment wash utilises enzymatic processes. By using cellulases the fibres at the surface of the fabric are weakened by hydrolysis and abraded by the friction during drum washing. As a result indigo is released from the surface more in the form of indigo-dyed fibres and fragments than as dispersed dyestuff. Representative conditions are pH 5.5–7 and temperature of 30–55 °C

The structure of the cellulose fibre is of distinct importance for the rate of hydrolysis. Thus by modification of the cellulose fibre, for example, through local application of swelling agents, a controlled local hydrolysis can be achieved [22].

Laccases are oxidoreductases, which can be used to oxidise indigo. Usually oxygen serves as the primary oxidant, which is utilised by the laccase to transfer a soluble redox system, called mediator, into its oxidised form. The reduced form of the

mediator is regenerated into the oxidised form by the action of the laccase. A representative system for a mediator compound is syringic acid. The complete scheme is
shown in Figure 12.23. Compared to the conventional bleach systems, the use of laccases still is rather expensive.

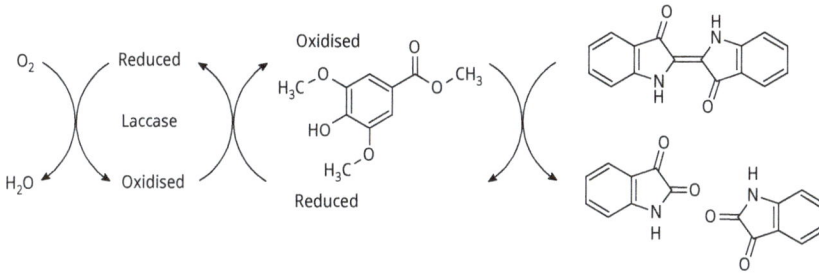

Figure 12.23: Reaction scheme for the indigo oxidation with laccases.

Laser processes: Instead of a chemical attack to destroy the indigo dye, thermal decomposition and sublimation of indigo by direct impact of a laser beam is also technically used.
On the contrary to the conventional garment wash, in laser treatment the pattern of used
look is developed on the garment by modulation of the laser energy. The piece of garment
is mounted on a manikin to define position and distance to the laser very accurately and
the desired picture of used look then is burned onto the surface of the garment. As a result of the thermal treatment, the colour of the laser-treated surface looks yellow-brown.
The coloured by-products from the laser treatment have to be removed in a following
rinse; thus in this case, this process cannot be claimed to be fully "water-free".

Alternative processes that are less polluting are waterjet fading, where the indigo is
removed by hydrojet treatment. As long as no chemicals are used, pollution will result
only from sludge formed, which consists of the removed indigo and fibre fragments. No
bleach effect of the weft yarn is obtained.

Developing the classical vintage look for denim products is coupled to woven fabric as a coloured warp and a white weft are required to develop the desired look. This
makes the application on highly elastic knitwear technologically impossible. For such
products, a different approach is used, with printing a photographic pattern of used
garment on the elastic material.

12.15 Finishes that influence thermal regulation

Considering thermal insulation, the resistance for heat transfer R_{ct} (thermal resistance
in Km^2/W) characterises the ability of a fabric to reduce heat flow from the body surface to the environment. The height of R_{ct} depends on the construction of a textile
layer. The major aspect is the amount of air that is trapped within the structure, for

example, a fine web of fibres and feathers (e.g., down). The conductivity of the fibre material itself is of minor relevance and becomes important only in case of very dense materials.

Under normal conditions, a dynamic equilibrium between heat generated by the body and the environment establishes. The surface temperature near the skin thus depends on the amount of heat flow generated and the heat flow to the ambient air. Hyperthermy (overheating, an increase in core temperature) or hypothermy (reduction in core temperature) appears when the thermal equilibrium is shifted outside of the usual range of comfort.

For a limited period, the heat capacity of melting (latent heat of fusion ΔH_{fus}) of the so-called phase change materials (PCM) can be used to stabilise thermal conditions [22].

Higher thermal energy production during physical activity leads to melting of the PCM, while during cooling the freezing of the substance stabilises the temperature against rapid drop. The melting temperature of suited substances should be in the range of the skin temperature.

Typical substances used as PCM are paraffin waxes (ΔH_{fus} = 140–260 kJ/kg) and polyethyleneglycols (ΔH_{fus} = 150–190 kJ/kg). Mixtures of fatty acids or polyalcohols are also possible PCM material.

For an application on textile materials, the PCM has to be microencapsulated or integrated into the fibre during manufacturing. A coverage of the material is required to seal the PCM in molten state and to prevent mobility of the oily substances. The coating of the PCM can be applied by physical methods (e.g., during spray drying), physicochemical methods (e.g., coacervation) or by chemical methods through interfacial polymerisation. Coacervation is the separation of a mixture into two phases, for example, in the case of PCM an oil-in-water system.

The fixation of the few micrometre-sized PCM particles on the fibre surface is achieved by using binder systems in a similar approach as applied in pigment dyeing. While considering PCM as a concept in thermoregulation, it is important to consider the use case and to estimate the total heat capacity of the PCM used. While a contribution to thermoregulation can be achieved for a limited period of time, the overall heat capacity will be limited. In particular in application of high-performance sports, the maximum function per weight should always be kept in mind.

12.16 Sorption of fragrances and functional substances

Deposition of fragrances, for example, for cosmetic textiles and functional substances, can be achieved by a similar approach through microencapsulation and binding at the surface. Friction leads to release of substances from the microcapsules.

Another approach is the use of covalently bound adsorbent systems, for example, cyclodextrins, calixarens or compounds with dendritic structure (Figure 12.24).

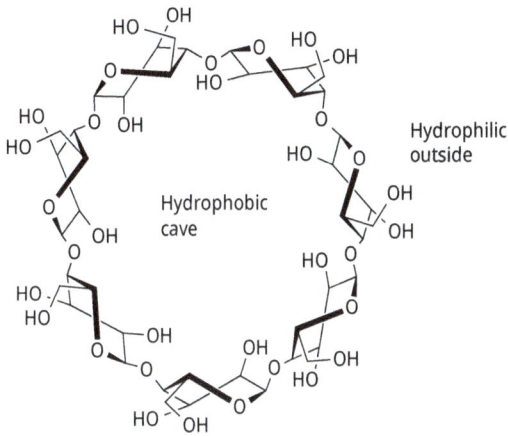

Figure 12.24: Structure of a cyclodextrin.

These compounds exhibit a cavity with typical diameters of 0.50–0.85 nm; their size being dependent on the number of rings in the cyclodextrine.

The inner side of the cavity exhibits low polarity; thus, fragrances and other functional substances with low polarity can be trapped inside the cavity. Based on the equilibrium constant for sorption/desorption, the active substance is released from the cavity.

In case cyclodextrin is attached to the fibre through a covalent bonding, the cyclodextrin remains on the fibre surface and could be recharged with the functional compound. Functional components other than fragrances could be antimicrobial compounds, pharmaceutical drugs, insecticides and repellents.

12.17 Plasma chemistry in textile treatment/modification

12.17.1 Introduction

One of the most efficient technologies for surface modification is the plasma modification technique. Starting with the usage in the microelectronic industries in the 1960s, plasma technology has been developed over last decades. Plasma is partially ionised gas, consisting of ions, electrons and neutral particles. These gases are produced by electrical discharges, providing very high concentrations of energetic and chemically active species. This feature makes plasma suitable as an additional tool to intensify traditional chemical processes, which essentially increases their efficiency [25]. Nowadays plasma is not only used for surface modification of a wide range of materials, including metals and ceramics, but also for polymers and bio-based materials. In the field of textile modification, considerable efforts have been taken since the 1980s to develop plasma technology, which being an appropriate alternative to classical wet chemical techniques. Typically, non-thermal plasma (also referred as cold or non-equilibrium plasma) is used for the treatment of fibres and textiles. Non-thermal plasma is characterised by high-energy electrons and low-energy ions and neutral species. Plasma with such characteristics interacts with textile surfaces, initiating reactions that would otherwise occur only at elevated temperature. Hence, non-thermal plasma can initiate reactions without causing excessive heat and degradation of textile substrates. In addition, it is a versatile technique, where a large variety of chemically active functional groups can be incorporated into the textile surface.

12.17.2 Effect of plasma treatment on fibre and textile surfaces

According to general understanding, plasma chemistry is associated with chemistry organised in or with plasma. Chemically active plasma is a multi-component system with high concentrations of charged particles (electrons, negative and positive ions), excited atoms and molecules, radicals and UV photons. Electrons, for example, are usually first to receive the energy from an electric field and then distribute it among other plasma components. Changing parameters of the electron gas (density, temperature and electron energy distribution function) permit control and optimisation of plasma-chemical processes.

The vibrational excitation of molecules often makes a major contribution to plasma-chemical kinetics as plasma electrons primarily transfer most of the energy in gases such as N_2, CO, CO_2 and H_2 into vibrational excitation [25].

In case of polymer substrates, a *high density of free radicals* is created by the dissociation of chemical bonds in polymers (Table 12.2).

Table 12.2: Dissociation energies of different covalent bonds in organic molecules [26].

Bond	Dissociation energy eV
C–C	3.6
C–H	4.3
C=O	7.8
C–N	3.2
C=N	9.3

The high-energy plasma-emitted species and UV photons are sufficient to react with the polymer molecules on the substrate surface resulting in the formation of active sites, that is, radicals by hydrogen abstraction or ablation of side chains. The formed free radicals can be utilised to functionalise fibre/textile surfaces by reacting with plasma active species and/or other functional groups. The effect, however, is commonly accompanied by C–C bond scission and the formation of short chain polymer fragments, that is, polymer degradation. The free-radical chemistry builds the basic for plasma-initiated chemical reactions on polymer fibre and textile surfaces.

12.17.3 Free radical formation

The initiation reaction for free radical formation is the abstraction of hydrogen from the polymer chain to form free alkyl radicals (eqs. (12.12) and (12.13)). Oxidation of polymers occurs by the combination of oxygen with the free-radical species to create new reactive species, including peroxy radicals and hydroperoxides (eqs. (12.14) and (12.15)). Since the oxidation takes place in the polymer cage, the products may recombine to form hydroperoxides (eq. (12.16)). The hydroperoxides, in turn, are themselves reactive, creating new free-radical species, such as hydroxy and alkoxy radicals.

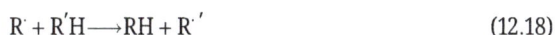

$$RH \xrightarrow{e,\,h\nu} R^{\cdot} + H^{\cdot} \tag{12.12}$$

$$RH + O_2 \longrightarrow R^{\cdot} + {}^{\cdot}OOH \tag{12.13}$$

$$R^{\cdot} + O_2 \longrightarrow ROO^{\cdot} \tag{12.14}$$

$$ROO^{\cdot} + RH \longrightarrow ROOH + R^{\cdot} \tag{12.15}$$

$$R^{\cdot} + {}^{\cdot}OOH \longrightarrow ROOH \tag{12.16}$$

$$RH + H^{\cdot} \longrightarrow H_2 + R^{\cdot} \tag{12.17}$$

$$R^{\cdot} + R'H \longrightarrow RH + R^{\cdot'} \tag{12.18}$$

Based on the free radical formation and reactions, the major effects of plasma on fibre and textile surface are surface cleaning, etching and modification of surface chemical structure [27].

12.17.4 Etching/cleaning

The etching and cleaning effect in polymer fibre/textiles is due to the ablation of the polymer outer layer and/or organic contaminants when exposed to plasma, which favours alterations of polymers. Etching effect is mainly based on the degradation of polymer, resulting in the formation of low molecular weight oxidised materials and release of volatised species, initiated by free radical active sites on fibre surface (eqs. (12.15)–(12.18)). Compared to inert polyolefins, polymers bearing functional groups containing oxygen (ethers, esters, ketones and carboxylic acids) are much more sensitive to etching. This feature of plasma is often used for surface decontamination to remove organic micro-pollutants from the surface (cleaning).

Applications on fibres and textiles are homogenisation of fabrics by removing a previously required finish, surface cleaning prior to finishing operations or increasing the colour shade by increasing the micro-roughness. Examples discussed in literature include desizing of cotton fabrics [28, 29], sterilisation [30] or anti-shrink treatment of wool by smoothing of the scales of the wool fibres [31].

12.17.5 Surface activation and functionalisation

Plasma is often used to activate the fibre/textile surface resulting in increased surface energy by the formation of, for example, –OH, –C=O, –COOH, –NH_2, –F and –Cl, on the substrate surface. Surface activation and modification is typically needed for materials having a low intrinsic surface energy, for example, polyolefins. The mechanism of the surface activation is described by the example of CO_2 plasma for the formation of –COOH groups on polymer fibre surface.

The preferential dissociation of CO_2 in plasma gives the following:

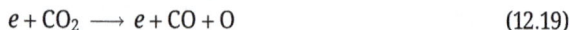

$$e + CO_2 \longrightarrow e + CO + O \tag{12.19}$$

However, CO_2, CO and O have ionisation potential in the range of 13–14 eV. Therefore, the average electron energy level needs to be high and exceeds the dissociation energies of C–C and C–H bonds (Table 12.2), leading to radical formation and polymer degradation [26]. As the insertion of CO_2 into an unactivated C–H bond by nucleophilic substitution is improbable, following free radical reactions are suggested (eqs. (12.20)–(12.23)):

$$R^\bullet + CO_2 \longrightarrow RCO_2\bullet \tag{12.20}$$

$$RCO_2\bullet + RH \longrightarrow RCOOH + R^\bullet \tag{12.21}$$

or

$$CO_2 + H^\bullet \longrightarrow {}^\bullet COOH \tag{12.22}$$

$$R^\bullet + {}^\bullet COOH \longrightarrow RCOOH \tag{12.23}$$

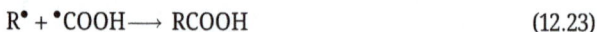

Plasma surface activation of fibres/textiles leads to improved wetting behaviour, often resulting in shorter absorption time and better wicking [32, 33] and higher dyeing and printing capability [34, 35].

12.17.6 Surface coating by plasma-induced polymerisation

Utilising suitable precursors/monomers, polymerisation and deposition of very thin polymer layer on fibre/textile surface can be realised by utilising the formed active sites. Most common types of polymer structure formation on fibre/textile surface are as follows:

- Radical chain growth polymerisation
- Recombination of macro-radicals, often leading to branching/crosslinking structure
- Grafting of polymers on to activated functional groups

Possible mechanism for polymerisation is as follows (eqs. (12.24)–(12.27)):

Start reaction:

$$R^\bullet + H_2C = CHX \rightarrow R - CH_2 - \overset{\bullet}{C}HX \tag{12.24}$$

Propagation:

$$R - CH_2 - \overset{\bullet}{C}HX + H_2C = CHX \rightarrow R - CH_2 - CHX - CH_2 - \overset{\bullet}{C}HX \tag{12.25}$$

Termination through recombination or disproportionation

$$\sim\!\sim\!H_2C - \overset{\bullet}{C}HX + XH\overset{\bullet}{C} - CH_2 \sim\!\sim \longrightarrow \sim\!\sim CH_2 - CHX - CHX - CH_2 \sim\!\sim \tag{12.26}$$

$$\sim\!\sim\!H_2C - \overset{\bullet}{C}HX + XH\overset{\bullet}{C} - CH_2 \sim\!\sim \longrightarrow \sim\!\sim H_2C - CH_2X + XHC = CH \sim\!\sim \tag{12.27}$$

X: Acrylic acid, methyl methacrylate and so on.

Examples of plasma-initiated polymerisation and coating on textiles are polymerisation of acrylamide on polypropylene [36], on cotton fibres [37] and deposition of polysiloxane on cotton fabrics [38].

12.17.7 Industrial plasma technologies and applications on textiles

Nowadays, batch-operated low-pressure plasma systems are still widely used for textile treatment. However, in the last decades, efforts were made to develop continuous plasma technology for fibre and textile treatment where atmospheric plasma gets much more attention [27].

Low-pressure plasmas are operated typically in the pressure range of 0.01 kPa, equipped with vacuum chamber and pumps. They can be generated not only using noble gases but also with O_2, N_2 or air. Roll-to-roll systems are used for textile transport and are placed in the plasma vacuum chamber. Low-pressure plasmas are characterised by their good uniformity but also by high investment and operation cost. Atmospheric plasma systems are, on the other hand, operated at standard atmospheric pressure of about 100 kPa. Modern atmospheric plasma systems offer air, N_2 and noble gases as process gas. The possible integration of atmospheric plasma into existing finishing lines exhibits a major advantage from industrial point of view.

With respect to textile treatment, the potential of plasma technology can be exploited in several areas, for example:
- facilitating the removal of sizing agents, scouring of fibres and fabrics [28, 29];
- sterilisation of textiles [30];
- anti-shrink treatment of wool [31];
- imparting hydrophilic properties and increasing adhesion [32, 33];
- enhancing printability and dyeability [34, 35];
- imparting hydrophobic and oleophobic properties [38] and
- coating/deposition [36–38].

References

[1]   Schindler, WD., Hauser, PJ. Chemical Finishing of Textiles, Woodhead Publishing, Cambridge, England, 2004.
[2]   Schimper, C., Bechtold, T. Mobile and hydrolysable formaldehyde in low-formaldehyde of cellulose textiles. Cell Chem. Technol. 2005,39/5-6,593–560.
[3]   AATCC Technical Manual 70, American Association of Textile Chemists and Colorists, Research Triangle Park, NC, USA, 1995.
[4]   Schramm, C., Rinderer, B., Bobleter, O. Nonformaldehyde durable press finishing with BTCA - Evaluation of the degree of esterification by isocratic HPLC. Textile Chemist and Colorist. 1997,29,37–41.
[5]   Welch, CM. Formaldehyde-free durable-press finishes. Rev. Prog. Color. 1992,22,32–41.
[6]   Abu Rous, M., Lenzing, AG. Lenzing, Austria. internal communication, 2018.
[7]   OECD(2013), OECD/UNEP Global PFC Group, Synthesis paper on per- and polyfluorinated chemicals (PFCs), Environment, Health and Safety, Environment Directorate, (accessed 14.02.2018) at http://www.oecd.org/env/ehs/risk-management/PFC_FINAL-Web.pdf.
[8]   Horrocks, AR., Price, D. Fire Retardant Materials, CRC Woodhead Publishing, Cambridge, England, 2003.
[9]   Reumann RD. (ed.), Prüfverfahren in der Textil- und Bekleidungstechnik, Springer Verlag, Berlin, 2000.
[10]   Berger, W., Faulstich, H., Fischer, P., Heger, A., Jacobasch, H-J., Mally, A., Mikut, I. Textile Faserstoffe - Beschaffenheit und Eigenschaften, Bobeth w. (ed.). Springer Verlag, Heidelberg, 1993, ISBN 978-3-642-77655-7.
[11]   Bechtold, T. Verfahren und stoffgemisch zur erhöhung der oberflächenleitfähigkeit von hydrophoben polymeren. PCT/AT/2006/000052, A 241/2005, 15.Februar 2005.
[12]   Bernhard, A. Characterisation of the Ultraviolet Protection of Wearable Textiles, Master Thesis, University Innsbruck, 2017.
[13]   Teufel, L., Schuster, KC., Merschak, P., Bechtold, T., Redl, B. Development of a fast and reliable method for the assessment of microbial colonisation and growth on textiles by DNA quantification. J. Mol. Microbiol. Biotechnol. 2008,14,193–200.
[14]   Rennie, PJ., Gower, DB., Holland, KT. In vitro and in vivo studies of human axillary odour and the cutaneous microflora. Br. J. Dermatol. 1991,124,596–602.
[15]   Emam, HE., Manian, AP., Siroka, B., Duelli, H., Redl, B., Pipal, A., Bechtold, T. Treatments to impart antimicrobial activity to clothing and household cellulosic-textiles - Why "Nano"-silver?. J. Cleaner Production. 2013,39,17–23.
[16]   Turalija, M., Merschak, P., Redl, B., Griesser, U., Duelli, H., Bechtold, T. Copper(I)oxide microparticles – synthesis and antimicrobial finishing of textiles. Journal of Material Chemistry B. 2015,3,5886–5892.

[17] ASTM, E2149-01, Standard Test Method for Determining the Antimicrobial Activity of Immobilized Antimicrobial Agents Under Dynamic Contact Conditions, Annual Book of ASTM Standards, ASTM, West Conshohocken, PA, 2004,p.1629.

[18] Roshan P, (ed.). Denim – Manufacture, Finishing and Applications, Woodhead Publishing, Cambridge, UK, 2015.

[19] Receveur, V., Czjzek, M., Schülein, M., Panine||, P., Henrissat B., Dimension. Shape, and conformational flexibility of a two domain fungal cellulase in solution probed by small angle X-ray scattering. J. Biol. Chem. 2002,277,40887–40892. doi: 10.1074/jbc.M205404200.

[20] Ostlund, A., Idström, A., Olsson, C., Tomas L.P., Nordstierna, L. Modification of crystallinity and pore size distribution in coagulated cellulose films. Cellulose 2013,20,1657–1667; DOI: 10.1007/s10570-013-9982-7.

[21] Oeztuerk, HB., Potthast, A., Rosenau, T., Abu-Rous, M., MacNaughtan, B., Schuster, KC., Mitchell, JR., Bechtold T. Changes in the intra- and inter-fibrillar structure of Lyocell (TENCEL) fibers caused by NaOH treatment. Cellulose 2009,16/1,37–52. DOI 10.1007/s10570-008-9249-x.

[22] Roshan, P., (ed.). Functional Finishes for Textiles, Improving Comfort, Performance and Protection, Woodhead Publishing, Cambridge, UK, 2015.

[23] Maier, P., Krüger, R., Grüninger, G. Yellowing of indigo-dyed jeanswear. Melliand Textilberichte. 1996,77,786–787.

[24] Schimper, C., Ibanescu, C., Bechtold, T. Effect of alkali pre-treatment on hydrolysis of regenerated cellulose fibers (Part 1: viscose) by cellulases. Cellulose. 2009,16/6,1057–1068.

[25] Fridman, A. Plasma Chemistry, Cambridge University Press, UK 2008.

[26] Friedrich, J. The Plasma Chemistry of Polymer Surfaces: Advanced Techniques for Surface Design, Wiley-VCH Verlag & Co. KGaA, Weinheim, Germany, 2012.

[27] R. Shishoo, (Ed.). Plasma Technologies for Textiles, Woodhead Publishing Limited, UK 2007.

[28] Riccobono, PX., Rolden, L. Plasma treatment of textiles: A novel approach to the environmental problems of desizing. Text. Chem. Color. 1973,5,239–248.

[29] Cai, Z., Qiu, Y., Zhang, C., Hwang, Y., McCord, M. Effect of atmospheric plasma treatment on desizing of PVA on cotton. Text. Res. J. 2003,73, 670–674.

[30] Roth, JR., Chen, Z., Sherman, DM., Karakaya, F., Tsai, PPY., Kelly-Wintenberg, K., Montie, KTC. Increasing surface energy and sterilization of nonwoven fabrics by exposure to an atmospheric uniform glow discharge plasma. Int. Nonwoven J. 2001,10,34–47.

[31] Canal, C., Erra, P., Molina, R., Bertran, E. Regulation of surface hydrophilicity of plasma treated wool fabrics. Text. Res. J. 2007,77,559–564.

[32] Temmerman, E., Leys, C. Surface modification of cotton yarn with a DC glow discharge in ambient air. Surf. Coat. Technol. 2005,200,686–689.

[33] Costa, THC., Feitor, MC., Alves, C., Freire, PB., de Bezerra, CM. Effects of gas composition during plasma modification of polyester fabrics. J. Mater. Process Technol. 2006,173,40–43.

[34] Ferrero, F., Tonin, C., Peila, R., Pollone, FR. Improving the dyeability of synthetic fabrics with basic dyes using in situ plasma polymerisation of acrylic acid. Color. Technol. 2004,120,30–34.

[35] El-Zawahry, MM., Ibrahim, NA., Eid, MA. The impact of nitrogen plasma treatment upon the physical-chemical and dyeing properties of wool fabric. Polym. Plast. Technol. Eng. 2006,45,1123–1132.

[36] Chen, KS., Lin, HR., Chen, SC., Tsai, JC., Ku, YA. Long Term Water Adsorption Ratio Improvement of Polypropylene Fabric by Plasma Pre-treatment and Graft Polymerization. Polym. J. 2006,38,905–911.

[37] Castelvetro, V., Fatarella, E., Corsi, L., Giaiacop,i S., Ciardelli, G. Graft polymerisation of functional acrylic monomers onto cotton fibres activated by continuous Ar plasma. Plasma Process Polym. 2006,3,48–57.

[38] Goodwin, AJ., Ward, L., Merlin, P., Badyal, JPS. Method and apparatus for forming a coating. EP. 1326718, 4. Oct. 2000.

Take home messages

Finishing processes intend to strengthen or develop a functional property of the product. Many finishing operations are performed to improve performance during garment production (sewing and softness), durability of the product (fastness and light degradation) or to meet customer expectations (softening, crease recovery, low shrinkage and water repellency).

A considerable number of finishes is performed to add a special functionality designed for a certain application, such as antimicrobial properties, flame resistance, insect repellency and antistatic properties.

The definition of an appropriate finishing concept has to follow the expected use and the functional requirements the product has to meet.

Quiz

1. Why wet crosslinking leads to higher water retention when compared to dry crosslinking?
2. What is the principle of an anti-felt treatment for wool?
3. What are important principles to reduce flammability of a product?

Exercises

1. Define an example for finishing concept (properties) for a business shirt.
2. Define an example for a finishing concept for a pilot overall.
3. Define an example for a finishing concept for a towel.

Solutions

1. (Question 1): The wet crosslinking stabilises the swollen state of a fibre, while in dry crosslinking the dry state is stabilised; thus, swelling in water will be higher and the water retention is higher.
2. (Question 2): By the combination of an oxidative and reductive treatment, the structure of the scales of wool is weakened. This leads to lower unidirectional friction and shrinkage. In addition, the polymer coating smoothes the fibre surface.
3. (Question 3): Selection of appropriate fibres (high thermal stability), formation of char, intumescent material to reduce access of oxygen, formation of volatile components that reduce access of oxygen (water, CO_2) and interruption of radical chain reaction as part of flame propagation.

4. (Exercise 1): Softening, easy-care finish (e.g., liquid ammonia treatment and moist crosslinking) and compressive shrinkage (for improved dimensional stability).
5. (Exercise 2): Selection of fibres with high thermal stability or flame retardancy, compressive shrinkage (dimensional stability) and optional: insect repellency and antimicrobial finish.
6. (Exercise 3): Mainly hydrophilic softening.

List of abbreviations/symbols

PP	polypropylene
DMU	*N,N'*-dimethylol-urea
DMDHEU	*N,N'* dimethylol-dihydroxyethylene-urea
DMeDHEU	*N,N'*-dimethyl-dihydroxyethylene urea
BTCA	1,2,3,4-butanetetracarboxylic acid
LOI	limiting oxygen index
DBDPO	decabromodiphenyl oxide
CVD	chemical vapour deposition
UVPF	Ultra-violet protection factor
PHMB	polyhexamethylene biguanide
CFU	colony-forming units
ASTM	American Society for Testing and Materials
AATCC	American Association of Textile Chemists and Colorists
AOX	adsorbable halogenated organic compounds
R_{ct}	thermal resistance

13 Technical approaches in dyestuff/chemical application

13.1 General aspects

In this chapter, relevant aspects of textile machinery and processing are highlighted. From the point of view of textile chemistry, we can classify machinery and processes with regard to different characteristics:

Dry processing: Such techniques include sanding, brushing and shearing, which are mechanical processes, as well as plasma processes, singing, laser treatment of denim, which also include chemical/physicochemical processes.

Wet processing: This involves techniques that use a solvent-based system for the textile chemical processes, for example, water, steam (condensate), printing pastes, supercritical CO_2 and solvent systems (e.g., petrol and halogenated hydrocarbons), ionic liquids and foam.

For classification of machinery with regard to applications of processes, it is important to understand the structure of a textile dyehouse and finishing plant (Figure 13.1).

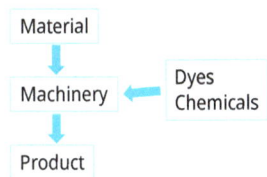

Figure 13.1: Structure of a textile dyehouse and finishing plant.

The machinery for wet processing of textiles with few exceptions is designed for the treatment of a certain group of textile materials, for example, fibres, yarn and fabric, either in discontinuous (batchwise) or continuous processes.

The design of the machinery allows applications of a wide range of processes. In a similar manner, the suppliers of dyestuff and auxiliaries form groups of dyes and chemicals that have been optimised for a certain application.

Because of the enormous range of variations in products, a textile dyeing and finishing company has to specialise on a certain group of products, which can be processed on the available equipment. The recipes for chemicals and dyes are fitted to the machinery available, with the aim to achieve an optimum quality of the product.

In this chapter, some general aspects of selected representatives of machinery for textile chemical processing are explained by using a few representative examples. However, it has to be kept in mind that many other processes could be run on the same machinery and a huge number of technical variations in equipment exist [1].

The major emphasis of this chapter will be on the use of equipment for water-based textile chemical processes.

https://doi.org/10.1515/9783110795738-013

13.2 Batchwise operation

In this type of equipment, the material is treated in a form of defined lots, with a series of consecutive steps. A general restriction in the use of batch-dyeing apparatus comes from the design of the machine for a given liquor ratio; thus, a machine that has been built for a capacity of 100 kg achieves the expected reproducibility of a recipe only with the exact amount of material. In modern machinery, the use of replacement bodies permits a reduction in capacity to 50%, without any significant change in the liquor ratio.

A reduction in amount of material processed often cannot be directly compensated by a reduction in the volume of processing baths, as a minimal volume is required to operate the machine. A certain compensation for the reduction in volume can be achieved by filling unused space with dummy bodies.

A change in liquor ratio will lead to need for adaptation of the recipes as, for example, dyestuff fixation will be directly dependent on the change in liquor ratio. Difficulties will arise from dyestuff combinations that react in different intensities on the changes in the dyebath volume.

13.3 Yarn dyeing apparatus

A schematic representation of a package dyeing apparatus is shown in Figure 13.2.

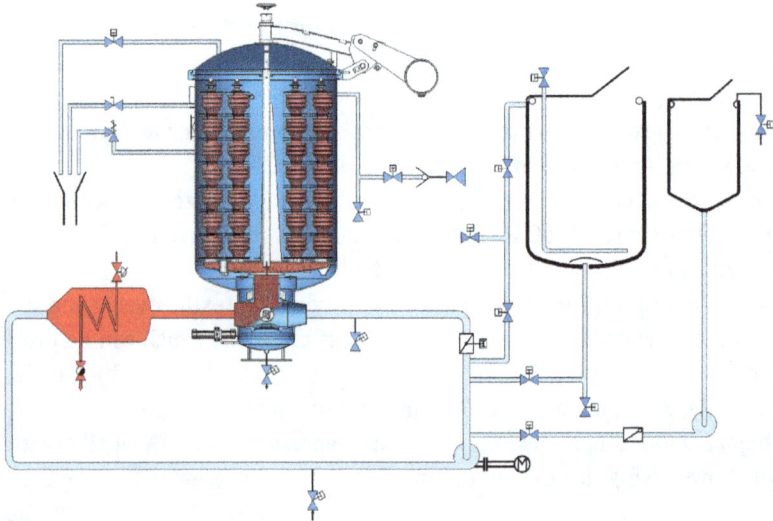

Figure 13.2: Schematic presentation of a yarn dyeing apparatus (courtesy of Thies AG, Coesfeld, GER).

The processing of the material (fibres, yarn and warp beam) is performed in a closed vessel. In the case of yarn dyeing, the yarn is wound on perforated tubes (conical and cylindrical), and fibres are pressed into perforated compartments. Such a machine also permits the dyeing of a warp beam; in this case, the warp has been rolled on a perforated axis. In a similar approach, pieces of fabric can also be dyed.

The processing liquor is transported through the material by means of powerful circulation pumps. A typical value for the circulation flow is about 35 L/hkg, which means that the circulation power in a 100 kg dyeing apparatus has to be at 3,500 L/h. The direction of the bath flow is reversed inside/outside and outside/inside within regular intervals of time, which are not necessarily equal in duration.

When the dyeing is performed at a liquor level of 1:6–1:7, the dyebath volume in the dyeing apparatus will be 600–700 L and the full dyebath will be circulated through the material approximately 5–6 times per minute.

Lower circulation flow increases the risk of local variations in process conditions; substantially higher flow rates lead to high-pressure drop on the material, which increases the risk of mechanical damage of the goods.

The need for such a high flow in bath circulation stems from the request that bath circulation must be sufficiently high to maintain differences in conditions between inside /middle/outside of a 1 kg yarn cone at a very low level. For example, reduction in dyestuff concentration during one full bath exchange should be less than 2%. A combination of temperature control and dosage of chemicals is used to steer dyestuff exhaustion in dyeing. The bath circulation is set to an appropriate maximum as lower values automatically would require an increased dyeing time.

The use of closed vessels permits the application of conditions of elevated pressure. Such conditions are required for proper function of the circulation pump at elevated temperatures and to avoid cavitation. A representative value for the pressure drop through a yarn cone is in the magnitude of 1 bar; thus in particular for the flow direction from outside to inside of a yarn cone, an overpressure in the apparatus is required so as to avoid conditions of critically low pressure inside the yarn cones. As a result of the high processing temperatures and the need for an overpressure inside the machine for safety reasons, the manual opening of valves is blocked automatically at temperatures above 80 °C.

The machines are designed for rather universal use; thus, prewashing, scouring, bleaching, dyeing and after treatment can be performed in the same machinery.

For example for dyeing of cotton yarn with vat dyes, the following steps could be arranged in a consecutive order:

Scouring – rinse – bleach – rinse – vat dyeing – reductive rinse – oxidation and soaping – rinse – softening. Then the yarn cones are dried in a separate machine.

Dyeing of warp beam and fabric requires the use of adapted inserts; however the, textile chemical principles are the same.

13.4 Overflow dyeing machines/jet dyeing machines

In this group of dyeing machinery, the goods are moved by action of a hydraulic stream. In the case of a jet dyeing machine, the dyebath is circulated through a nozzle that transports the goods. In the case of an overflow dyeing machine, the hydraulic transport is more gentle and thus the machine is better suited for delicate material. Representative schemes of a jet dyeing machine and an overflow dyeing machine are shown in Figure 13.3.

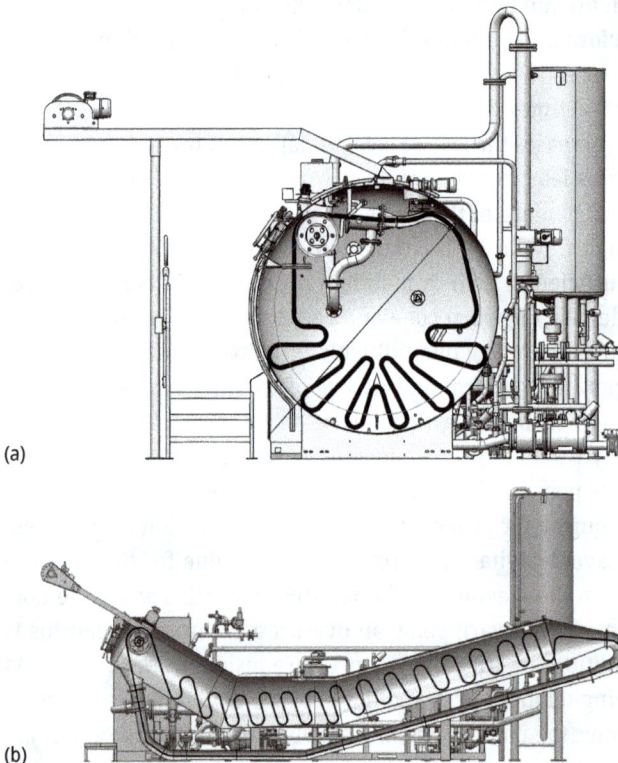

(a)

(b)

Figure 13.3: Representative schemes of (a) a jet dyeing machine and (b) an overflow dyeing machine (courtesy of Thies AG, Coesfeld, GER).

Typical liquor ratios are between 1:5 and 1:20; however with a partial filling or fully flooded operation, the liquor ratio increases. A jet dyeing machine operates at a material velocity of up to 200–500 m/min, and in the overflow dyeing machine material velocity of up to 500 m/min is used. The maximum velocity depends on the quality of fabric to be processed. The capacity of a jet is limited by two conditions: the maximum velocity of the textile and the limitation for the dwell time in the storage. The exhaustion of dye per cycle should not be higher than 1–3% of the total amount used, which defines a useful time for one cycle t_R with typically 0.2–1 min. The capacity m_{load} of a jet dyeing machine thus is defined by the linear meter weight m_L of a material, the speed of operation v_{max} (e.g., 200 m/min) and the request for a maximum resident time t_R (e.g., 1 min; eq. (13.1)).

$$m_{load} = v_{max} m_L / t_R \tag{13.1}$$

Thus, batch sizes usually are below 1,000 m per jet; larger capacities have to be achieved by coupling several units (e.g., 2 or 4) and circulation of a common dyebath.

Several critical issues have to be considered to achieve good results:
- The hydraulic stress in jet dyeing is considerable and the material quality with regard to fibre type and construction (knitted material or fabric) must be resistant enough to withstand the applied tensional and frictional effects. The surface of a sensitive material can be damaged irreversibly through abrasion, fibrillation and crease formation.
- The tendency to mark creases during the period of low movement in the storage zone must be low to avoid development of creases, which become fixed in the fabric or marked by the dyestuff. Similar to yarn dyeing apparatus, the exhaust rate has to be controlled carefully, for example, through temperature control (polyester dyeing) or pH control (alkali dosage in reactive dyeing).

Substantially reduced liquor ratio has been achieved through a combination of a liquid jet with pneumatic transport. In the "airflow" system, the effective liquor ratio applied can be reduced to 1:2–1:4. The process is demanding with regard to the properties of the material and the levelling behaviour of the dyes, however, offers substantial reduction in consumption of chemicals and dyestuff [2].

13.5 Continuous pretreatment

Continuous processes are highly efficient for processing larger amounts of goods or for performance of standardised pretreatments. Representative examples in pretreatment are as follows:
- anti-felt treatment of wool sliver with the oxidant/sulphite/polymer treatments;
- scouring and bleaching of woven and knitted cotton fabric and
- mercerisation and alkalisation of cotton-based fabric.

The fabric can be processed in open width and in form of a rope. Tension applied on the fabric is a critical issue in the continuous processing of fabric. A certain tension is required to permit crease free transportation; however, careful control of stretch is required during processing of highly extensible elastic knitwear.

In the following examples, continuous plants for pretreatment of cotton fabric are shown.

Figure 13.4 shows a continuous desizing and one-stage bleach plant. The fabric is dipped into water to swell the water-soluble size, which is then washed off. At the end of every washing compartment, the excess liquid is squeezed off to minimise unwanted carry over from the washing bath into the next wash compartment. The application of the bleach chemicals is performed as wet-in-wet application. The bleach process usually is based on hydrogen peroxide as an oxidant and requires sufficient time and heat for the reaction to proceed. Thus, a steamer is used to execute the reaction. In the first phase of the steaming process, the fabric is guided over rolls to achieve a controlled heat transfer and to achieve a uniform bleach effect. The material then is rested on a roller bed, which slowly relocates the material to avoid crease marks and moves the material to the exit of the steamer, where it is brought into open width form again. The chemicals are washed off the fabric in a wash box (roller vat), where the flow of water is in counter-current direction. The material is then dried on a drum dryer.

The combination of aggregates shown in Figure 13.5 permits a continuous desizing, alkaline scouring and peroxide bleach of fabric in a combination of applicators, steamer boxes and wash boxes. The machine shown in Figure 13.5 delivers the bleached material in wet state. The drying step is performed separately.

In such an installation, the duration of all processing steps has to be synchronised to achieve maximum capacity of the installation. For an efficient operation, a careful setting of all conditions and recipes has to be worked out to guarantee completion of the respective process within the process time permitted.

An example for a continuous alkalisation plant is shown in Figure 13.6. The material is treated open width and in a first step is impregnated with concentrated alkali (e.g., 300 g/L NaOH). The swelling of the cotton fibres in alkali leads to an increase in fibre diameter that then causes substantial shrinkage forces. Attention must be paid to a control of shrinkage forces as otherwise the dimensions of the fabric (width and mass per area) will change. The machine in Figure 13.6 shows two different technical solutions to process fabric under tension. In the first section of the machine, the fabric is guided along rollers and transported without any air gap between the rollers. Thus, shrinkage along the warp direction is prevented by the direct contact of the fabric to the rollers. As the fabric is not held at the edges, shrinkage in the direction of the weft can occur, beginning from the sides of the fabric. As the middle part of the fabric is under tension both in the warp and weft direction, this part of the fabric maintains its initial dimensions, while the fabric tends to shrink at the edges. This effect leads to a slightly higher mass per area at the sides of the fabric when compared to the middle zone.

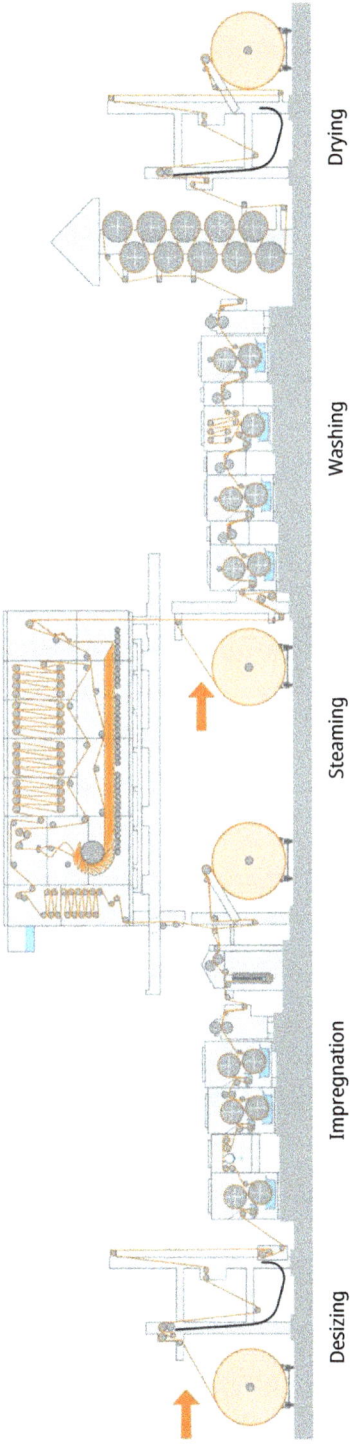

Figure 13.4: Continuous, desizing and one-stage bleach plant for elastic fabric (courtesy of Benninger AG, Uzwil, CH).

Desizing — Impregnation — Steaming — Washing — Drying

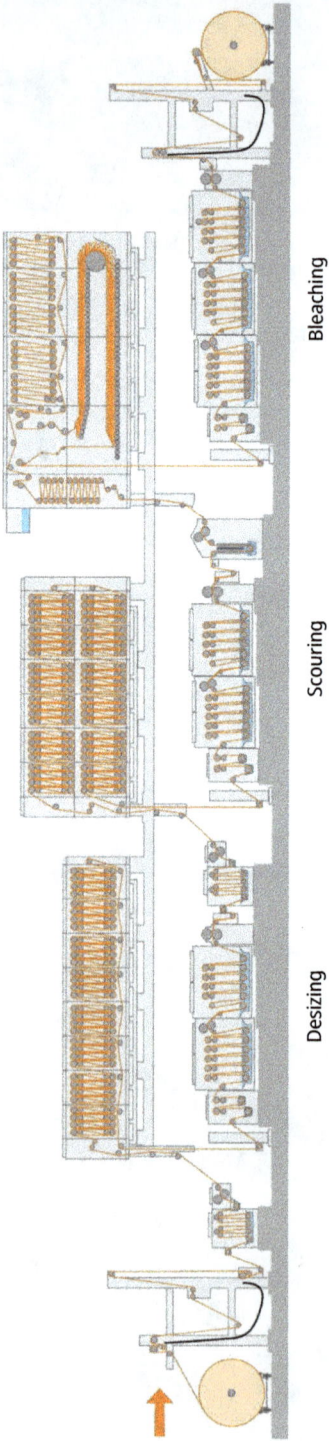

Figure 13.5: Three-stage pre-treatment plant for fabric (desizing, scouring and bleaching) (courtesy of Benninger AG, Uzwil, CH).

Desizing   Scouring   Bleaching

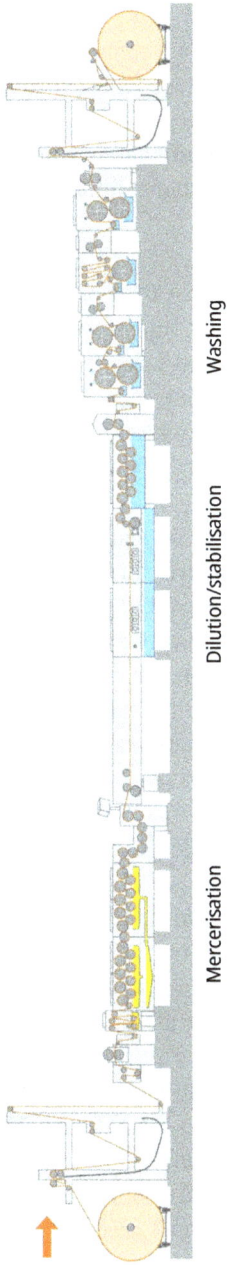

Figure 13.6: Mercerisation plant, combination of chain mercerisation and roller mercerisation (courtesy of Benninger AG, Uzwil, CH).

Mercerisation   Dilution/stabilisation   Washing

Such effects are undesirable for several reasons:

– The differences in yarn density can become visible in case it is a colour-woven fabric. The differences in pattern will be visible latest at the stage where a garment is produced from different parts of the fabric.

– In the case of a continuous dyeing, the pick-up of dye containing liquor will be different between middle part and sides of the fabric. Thus, colour differences along the weft direction will occur.

In the second section of the machine in Figure 13.6, the fabric becomes fixed on a needle frame that then expands the fabric to the original width. By this process the shrinkage at the sides of the fabric can be minimised. The required forces are high and can come into the dimension where risk of rupture of the fabric begins. A reduction in lye concentration is already initiated on the needle frame to reduce the shrinkage forces and stabilise the dimensions of the material before it is transported to the continuous drum washers.

The deep penetration of the concentrated alkali into the cellulose structure and the swelling of the fibres make the removal of alkali form cellulose fibres a rather slow process. Thus, after a series of rinsing processes, a hot rinse is introduced to remove alkali from the core of the material. Based on the following processes, a water rinse is sufficient or a neutralisation step is required. Careful neutralisation, for example, is necessary in case a reactive dyeing should follow, as residual alkali brought into the dyebath with the material would initiate early and uncontrolled dyestuff fixation.

13.6 Continuous dyeing processes – cold-pad-batch dyeing

Cold-pad-batch dyeing with reactive dyes belongs to the highly efficient dyeing processes with regard to dyestuff fixation, salt and energy consumption. In addition, other dyes permit application of the cold-pad batch technique; however, the use of reactive dyes is of highest importance and thus will be discussed as a representative case (Figure 13.7).

The dyestuff containing liquid is added to the fabric through an immersion and squeezing step. Representative values for a pick-up of dye solution are between 70 and 80 wt%. The uniformity of the add-on must be controlled carefully as the amount of added dyestuff solution will directly determine colour fixation and colour depth.

Affinity of the dyes must be low; otherwise, slight exhaustion will occur during the padding process. This will lead to a reduction of the dyestuff concentration in the pad box and as a result a reduction in colour depth along the production length will be observed (ending and tailing).

In particular at higher width of the fabric, the rollers of the padder can begin to bend, which leads to substantial differences in pressure and thus in pick-up.

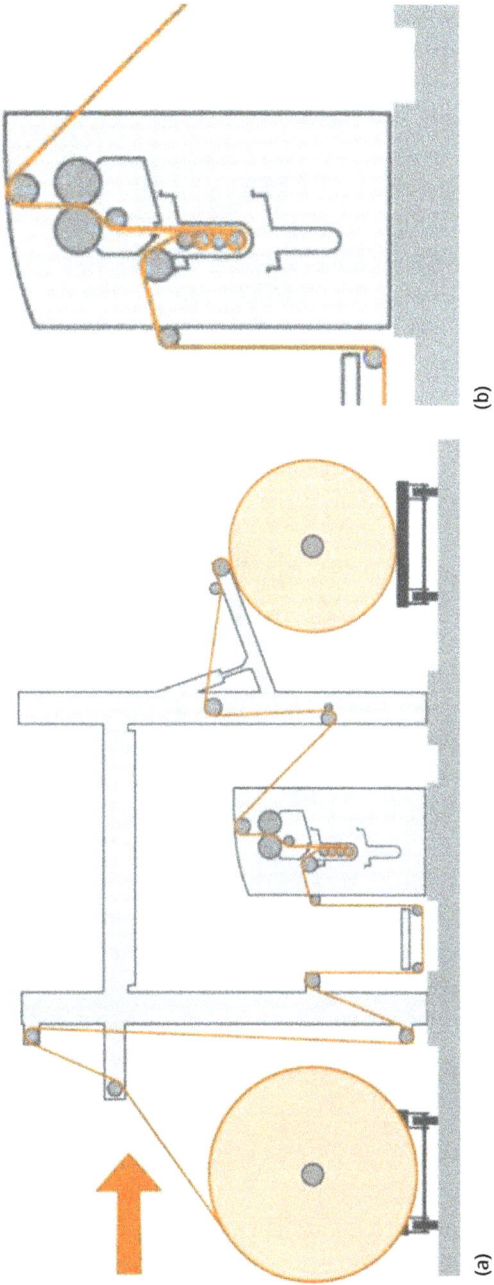

(a)

(b)

Figure 13.7: Cold-pad-batch (a) complete installation and (b) padder (courtesy of Benninger AG, Uzwil, CH).

The add-on can be controlled by means of a microwave absorption detector. When a padder with adjustable squeeze pressure is used, differences in add-on between middle and sides can be compensated. Such padders, for example, use a hydraulic pressure system or an air cushion, which then permits the adjustment of pressure in different zones of the padder.

For padding of the material, careful pretreatment is required to guarantee rapid wicking within the short length of immersion in the pad box and in particular if nip dyeing is intended. The time for full wetting is then within a few seconds. Equal penetration with dye liquor has to be achieved as a prerequisite for a uniform dyeing.

The impregnated fabric is then transferred on a skein and rotated during the period of dyestuff fixation. Rotation is necessary to prevent settling of water in the rolled material, which then would cause serious problems during rewinding due to unbalances inside the fabric roll.

Time for fixation has to be substantially longer than the minimum time required for fixation to avoid any difference in colour fixation, as the last part of the batch will be rinsed as first part of the batch.

In case of a reactive dyeing, a metering pump doses the dyestuff solution and the alkali solution required for fixation into the padder box. The running speed, length of immersion, temperature of the dye liquor and concentration of dye in the pad box must be kept constant during the continuously running process.

With a pick up of 75%, the formal liquor ratio is 1:0.75; thus, no salt is required for dyestuff fixation of reactive dyes. Dyestuff solubility can be an issue at darker shades; a possible solution to increase dyestuff solubility is the addition of urea. The content of the padder box has to be exchanged within a few minutes to avoid tailing effects due to dyestuff hydrolysis.

Pick-up in a padder depends on the applied pressure between the rolls, the type of material and on the running speed of the padder. The pick-up for a material has to be determined in laboratory experiments, and the results then are transferred on the large-scale dyeing.

The dyed material is then washed to remove hydrolysate and alkali either on continuous roller boxes, in drum washers in open width form or in form of a rope.

13.7 Continuous dyeing processes – pad-dry/pad steam plants

A highly efficient installation for continuous dyeing of fabric uses rapid fixation of dyes at elevated temperatures. Such an installation can be applied for a wide range of processes, dyestuff systems and combinations of dyes, for example, for fibre blends (e.g., cotton/polyester). A representative case for a full installation is shown in Figures 13.8, 13.9 and 13.10.

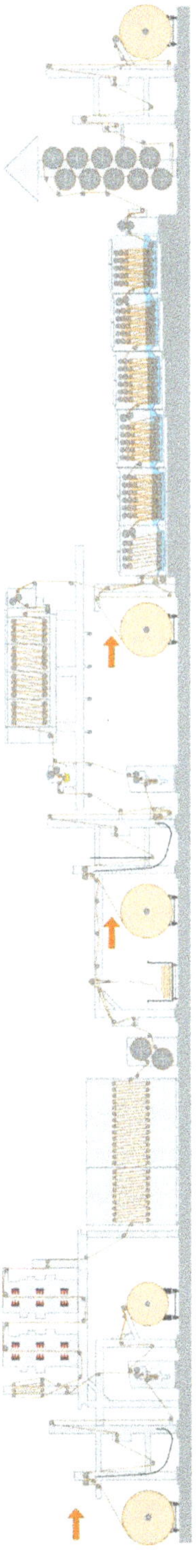

Figure 13.8: Continuous plant for pad-dry and pad-steam fixation (courtesy of Benninger AG, Uzwil, CH).

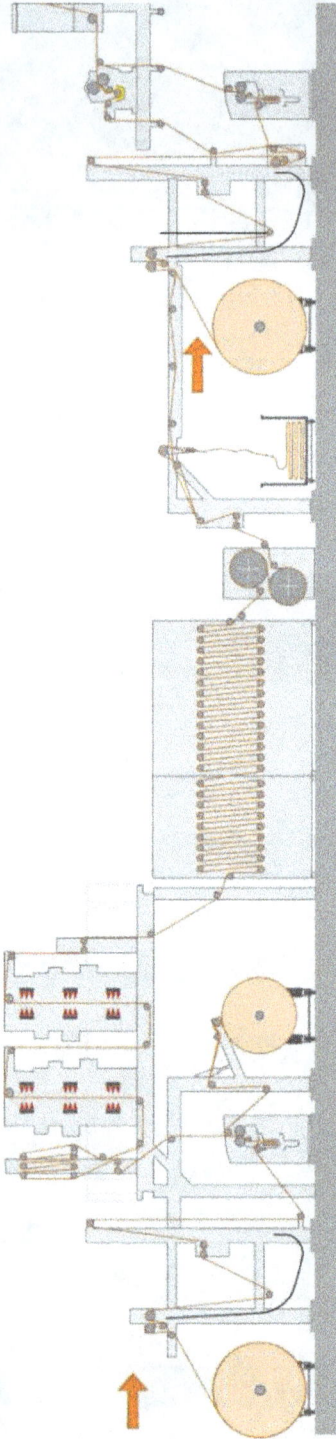

Figure 13.9: Magnified representation of a continuous pad-dry, pad-steam dyeing plant, first part (courtesy of Benninger AG, Uzwil, CH).

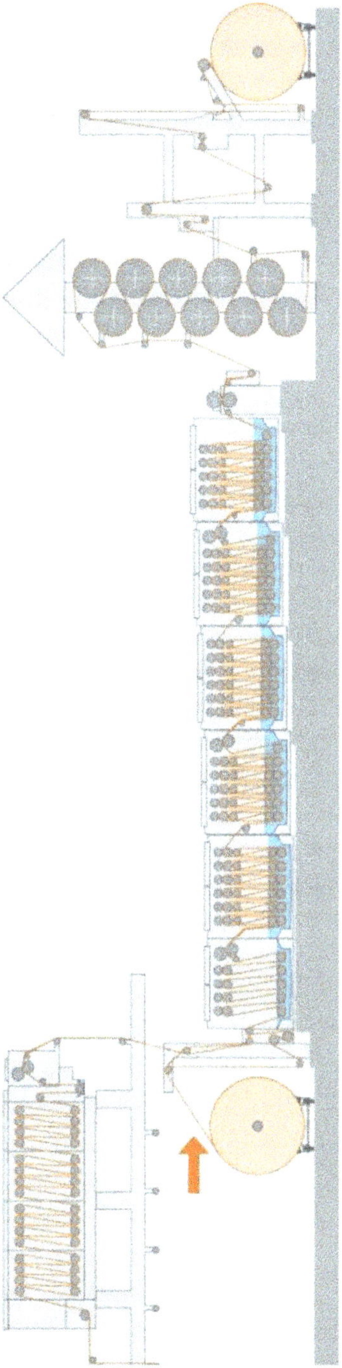

Figure 13.10: Magnified representation of a continuous pad-dry, pad-steam dyeing plant, second part (courtesy of Benninger AG, Uzwil, CH).

Elements of such an installation can also be used separately, which widens the field of applications. In the following explanation of the different units, dyeing of a fabric made form an intimate cotton/polyester fibre blend is discussed as a representative case.

The pretreated cotton/polyester fabric at first is impregnated with the pad liquid that contains both dyes in dispersed form: a polyester dye and a vat dye. The dye liquor is padded on the fabric and then following to a short air passage pre-dried in a vertical infrared dryer. The drying is not performed completely to the dry state and to avoid overheating of the fabric in the infrared dryer. A rapid pre-drying, however, is necessary to avoid redistribution of dye pigments on the fabric through migration effects. In addition, auxiliaries are used to reduce dyestuff migration by increasing the viscosity of the padded liquid (antimigration auxiliaries).

After pre-drying the material is guided into the hot-flue, a heated chamber equipped with rollers. At a temperature of 180–220 °C, the fixation of the disperse dye proceeds inside the hot-flue in hot air within 90–15 s (*thermosol* process).

The material is then cooled down through cooling rollers and then guided to a next padder that contains the alkaline-reducing agent for the vatting of the vat dye.

The impregnated material is transported into a steamer where dyestuff reduction is completed under the condition of an oxygen-free saturated steam atmosphere. At the same time, a reductive cleaning of the disperse dye is achieved.

Dyestuff fixation of the reduced vat dye is achieved at this temperature (102 °C) within a short time. The material leaves the steamer through a water seal, which reduces temperature of the goods and lowers the chemical concentrations on the fabric.

In the following continuous roller boxes, the vat dyeing is finished through rinsing, oxidation, soaping near to the boil and another final rinse.

Similarly cotton/polyester blend can be dyed with disperse dyes and reactive dyes. In this case, the first part of the unit is used for the thermosol dyeing of the polyester fibres, and the second part of the plant executes the reactive dyeing, for example, as a pad-steam process. Here the dye liquor that is padded on the material contains the reactive dye in combination with alkali. Fixation is achieved through treatment in saturated steam. The continuous open width washers are then used to rinse off unfixed reactive dye and chemicals.

At the end of the dyeing processes, the material is dried by means of a drum dryer. Such a drying technique does not permit adjustment of the dimensions (width) of the material, which is done in a later stage during finishing on a tenter.

13.8 Drying, fixation and finishing

Drying of wet textile fabric can be achieved with different machinery. *Drum dryers* permit an efficient drying through direct contact of the material with heated drums. Pre-drying can be achieved through direct *IR-radiation*, which is produced, for example, by electrical heaters and gas heating. Such devices achieve very high energy transfer and thus are used for rapid evaporation of a major part of the water; however, they are not used for full drying as the temperature of the material mainly is controlled by the evaporation of water and rises rapidly to ignition if the material begins to dry completely.

Drying without dimensional control can be achieved in heated chambers with intensive air circulation. In the hot-flue, the material is guided by rollers; in the sieve band dryer, the material is transported in loose and relaxed form by means of air-permeable sieve and dried by circulation of hot air.

A controlled adjustment of dimensions (mass per area and width) is achieved by drying in a tenter. The material at first is fixed on moving chains either by needles or clamps and then transported through a series of heated chambers (fields). Rapid drying is achieved by intensive air circulation. A padder at the entry of the heated tenter permits addition of crosslinkers for finishing, which are then reacted in the last fields of the tenter frame.

Dry state of the material is monitored at the exit of the tenter frame by means of electrical conductivity, which is low in the case of completely dried material. Storage of wet material bears the risk of microbial growth (fungi), thus leading to dark spots and colour changes. Overdrying to very low residual humidity is undesirable as static charge and stiffness of the material increase considerably.

13.9 Minimum pick-up applications

Drying is a highly energy-consuming process step. As a rough estimate the evaporation of 1 kg of water will require 1 kWh of energy. Thus, it is important, as much as possible in processing steps, to keep the material in wet state without drying, for example, singing – desizing – scouring/washing – bleaching, first drying, continuous dyeing – rinsing, second drying, finishing, third drying.

In such a case, at least three dyeing steps are required and the total energy consumption for drying can be estimated with 3 kWh/kg of fabric. In particular for the processing of natural fibres, the high water uptake requires more energy than in the case of synthetic fibres where low swelling and appropriate mechanical methods for dewatering reduce the energy consumption of drying considerably.

Another group of techniques, the so-called low-add on techniques, permit application of chemicals from more concentrated solution and thus reduce energy for subsequent drying.

In these processes, only an add-on of, for example, 30 wt%, is brought on the fabric. These techniques avoid dipping of the fabric into the solution as dewatering to a pick-up of 30% will not be possible in many cases.

Typical techniques include the following:
- Application of process liquids in form of a *foam*, which collapses after impregnation.
- *Kiss-roll techniques*: An engraved application roll dips into the chemical solution and transfers only a thin layer of chemical solution on the fabric.
- *Spraying* of solutions on the surface.

The use of low-add on techniques is favourable for the application of finishing chemicals as any deviation in the uniformity in the distribution of chemicals will be less visible compared to a dyeing process. In these techniques, chemicals are often applied only from one side, which thus also permits the production of two-sided materials.

13.10 Coating and laminating

Many functional and technical textiles have been covered by a polymer layer (coating) or consist of a combination between a membrane and at least one textile fabric [3].

For coating process, the viscous polymer mass is distributed on the surface of the material by different methods:

- In knife coating, an adjustable blade is positioned at a short distance to the fabric. The material is transported under the blade that distributes the viscous mass on the fabric in a desired thickness. The viscosity of the mass must permit intensive contact to the surface, but prevent soaking into the fabric. Such a process permits one-sided coating, for example, for the production of artificial leather, protective clothes and rain coats.
- The mass is placed on the fabric through rollers by means of a kiss roll, a modified calender or by padding applied. Such processes also permit double-sided coating.
- Also modified printing techniques can be used to obtain coated surfaces, for example, by using an open screen in a screen printing machine.

Almost every flexible polymer is used for coating: polyvinylchloride (PVC), polyurethanes (PUR), silicones, natural rubber, styrene–butadiene rubber (SBR), polyolefins (low-density polyethylene (LDPE), high-density polyethylene (HDPE) and polypropylene (PP)).

By laminating combinations of fabric and membranes can be produced, which, for example, are of high interest as water-impermeable breathable material for outdoor applications.

In continuous processes the fabric and the membrane are pressed together and bound through an adhesive system, which can be based on melt glue, reactive polyurethanes and so on.

Important representatives for membrane materials are polytetrafluoroethylene (PTFE) and polyurethane (PUR).

A particular challenge results from laminating of durable water repellent (DWR) finished textiles, as the adhesion of the adhesive must be sufficient to achieve a stable and durable lamination. A later finish with DWR chemicals is not possible as this would block the pores of the breathable membrane.

References

[1] Fischer, K., Marquart, K., Schlüter, K., Gebert, K., Borschel, EM., Heimann, S., Kromm, E., Giesen, V., Schneider, R., Wayland, RL. Textile Auxiliaries. in ULLMANN's Fibres, Volume 2, Textile and Dyeing Technologies, High Performance and Optical Fibres, Wiley VCH Verlag, Weinheim, Germany, 2008.

[2] Nur, M. THEN-airflow® dyeing machine (accessed 25.10.2018 at .https://www.slideshare.net/imnur/then-airflow-dyeing-machine)

[3] Singha, K. A review on coating & lamination in textiles: processes and applications. Am. J. Polym. Sci. 2012,2(3),39–49. DOI: 10.5923/j.ajps.20120203.04.

Take home messages

Machinery for pretreatment and dyeing of textiles is designed as universal machines suited for the treatment of a wide range of materials and different processes. Restrictions arise from specific conditions required for a certain process (e.g., dyeing in aqueous dyebath at temperatures over 110 °C) or from the material (sensitive surface, elastic and fibre blends).

As a result the existing equipment of a textile dyehouse also defines the major products to be processes. The use of inappropriate procedures or not optimal apparatus would increase the risk of low quality of the final product.

Quiz

1. Why the circulation of a package dyeing apparatus has to reach a level of 35 L/kg min?
2. Why the speed of transport and the filling capacity of a dyeing jet are linked together?
3. What could you check if a continuous dyeing shows variations in shade between the sides and the middle of the fabric? What is tailing and which parameters should you check if your material shows tailing?

Exercises

1. In an exhaust dyeing of 100 kg cotton yarn with vat dyes, a combination of three dyes with different bath exhaustion is used. Each dye is used at 2 wt% colour depth. At a liquor ratio of 1:10, the exhaustion of the yellow dye is at 85% of the initial concentration, exhaustion of the blue dye at 90% and exhaustion of the red at 70%. Calculate the initial concentration in the dyebath and the concentrations in dyeing equilibrium.

   What happens to the concentrations if the dyeing is performed with a lower amount of material and the liquor ratio increases to 1:12. Compare the concentrations of dyestuff on the material at 1:10 and 1:12. Is there any chance to compensate for the changed conditions by using different amounts of dye? Can we formulate a common correction factor? Yes/no and why?

2. The same reactive dyeing is performed in an overflow (LR 1:20), a jet (1:10) and an "airflow" (LR 1:4). Reference dyeing: a 5 wt% dyeing is performed in the overflow, dyestuff fixation is at 85% and 50 g/L NaCl are added to the dyebath. The transfer TR of a used bath into the next bath by the wet material is 200 wt%. The material is rinsed four times with a liquor ratio of 1:20. (A) Correct the recipe in the other machines for the same colour depth (colour fixation per kg goods). (B) Calculate an estimate of the savings in dyestuff, salt and water. For the calculation of water consumption, it is important to consider transfer of dyestuff into the next bath and to rinse the goods so often that the same final dyestuff concentration as in the reference process is obtained. (C) Draw a graphical representation of the decrease of dyestuff concentration as a function of number of rinsing baths for the three processes. Also draw a representation of a decrease of dyestuff concentration as a function of total volume of water spent.

3. In a wet-in-wet application, the fabric leaves the preceding stage with an add-on (pick-up) of 65 wt%. The chemical concentration of hydrogen peroxide on the fabric should be at 15 mL/kg of dry goods. The pick-up in the applicator is 95 wt%. Assume a full exchange equilibrium in the applicator (padder). What concentration of peroxide has to be used in the padder box.

4. A mercerisation machine operates at a speed of 30 m/min. The alkali concentration is set at 300 g/L NaOH. The pick up at the end of the alkalisation step is 150%. What consumption of NaOH in kg/day can be estimated at the following conditions: Fabric width 1.8 m, mass per area 160 g/m^2 and 20 h production per day. The average consumption of water is at 11 L/kg material. What is the total water consumption per day, and what is the average NaOH concentration in the waste water stream?

5. A padder box has a content of 25 L, and another box only contains 12 L. Material is processed at a speed of 30 m/min, width 1.8 m, 120 g/m^2 and pick-up 75 wt%. What is the average resident time of the dyebath in the pad box. When 5 min is set as critical limit for the resident time, what is your recommendation?

   What happens to the resident time if lighter, more heavier material with 100 g/m^2 or 220 g/m^2 is dyed.

6. Calculate the time of immersion in continuous dyeing using different length of immersion in the dye box/padder nip. Indigo dyeing: length of immersion 2 m, speed 25 m/min; padder: length of immersion 1 m, speed 30 m/min and nip dyeing: length of immersion: 25 cm, speed 30 m/min.

7. Calculate the maximum capacity of a jet as a function of the velocity of the textile in the range of 50–600 m/min for the following conditions: (a) longitudinal weight of the material 150 g/m and maximum dwell time 1 min; (b) longitudinal weight of the material 300 g/m and maximum dwell time 1 min. Draw a graphical representation of the data in the form of a diagram of capacity as a function of

material velocity. Assume the critical velocity is 300 m/min for (a) and 200 m/min for (b) and indicate the technical maximum in the drawing.

Solutions

1. (Question 1): The differences in the composition of the treatment bath, for example, chemical concentration and dyestuff concentrations, must be kept low (less than 2%); thus, either circulation has to be in this level, or the processes have to be performed very slowly.
2. (Question 2): Similarly as in the case of Question 1, the dwell time of the goods in the storage of the machine has to be kept below a critical limit, which would otherwise cause differences in colour, creases and so on. Thus, either speed of goods has to be high or the mass of goods in the storage is limited.
3. (Question 3): Variations in colour depth and shade between middle of a fabric and sides can be due to differences in pick-up due to differences in mass per area or bending of the roller of the padder, also differences in heat-flow and differences in drying can lead to migration of pigments and visible colour differences. Tailing of goods can be due to a number of reasons: slight exhaustion of the dye during immersion in the pad box, hydrolysis of a reactive dye in the pad-box, temperature variation in steaming or dry heat fixation and variations in pick-up of the padder.
4. (Exercise 1): A mass of 100 kg of goods will be dyed in a 1,000 L dyebath. A 2% dyeing corresponds to 2 kg of dye; thus, the initial dyebath concentration of each dye will be at 2 g/L. The final concentration of the yellow dye will be $c_y = 2 \times 0.15 = 0.3$ g/L, $c_B = 0.2$ g/L and $c_R = 0.6$ g/L. For a change in liquor ratio to LR = 1:12 every dye has to be corrected individually. We assume that the final concentration will be the same, thus the correction has to consider the increased volume of the bath. The bath volume will be 1,200 L; thus, the equilibrium concentration for 200 L has to be added to the recipe, for example, for the yellow dye $m_y = 2,000$ g + 200 L × 0.3 g/L = 2,060 g/L.
5. (Exercise 2): (A) A mass of 50 g dyestuff are used to dye 1 kg. Dyestuff fixation will be $m_{fix} = 50 \times 0.85 = 42.5$ g, and the dyestuff concentration in equilibrium will be $c_D = 7.5/20$ g/L = 0.375 g/L. As an assumption the same concentration thus will be present in all dyeing processes. The amount of dyestuff required for the jet thus will be $m_{fix} + c_D \times V_{dyebath} = 42.5$ g + 0.375 × 10 g = 46.25 g, the amount for the "airflow" will be 42.5 g + 0.375 × 4 = 44 g. (B) The savings of dye can be seen at the reduced amount of dyestuff required, savings of salt: at a liquor ration of 1:20 we use 20 × 50 g = 1,000 g salt per 1 kg of goods, at 1:10 500 g NaCl and at 1:4 200 g of NaCl are spent. Water consumption: we have to calculate the amount of dye in a bath as a function of liquor ratio and transfer of a bath into the next one. The concentration in the first bath= c_{D1}, the concentration in the next bath $c_{D2} = TR \times c_{D1}/$ ((100 × LR + TR). Thus we can formulate a general scheme in Excel as a function of TR, LR and c_{Di}.

	TR	200 %		
	c_{D1}	0.375 g/l		
	LR 1:	20	10	4
Dyestuff concentration				
Bath number	1	0.3750	0.3750	0.3750
	2	0.0341	0.0625	0.1250
	3	0.0031	0.0104	0.0417
	4	0.0003	0.0017	0.0139
	5	0.0000	0.0003	0.0046
	6	0.0000	0.0000	0.0015
Water consumption		litres		
Bath number	1	20	10	4
	2	40	20	8
	3	60	30	12
	4	80	40	16
	5	100	50	20
	6	120	60	24

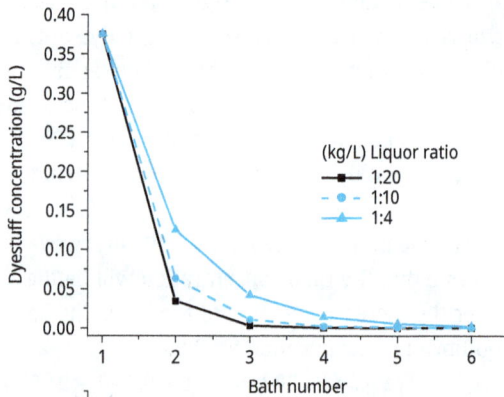

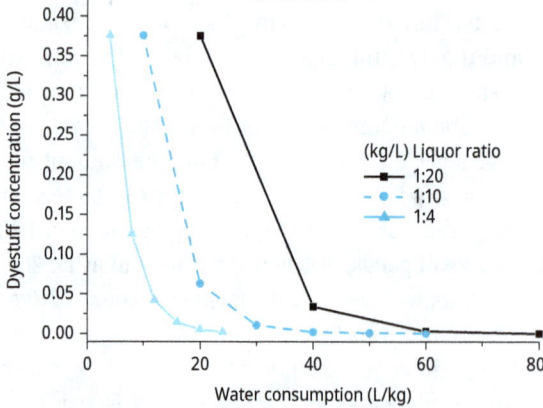

6. (Exercise 3): Only the difference in pick-up will be added to the fabric. Thus, the amount of 15 mL/kg has to be added with a pick-up of 95–65% = 30%. The peroxide concentration in the pad box thus has to be c_{pad} = 15 mL/kg/0.3 L/kg = 50 mL/L.

7. (Exercise 4): At first we have to calculate the production capacity of the machine per day: Capacity = mass per area × width × speed × time of operation = 0.160 kg/m² × 1.8 m × 30 m/min × 60 min/h × 20 h/day = 10,368 kg/day.

   The alkali consumption thus is = 6,912 kg/day × (150/100) L/kg × 300 g/L/1,000 g/kg = 4666 kg of NaOH. The concentration in the waste water will be c_{NaOH} = 4666 kg NaOH/(10368 kg × 11 l/kg) = 40.9 g/L.

8. (Exercise 5): The resident time of the dyebath in the padder can be calculated from the pick-up: $V_{pick-up}$ = mass per area × width × speed × pick-up/100 = 0.120 kg/m² × 1.8 m × 30 m/min × 0.75 L/kg = 4.86 L/min. For a 25 L pad box the resident time will be t_{res} = 25/4.86 min = 5.1 min (critical!) and for the smaller box t_{res} = 12/4.86 min = 2.47 min (fine). We should use the smaller pad box. If more heavy material is used, the resident time will be shorter and we can use also the 25 L pad box (t_{res} = 2.78 min). If lighter material is used, only the smaller pad box is recommended.

9. (Exercise 6): The time of immersion t_i = length of immersion/speed. Indigo dyeing: t_i = 2 × 60/25 = 4.8 s; padder: t_i = 1 × 60/30 = 2 s; nip dyeing t_i = 0.25 × 60/30 = 0.5 s (!).

10. (Exercise 7): The capacity corresponds to the mass of goods that are transported at given speed and maximum dwell time.

    Capacity = speed × longitudinal weight × dwell time

Dwell time	1 min		Dwell time	2.5 min	
Velocity m/min	Long. mass g/m		Velocity m/min	Long. mass g/m	
	150	300		150	300
50	7.5	15	50	18.75	37.5
100	15	30	100	37.5	75
150	22.5	45	150	56.25	112.5
200	30	60	200	75	150
250	37.5	75	250	93.75	187.5
300	45	90	300	112.5	225
350	52.5	105	350	131.25	262.5
400	60	120	400	150	300
450	67.5	135	450	168.75	337.5
500	75	150	500	187.5	375
550	82.5	165	550	206.25	412.5
600	90	180	600	225	450

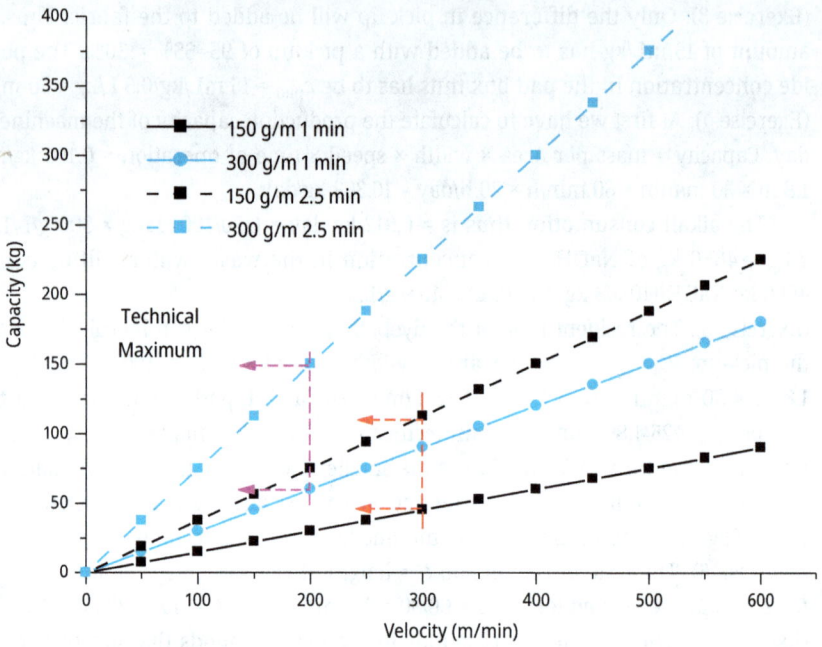

List of abbreviations/symbols

LR	liquor ratio
m_L	metre weight
v_{max}	velocity of operation
t_R	resident time
m_{load}	capacity a dyeing machine
PVC	polyvinylchloride
PUR	polyurethanes
SBR	styrene–butadiene rubber
LDPE	low-density polyethylene
HDPE	high-density polyethylene
PP	polypropylene
PTFE	polytetrafluoroethylene
DWR	durable water repellent

14 Surfactants, detergents and laundry

14.1 Surfactants

Surfactants contain two different structural elements:
- A hydrophilic part that is polar and responsible for molecular interactions with water molecules. Polar groups can bear a positive or negative charge, or contain a number of polar groups (e.g., polyethylene oxides).
- A hydrophobic part that exhibits low polarity and thus is not able to interact with water molecules. Representatives for nonpolar groups are long chain hydrocarbons, perfluoroalkyl groups and alkylaromates.

The amphiphilic character of a surfactant leads to a preferred concentration of the surfactant on the water surface, thus causing significant effects:
- reduction in surface energy between water and the adjacent phase (e.g., air, glass and textile fibre);
- modification of the wetting properties between water and solid surfaces
- formation of an electrical double layer at the boundary layer.

Above a certain concentration in solution, the formation of larger aggregates is also possible, which then change the rheological properties of the solution [1].

14.2 Classification of surfactants

Anionic surfactants: In anionic surfactants, the surface-active molecule contains at least one negatively charged group. The positive counter ion can be, for example, Na^+, K^+ or a quaternary ammonium ion.
    Important representatives are as follows:

Soap	$R-COO^-Na^+$
Alkyl sulphate	$R-O-SO_3^-Na^+$
Alkyl sulphonate	$R-SO_3^-Na^+$
Alkyl benzene sulphonates	$R-Ar-SO_3^-Na^+$
Fatty alcohol ether sulphate	$R-O-(CH_2-CH_2-O)_x-SO_3^-Na^+$
Alkyl phosphate	$R-O-PO_3^{2-}2Na^+$
Alkyl phosphonate	$R-PO_3^{2-}2Na^+$

Cationic surfactants: These surfactants bear a positively charged group in their active molecule part. Counter ions can be Cl^- and Br^-. Examples are as follows:

https://doi.org/10.1515/9783110795738-014

| Quaternary ammonium salts | $R^4 N^+ Cl^-$ |
| Alkyl pyridinium salts (Py, pyridinium) | $R^3 PyN^+ Cl^-$ |

Zwitterionic/amphoteric surfactants: Surfactants that bear a positive and a negative charge in the surface-active part are called zwitterionic and amphoteric, respectively. Such surfactants do not require a small counterion for reasons of charge neutrality as charges of both sign are present in the molecule.

Representative examples are as follows:

| Aminoxides | $R_3N^- -O^+$ |
| Betaines | $R_3N^+ -CH_2 -CO_2^-$ |

Nonionic surfactants: Surface-active properties also can be achieved by the presence of larger molecules that exhibit a certain polarity.

Representative examples are as follows:

Alkyl polyethylene glycol ethers	$R-O-(CH_2-CH_2-O)_x-H$
Alkyl polyalcohols	$R-O-CH_2-CHOH-CH_2-OH$
Akyl (poly)glycosides	$R-O-(Glyc)_x$
Alkyl polyamines	$R-NH-(CH_2-CH_2-NH-)_x-H$
Alkyl phenol polyethylene glycol ethers (APEO)	$R-Ar-O-(CH_2-CH_2-O)_x-H$
(Ar, aromatic hydrocarbon; in case of APEO, Ar, phenylene)	

The group of APEO-based surfactants is mentioned here because of their relevance as banned surfactants, which is due to their persistence, bioaccumulation and endocrine activity.

14.3 Solubility of surfactants

The solubility of a surfactant in water is the product of two effects: The polar group intends to dissolve in water, with a negative free energy ΔG_{polar}. The nonpolar part of the surfactant, however, does not dissolve in water; the major reason for the positive contribution to $\Delta G_{nonpolar}$ results from the negative entropy $\Delta S_{nonpolar}$.

The negative contribution of $\Delta S_{nonpolar}$ results from the inability of the nonpolar molecule part to interact with the surrounding water molecules. Thus, adjacent water molecules have to arrange into an ice-like structure of higher order.

Solubility of a surfactant is achieved as long as the sum of $\Delta G_{polar} + \Delta G_{nonpolar}$ yields a negative value. Thus, ionic surfactants with a shorter chain length, for example, a C_{12} hydrocarbon group will be soluble in water. Solubility decreases with increasing length of the nonpolar group as the contribution of $\Delta S_{nonpolar}$ increases.

A compromise in the solubility of a surfactant is achieved when the molecules accumulate at the surface of the liquid, with orientation of the polar group in the direction of the aqueous phase. In this state, the polar group still remains in "dissolved" state, while the orientation of the nonpolar group out of the liquid keeps the contribution of $\Delta S_{nonpolar}$ low. As a result of this orientation effect, the surface energy of the liquid is decreased.

This physicochemical basis leads to the characteristic solubility behaviour of surfactants, which is shown in a general form in Figure 14.1 in the form of a phase diagram. Below a certain concentration, the surfactant is present in state of molecular dissolution. The solubility increases with temperature. At lower temperature, the maximum solubility is defined by the solubility curve. At concentrations above this curve, the dissolved surfactant is present in its saturated solution, which is in equilibrium with an excess of insoluble material.

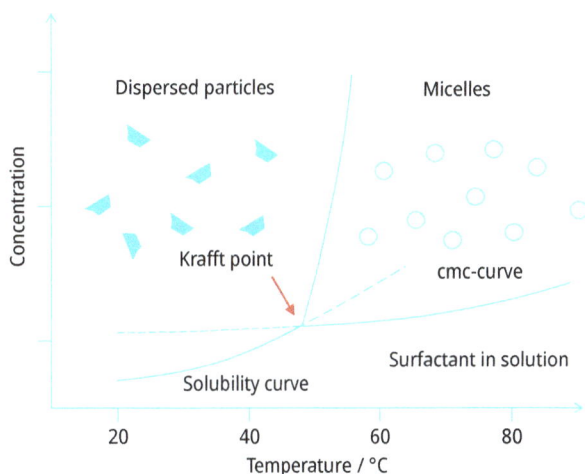

Figure 14.1: Phase diagram of a surfactant.

At higher temperatures, the insoluble material dissolves and molecular aggregates (micelles) are formed. The critical concentration for the formation of micelles is indicated by the cmc curve (critical micelle concentration). The solubility curve and the cmc curve intersect at the so-called Krafft point.

When a surfactant is added to a solution in a concentration that exceeds the monomolecular solubility at temperatures below the Krafft point, an equilibrium between the insoluble excess and the molecular solution exists. At temperatures above the Krafft point, the insoluble parts dissolve under formation of micelles. In aqueous phase micelles are aggregates that consist of a double layer formed by surfactant molecules. The inner part of the layer contains the hydrophobic parts of the surfactant molecules, while the surface of the aggregate is formed by hydrophilic groups. Thus formally the micelle can be understood as a folded additional surface layer, which has been formed as a result of the need to position more surfactant at the water surface.

Different shapes of micelles can be observed, for example, spherical micelles, cylindrical micelles and lamellar phases. A special case is a vesicle, which is a spherical structure formed by a bi-layer of surfactants and which can contain a solution of different composition in the core of the vesicle.

Below the solubility curve or the cmc curve, the surface tension of a surfactant solution decreases with concentration of the surfactant. At concentrations above the solubility limit or the cmc concentration, the surface tension does not decrease further. The formation of micelles does not contribute to the reduction of surface tension, but will be of significant relevance for dispersing and emulsifying capacity of the surfactant solution. Figure 14.2 shows a typical curve for the reduction in surface energy as a function of concentration. Above the solubility maximum, the reduction in surface energy stops and micelles or the formation of insoluble parts of the surfactant appear.

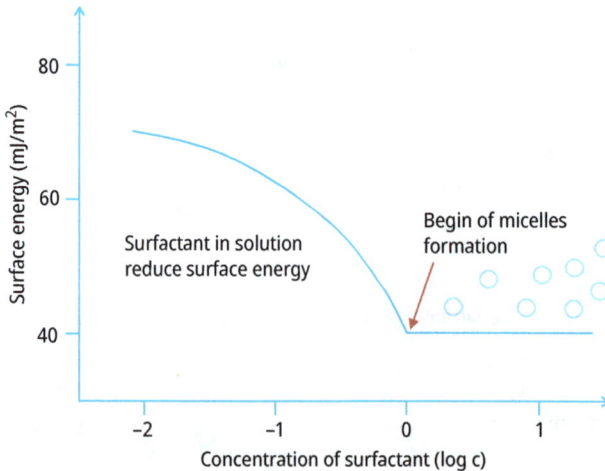

Figure 14.2: Reduction in surface energy as a function of surfactant concentration.

14.4 Contact angle

The surface energy of water is rather high with 72.8 mN/m. Olive oil as well as sweat exhibit lower surface energies about 30–32 mN/m. Representative values for synthetic polymers are as follows:

Polypropylene 28–30 mN/m
Polytetrafluoroethylene (PTFE) 18 mN/m
Polyethylene terephthalate 43 mN/m
Polyamide 6,6 46 mN/m.

While wetting of usual synthetic fibres can be achieved with standard surfactants, the wetting of PTFE requires special determents (perfluoro surfactants) to lower the surface energy for proper wetting. For the explanation of the function of a surface-active substance, we will consider the concepts of surface energy and surface tension in parallel, as explanation of effects sometimes is more simple to describe with the use of vectors representing the surface tension.

When a drop of liquid is in contact with a plane surface, an equilibrium between the surface tensions at the interfaces between γ_S solid surface/air, γ_L liquid surface/air and γ_{SL} solid surface/liquid establishes. The equilibrium state is described by the Young equation (eq. (14.1)). Θ represents the contact angle between liquid and surface. A representation of the different force vectors involved in the formation of an equilibrium state is shown in Figure 14.3.

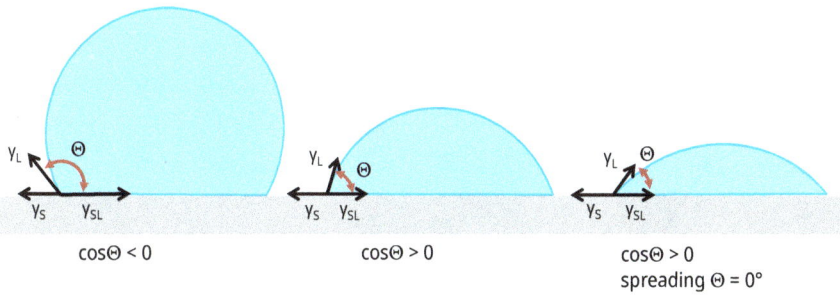

Figure 14.3: Equilibrium of forces due to surface tension for wetting of a solid.

$$\gamma_S - \gamma_{SL} = \gamma_L \cos\Theta \tag{14.1}$$

γ surface tension (N/m, mN/m), γ_S surface tension solid/air, γ_{SL} surface tension solid/liquid, γ_L surface tension liquid/air and Θ contact angle between liquid and solid surface.

A contact angle between water and a highly hydrophobic surface will be at 150°, ideally at the theoretical value of 180°. A hydrophilic surface will have contact angles of, for example, 30°. Spontaneous spreading of a liquid on a surface will be observed when the surface tension of the solid is higher than the sum of the surface tension of the liquid and the liquid/solid interfacial tension $\gamma_S \geq \gamma_L + \gamma_{SL}$. Analogous surface energies can be used to replace the concept of surface tension.

As the surface energy of the solid material usually is at a given value, the wetting or spreading of a textile material has to be achieved by the reduction of the interfacial energy of the interface liquid to air and the interfacial energy between the solid substrate and the liquid phase.

Thus in wetting and washing processes, a reduction of the interfacial energy to a critical minimum has to be achieved by using surfactants. A similar situation is observed in washing processes where oily residues should be removed during washing.

In this case, two liquid phases (water and oil) and a solid phase (fibre surface) must be included in the assessment of the detergent to remove an oily impurity. Figure 14.4 indicates the different vectors representing the respective interfacial surface tensions. In the example given in Figure 14.4, the replacement of an oil/solid surface by a water/solid surface is favourable, a net force in direction of the oil drop appears and as a result the "roll-up" of an oily stain into a drop is initiated. However, usually mechanical agitation is required to complete the drop formation and achieve a final lift off.

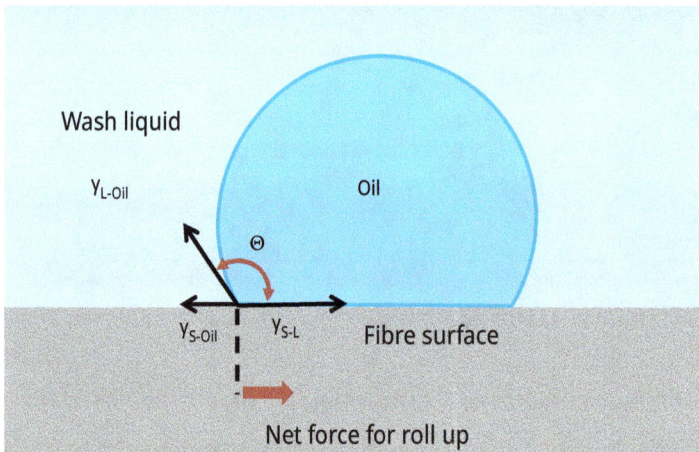

Figure 14.4: Representative forces at the interfaces of an oily residue present during a wash process.

The stabilisation of removed oily liquids and greases in a wash solution can be achieved through two different processes:

Emulsification: Here the surfactants support formation of small droplets and stabilisation in the wash liquid. Redeposition can occur in case the stabilisation of the emulsion is insufficient, in particular during rinsing, where the surfactant concentration is lowered by dilution.

Solubilisation: The formation of micelles leads to stabilisation of oil droplets and other water-insoluble material and thus reduces the risk of redeposition of previously removed material.

14.5 Cloud point

Non-ionic surfactants, for example, alkyl polyethyleneglycols, exhibit reverse temper-ature-dependent solubility behaviour. With increasing temperature, solubility de-creases and at the temperature of the cloud point a two-phase system of surfactant and water is formed. The solution becomes turbid.

The cloud temperature is dependent on the structure of the surfactant:
– length and form of the nonpolar hydrocarbon chain;
– number and type of ether groups (ethylenoxides and propylenoxides) and
– additives present in solution (e.g., presence of a ionic surfactants, salt).

Different models are used to explain the behaviour: Formation of hydrogen bonds be-tween the ether group and water leads to solubility of the surfactant. At higher tem-perature, the hydrogen bonds weaken and insolubility appears when the stabilisation of the surfactant in solution becomes insufficient (eq. (14.2)).

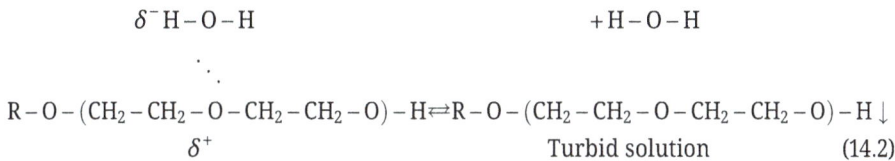

$$\delta^- H-O-H \qquad\qquad\qquad +H-O-H$$

$$R-O-(CH_2-CH_2-O-CH_2-CH_2-O)-H \rightleftharpoons R-O-(CH_2-CH_2-O-CH_2-CH_2-O)-H \downarrow$$

$$\delta^+ \qquad\qquad\qquad\qquad \text{Turbid solution} \qquad\qquad (14.2)$$

Another explanation for the unexpected behaviour is based on the higher conforma-tional energy of the polyoxyethylene chains, thus leading to conformational states that exhibit less hydration energy. Above the cloud point, the hydration becomes less favourable compared to self-assembly and tendency to separate from solution [2].

14.6 Surfactant adhesion at interfaces

The tendency of surfactants to concentrate at the boundary region between liquid and solid phases leads to an adsorption at the surface of solid phases. Based on the polarity, the phase-oriented adsorption occurs:

On a nonpolar surface, for example, polyolefin fibres, the surfactants adsorb with the hydrophobic side directed to the nonpolar solid material.

On a polar surface, the polar side of the surfactant interacts with polar groups of the surface and the nonpolar side is directed towards the aqueous phase. This nonpo-lar layer is covered with another layer of surfactant. As a result a double-layer ad-sorption occurs with both polar sides oriented to the surface and the aqueous phase (Figure 14.5).

In case of sorption of a ionic surfactant on a surface of low polarity, the charge density of the surface will increase, which leads to an increase in zeta potential. The increase in surface charge density will lead to an increased repulsion between fibre

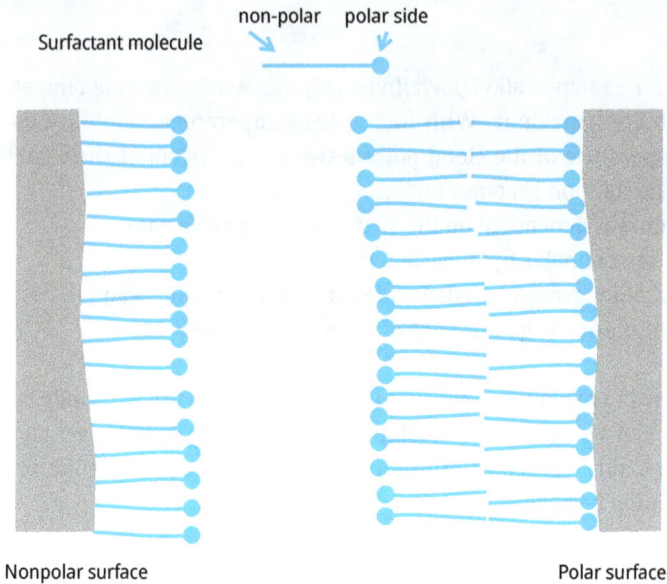

Figure 14.5: Adsorption of surfactants on polar/unpolar surfaces.

surface and an adhering particle and thus stabilises the dispersed state and supports removal of particles from the surface.

Adsorption of charged surfactants on the fibre surface will also increase repulsive forces to particulate material, an effect that can be obtained by adsorption of charged macromolecules, for example, carboxymethyl cellulose (CMC).

Besides anionic surfactants, cationic products could also be used to increase the repulsive forces of dispersed soil particles, which in principle also would work for the washing step where higher concentrations of surfactant are applied. The usually negative zeta potential, both of the fibres and the pigment, shifts through adsorption of cationic molecules to positive values. During rinsing however, the concentration of the cationic surfactant reduces, the fibre surface potential reverses again and also the stability of the dispersed pigments decreases. As an overall effect, redeposition of already removed pigments is observed during rinsing.

In the case of sorption of non-ionic surfactants, no substantial change in zeta potential is observed, as these surfactants do not bear charged sites. For the removal of pigments from fibre surfaces in many cases, the pure action of surfactants is not sufficient and a certain mechanical force is required as well. This effect is achieved through regular movement of goods during the wash cycle. However, it must be considered that such a principle will be effective only for particulate materials, which are sufficient in size to reach through the stagnant diffusion layer near the surface. The higher the velocity of the wash liquid, the smaller will be the thickness of the stagnant layer and the smaller will be diameters of particles, which will be affected from the agitation. Very small particles (e.g., <0.1 μm) will be very difficult to be displaced by agitation. This explains repair of an accidentally low rubfastness of a vat dyeing or an indigo dyeing will be difficult or even impossible to be achieved by only washing and soaping. The small dye particles at the fibre surface in such cases have to be removed by chemical reduction and transferred into the dissolved state. In addition, removal of non-fixed disperse dye is performed by a reductive cleaning and not by simple washing with detergent.

14.7 The role of multivalent ions

The presence of multivalent ions near the interface of the textile fibres plays a signifi-cant role in the release of soil. Important representatives for such ions are calcium, magnesium and iron. Calcium and magnesium are present as water hardness, and iron oxides also are released from soil. These multivalent ions can bind at the charged sites on the fibre surface and thus form cationic bridges to soil particles. As a result removal of soil is difficult as long as these multivalent ions are not removed or masked.

In cellulose, the presence of carboxylic groups permits a stoichiometric binding of multivalent ions [3]. In the case of calcium, an additional counterion is required to neutralise the charge of the Ca^{2+}, which, for example, can be a HCO_3^- ion or a charged soil particle (eq. (14.3)).

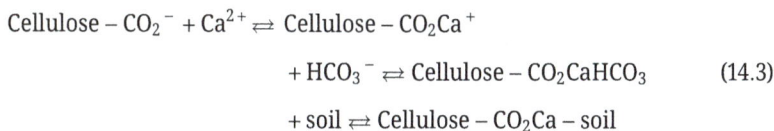

$$Cellulose - CO_2^- + Ca^{2+} \rightleftharpoons Cellulose - CO_2Ca^+$$

$$+ HCO_3^- \rightleftharpoons Cellulose - CO_2CaHCO_3 \qquad (14.3)$$

$$+ soil \rightleftharpoons Cellulose - CO_2Ca - soil$$

In the case of adsorbed coloured soil pigments, the wash performance is reduced as Ca^{2+} ions are not removed by appropriate systems. The presence of hydrogen carbon-ate ions will lead to formation of $CaCO_3$ on the fabric during drying and ironing of the washed goods [4].

14.8 Detergent ingredients

Detergents are complex combinations of surface-active substances and components designed for a certain application (e.g., household laundry, industrial laundry, dish washing etc.).

Formulations of products for industrial use or textile chemical applications (e.g., dispersing agents and wetting agents) are rather straightforward and contain a re-duced amount of components. The specialised purpose of a certain technical applica-tion and the technological environment of use (e.g., in a textile dyehouse) permit the use of different concentrated products in combination to obtain the final recipe of application. Thus, a scouring or bleaching recipe in a dyehouse consists of a set of individual components (dispersing agent, alkali, complexing agent, stabiliser and bleach chemicals). For household applications, an even higher number of components are formulated together into one product.

The composition and the formulation of a detergent depends on
- the purpose (heavy duty, wool and synthetic fibres);
- the equipment used (drum washer, top-loader and hand wash) and
- applicatory conditions (temperature, concentration and liquor ratio).

Thus, signification differences in formulations exist between continents and countries.

Major groups of ingredients in a household laundry product are as follows:
- surfactants
- builders
- bleaching agents
- auxiliary agents.

Surfactants: In the vast majority, a combination of nonionic and anionic surfactants is used. Examples are soap, alkyl benzene sulphonates, fatty alcohol sulphates, alkyl sulphonates and fatty alcohol ether sulphates.

Technological aspects for the selection of suitable surfactants include wash performance (reduction in surface tension, specific adsorption, low sensitivity to water hardness, readily soluble, dispersing and emulsifying properties, soil anti-re-deposition properties, low toxicity and biodegradable).

The use in a liquid product requires appropriate solubility in a concentrated formulation. For use in a powdered product, the requested flow behaviour and consistency of the powder have to be achieved. Representative values for the content of surfactants in a heavy duty detergent are between 15 and 20 wt%.

Builders take a complex role in the function of a detergent. Important criteria are as follows:
- elimination of multivalent ions from the wash liquor present due to the water hardness (Ca^{2+} and Mg^{2+}) and ions brought into the wash bath with the textile goods and soil contaminants (Fe^{3+});
- adjustment of appropriate pH value in the wash liquor;
- contribution to an increase in negative zeta potential of the fibre surface and solid impurities (pigment and soil), thus increasing wash performance and
- stabilisation of dispersed impurities in the wash liquor through charge repulsion.

Three different principles to reduce the water hardness are utilised in detergents:
- removal through precipitation (such principles are not able to remove bound Ca^{2+} ions from the textile surface);
- complex formation, which requires use of at least a stoichiometric amount of complexing agent and

– ion exchanger: Soluble systems can remove Ca^{2+} from the surface of the textile material. In case an insoluble zeolith is used as ion exchanger, another Ca^{2+}-transporting system is required to remove any surface bound Ca^{2+}.

Important representatives are as follows:

Alkalis, to increase pH and partly also precipitate Ca^{2+} ions: soda Na_2CO_3, sodium metasilicate Na_2SiO_3 and trisodium phosphate Na_3PO_4.

Complexing agents to bind Ca^{2+}, to support the wash process and to prevent deposition of $CaCO_3$ on laundry and washing machines. These agents have to be used in stoichiometric quantities; thus, the share of such products in a formulation is considerable and determines the amount to be dosed per wash cycle.

Important representatives are ethylenediamine tetraacetic acid (EDTA), nitrilotriacetic acid (NTA), sodium triphosphate (STP) and phosphonates, citric acid.

Ion exchangers to increase Ca^{2+}-binding capacity and to reduce the amount of complexing agents required. Important representatives are insoluble aluminosilicates (zeolite, $xNa_2O\ Al_2O_3\ ySiO_2\ zH_2O$), soluble polyacrylates ($-[CH_2-CHCO_2Na]_n-$).

As the zeolites are insoluble in water, the removal of adsorbed Ca^{2+} ions requires the presence of a soluble carrier system; thus, zeolites are often used in combination with a soluble complexing agent (e.g., polycarboxylate). The amount of builders required in the wash liquor thus depends on the water hardness and the level of soiling of the material.

In any case, substantial amounts of builders are required; thus the ecological profile of these components also has to be considered. Important aspects are biodegradability in wastewater treatment plants, contribution to eutrophication, no accumulation in the environment or detrimental effects on drinking water. In addition, the temperature stability of the complexes formed has to be considered, as, for example, the calcium-binding capacity of citric acid decreases form 195 mg CaO per g of citric acid at 20 °C to 30 mg/g at 95 °C, while values for EDTA or NTA reduce only from 219 to 154 mg/g and from 285 to 202 mg/g, respectively. The capacity of an insoluble zeolite, however, increases from 165 mg/g at 20 °C to 190 mg/g at 95 °C. This increase is a result of kinetic effects for Ca^{2+} binding in the heterogeneous system. Representative values for the content of builder in a heavy duty detergent are between 20 and 50 wt%.

An important effect to mention is the so-called threshold effect. Phosphate-based systems and phosphonates retard and interfere to the formation of precipitates e.g. $CaCO_3$ in the wash liquor even at understoichiometric amounts. Thus, low amounts of phosphonates are found as additives in liquid detergents due to their ability to minimise incrustations on textiles and machinery.

Bleach systems: These are effective in the removal of coloured stains, which come from coloured plant dyes, chlorophyll, blood and food colorants, as these chemicals destroy the coloured chromophores of these substances.

For an oxidative bleach, usually *peroxide*-based systems are used, for example, sodium perborate monohydrate $NaBO_3 \times H_2O$, sodium percarbonate $Na_2CO_3 \times 1.5H_2O_2$.

There are other bleach chemicals, for example, sodium perphosphate $Na_4P_2O_7 \times 3H_2O$, sodium persulphate and potassium peroxidisulphuric acid; however, the relatively high amount of chemicals used for one equivalent of peroxide makes these systems unfavourable for use.

For liquid formulations (e.g., commercial laundry in hotels and hospitals), a concentrated solution of H_2O_2 (e.g., 20 wt%) can be dosed directly into the wash liquor. The sensitivity of H_2O_2 to decompose in liquid formulations makes the integration of H_2O_2 into liquid formulations difficult.

The limited reactivity of H_2O_2 requires the addition of an *activator system* to achieve a bleach effect at low-temperature washing (e.g., 60 °C). A representative case for an activator is the well-known TAED (tetra-acetyl-ethylenediamine; Figure 14.6). As a common reaction for many peroxide activator chemicals, a reaction of peroxide with the formation of the corresponding peroxoacid (e.g., peroxoacetic acid CH_3CO_3H) is formulated.

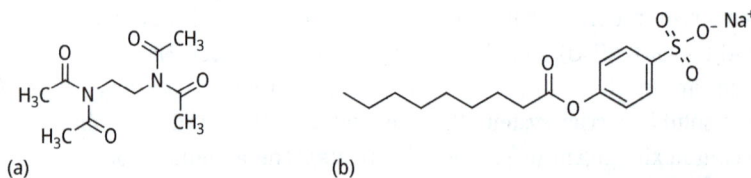

(a)   (b)

Figure 14.6: Representative formulas of activators for peroxide-based oxidants TAED and NOBS (sodium nonanoyloxybenzenesulphonate).

The higher reactivity of the peracids increases the bleach performance substantially. In addition, organic peracids (in particular, peroxoacetic acid) are highly effective as antimicrobial agents and thus are used as disinfectants in laundry. In comparison to the bleach process, which is intended to occur during the main washing step, the disinfectant is added in low concentration to the final rinse.

In addition, other systems (e.g., metal complexes) have been proposed as catalytic activators for peroxides; however, the reactivity of these systems was difficult to control.

Besides activators, stabilisers are also used in peroxide bleach baths. Similar to the textile bleach processes, magnesium silicates and complexing agents (EDTA, phosphonic acids) are added, with the aim to bind or mask iron and manganese ions.

Chlorine-based bleach systems permit bleach processes at lower temperature. In alkaline solution, the bleaching agent is the hypochlorous ion OCl^-. Sodium hypochlorite is added to the wash bath as a separate solution of NaOCl. In addition, solid chlorine forming agents are in use; for example, sodium dichloroisocyanurate (NaDCC) releases two moles of NaOCl per molecule NaDCC, and thus is a very effective substance for hypochlorite formation (Figure 14.7).

Dyestuff stability has to be considered with care, as many reactive dyes and also fluorescent whitening agents exhibit limited stability against hypochlorite. Chlorine-based bleach systems also exhibit strong antimicrobial activity.

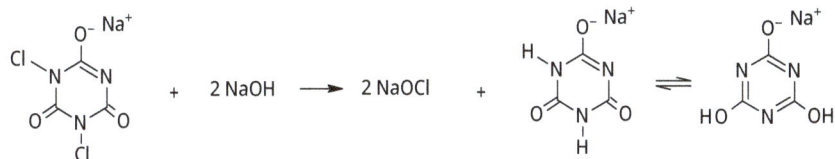

Figure 14.7: Formation of hypochlorite by NaDCC.

Reductive bleaches are rare in textile laundry, as the machines agitate a low volume of wash liquor with high intensity and thus rapid oxidation of the reducing agents (e.g., sulphite and dithionite) proceeds. Such chemicals are used only in special products, with the aim to remove coloured stains on white material, which had been transferred from products with low colour fastness.

Auxiliary additives: Based on the use of a detergent, a number of minor components are present in a commercial washing powder. Despite the comparable low content, these components make a substantial contribution to the overall wash performance as well as to the production costs of a product.

Enzymes: Protein-based stains arise from a number sources: milk, cacao, egg yolk, blood and grass. For removal from the fibre surface, a hydrolysis of the denaturated proteinaceous material by the action of proteases is essential.

Because of the alkaline washing conditions and the elevated temperatures, such proteases must be stable in the pH region of 10–11 and up to temperatures of 60–70 °C. Representative examples are bacillopeptidases, which are obtained by fermentation of genetically modified bacteria (*Bacillus subtilis* and *bacillus licheniformis*).

In solid formulations, the proteins are added in form of coated granules, prills. The coating protects the proteases against damage by alkaline ingredients. In liquid formulations, an appropriate stabilisation of the proteases has to be achieved to avoid autohydrolysis.

Because of the temperature sensitivity of the proteases, the activity of the enzymes decreases at wash temperatures above 70 °C; thus in the design of a wash process, the proteases have to be effective during the period of heating.

Other enzymes of interest for laundry applications are lipases and enzymes, which support hydrolysis of greasy stains. At higher temperature (e.g., above 40 °C), the performance of anionic surfactants and alkali concentration makes lipases unnecessary.

Remarkably the addition of amylases is not of high interest for textile laundry; however, it is useful in dishwasher product as well as for removal of starch-based sizes during textile pretreatment.

Soil anti-redeposition agents: This group of additives supports the stability of dispersed and emulsified soil in the wash liquor. Insufficient soil anti-redeposition properties will lead to a pronounced redistribution of soil on the material, thus causing undesirable greying. The anti-redeposition capacity is a sum of action of builders, detergents and special additives, for example, CMC or CMS (carboxymethylstarch) that adsorb on cellulose fibres and thus increase charge repulsion between fibre surface and already dispersed soil.

Foam control: Based on machinery used, foam control is required (drum washer) or foam boosters are added.

In washing machines other than drum washers, consumers expect development of a certain level of foam; thus, foam boosters are required (e.g., fatty acid amines and fatty acid alkanolamines) to demonstrate the proper function of the detergent.

In Europe, mainly drum-type washing machines are in use. This type of machine agitates the goods and the liquid intensively; thus, the tendency to form excessive foam with high stability has to be controlled. Foam regulators have to be added to limit the level of foam formed in the drum.

The chemical concept of a foam regulator has to be adapted to the overall formulation of the detergent.

Soap: Soap acts as a foam-controlling substance by the formation of insoluble precipitates of Ca-soap. Such a concept is effective as long as a sufficient concentration of Ca^{2+} ions is available in the solution. In the presence of a strong chelating agent for Ca^{2+} (e.g., NTA) in over-stoichiometric amount, formation of Ca-soap is impossible and foam regulation fails. This can also happen when the detergent is used in high dosage or in regions of low water hardness.

Hydrocarbons/silicone oil: Non-polar insoluble oily substances enrich at the water surface and thus reduce foam stability. These components often are added by means of a carrier substance, for example, on zeolite or silicates.

Nonionics with low solubility in water: In liquid detergents, the addition of nonionic surfactants with low water solubility also permits the formulation of low foaming detergents.

Fluorescent whitening agents/optical brighteners: Similar to the pretreatment of white textile materials, fluorescent whitening agents are added to detergents. The products have been designed for washing of white textiles. The optical brightener should be able to stain all types of material present in the wash bath; thus, adsorption at the fibre surface is required.

Fragrances: Fragrances are added to the detergent for two reasons:
- to cover the unpleasant smell of the laundry process, in particular, during periods of hot wash and
- to provide a fresh smell to the washed goods when the material is removed at the end of the wash process.

Fillers: These are added to adjust the density of the powder detergent to the defined level, as dosage usually is done by volume.

The substances are intended to support the handling, the flow and storage behaviour of the powder (no caking and no dusting). Examples for formulation aids are salt and polyethylene glycols.

The addition of short chain alkyl-benzensulphonates (e.g., cumolsulphonate) intends to support rapid dissolution of other surfactant components. A representative example for a formulation of a solid phosphate free detergent is given in Table 14.1 [5].

Table 14.1: Major and minor constituent parts of a phosphate free detergent.

Component	Major wt%	Minor wt%
Surfactants (anionic and non-ionic)	15–20	
Foam controlling substances		0.1–3
Foam boosters		(0–2)
Chelating agents	5–10	
Ion exchangers	20–25	
Alkali and builders	5–15	
Bleaching agents	10–25	
Bleach activators	0–5	
Beach stabilisers		0.2–0.5
Antiredeposition agents		0.5–1.5
Enzymes		0.3–0.8
Optical brighteners		0.1–0.3
Anticorrosion agents (sodium silicate)	2–6	
Fragrances		< 1
Formulation aids and fillers	ad 100	

Special detergents for coloured material do not contain optical brighteners and bleaching agents. Detergents for wool have been formulated to perform at neutral pH and thus avoid alkali damage of the wool or felting due to high fibre swelling.

Liquid household detergent formulations usually do not contain bleaching agents, as the stabilisation of peroxide or hypochlorite in the presence of a complex surfactant formulation is difficult. In addition, solid ingredients (e.g., zeolites) are difficult to use in liquid formulations as settling can occur during storage.

Large volume consumers (e.g. hotels and care homes) often use a modular combination of liquid products that are automatically dosed according to the wash programme chosen.

Major components are as follows:
- neutral formulation containing surfactants, enzymes, fluorescent whitening agents and fragrances;
- bleaching agent (mainly H_2O_2 based) and
- concentrated alkaline solution containing complexing agents, sequestering agents.

Additional components can be as follows:
- concentrated alkali solution for washing of heavily soiled clothes and kitchen textiles;
- disinfectant rinse to be added into the last rinsing bath (e.g., peracetic acid) and
- softeners to be added to the last rinse.

14.9 Washing and environment

Resources consumption: The washing of clothing contributes substantially to the resources consumption that is allocated with textiles and garment production, use and maintenance. A representative case for automated washing of clothing in a drum-type washing machine is given as follows:

The liquor ratio for a washing step is at 1:4–1:5, which means that 1 kg of goods required approximately 4–5 L of water. The concentration of a detergent is in the magnitude of 5 g/L. When four rinse steps with a similar liquor ratio are considered, then the total water consumption is at 20–25 L/kg for one wash step.

Based on the type of textile in many cases, the water consumption of two-wash cycles already exceeds the water consumption of the textile chemical processing.

A similar situation arises in case dryers are used. The energy consumption for evaporative drying of the textiles depends on the water retention value of the material. After centrifugation of the washed goods at the end of a wash process, wet cellulose fibre-based textiles carry approximately their own weight water. This has to be removed with external energy. As an estimate evaporation of 1 kg water will require 1 kWh of electrical energy. The water retention of textiles form synthetic fibres is much lower (e.g., 30%); thus, such textiles require substantially less energy for drying. The use of heat-pump tumble dryers reduces the energy for mechanical drying substantially through condensation and energy recovery from evaporated moisture.

For 25 wash cycles of a simple T-shirt, reduction in washing temperature from 60 °C to 40 °C contributes to a reduction in 10% of total environmental impact, while tumble drying contributes to 50% of the total environmental impact [6]. Thus in all cases of environmental considerations, the full lifecycle of a textile product has to be considered, including the resources consumption during use and care, instead of a simplified focus on its production alone.

References

[1]   Holmberg, K., Jönsson, B., Kronberg, B., Lindman, B. Surfactants and Polymers in Aqueous Solution, 2nd ed. John Wiley & Sons, Chichester, England, 2007.
[2]   Stache, H. Tensid Taschenbuch, Carl Hanser Verlag, München, 1981.
[3]   Bechtold, T., Fitz-Binder, C. Sorption of alkaline earth metal ions Ca^{2+} and Mg^{2+} on lyocell fibers.. Carbohydr. Polym. 2009, 76/1, 123–128.
[4]   Fitz-Binder, C., Bechtold, T. Ca^{2+} sorption on regenerated cellulose fibres.. Carbohydr. Polym. 2012, 90/2, 937–942. DOI/10.1016/j.carbpol.2012.06.023.
[5]   Jakobi, G., Löhr, A. Detergents and Textile Washing, Principles and Practice, VCH, Weinheim, Germany, 1987.

[6]   Allwood, JM., Laursen, SE., de Rodriguez, CM., Bocken, NMP. Well dressed, The present and future
      sustainability of clothing and textiles in the United Kingdom,Univ. Cambridge, Cambridge, United
      Kingdom, ISBN 1-902546-52-0. 2006.
[7]   Solid surface energy data (SFE) for common polymers (accessed at 1. Nov. 2018) http://www.sur
      face-tension.de/solid-surface-energy.htm

Take home messages

Surfactants are amphiphilic compounds that exhibit a polar and a non-polar site. This property leads to an accumulation of these compounds at the interface between the liquid phase and solids or air. A substantial reduction of the surface energy of water occurs in the presence of surfactants, which thus permits wetting of surfaces with low surface energy. The maximum reduction in surface tension is achieved when the surfactant reaches its solubility limit.

At a concentration and a temperature above the Krafft point, the formation of aggregates (micelles) occurs. The presence of micelles is the reason for an increase in the dispersing and emulsifying capacity of a surfactant solution.

Household detergents are complex mixtures containing a number of components: surfactants, builders, bleaching agents and a lot of additives (e.g., fluorescent whitening agents, enzymes, stabilisers and fragrances).

For an assessment of the ecological profile of a product, besides the production, use and care (washing, drying) as well as disposal and recycling also have to be considered. Often the resources consumption during use exceeds the consumption allocated to production by far.

Quiz

1.   What is the difference between Krafft point and cloud point?
2.   Which groups of surfactants can be distinguished? What happens if two types of surfactants are mixed?
3.   Explain the negative influence of multivalent cations on solid soil removal.

Exercises

1.   Calculate the contact angle between a drop of liquid (water 72 mJ/m^2, a surfactant solution 40 mJ/m^2 and sweat 31 mJ/m^2) and the following polymers (PA6,6 46.5 J/m^2, PET 44.6 J/m^2, PP 30.1 mJ/m^2 and PTFE 20 mJ/m^2) [7]. Assume γ_{SL} as difference between γ_L and γ_S, $\gamma_{SL} = \gamma_L - \gamma_S$.

Solutions

1. (Question 1): The Krafft point defines conditions above which the formation of micelles is possible; thus, it is a general characteristic for surfactants. The cloud point describes the behaviour of non-ionic surfactants that can exhibit decreased solubility at elevated temperature (above the cloud point).

2. (Question 2): We can distinguish between anionic, non-ionic and cationic surfactants. A mixture of anionic and cationic surfactants is not recommended as precipitation of the surface active cationic molecule together with surface active anionic molecule would occur.

3. (Question 3): The multivalent cations can form charged bridges between the negatively charged surface of the fibres and the negatively charged surface of the soil particle. As a result repulsion between soil and fibre is lowered and removal becomes more difficult.

4. (Exercise 1): From eq. (14.1) $\gamma_S - \gamma_{SL} = \gamma_L \cos\Theta$, we can calculate $\cos\Theta = (\gamma_S - \gamma_{SL})/\gamma_L$.

Polymer	Contact angle/°		
	Water	**Surfactant solution**	**Sweat**
PA6,6	73.0	spreading	spreading
PET	76.2	spreading	spreading
PP	99.4	59.7	19.6
PTFE	112.9	84.3	65.2

List of abbreviations/symbols

Py	pyridinium
Ar	aromatic hydrocarbon
APEO	alkyl phenol polyethylene glycol ethers
Glyc	glycoside
CMC	carboxymethyl cellulose
CMS	carboxymethylstarch
EDTA	ethylenediamine tetraacetic acid
NTA	nitrilotriacetic acid
STP	sodium triphosphate
NaDCC	sodium dichloroisocyanurate

15 Environmental aspects of textiles

15.1 Waste water

Textile industry is one of the largest consumers of high-quality water. The vast majority of all textile chemical processes are performed with aqueous solutions.

Taking the water consumption of 100–200 L per 1 kg of processed textile material as a representative value and a processed volume of 50 million tonnes as an estimate, the total water consumption of textile dyeing and finishing operations will be in the dimension of 5,000–10,000 million cubic meters of water.

A considerable problem in waste-water treatment thus arises from the huge amounts of wastes, minor contents can also be of serious impact when the total amount is considered.

Industrial interest in waste-water treatment is limited as the processing of effluents is directly linked to costs, but indirectly related to the value of a commercial product.

In Europe, industrial activities are regulated by the Council Directive 96/91/EC for integrated pollution prevention and control (IPPC) [1]. This document defines a number of relevant terms:

IPPC plant: An IPPC plant is a plant which exhibits a theoretical production capacity of more than 10 tonnes per day. Here, the theoretical maximum capacity is taken as a measure and not the real production achieved.

BAT (best available technology): BATs are used to minimise environmental pollution. A BAT technique defines the best and most efficient technique to run a process and handle the respective emissions. The choice of a BAT technique also considers economical and technical viability, thus implementation should be possible without extensive research and development activities.

BREF: The BREF document represents the technology reference document which describes the best available technologies for selected textile processes, for example, washing, bleaching, mercerisation or dyeing of fibres/textiles [2].

The selection of applicable processes for wastewater treatment must also consider a careful analysis of existing production processes with regard to BAT and BREF documents to optimise the production conditions before considering the end of pipe solution.

General strategies in optimising the problems in waste water are as follows [3]:
– *Replacement and minimisation* of product use:

A number of products have already been banned (e.g. APEO) or been restricted for certain applications (EDTA, phosphonates), others (e.g. CMC) exhibit poor biodegradability and thus should be replaced by alternative products. The extensive use of chlorine for bleach processes leads to problems of AOX in the waste water and has been successfully substituted in many processes by hydrogen peroxide as the greener oxidant. Hypochlorite-based processes, however, are still extensively used in washing jeans.

https://doi.org/10.1515/9783110795738-015

The substitution of chemicals in a process often requires a complete redesign of the technology. As an example, the reaction of indigo with hypochlorite is completely different than with hydrogen peroxide, thus a direct exchange is not possible.

- *Optimisation of processes*: Substantial reduction in energy consumption, chemical consumption and wastewater load can be achieved through reconsideration of the processes used and optimisation of processes, for example, use of reactive dyes with higher fixation will reduce the colour in the effluents, use of machinery with lower liquor ratio will save water, chemicals and dyestuff, optimisation of the volume to fill the pad-box will reduce the amount of excess padding liquid required.
- *Separation and Recycling*: In an ideal case, valuable components which still exhibit their technical/chemical functionality are separated from the wastewater stream and recycled. Relevant examples are reconcentration of spent caustic soda solutions which are released from mercerisation through reboiling and recovery of water soluble sizes from the desizing process.
- *Separation, disposal and drain*: In many cases, the waste water contains a complex mixture of added chemicals, reaction products and impurities brought into the treatment bath by the fibres. At representative case is the waste water from scouring of cotton, which contains surfactants, complexing agents, alkali and fibre components (waxes, pectines, hemicelluloses etc.). Part of theses substances can be precipitated and collected by settling or filtration.

  If sludge or concentrates are formed during wastewater processing, the disposal of such products may cause substantial costs. Thus, in case of biodegradable products the biological treatment in a communal wastewater treatment plant (CWWT) may be preferable.

15.2 General wastewater treatment

In the vast majority, textile dyehouses release their waste water into a CWWT.

Waste water has to fulfil a series of defined limits before it may be released into a CWWT. As an example, some important aspects of the Austrian legal regulations are summarised in Table 15.1 [4].

Examples for substances which may not be released into the waste water are:
- Chromium(VI) compounds released from sulphur or vat dyeing
- Chloroorganic carriers in PES dyeing
- Arsenic, mercury or tin containing preservative agents
- Akyl phenolethoxylates (APEO) from washing
- Printing pastes, finishing chemicals

Table 15.1: Selected limits for release of textile waste water into a CWWT.

General parameters	Limit for release into a CWWT
Temperature	40 °C
Toxicity	No hindrance of biodegradation
Filter residue	300 mg/L
pH	6.5–10.0
Colour	
Spectral coefficient of extinction	
436 nm (yellow)	$28.0 \ \mathrm{m}^{-1}$
525 nm (red)	$24.0 \ \mathrm{m}^{-1}$
620 nm (blue)	$20.0 \ \mathrm{m}^{-1}$
Inorganic parameters	
Aluminium	Limited by filter residue
Lead	0.5 mg/L
Chromium – total	0.5 mg/L
Chromium(VI)	0.1 mg/L
Cobalt	0.5 mg/L
Iron	Limited by filter residue
Copper	0.5 mg/L
Nickel	0.5 mg/L
Zinc	2.0 mg/L
Tin	1.0 mg/L
Total chlorine (Cl_2)	0.3 mg/L
Sulphate	200 mg/L
Sulphide	1.0 mg/L
Sulphite	10 mg/L
Organic parameters	
Total hydrocarbons	20 mg/L
Phenol index calculated as phenol	10 mg/L

The maximum concentration of sulphate has been limited with regard to potential corrosion of concrete tubes in the ground and not for reasons of toxicity. A number of mineral waters contain higher concentrations of sulphate.

15.3 Processing steps and relevant aspects of wastewater treatment

15.3.1 General

The choice of a certain wastewater treatment is specific for the process steps performed in textile processing. Different technologies are applied depending on the substances, chemicals and dyes to be removed. In many cases, a selective treatment of a certain waste-water stream is favourable, instead of dilution with effluents from other steps. Thus, in this chapter representative examples for process steps and appropriate technologies will be discussed [5].

15.3.2 Wool processing

Washing of raw wool: A substantial amount of organic material is removed during the first wash of raw wool. The major problem arises from the high load in fat (lanolin) which is removed together with soil, salt, proteins and plant materials.

From washing 1 kg of raw wool, approximately 3–5 l of water with a total COD of 80,000 mg/L O_2 is released.

In the first step, the wash liquor is destabilised with an addition of auxiliaries to coagulate and separate the lanolin through sedimentation or flotation. Soluble impurities remain in the aqueous phase, while the oily phase (lanolin) is removed through centrifugation. A further purification can be achieved by ultrafiltration.

The composition of waste water is complex and difficult to control, thus membrane fouling is a critical issue.

Anti-felt treatment: The classical superwash treatment of wool uses a chlorine treatment as the first step, followed by a sulphite treatment and polymer addition. Due to the reaction of chlorine with the protein fibre, substantial concentrations of halogenated compounds are produced (AOX). Representative concentrations of AOX in the waste water of the Chlorine–Hercosett process are in the dimension of 20 mg/L (wastewater limit for this particular case is 2.5 mg/L).

An optimisation of the process does not solve the problem of AOX formation, as this is part of the chemical process performed.

Only the substitution of the process with non-chlorine–based oxidation, such as persulphate techniques, solves the AOX problem in anti-felt treatment of wool [6].

Chromium from dyeing with metal comple: Considerable amounts of chromium can be released from wool dyeing with chromium-containing metal complex dyes. In particular, from dyeing of dark shades like black with metal complex dyes, considerable amounts of chromium can be released into the wastewater stream. Through optimisation of the dye exhaustion and use of dyes which contain low amounts of non-complexed chromium, substantial improvement of the chromium concentration in the waste water is possible.

15.3.3 Sizing/desizing

For production of woven fabric, usually the warp has to be strengthened by a polymer coating, which has to be removed after weaving.

Depending on the quality of the yarn, the add-on of polymer on the warp yarn is in the dimension of 8–20% of the yarn weight. Thus, during desizing this amount has to be removed.

When native starch is used as size, the desizing requires an enzymatic degradation of the starch polymer. This prevents a possible reuse, as the film-forming properties and the adhesive strength of the starch size is lost.

Synthetic sizes such as polyacrylates, CMC (carboxymethylcellulose), PVA (polyvinylacohol) or CMS (carboxymethylstarch) are water soluble and thus can be removed through intensive washing with water. The diluted solution then contains the size in almost an unchanged chemical form. However, a number of impurities are removed from the cotton during the desizing step and thus are present in the waste water.

Desizing causes two major problems in the waste water:

- The high load of organic material such as hydrolysed starch leads to a substantial contribution to the overall COD load released from textile processing. This can cause problems in the following treatment in a CWWT due to the limited capacity of the aerobic/anaerobic stages of the CWWT plant.
- Synthetic sizes such as polyacrylates, PVA and CMC are only slowly biodegradable, a part of the sizes is removed in the CWWT through precipitation and adsorption at the sludge, however, another part of the sizes passes through the CWWT which is detected as refractory substances in the effluents from the CWWT.

Thus, treatment of such waste streams is necessary.

For recovery of synthetic sizes, two process stages are required:

- In the first stage, the wash process to remove the size is optimised to deliver a concentrated aqueous regenerate with concentration of 10–15 g/L.
- In the second step, a reconcentration to 80–150 g/L of size is performed by ultrafiltration. Higher concentrations are difficult to realise as viscosity of the solution increases and thus reduces the permeate flow substantially.

A concentration of the regenerate is required to achieve useful concentrations for reuse in sizing.

Assuming a size add-on of 10% and a pick-up of 75%, the concentration of the size in the pad-box has to be of at least 133 g/L.

For a successful introduction of size recovery and reuse, a number of pre-conditions have to be fulfilled:

- The reconcentrated regenerate is still a diluted aqueous solution, which has to be reused at a short distance to the site of generation (desizing). Thus, the weaving process has to be located near to the site of desizing.
- Through UF, the molecular distribution of the size is modified depending on the cut-off characteristics of the membrane. This modifies the technological properties of the regenerated size and measures have to be taken to control and maintain a constant quality of the regenerate.

- The regenerates contain impurities washed off from the raw material, which can change the properties of the size and reduce efficiency in weaving. As examples, in case of coloured fabric the regenerate will also contain a released dyestuff (dissolved reactive dyes, dispersed vat dyes). Singed material will release dust into the regenerate.
- A technological requirement results from the need to use similar sizing recipes for a major part of the weaving processes, otherwise separation of the desizing processes with regard to the size used becomes necessary and handling of regenerates will be complex.
- The regenerate contains biodegradable components, thus growth of microorganisms has to be controlled to avoid a reduction in the degree of polymerisation and odour development.

15.3.4 Scouring, bleaching, pre-washing

In scouring and bleaching of cotton, a substantial part of soluble fibre components is removed through the action of heat, alkali and oxidants.

The removed substances do not cause substantial problems with regard to biodegradability as these are mainly pectins, hemicelluloses, waxes, proteins and minerals. The released amount of 6–8 wt% of the processed cotton leads to a substantial COD load in the waste water. The contribution of auxiliaries and surfactants to the total COD is limited, compared to the amount of the removed cotton impurities.

The treatments are performed near to the boil, thus the temperature and alkalinity of the released waste-water stream require care. Usually these parameters are adjusted during collection and mixing in a buffer tank, which levels out the peaks in temperature and alkalinity.

The use of peroxide bleach requires an addition of complexing agents and stabilisers, which must be biodegradable. Typical examples for complexing agents used are polyacrylates, phosphonates, EDTA (ethylenediaminetetraacetate acid), NTA (nitrilotriacetic acid), organic acids (heptagluconate, citric acid).

Through the replacement of hypochlorite bleach by peroxide-based systems, the AOX release from the bleaching processes stops.

Fabrics from man-made fibres require pre-washing/relaxation stabilisation as pretreatment before dyeing. In these processes, substantial amounts of spin finishes, knitting oils and other yarn preparations are removed. Thus, waste water released from such steps transports a substantial load of greasy substances: derivates of fatty alcohols, fatty amines, hydrocarbons, silicone oil. In particular, the use of elastomer fibres requires high amounts of such auxiliaries as control of friction is highly important to achieve reproducibility in the elastic properties during knitting. In case of elastomer fibres, the added amounts of silicone oil can reach 10 wt% and more.

Attention has to be directed to the concentration of hydrocarbons and silicones as biodegradation is very low and elimination mainly occurs through adsorption at the activated sludge in the CWWT.

15.3.5 Mercerisation/causticising

Alkalisation/mercerisation of cellulose-based materials, in particular cotton fabric, is performed with highly concentrated alkali solutions. Representative values for the alkali concentration of the treatment bath are between 120 g/L NaOH (alkalisation of lyocell-type cellulose fibres) [7, 8] and 300 g/L mercerisation of cotton. Also, concentrated KOH solutions are in use, thus concentrations are typically near 250 g/L KOH.

Assuming a pick-up of 120 wt% at the passage from the mercerisation bath to the stabilisation bath, 1,200 g of alkali solution is washed out per 1 kg of processed fabric. The water flow in the wash baths and stabilisation baths is organised as counterflow, thus concentration of alkali increases stepwise in the opposite direction to the material flow. From the first stabilisation bath, rather concentrated alkali solutions can be recovered (e.g. 40–60 g/L NaOH). This concentration is high enough to permit an economic reconcentration through reboiling.

Reboiling is an energy-intensive process. The amount of water which is to be evaporated depends on the initial concentration of the diluted caustic soda solution and, to a minor extent, on the final concentration, which is more limited by technical aspects such as viscosity of the concentrate, corrosion of the distillation unit and a standardised concentration of recovered NaOH. Reconcentration of a diluted NaOH solution with 50 g/L NaOH requires evaporation of 7.8 l of water to achieve a final concentration of 440 g/L NaOH. Evaporation of 1 l of water requires approximately 1 kWh of energy. Thus, evaporation plants with several stages (three and more) are used to reduce the energy consumption of NaOH reboiling units to values of 0.2–-0.3 kWh/kg. Such plants recover the condensation energy of the steam generated in the preceding stages to heat the following evaporative stage.

The condensation energy of the last stage is recovered in the form of warm water, which must be used in the textile dyehouse as part of the overall energy concept. The use of the heat recovered in form of warm water is an important element of the overall profitability of such an installation.

As reboiling does not remove impurities, accumulation of impurities (e.g. dyes, residual size, fibre fragments, auxiliaries) has to be considered in reuse.

Partial removal of dispersed material can be achieved through filtration (e.g. fibre fragments), centrifugation and flotation processes [9, 10].

15.3.6 Dyeing processes

In textile dyeing operations, fibre and dyestuff-specific chemicals are used to achieve binding of dyestuff to the fibre. Relevant substances found in the waste water of dyehouses are:

– Surface-active auxiliaries (e.g. dispersing agents, wetting agents, levelling agents)
– Chemicals to adjust pH (alkali, acids)
– Reducing agents (vat dyes, sulphur dyes and indigo)
– Unfixed dyestuff, hydrolysate and dyestuff by-products

Different problems can arise from spent dyebaths. A major problem with used dyebaths results from intensively coloured waste water, which easily exceeds the legal limits set for absorbance at three different wavelengths. Also, heavy metal content or contribution to AOX values can result from effluents from dyeing processes.

Quite often, recycling and possible reuse of dyestuff is discussed; however, such an approach is not possible for dyes where a chemical reaction alters the dye molecule, for example, in reactive dyeing and naphtol dyeing.

Often dyes are used in combinations, which would thus require concentration determination and replenishment of exhausted dyes. Such approaches are useful only for processes where standard dyeing recipes with few variations in the dye recipes such as indigo dyeing and dyeing with sulphur black are used. Also, dyeing on a standing dyebath with replenishment of chemicals and dyes can reduce chemical load fed into the waste water. However, it must be considered that such concepts can only be used for processes which are controlled via temperature adjustments or minor changes in the pH. For dyeing processes which use controlled dyestuff exhaustion through dosage of chemicals or dyestuff, it will be difficult to implement recycling strategies.

Representative examples of dyeing processes and possible recycling strategies are discussed here.

Cellulose fibres

– Direct dyes: Direct dyes exhibit a high degree of exhaustion, thus the residual colour is not a big problem. Metal content (Ni, Cu) in the waste water from metal complex dyes and AOX from halogen-containing chromophores can be optimised through an increase in dyebath exhaustion and dye fixation.
– Reactive dyes: Due to the hydrolysis of the anchor group, waste water contains inactive hydrolysate which is not regenerable. Considerable amounts of unfixed dye are released from exhaust processes and also in continuous pad-batch dyeing with the residual filling of pad-boxes at the end of the impregnation step. In exhaust dyeing, substantial amounts of salt are released in addition, such as from a 1:10 dyeing with 50 g/L NaCl, a total amount of 0.5 kg salt is released for each kg of dyed goods. Recycling approaches thus have to focus on the recovery of water and salt solutions from washing processes through membrane filtration processes.

Decolourisation is usually performed as part of the end-of-pipe treatment of the textile effluents through precipitation, flocculation with iron salt, adsorption. Treatment with hypochlorite/chlorine-producing chemicals should be avoided as AOX problems will occur.

–   Vat dyes: Effluents from vat dyeing contain non-exhausted dyestuff in reduced form such as dispersing agents, levelling agents, alkali and an excess of reducing agents. Theoretically, the dyestuff could be regenerated through oxidation into its pigment form and concentration through filtration, however, the use of dye combinations prevents such an approach. The reuse of the dyebath chemicals is difficult as the follow-up products of reducing agents will change the salt content of the solution, and in case of organic reducing agents (e.g. hydroxyacetone) it will be difficult to oxidise completely before a bath reuse is possible. Dyestuff oxidation and soaping of the dyeing do not cause particular problems with regard to wastewater composition.

–   Indigo dyeing: Indigo dyeing for denim represents a unique case, as the recovery of dispersed indigo from wash baths following the dyestuff fixation with ultrafiltration has been proposed as a technical solution. A difficulty arises from the impurities from topping or bottoming with sulphur dyes. In indigo dyeing, constant addition of chemicals into the dyebath leads to an accumulation in the dyebath, thus rather high concentrations of chemicals are found in the waste water, such as sulphate concentration and COD from organic reducing agents. A substantial improvement of wastewater load could be achieved by the introduction of hydrogenated indigo, which can be directly fed into the dyebath without the use of a stock vat. By this approach, consumption of reducing chemicals (e.g. dithionite) is lowered to approximately 50%, which also reduces the resulting sulphate load in the waste agent to half the value [11].

–   Sulphur dyes: Sulphur dyes require a less negative redox potential for dyestuff reduction. Nowadays, reducing sugars are mostly used for dyestuff reduction instead of sulphides and polysulphides, which should be replaced by the easily biodegradable sugar-based products. Also, direct electrochemical reduction can be used to produce low sulphide-containing dyes [12, 13]. Oxidation of leuco dye should be performed with the use of peroxide-based oxidation agents. The use of halogen-based oxidants ($NaOCl$, $NaBrO_3$) results in the formation of AOX.

Protein fibres (wool, silk)

These fibres exhibit a high absorbency for acid dyes and metal complex dyes, and thus dyeing of such fibres does not exhibit inherent sources for pollution. Considerable concentrations of heavy metals (e.g. chromium) can result from dyeing of dark shades (black) with chromium-containing metal complex dyes.

Synthetic fibres

In dyeing of synthetic fibres such as polyethylene terephthalate (PET) or polyamides (PA6, PA6,6), only minor amounts of chemicals are released from the dyebath, for example, dispersing agents in case of disperse dyes and organic acids, levelling agents in case of polyamide dyeing with acid dyes and metal complex dyes.

Attention is required for low-temperature dyeing of polyester fibres as in this case carriers are used to achieve dye uptake at temperatures below 100 °C. These auxiliaries nowadays are aromatic esters and substituted phenols.

15.3.7 Printing

Screen printing techniques require the use of a thickened printing paste. The thickener in these pastes causes high COD values of up to 150,000–350,000 mg O_2/kg of paste. Thus, excess pastes must be collected and disposed instead of releasing into the waste water. Modern printing machinery has been optimised to reduce the volume of paste required for proper function. Depending on the cover factor of the print and the length of the print, 20–30% of the paste might be required for filling of the tubes and printing screen.

Also, high dye content (up to 100 g/kg), emulsifiers, pigment binders, melamine-based fixation agents, oxidants (m-nitro-benzene-sulfonic acid) and preservatives against microbial growth contribute to the chemical load.

Only the release of printing pastes which are removed during washing of printed goods is acceptable.

Due to the complex composition of printing pastes, their recycling or reuse is difficult. Usual methods for disposal include incineration, anaerobic degradation and binding in concrete.

Depending on the fibre material used, digital printing techniques permit direct printing of pigment dyes and disperse dyes, or require a pre-treatment of the fabric with chemicals for fixation of reactive dyes. In most cases, chemical consumption is lower compared to the screen printing techniques.

15.3.8 Finishing

In many cases, finishing is achieved through pad/dry/cure operations, the chemicals applied thus remain on the goods. In such a case the only wastewater load results from the residual filling of the padder, which is in the dimension of 15–20 l and a small surplus of finishing solution which is required to compensate for an unexpected slightly higher pick-up.

Critical components in such liquors are mainly hydrocarbon or silicone-based softeners and polymers obtained from dispersion (polyacrylates).

Specialty finishes such as antimicrobial agents (silver), insecticides (permethrin), fluorocarbon finishes are collected and either disposed or pre-treated before release into the waste water is permitted.

In case of exhaust processes as first step optimisation with regard to exhaustion/ fixation of chemicals should be undertaken. Based on the achievements and the concentrations in the waste water it has to be decided then if further processing is required.

15.4 End-of-pipe technologies

15.4.1 Neutralisation

From different process steps, effluents are released into a common levelling basin with the aim to level out peaks in pH, temperature and colour before release of waste water to the CWWT.

Depending on the composition of the waste water, additional less specific treatments may be required.

Neutralisation of alkaline waste water: Dyehouses with a specialisation in cellulose fibres often release alkaline waste water, which can be neutralised through introduction of CO_2-containing waste gases from the heat/steam generating plant.

15.4.2 Filtration techniques

Microfiltration permits removal of particles with rather bigger size [14]. The pore size of the filter is of the dimension of (0.2–0.1 μm). The transmembrane pressure is between 0.2–5 bar. Such filters permit removal of suspended substances.

In *ultrafiltration* (UF), ceramic membranes with a diameter in the dimension of μm (0.1–0.01 μm) finely dispersed substances and high molecular weight polymers can be collected through filtration by membrane filters. The surface of the membrane is flushed continuously with the filtration solution to avoid concentration polarisation. Transmembrane pressure is between 1–10 bar. The separation characteristics of a membrane are described by the molecular weight cut off (MWCO), which describes the size of molecules which are retained. A typical MWCO for a UF membrane is near 500,000–20,000.

Examples are the removal of dispersed dyestuff particles (indigo), separation of oily impurities (lanolin), separation of water soluble sizes (CMC, PVA) and recovery of thickeners in printing pastes.

The use of chemically resistant ceramic membranes has widened the range of application to solutions of higher alkalinity such as 40 g/L NaOH.

Membrane fouling is the loss in membrane flux (l/m^2 h) due to a coverage with impurities, small particles, deposits, and also through physical compression of the membrane. As membrane fouling reduces the filtration efficiency, measures have to be taken to control the composition of the effluent with regard to the following treatment e.g. nanofiltration.

Nanofiltration (NF) uses membranes with lower pore diameter such as 0.01--0.001 μm (10–1 nm). Thus, the already diluted solutions of larger molecules can be reconcentrated. Organic filtration membranes or ceramic membranes are used. Ceramic membranes offer a wider range of chemical stability (pH 0–14), organic membranes also exhibit good chemical resistance depending on the membrane polymers (pH 1–14) [14]. As an example, the removal of organic dyes from coloured waste water is possible through NF. Treatment of dilute alkali solutions (e.g. wash water) from reactive dyeing is possible. Transmembrane pressure is typically between 5 and 10 bar. A typical MWCO for a NF membrane is nearly 20,000 to 200.

In *reverse osmosis* (RO), membranes with lower pore diameter such as below 1 nm are used. Thus, the concentration of salt solutions is also possible. RO is used in sea water desalination and reconcentration of salt solutions. The membranes are quite sensitive with regard to the quality of the solution and its chemical conditions. Thus, the applicable range of pH is limited between 3 and 12. A typical MWCO for a RO membrane is below 200.

Filtration pressure needs to overcome the osmotic pressure between retentate and filtrate, thus RO operates at rather high pressure of up to 10–150 bar.

Often, a pre-filtration through UF or NF membranes is performed to remove impurities before treatment in the RO membrane.

The use of membrane filtration technique permits the recovery of substantial amounts of water.

Major disadvantages of such processes are:
- Investment costs are high for full installation and considerable energy costs for pumping and filtration.
- A concentrate is formed, which requires further treatment (disposal, incineration).
- The risk of irreversible membrane fouling includes a substantial financial risk as the costs for the membranes represent a major share of the total investment.
- A limited value of the recovered chemicals leads to a low contribution to amortisation.

15.4.3 Flocculation and sedimentation

An unspecific removal of dissolved, dispersed components can be achieved through flocculation and sedimentation. Through addition of flocculants ($Fe^{2+/3+}$ salts, Al^{3+}, Ca^{2+} salts, cationic polyelectrolytes), the formation of insoluble aggregates is initated, which then settle to the ground and can be removed/collected by filtration, centrifugation, flotation.

The formation of sludge leads to the next problem, as this product often exhibits low biodegradability and has to be disposed (e.g. by incineration, anaerobic digestion) which can cause substantial costs.

15.4.4 Oxidative processes

Oxidative processes can be used to destroy coloured substances and to degrade refractory substances. Representative processes can use Fenton's reagent (a mixture of $Fe^{2+/3+}$ and H_2O_2), ozone, as well as combinations of oxidants with UV irradiation. Also, low-pressure wet oxidation (oxygen and catalyst, 5–20 bar, < 200 °C) has been proposed to degrade polymers like PVA and CMC [15].

In case the addition of oxidants is considered as a possible method, it has to be kept in mind that a stoichiometric addition will be required for a total oxidation, thus treatment of separated waste streams is preferable to limit the costs of chemicals.

Possible formation of chlorine and an increase in AOX concentration as the result of oxidation of chloride ions also have to be evaluated when introduction of oxidative treatments is considered for wastewater processing.

15.4.5 Adsorption

Adsorptive processes with the use of charcoal or other sorbents can remove selected critical components. Such processes permit the removal of dyes to reduce the colour of waste water or of specialty chemicals. However, it is important to use such processes in side streams to avoid the formation of huge amounts of spent adsorbent, which then have to be disposed or regenerated with substantial costs.

15.4.6 Biological (aerobic, anaerobic) treatment

Biological treatment of textile effluents is of interest for large facilities which produce high volumes of waste water and release a substantial load of organics which serve as nutrients for bacterial consortium. It is important to consider that adaptation of bacteria to a certain composition of the effluents to be digested in an aerobic or/and anaerobic process will require time. Maintenance of a relatively constant composition and volume flow is required to feed the biological stage and maintain activities. This requirement can cause serious problems with usual textile production as variations in composition as well as in volume flow are considerable, and shut down of production during plant holidays can lead to instabilities in the biological stage.

In an ideal case pH and temperature buffering is performed in the levelling tank at the site of the textile plant. Biological treatment is performed in a CWWT, which also handles communal waste water. Thus, variations in the load or amount of waste water will not be as critical as in the case of a separate biological treatment of the effluents in the textile dyehouse.

15.5 Recycling and disposal

Design and production of sustainable apparel is still a challenge [16]. Used textiles are collected, sorted and separated according to the following:
–   Value of the material to be recycled: Fibres with higher value such as silk or wool are of higher interest for recycling due to higher value of the fibre material.
–   Condition of the used product: A considerable part of a garment is used in very little or even in negligible amount. Such a product can be recycled with the mark of second-hand apparel. Heavily used clothes can still be used as disposable wipes in technical cleaning processes or are processed further to form noise absorbing structures. A substantial part of the used garment, however, is disposed and incinerated.

Several technical problems make recycling of textiles, fibres and polymers difficult.
–   At present many textile products are constructed from a mix of materials. The fibres can be blended at the level of a yarn such as a cotton/polyester or wool/ polyamide yarn, polyurethane filaments wrapped by polyamide fibres. Different threads can be combined at the level of knitting, weaving, embroidery and finally during the production of a garment. Separation of a multi-material mix is difficult, often impossible.
–   The separation of fibres from a fabric requires substantial mechanical energy and the fibre length is shortened considerably. Often a break down of fibres of sufficient length to be spun to yarn is not possible.
–   A number of accessories have to be removed and separated: buttons, zips, velcro, rivets.
–   Used textiles in garment form do not consist of fibre polymers alone. There is a complex mixture of additives, dyes, finishes and auxiliaries present in the material. Also, the fibre polymers may be degraded through the impact of light, chemicals, oxygen and detergents. Thus, a simple separation and reuse will be difficult.

Besides the technical problems, many consumers do not accept products from recycled fibres as competitive to new products. While consumer awareness for recycling is considerable, readiness to buy recycled goods is low.

A substantial reduction in the volume of wastes could be achieved through longer lifetime of products and willingness of consumers to reduce the frequency of garment replacement.

15.6 Approaches and challenges

Remarkably, only a limited number of concepts for recycling of used textiles have reached the technical scale. Representative examples are:

- Recycling of production wastes from cutting of materials during garment production is a state of the art. These wastes are well defined in compositions, technically new and unused. The waste material is collected, mechanically opened and used to produce relatively coarse yarns for weaving of blankets.
- Recycling of pure cellulose fibres in manufacturing of regenerated cellulose fibres can be used to substitute fresh pulp.
- Polyester collected from bottles can be used for manufacturing of recycled polyester yarns.
- Depolymerisation of PA6 collected in the form of used fishing nets made of PA6 permits recycling of the ε-amino-caprolactam and production of new polyamide fibres [17].

To support future recycling of textile materials, some general concepts should be considered:

- The number of different materials used in a garment must be reduced to few main components.
- The design of a product must consider later separation and recycling (e.g. thermal removal of seams).
- Recycling of fabric, or threads, is preferable but more difficult when compared with recycling of fibres and depolymerisation of polymers.
- Additives, dyes and finishes must not interfere with the planned recycling process.
- A concept for handling of by-products of the recycling process is required, as considerable amounts of additives, dyes etc. will be separated during polymer recycling.
- There is a substantial need for appropriate separation processes which will permit a disintegration and separation of textile structures without an extensive use of resources.

In any case, appropriate logistics will be required to collect and separate the material and to avoid long-distance transportation to sites of further treatment.

References

[1]   European Parliament – Committee on the Environment, Public Health and Consumer Policy 2004. Report on progress in implementing Council Directive 96/91/EC concerning integrated pollution prevention and control, A5-0034/2004 final, 28 January 2004.

[2]   Integrated Pollution Prevention and Control (IPPC), Reference Document on Best Available Techniques for the Textiles Industry, BREF 07/2003; European Commission, http://eippcb.jrc.ec.eu ropa.eu/reference/txt.html, July 2003.

[3]   Schönberger, H., Kaps, U. Reduktion der Abwasserbelastung in der Textilindustrie, Umweltbundesamt, Berlin, Germany 1994.

[4]   AEV Textilveredelung und -behandlung – Verordnung des Bundesministers für Land- und Forstwirtschaft, Umwelt und Wasserwirtschaft über die Begrenzung von Abwasseremissionen aus der Textilveredelung und Behandlung StF: BGBl. II Nr. 269/2003, http://www.ris.bka.gv.at/Geltende Fassung.wxe?Abfrage=Bundesnormen&Gesetzesnummer=20002744, 2003

[5]   Bechtold, T., Burtscher, E., Hung, Y. Chapter "Treatment of textile wastes". in "Handbook of Industrial and Hazardous Wastes Treatment", Second Edition, Wang, LK., Hung, Y-T., Lo, HH., Yapijakis, C. (ed.) Marcel Dekker, Inc. New York, ISBN: 0-8247-411-5, 2004.

[6]   Bechtold, T., Mahmud-Ali, A., Siroky, J., Riehl, M., Krüger, M. Superwash ohne chlor – ein neuer standard wird technische realität. Melliand Textilberichte. 2012, 3, 148–150.

[7]   Manian, AP., Abu-Rous, M., Schuster, KC., Bechtold, T. The influence of alkali pre-treatments in lyocell resin finishing. J. Appl. Polym. Sci. 2006, 100/5, 3596–3601.

[8]   Siroky, J., Manian, AP., Siroka, B., Abu-Rous, M., Schlangen, J., Blackburn, RS., Bechtold, T. Alkali treatments of lyocell in continuous processes. part 1: Effects of temperature and alkali concentration in treatments of plain-woven fabrics. J. Appl. Polym. Sci. 2009, 113, 3646–3655.

[9]   Bechtold, T., Gmeiner, D., Burtscher, E., Bösch, I., Bobleter, O. Flotation of particles suspended in lye by the decomposition of hydrogen peroxide. Sep. Sci. Technol. 1989, 24, 441–451.

[10]   Bechtold, T., Burtscher, E., Sejkora, G., Bobleter, O. Moderne verfahren zur laugenrückgewinnung. Internationales Textilbulletin – Veredlung. 1985, 31, 5–26.

[11]   Blackburn, RS., Bechtold, T., John, P. Indigo reduction methods and pre-reduced indigo products: A historical review and recent developments. Color. Technol. 2009, 125, 193–207.

[12]   Heid, C., Holoubek, K., Klein, R. 100 Years of sulfur dyes. Melliand Textilber Int. 1973, 54, 1314–1327.

[13]   Bechtold, T., Burtscher, E., Turcanu, A., Bobleter, O. Continuous sulfur dyeing without reducing agents: Fully reduced sulfur black 1 by cathodic reduction. Text. Chem. Color. 1998, 30/8, 72–77.

[14]   Melin, T., Rautenbach, R. Membranverfahren – Grundlagen der Modul- und Anlagenauslegung, Springer-Verlag, Berlin Heidelberg, 2007.

[15]   Horak, O. Katalytische naßoxidation von biologisch schwer abbaubaren abwasserinhaltsstoffen unter milden reaktionsbedingungen. Chem. Ing. Tech. 1990, 62, 555–557.

[16]   Blackburn, R. (ed.), Sustainable Apparel – Production, Processing and Recycling, Woodhead Publishing Ltd., Cambridge, UK, 2015.

[17]   ECONYL® Regenerated Nylon. https://www.econyl.com (accessed 5.11.2018).

Take home messages

Textile waste water and textile wastes create substantial environmental problems. The major reason for the need of wastewater treatment or increased activity in recycling does not come from the toxicity of the material, but from the huge amounts which are generated.

The use of critical components such as alkyl phenol ethoxylates, chloroorganic carriers and fluorocarbon finishes have already been banned or regulated.

Wastewater limits have been defined in many countries. The justification of such limits comes from the toxicity of the substances, like heavy metal, from the unpleasant appearance of the effluents in case of coloured waste water or from the risk of corrosion of concrete installations in case of high sulphate concentrations.

Wastewater processing has to be considered with care as considerable costs can arise from an inefficient technological concept. Optimisation of processes must be the first step, then in many cases collection and treatment of concentrated streams of waste water is favourable.

End-of-pipe technologies in many cases are limited in efficiency and rather expensive as high volumes of diluted solutions must be processed.

Biological treatment of textile effluents at the site of a textile dyehouse may be of interest for larger plants; however, in this case a relatively constant composition of waste water fed into the biological state is important to achieve rapid biodegradation and stable results.

In an ideal case, waste water contains biodegradable organic substances and the critical components remain below the limits set for release into a CWWT.

Quiz

1. Which important aspect should be considered critically when the following techniques are discussed for wastewater treatment:
    membrane processes; evaporation/reboiling; settling/filtration; decolourisation by addition of oxidants
2. Which important aspect should be considered critically when the following processes are discussed:
    desizing; scouring/bleaching; pre-washing; mercerisation/alkalisation; dyeing; printing; finishing

Exercises

1. Based on the total amount of wool produced per year, an estimation of the water consumption for raw wool washing should be made. Assuming a lanolin content of 14 wt%, estimate the total amount of lanolin available for recovery and use in cosmetic products.
2. Calculate the concentration of a size regenerate which is required as a minimum to achieve a size add-on of 6%, 10%, 14% at a pick-up of 75% and 85%. How much energy demand will increase in case the pick-up is changed from 75% to 85% when we assume an energy demand of 1 kWh to dry 1 kg of water?
3. Calculate the COD in mg O_2/L of waste water from scouring, assuming the following process: water consumption is 8 l/kg, removal of 5 wt% of impurities from cotton, molecular basis for calculation, average formula of hemicelluloses.

4. Calculate the overall transfer of alkali in the form of 50 wt% NaOH into the stabilisation compartment of a mercerisation unit. The material is introduced into the alkali bath (300 g/L NaOH, ρ = 1.263 g/mL) in dry state. Pick-up of swollen cotton is at 150 wt%, fabric data is 150 g/m^2, 1.8 m width, speed is 25 m/min.

   In case the material is mercerised in wet state (e.g. pre-bleached material, water add-on 70 wt%), what is the minimum concentration of the NaOH solution to be dosed into the mercerisation bath to maintain the bath concentration at 300 g/L without overflow in bath volume? What happens if the concentration of the recovered reboiled NaOH solution is only 440 g/L (density 1.357 g/L)? What volume of mercerisation bath will be produced in excess per 1 tonne of processed material if c(NaOH) is maintained at 300 g/L?

5. Calculate a graph which represents energy consumption of a three-stage evaporation plant as function of NaOH concentration (in g/kg) to be reboiled (c(NaOH) = 50 – 400 g/kg NaOH). Range of diagram is 5 g/kg NaOH–60 g/kg NaOH, destillation of 1 kg of water requires an energy consumption of 0.3 kWh.

6. Calculate the total amount of dyestuff released per day (24 h) in cold pad-batch dyeing with reactive dyes from the excess of dye solution prepared to fill the pad-box. Average dye concentration is 50 g/L, filling of pad-box plus surplus of dye solution: 20 L; two padders work in three shifts, average time of production is 2.5 h per batch.

7. An indigo dyeing plant washes the dyed yarn with a specific water consumption of 3 L/kg. What amount of indigo can be regenerated per day from the waste water in case the indigo concentration is at 0.1 g/L indigo and the daily production is at 15,000 kg yarn per day? What is the estimated indigo consumption of the plant in case a 2 wt% indigo dyeing is performed without recycling?

Solutions

1. (Question 1): Membrane processes: membrane costs, stability and fouling; Evaporation/Reboiling: energy recovery; Settling/Filtration: disposal of sludge and filter cake; Decolourisation by addition of oxidants: formation of AOX, by-products.

2. (Question 2): Desizing: high COD content, PVA and acrylates exhibit limited biodegradability; Scouring/bleaching: complexing agents, COD load; Pre-washing: substantial load in oils, paraffines and silicones; Mercerisation/alkalisation: concentrated alkali solution for recovery; Dyeing: highly coloured effluents; Printing: pastes with high COD and preservatives; Finishing: specialty finishes require attention (e.g. antimicrobial finishes), considerable hydrocarbon or silicone content.

3. (Exercise 1): Total amount of wool produced per year 1.5 million tonnes. Thus 14 wt% correspond to 210,000 tonnes of lanolin available for recovers. Only a share of this amount will be useable for cosmetic products after purification.

4. (Exercise 2): The concentration of size in the regenerate can be calculated by dividing the add-on with the pick-up. An add-on of 6 wt% corresponds to 60 g/kg. Assuming the density of the size with 1 g/mL per litre of size corresponds to 1 kg, the concentration can be calculated with 60/0.75 = 80 g/L. Thus, the concentrations in the regenerate must be 80 g/L, 133 g/L and 187 g/L for a pick-up of 75 wt%, and 71 g/L, 118 g/L, 165 g/L for a pick-up of 85%. The energy demand will increase for drying off 0.85 kg of water instead of 0.75 kg, thus an increase by a factor of 0.85/0.75 = 1.13 (+13%) will take place.

5. (Exercise 3): We assume a formula: $C_5H_8O_4$ for the hemicelluloses. The oxidation will follow the reaction:

$$C5H8O4 + 5O2 \rightarrow 5CO2 + 4H2O$$

mol of hemicelluloses M = 132 g/mol will require $5O_2$ (5 mol × 32 g/mol = 160 g). For oxidation of 50 g hemicelluloses in 8 L of water we will need 50 × 160/(132 × 8) = 7.5 g O_2 this corresponds to a COD of 7,500 mg/L O_2.

6. (Exercise 4): At first we calculate the total production of mercerised goods per hour: m_{goods} = 25 m/min × 60 min/h × 1.8 m × 0.150 kg/m² = 405 kg/min. The transfer of NaOH m_{tr} into the next bath will be: m_{tr} = m_{goods} × 1.50 × 300 /1.263 = 144 kg/h. In case of wet mercerisation, the difference in pick-up defines the concentration of the alkali solution. Δpick-up = 150–70% = 80%. The concentration of reboiled alkali thus must be 144 kg/(405 × 0.8) kg = 0.444 or 44.4% (from density tables: approximately 645 g/L NaOH, 1.475 g/mL). The bath volume will increase constantly and this will lead to an overflow.

   We calculate on the basis of 1 kg of goods. Due to the differences in density we have to perform the calculations on the basis of mass and not with concentrations. The intake of the amount of water has to be compensated by addition of concentrated NaOH solution (c = 440 g/L).

   With the mercerised material 150 %wt of NaOH solution (c = 300 g/l) leave the bath. Considering the density 1.263 g/mL this corresponds to a mass of 356.3 g NaOH and a volume of 1.18 L.

   The intake of 0.7 kg of water with the wet goods must be concentrated to 300 g/L by addition of NaOH solution (440 g/L). By mixing 0.7 kg of water with 1.891 kg NaOH (c = 440 g/l; 613.2 g NaOH and 1278 g water) we obtain 2.591 kg of NaOH (c = 300 g/L). 1.5 kg of solution are removed with the goods, thus an excess of 1.091 kg NaOH (c = 300 g/l) remains. This corresponds to a volume of 1.091/1.263 = 0.864 L which will overflow per kg (864 L per 1000 kg) of mercerised goods.

7. (Exercise 5): The general formula for energy consumption is: Energy per kilogram of concentrate = $(c_{End}/c_{Start} - 1) \times 0.3$ kWh/kg.

   The diagram for 5, 20, 40, 60 g/kg NaOH is shown below:

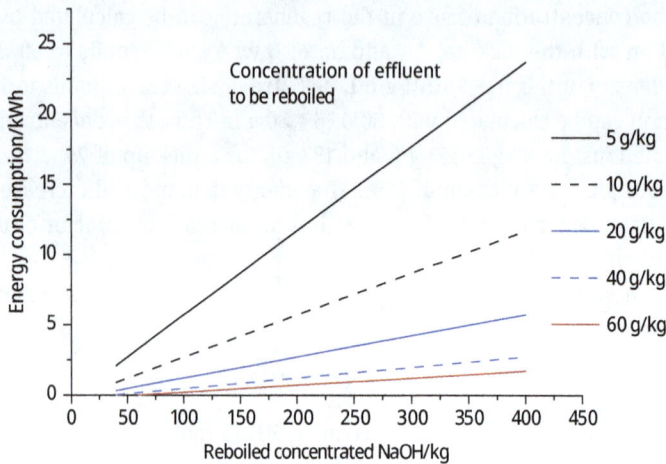

8. (Exercise 6): The total volume of excess dye solution released will be:
    V_{Total} = 20 L × 24 × 2/2.5 L = 384 L. This corresponds to a total mass of dyestuff
    $m_{dyestuff}$ = 384 × 50 /1,000 = 19.2 kg (!) per day.
9. (Exercise 7): Total indigo recovery m_{recov} = 15,000 kg × 3 L/kg × 0.1 g/L = 4,500
    g = 4.5 kg. Total indigo consumption: m_{indigo} = 15,000 kg × 0.02 = 300 kg.

List of abbreviations/symbols

APEO	alkyl phenolpolyethyleneglycolethers
EDTA	ethylenediaminetetraacetic acid
NTA	nitrilotriacetic acid
AOX	adsorbable halogenated organic compounds
STP	sodium triphosphate
CWWT	communal wastewater treatment plant
UF	ultrafiltration
NF	nanofiltration
RO	reverse osmosis
MWCO	molecular weight cut-off
CMC	carboxymethylcellulose
CMS	carboxymethylstarch
PVA	polyvinylalcohol
PET	polyethylene terephthalate

16 Circularity, recycling and disposal

Circularity requires integration of all stages of production as well as consumer use and recycling into the design of a new textile-based product. During the last decades, the majority of textile products had been designed to meet fashion demands and to achieve optimum functionality per money. Today global warming and tremendous growth of amounts of disposed textile wastes put pressure on the existing production. Concepts including reuse, repair and recycling have been estimated as primarily cost-generating activities, which apparently do not add value to a product.

A thermodynamic analogue to the current situation could be described as follows:

Our current use of textiles ends up with a high increase in disorder (entropy change, ΔS, by wastes and pollution) and consumes a lot of resources (polymer, fuel and water) which are equivalent to energy/enthalpy change, ΔH. As a result, a considerable change in free enthalpy ΔG of the system is the driving force which makes such a system easy going (eq. (16.1)).

Circularity intends to reduce the overall entropy change, ΔS, of the textile production and use through reduction in amounts of waste and pollution, and at the same time requests to reduce the consumption of resources (less energy, less non-renewable raw materials and no petrol-based products) which are equivalent to ΔH in our model.

In total, we optimise the products and processes to run the circular economy at lower ΔG:

$$\Delta G = \Delta H - T\Delta S \qquad (16.1)$$

Any time we are considering new processes and strategies to lower the disorder, for example separation of fibre blends, depolymerisation/repolymerisation of macromolecules, decolourisation of dyes and removal of chemicals for purification of feedstock material we have to assess the overall impact of the process on the resource consumption of water, energy and chemicals. For a valuable concept, it will not be sufficient to demonstrate the principle of a process without any comparison to existing technologies.

16.1 The EU concept of circularity

The production and consumption of textile products are constantly growing and are still projected to grow further. Between 2000 and 2015, the global consumption of clothing and footwear doubled and is expected to grow further by 63% from currently 62 million tonnes (2022) to 102 million tonnes in 2030 [1].

The contribution of textile consumption is at position 4 with regard to climate change and negative environmental impact and at position 3 in global water consumption and land use.

https://doi.org/10.1515/9783110795738-016

As a result of the high consumption of textile-based products, a total of 5.8 million tonnes of textiles are discarded per year in the EU, which corresponds to an average of 11 kg per person and year.

The trend for fast fashion has led to overproduction and overconsumption. A major part of the used textiles goes into incineration and landfill.

The trend for rapid product production has led to cheap products of inferior quality, which had been produced with inefficient processes and a high consumption of non-renewable resources. Also social aspects, for example child labour as well as gender equality dimensions due to the high share of women in low-wage and unskilled workforce employment, have raised concerns.

In a circular economy of textile products, the priorities change. Short lifetime products must be replaced by products of high quality, durability and recyclability. Products must allow a longer time of use and reuse, and must be designed for repair.

Fibre-to-fibre recycling then replaces disposal and incineration.

In the 2020 Circular Economy Action Plan and the 2021 update of the EU Industrial Strategy, textiles were identified as a key product value chain for the transition to sustainable and circular production, consumption and business models.

A short summary of the vision for sustainable and circular textiles for 2030 includes the following main aspects:

- Textile products will be long-lived and made to a great extent of recycled materials.
- The products are free of hazardous substances and have been produced in accordance to social rights and the environment.
- Fast fashion will be replaced by high-quality affordable textiles, which are suited for repair and reuse.
- Fibre-to-fibre recycling replaces incineration and landfill.

Currently less than 1% of the textile post-consumer wastes are used to produce new textiles globally [1]. A share of 20% of the collected and separated wastes is used in downcycling, for example for wipes.

In the EU, a total of 2.1 million tonnes of post-consumer wastes from clothing and home textiles are collected at present for recycling and reuse, which corresponds to 38% of the total volume of textiles on the EU market, while 62% are discarded.

A growing source of concern is the release of microplastics from textiles. Approximately 40,000 tonnes of synthetic fibres are released from textile washing operations. The highest release of microplastics is observed during the first 5–10 washes; thus, longer lasting and durable textiles will reduce the release of microplastics. More gentle washing operations and use of filters to collect microfibres from the wash liquor are among the measures proposed to reduce the unintentional release of microplastics in the environment.

16.2 Terms and definitions used for assessment of products and processes

For an assessment of a certain product, a collection of data with regard to raw materials, resource consumption (chemicals, water and energy), quantity and quality of output and their impact on the environment (greenhouse gases, soil contamination and wastes) is required. The data set forms the life cycle inventory (LCI) database [2].

For product assessment over the full life cycle, a life cycle assessment (LCA) is formulated. The stages include creation, design, marketing, transport, consumer use and disposal. A careful definition of the system boundaries is required to achieve reliable results. As an example in the case of reuse of wastes, the environmental burden of the starting material is shifted to the preceding phase of product life; similarly, the environmental impact for generation of thermal energy from municipal solid waste incineration is allocated to the producer of the wastes [3]. This will require adaptation in the assessment of a circular concept.

Furthermore, a critical evaluation of the data with regard to their relevance has to be made. Important factors for assessment of a product over its full lifetime include:
–   Total energy use per produced unit (e.g. 1 tonne), non-renewable energy use and renewable energy use
–   Land use for production of biomass (e.g. cotton, plant material for pulping), fibre and textile production, storage, retail and transportation
–   Water use (irrigation, cooling and process water)
–   Global warming potential (CO_2 production)
–   Abiotic depletion (e.g. synthetic polyester), ozone layer depletion (e.g. halon emissions from crude oil production), human toxicity (e.g. polycyclic aromatic hydrocarbons), freshwater aquatic toxicity (e.g. pesticides and fertilisers), terrestrial ecotoxicity, photochemical oxidant formation (e.g. release of SO_2), acidification and eutrophication (e.g. fertilisers).

The different factors have to be combined and rated to obtain a final result. It is clear that dependent on the particular situation and the emphasis of certain factors, a considerable space for discussion will remain.

A serious problem arises from the use of false or deceptive claims with regard to green, eco-friendly, sustainable and environmentally friendly products, when these claims are not justified by facts. A similar situation is observed with the accuracy of claims for use of recycled polymers (mainly polyethylene terephthalate, PET). In majority, the polymer used for the fibre production stems from food packaging material (bottle-to-fibre) and not from fibre-to-fibre recycling. The fibre-to-fibre recycling of polyester textiles is substantially more difficult as these products contain dyes and finishes and are often used in combination with elasthane fibres.

16.3 Important steps in a circular concept

The conventional value chain in textile production and consumer use covers the steps of

- raw material extraction – (different sources can be distinguished) natural fibres, man-made fibres (synthetics, regenerated cellulosics and inorganic fibres) and recycled fibres – leading to pre-consumer wastes;
- fibre formation – for example as natural fibre or by chemical processes (man-made fibres) – leading to pre-consumer wastes;
- production of a textile fabric and garment/clothing – leading to pre-consumer wastes;
- retailing (conventional, e.g. stores and online, a difference arises from the energy use) – leading to pre-consumer wastes;
- consumer use – post-consumer wastes are generated, also care operations (washing and drying) contribute substantially to the overall balances of CO_2 emission, energy and water consumption, and release of microplastics – leading to post-consumer wastes.

Pre-consumer wastes can be collected at the site of formation; thus, the composition in terms of fibre material is known and the uniformity with regard to material composition is comparable high. Examples are short fibres and yarns from spinning, selvedges from weaving and textile wastes from cutting in garment production.

At the end-of-product lifetime, the textile products are collected as post-consumer wastes. These wastes require sorting with regard to their fibre composition and disassembly of mixed material.

16.4 Composition of textile material

The composition of textile wastes is complex, which is a result of the extreme number of products and variations on the market. Different fibre materials are used alone or in combination: natural fibres, regenerated cellulose fibres, synthetic polymers and so on.

The majority of textile products has been dyed, and special functions have been implemented by use of additives and finishes (softeners, cross-linkers, flame retardants, antimicrobial chemicals, etc.).

Besides the handling of the fibre material, also the role of dyes and finishes during recycling must be considered.

Appropriate labelling of material composition, dyes and finishes used in a product will be required to achieve an efficient sorting and separation of wasted textiles.

Pre-consumer wastes are collected at the site of textile production: dust from fibre spinning, selvedges from weaving and fabric from cutting of the pattern in garment production. These materials can be collected separately and are available in rather well-defined composition and comparable high purity. Post-consumer wastes are textile products which have been collected at the end of their usage. These products have aged during use and care (laundry). Sorting and facilitating of the collected products are required to remove impurities and hard parts (e.g. zippers) and to form batches with similar fibre composition.

16.5 Design for recycling

Activities in the design for recycling intend to extend the life of the products and also include the requirements coming from later recycling activities.

Major factors which lead to failure during use have been identified: poor fastness properties, low tear strength, malfunction of zippers and breakage of seams. An improvement of these factors is part of the product design for longer lasting lifetime.

At the stage of product design, the following aspects must be considered to reduce complications at the stage of product collection, separation and recycling:
- Correct and machine-readable labelling to identify fibre type, dyes and chemicals
- Reduction of material types combined in one product, for example fibre blends which require complicated separation processes and prevent economic separation and reuse
- Design of single material products wherever possible, for example use of same material for sewing threads
- Selection of dyes and finishes with regard to durability, reuse as well as recycling of fibres

According to the EU – a study on the technical, regulatory, economic and environmental effectiveness of textile fibre recycling – we can differentiate recycling techniques into four major groups (Figure 16.1) [4]:
- Mechanical recycling, where mainly physical forces are applied
- Thermal recycling, where thermoplasticity of polymers is utilised
- Chemical recycling, where use of solvents and chemical processes is applied for separation, degradation and recovery of polymers
- Biochemical recycling, where action of enzymes and microbial consortia is applied

16.6 Facilitating steps

For pre-consumer wastes, the material composition is known, and separated collection can be organised at the site of generation.

For post-consumer wastes, the first step after collection is sorting into fractions of the same or similar composition. This process is mainly done manually, however, instrumental methods, for example NIR (near-infrared spectroscopy), can also be used.

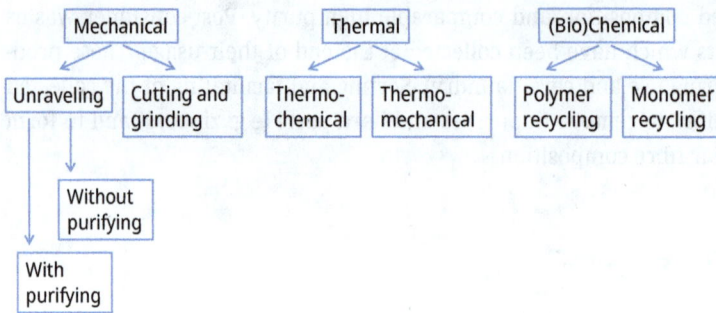

Figure 16.1: Textile recycling technologies (modified after [4]).

Complex instrumental methods which require sample preparation, for example differential scanning calorimetry, gas chromatography, microscopy and dissolution experiments, are too laborious and costly.

Limitations in automated identification result from low penetration depth of irradiation-based methods, for example NIR; thus, only the outer material is identified in a thicker garment which is composed of several layers. Further complications arise from coatings, finishes, presence of dyes in higher concentration (e.g. dark colours) and from ageing of materials during use.

In an ideal case, automated identification of products is achieved through intelligent labelling, which also provides access to information about possible recycling techniques.

Another important approach utilises the principle of design for disassembly. This includes destruction of seams, for example through microwave irradiation, which allows separation of hard parts and different materials. Also reversible debonding of coated and laminated textiles is under consideration.

16.7 Mechanical recycling

In mechanical recycling, physical forces are applied to separate fibre-based materials to a level which allows reuse of fibres, for example for spinning of yarn and preparation of non-wovens.

Fabric pieces are torn and separated into individual fibres; however, the input of mechanical energy leads to fibre shortening and formation of fluff and dust (Figure 16.2). The overall efficiency is dependent on the characteristics of the feedstock. Open-knitted material is more easy to open, woven fabric is difficult to open and a higher amount of fluff is obtained. Fluffs are short staple fibres <10 mm, which cannot be spun to yarn.

Coated, laminated products are bound together very firmly and not suited for direct mechanical opening.

```
┌─────────────────────────┐
│ Textile wastes          │
└─────────────────────────┘

┌ ─ ─ ─ ─ ─ ─ ─ ─ ─ ─ ─ ─ ┐
  Sorting-Facilitating
│ Formation of fractions  │
  of similar composition        ┌──────────────────────────┐
└ ─ ─ ─ ─ ─ ─ ─ ─ ─ ─ ─ ─ ┘     │ Purification / Washing    │
                                 │ Removal of contaminants   │
┌─────────────────────────┐     └──────────────────────────┘
│ Mechanical action       │
└─────────────────────────┘

┌─────────────────────────┐
│ Fluff                   │
│ Fibres                  │
│ Filling material        │
└─────────────────────────┘
```

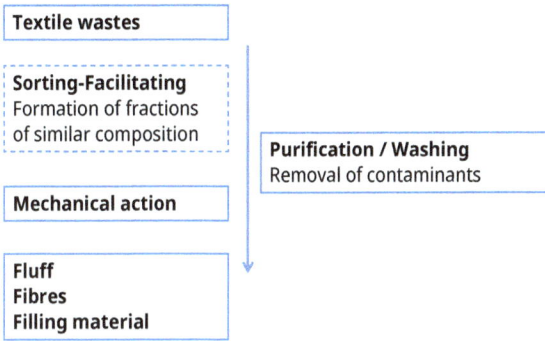

Figure 16.2: Mechanical recycling: main steps.

Many textiles contain polyurethanes (PUR, elastane) which can also generate problems and must be removed. Ideally one-component products are processed.

Pretreatment includes removal of hard parts, for example buttons, zippers and metal parts. Ideally at the stage of design for recycling, the products are optimised to allow rapid separation.

No removal of dyes or finishes occurs during the process, thus colour of the recycled yarn depends on the average colour of the material.

In case of contaminated textiles (unwashed clothing, dirty carpets and contaminated workwear), a prewash may be required. Any wet process has to be considered with care, as this will contribute to the overall energy and resource balance.

The technology is cheap and flexible; however, the overall material efficiency is limited. Mechanical processing may be an important pretreatment step for thermomechanical, chemical or biochemical recycling processes.

For natural fibres, that is cotton, an amount of only 5–20% of the input is obtained as spinnable fibres; for synthetic fibres (e.g. polyester fibres), 25–50% of the input is recovered as spinnable materials [4]. The lower quality of the recycled fibres limits their use to coarse yarns.

The energy consumption of mechanical processing is estimated with 0.3–0.5 kWh/kg input material, which is low, compared to other processes. However, in relation to the output fibre material, this value increases by a factor of 5–20.

As disadvantages, the technology delivers staple fibres with lower quality, and concerns exist about possible contamination by the presence of non-REACH conform chemicals, for example presence of polyaromatic hydrocarbons in used firefighter suits. Thus complete information about the history of wastes will be important.

Mechanical recycling will be the initial stage of material processing in recycling, which delivers the starting materials, for example polyester fibres and cotton dust, for a following processing by thermal or chemical recycling.

16.8 Thermal recycling

In thermal recycling, heat is applied to recover thermoplastic polymers or to break down polymers into smaller building blocks. Dependent on the main reactions, it is distinguished between thermomechanical and thermochemical recycling (Figure 16.3). Thermal recovery to deliver energy for other processes by combustion is not considered as recycling technology.

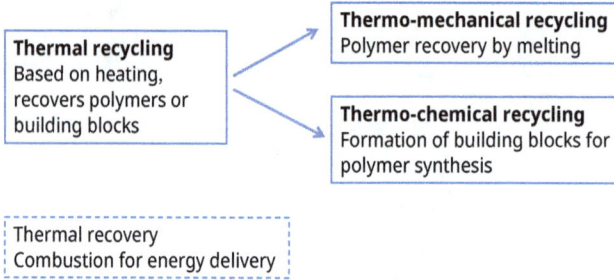

Figure 16.3: Thermal recycling.

Thermomechanical recycling: These processes are mainly applicable for thermoplastic polymers such as polyester (e.g. PET), polyamides (polyamide 6 (PA6) and polyamide 6.6 (PA6.6)) and polyolefins (e.g. polypropylene (PP)).

Raw materials are shredded, cleaned, degassed and thermally processed into granules which then can be extruded to fibres (Figure 16.4).

For fibre spinning, a comparable high purity of the recycled polymer is required to avoid problems during filament formation.

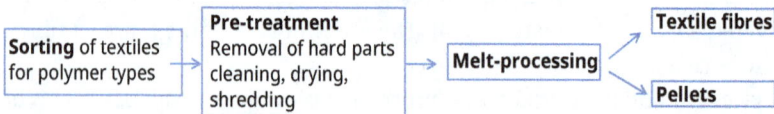

Figure 16.4: Flow scheme for thermomechanical recycling.

Thus, the quality of the sorting, cleaning and separation processes must be high. Both polyester and polyamide are sensitive to hydrolysis during melt processing. Consequently, attention must be paid during thermal processing to avoid hydrolytic or thermal chain degradation and reduction in the degree of polymerisation.

The presence of a few percentage of incompatible other polymers, for example PP in PET or mixtures of PA6 and PA6.6, will cause problems during fibre spinning and lower the quality of the recycled product. Also in case of polyester, different polymers are collected, for example PET, polybutylene terephthalate (PBT) and polytrimethylene terephthalate. Blending of these polyesters results in problems during processing and reduces the quality of products.

For fibre spinning, concentration of other polymers must be kept at a minimum, and a uniform and small distribution of dispersed polymers is required. A frequently used technique to remove smaller particles is melt filtration. At this stage of thermal processing, a degassing step helps to remove small molecular weight volatiles, which for example result from previous oxidative polymer degradation.

To increase the tolerance of a material for the presence of other polymers, the so-called compatibilisers are used. These additives are polymers which increase interfacial adhesion between the immiscible polymers and thus reduce dimensions of dispersed polymer phases and form a more uniform and stable morphology. Examples for compatibilisers are block/graft copolymers and non-reactive polymers with polar functionalities or reactive functionalised polymers (Figure 16.5). In case of reactive functionalised polymers, the precursors for block polymers are soluble in one phase of the polymer mixture and reactive to functional groups of the other polymer component. The block copolymer is produced in situ.

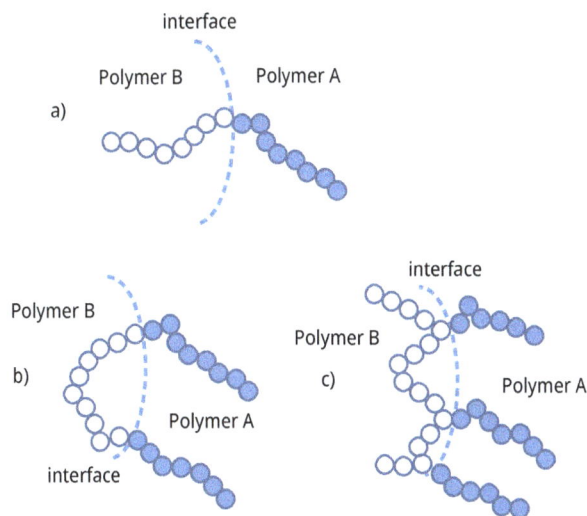

Figure 16.5: Compatibiliser at the interface of a heterogeneous polymer blend: (a) diblock, (b) triblock and (c) multigraft copolymers.

As an example, a compatibiliser for a polymer waste consisting of PET as a major component and PS (polystyrene) as a minor component, a block polymer from PS-b-PCL (poly-ε-caprolactone) can be used [5].

Figure 16.6 gives an example for a reactive functionalised polymer as compatibiliser. The poly(dimethylphenylene oxide) (PPO) serves as a compatibiliser for PBT. PPO-b-PBT block copolymers are formed at the interface during thermal processing.

X = Alkyl, cycloalkyl, aryl

PPO-EPOXIDE

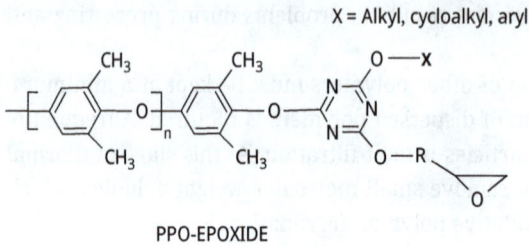

Figure 16.6: Epoxide end-capped poly(dimethylphenylene oxide), a reactive functionalised polymer, as compatibiliser.

The epoxide group can also react with H_2N and HO_2C groups of PA 6.6 during polyamide recycling.

In addition, reduction of the degree of polymerisation occurs during the phase of consumer use as well as during thermal recycling, which reduces the mechanical properties of recycled fibres.

Another substantial problem which has to be solved for textile recycling arises from the high number of different substances still present in polymer-sorted fraction of post-consumer wastes (Table 16.1).

Table 16.1: Contaminants (chemicals and impurities) present in post-consumer wastes.

Function of substance	Type of substance
Modification of lustre	TiO_2
Thermal and UV stabilisation	UV-absorbing organic substances
Colouration	Dyes Pigments
Product functionality	Finishing, e.g. crease resistance Softening silicones Quaternary ammonium products Flame retardant Antimicrobial properties
Coating	Polymer coatings, e.g. PUR and silicones
Seam taping for outdoor application	Adhesives (low melting polymer)
Consumer use	Impurities (soil, grease and wash residues)

At present, a wide range of chemicals and additives is used to optimise the performance of a textile product for its later use. The removal of all these substances will cause substantial problems during recycling. Also decolourisation of dyed products will lead to a substantial impact on the overall resource balance. A rigorous selection of chemicals and dyes will be required at the stage of product design to deliver a product which is directly recyclable after the wastes have been sorted into fractions of different colours.

Another important parameter to control is the molecular weight distribution in the polymer feedstock. Due to ageing processes, textile wastes can contain higher amounts of lower molecular weight fragments, and further degradation can occur as a result of the thermal stress applied during reprocessing. In case of PET, melt strength and mechanical properties of the recycled polymer can be improved by post-condensation processes. In this case, the polymer pellets are heated to a temperature above glass transition and below melting temperature (e.g. 200–240 °C).

In theory, a direct processing of the recycled material into fibres could be achieved, however due to the difficult feeding of fluffy material, the presence of impurities and the requirement to increase molecular weight, granulate production as an intermediate stage will be more easy. In addition the sorting of granulate with regard to colourants and pigments in the polymer will allow the formation of larger batches with constant colour for later fibre production.

Besides the supply with a defined feedstock for polymer recycling, also the presence of dyes, pigments and additives will be an important factor which determines the suitability of a polymer waste for thermomechanical recycling. The stage of product design will be of high importance to improve suitability of a product for recycling, for example by selection of thermostable dyes and replacement of additives and stabilisers.

Thermochemical recycling: In this category, processes are collected, which degrade the polymers by action of heat or oxidation processes. The products are used as feedstock for synthesis of polymers (monomers, syngas).

Use as fuel or combustion for energy production is excluded (Figure 16.7).

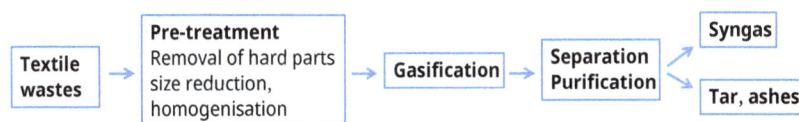

Figure 16.7: Thermochemical recycling of textile wastes.

The process is robust and could be of interest for fractions which cannot be recycled with more efficient techniques and for certain low volume fractions as contaminants are separated during the gasification process. Products that contain critical chemicals could be processed with such a technique, however, careful design of new products should reduce the total amount of textile wastes which require processing by thermochemical recycling. The overall efficiency for climate change is estimated to be 22% lower than the conventional syngas production, for example by gasification of coal. For an optimised process, the reduction in carbon footprint is estimated with 50%.

The type and amount of products depend on the fibre material used. From 100% cotton, 0.8–0.9 kg/kg$_{daff}$ (dry ash free fuel) as permanent gas (non-condensable fraction) is obtained, while 100% polyester yielded 0.7–0.8 kg/kg$_{daff}$ as permanent gas and also 0.1 kg/kg$_{daff}$ as aromatic BTXS (benzene, toluene, xylene and styrene) [4].

Besides the permanent gas fraction and the condensed oil/tar fraction, a solid inorganic residue "ash" (carbon soot, metals and minerals) is obtained.

The process is advantageous for impure and contaminated wastes, for example used carpets, and does not require high-quality sorted products. The output fractions can serve as feedstock for various applications, and their use is not restricted to textile fibre production. As disadvantages, energy consumption is relatively high and purification steps are required to deliver clean syngas for chemical synthesis.

16.9 Chemical recycling

In this section, chemical recycling processes are based on the dissolution of polymers in solvents for separation and regeneration or on chemical breakdown of the polymers into monomers followed by polymerisation and rebuilding fibres. The best suited process depends on the type of polymer and the textile waste (e.g. fibre blend).

Cellulose textiles – cotton recycling: With a total volume of 24 million tonnes cotton is the most important representative for cellulose-based fibres. Bast fibres such as jute, coir, hemp and flax hold a substantially lower share and sum up to 4–5 million tonnes per year. Regenerated cellulose fibres such as viscose, modal and lyocell fibres reach an annual production volume of 7 million tonnes.

Mechanical recycling of cotton gives favourable results in the LCA analysis [6]. Among the natural cellulose fibres, the elemental fibre of cotton is the longest with a typical fibre staple length between 30 and 35 mm. This relatively short fibre length requires a high number of turns to form a yarn with acceptable strength. As a result, mechanical opening of cotton-based textiles leads to considerable shortening in fibre length and also releases a high amount of dust and lint. This limits mechanical fibre recycling to production of coarse yarns with low strength [7].

In its chemical structure, cellulose is a homopolysaccharide built of β-D-glucopyranose units which are linked together via 1–4 glycosidic bonds. This complex structure hinders an economic and resource-saving recycling via depolymerisation and polymerisation [8]. Promising techniques for cellulose fibre recycling thus utilise the solubility of cellulose under appropriate chemical conditions. Most important processes for cellulose fibre recycling via the route of regenerated cellulose fibres are:

- Viscose process: Dissolution of cellulose after xanthation with CS_2 in diluted aqueous NaOH and regeneration of the cellulose in diluted sulphuric acid
- Lyocell process: NMMO (*N*-methyl-morpholine-*N*-oxide) as solvent allows direct dissolution of cellulose and fibre formation through spinning into a water bath.

– Ionic liquids: Dissolution and fibre formation can also be achieved with the use of ionic liquids and fibre regeneration in a non-solvent [9]. Examples for chemical structures are dialkylimidazolium chlorides. Also deep eutectic solvents, for example mixtures of choline chloride and malic acid, quaternary ammonium salts, can dissolve cellulose. In the Ioncell-F process, 1,5-diazabicyclo[4.3.0]non-5-ene acetate was used as solvent to process coloured textile wastes [10].
– Carbamate process: Cellulose, first, is converted into cellulose carbamate, which then is soluble in diluted aqueous NaOH solution. Regeneration of the fibre is achieved in diluted sulphuric acid.

More detailed information about the processes and fibres is given in Chapters 2 and 3 of this book.

The replacement of fresh pulp with recycled cellulose requires a number of adaptations:
– The degree of polymerisation of cellulose in cotton is substantially higher ($P_w >$ 1000 units) than required for the production of regenerated fibres (e.g. viscose P_w 200–250, modal P_w 400–600 and lyocell P_w 600–800). Thus, a controlled adjustment/reduction of the degree of polymerisation has to be performed before complete dissolution in the spin dope is achieved. Relevant methods utilise acid hydrolysis, electron beam treatment or enzymatic hydrolysis with an endoglucanase-rich enzyme.
– The majority of post-consumer wastes have been dyed and a high number of different dyes are in use. Sorting of wastes into groups of similar colour and knowledge about the dye class and selection of suitable dyes will be required to avoid uncontrolled colour change during fibre manufacturing and contamination of fibre products with dye fragments. Also release of dyes into the processing liquids, for example fibre regeneration bath, has to be minimised. In viscose fibre production reducing conditions can lead to reductive cleavage of azo groups in dyes. Oxidising conditions in the NMMO solution can attack the dye chromophore. The presence of Cu-ion traces, for example from metal complex-containing dyes (turquoise-reactive dyes with copper phthalocyanine groups), leads to the risk of increased rate of decomposition of NMMO–cellulose solutions [11].
– Fibre blends, for example with the use of PURs (elastane), for improved elasticity require separation from the second fibre component. In case the second component is present in a minor share, the cellulose dissolution and fibre separation could be combined.
– Besides information about dyes used also the finishes present in a material should be known. Cross-linking finishes, for example dimethyloldihydroxyethylene urea, in permanent press-finished textiles modify the dissolution behaviour of the cellulose.

Any chemical pretreatment of coloured post-consumer wastes in advance to the step of cellulose dissolution and fibre formation, for example chemical discolouration, requires careful consideration of the resource consumption (chemicals, energy and water).

Circularity of cellulose-based fibres will lead to a substantial shift in total cellulose fibre production. When cellulose-based wastes are fed into the production of regenerated cellulose fibres, the production capacity for these fibres will increase substantially on the costs of the cotton fibre production. The amount of cotton fibres will drop to the volume which is required to compensate for the losses in the circular system. The yearly production of regenerated cellulose fibres and the cotton fibre production will then be dependent on the efficiency of the cellulose recycling and fibre regeneration.

Polyamide/polyester: These two polymers represent the biggest share in the market of synthetic textiles. Often these fibres are found in combination with elastane (PUR); thus, additional separation processes to remove PUR have to be included. Two general strategies can be applied:
- Solvent-based separation of polymers without distinct change in molecular weight
- Depolymerisation to monomers, separation and repolymerisation

For separation of polymers by solvent systems, appropriate green solvents have to be available. As an example, the dissolution of PA6 and PA6.6 has been achieved in mixtures of $CaCl_2$/ethanol/water [12]. The polymer is then reprecipitated from solution by addition of non-solvent, for example water. The recovered polymer then goes into melt-spinning processes again. In an ideal case, the polymer dissolution process is combined with the separation of different fibres, for example separation of PA6.6 and elastane (PUR).

Depolymerisation of polyamides and polyesters to monomers or oligomers has been studied extensively, however, efficient processes are rarely found.

A general scheme for the steps in the fibre-to-fibre recycling via depolymerisation for polyester or PA6 is shown in Figure 16.8.

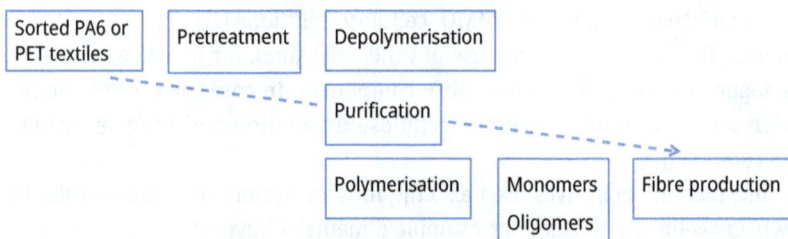

Figure 16.8: Process scheme for chemical recycling of polyester/polyamide 6 textile wastes via depolymerisation.

The hydrolytic depolymerisation of PA6, for example by high-pressure steam, yields ε-aminocaproic acid, which can then be repolymerised to PA6 [13]. Hydrolysis of PA 6.6 leads to formation of a mixture of two building blocks, hexamethylenedi-amine and adipinic acid, which complicates the later purification and repolymerisa-tion (Figure 16.9).

Polyester is broken down into its fragments such as glycol and terephthalic acid by hydrolysis (alkaline, acidic and steam), methanolysis or glycolysis (Figure 16.10).

The monomers or oligomers from the breakdown of the polymer require purifica-tion before repolymerisation can be initiated. At this stage, contaminants, for example colourants, additives, residues from finishing and coating as well as solvents added during hydrolysis/solvolysis, are removed. Techniques to remove contaminants in-clude filtration, decolourisation, for example by adsorption, and purification of the products, for example by distillation (e.g. ε-aminocaprolactam) or crystallisation (ter-ephthalic acid and bishydroxyethyl terephthalate).

Breakdown of PUR (elastane) can be achieved by reaction with glycol (glycolysis), however, a complex mixture is obtained, as elastane is a copolymer of PUR (hard seg-ments) with polyether or polyester segments (soft segments) (Figure 16.10). Use of a solvent for removal of PUR without degradation would be favourable [14].

Cotton/polyester blends: Fibre blends containing cotton and polyester are quite common, however, require appropriate separation processes. Again as a first step, pu-rification and removal of hard parts (buttons and zippers), followed by cutting and shredding, is undertaken. Three different approaches to process this group of textile wastes have been established (Figure 16.11).

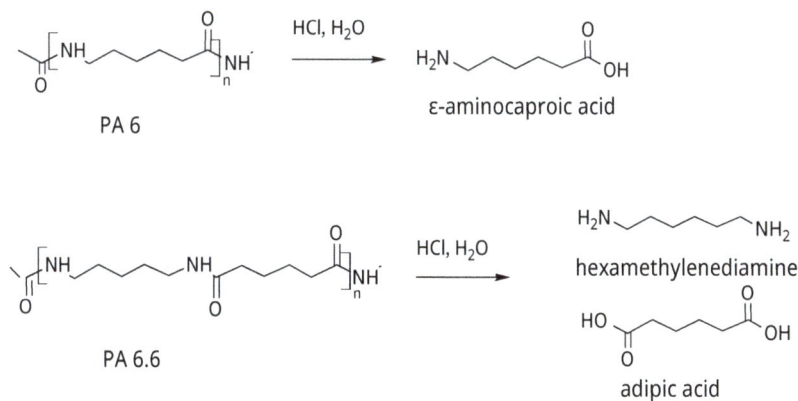

Figure 16.9: Hydrolytic depolymerisation of PA 6 and PA 6.6.

A straightforward approach is the removal of cellulose without hydrolysis by solvent-based systems to form spinnable cellulose solutions; however, attention has to be paid that contamination by hydrolysis products from the polyester fibres can be avoided. The

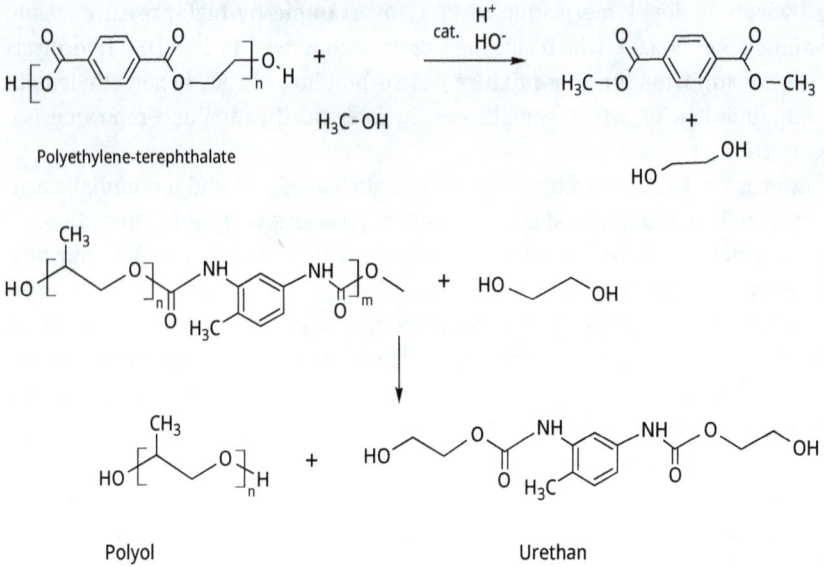

Figure 16.10: Alcoholysis – methanolysis of PET and glycolysis of PUR.

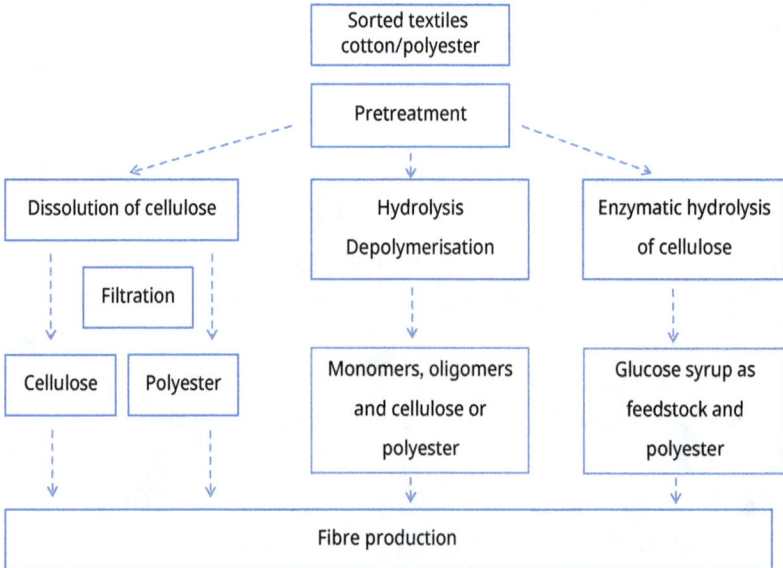

Figure 16.11: Approaches to separate and process cotton polyester blends.

polyester residue can then be used for fibre production by melt extrusion. A further complication arises when elastane has been added as a third component for improved elasticity as the tolerance of the processes for other polymers is limited.

Hydrolysis of cellulose and/or polyester can be used to separate the two polymers:

– either by hydrolysis of the cellulose under acidic conditions (e.g. organic acids and sulphuric acid) to lower molecular weight products which allow separation and collection of the polyester fraction,

– or by hydrolysis of the polyester fraction under alkaline conditions (NaOH and KOH) and recovery of the cellulose for pulp and fibre manufacturing. For improved efficiency and reduced energy consumption, a co-solvent may be added to the aqueous solution.

Also enzymatic breakdown of the cellulose to oligosaccharides and glucose has been studied as a method to separate cellulose fibres from other synthetic polymers. At first, a pretreatment to improve susceptibility to enzymatic hydrolysis is performed. Then a fungus is grown directly on textiles. The extracellular enzymes released by the fungi then lead to hydrolytic degradation of the cellulose. The hydrolysate is processed into a glucose-rich syrup, while the polyester remains intact. Again attention has to be paid to the presence of other polymer contaminants, for example PURs, dyes, finishes and in particular antimicrobial finishes. Biocides are frequently used in synthetic products to reduce bacterial growth and prevent odour formation during use.

16.10 Consumer use

During its phase of use, washing and drying of clothing lead to a significant impact to the overall balances of greenhouse gas emissions, energy and water consumption.

The frequency of washing, the efficiency of the washer and the washing temperature determine the water consumption and energy consumption of the processes. As an example for 45 washing cycles, a high-efficiency washer in combination with line drying requires 8.9 MJ energy and 614 L of water per kg of goods, and the greenhouse gas emissions sum up to 1.71 kg CO_2eq/kg. A regular efficient washer followed by a standard drier increases these values to 232.7 MJ energy and 1116 L of water per kg of washed goods, and 51.6 kg CO_2eq/kg [2].

Besides these values, household washing operations also represent a major source for release of microplastics into the environment, as the abrasive forces during the wash generate a high amount of short fibres which then are emitted with the spent wash liquor.

Some general parameters of the textile products design and of the general consumer use will determine these values significantly:

– The length of use of a garment between two washing operations determines the frequency of washing and also influences the lifetime of a garment.

– The fibre material and the textile construction determine the washing temperature, the required amount of detergents and the amount of water transported
with the material forward into the drying step. A non-swelling fibre can be dewa-
tered by centrifugation to contain only a few wt% of water, while cellulose fibres
and wool transport around 60 wt% and more into the drying step.
– Line drying requires only low amounts of energy, while machine driers require
substantial amounts of energy for water evaporation. Thus, a substantial difference in energy consumption for drying will arise from the amount of water retained in the wet textiles before drying.

During the design of clothing for a circular concept, the consumer behaviour and the
practical situation of use and care must be considered. Products which can be used at
sites where line drying is possible may then be designed different to products for consumers who live in areas of high population density where machine drying is expected to be the standard procedure.

References

[1] European Commission. EU Strategy for Sustainable and Circular Textiles. European Commission,
Brussels, 2022.
[2] Munasinghe, P., Druckman, A., Dissanayake, DGK. A systematic review of the life cycle inventory of
clothing. J. Clean. Prod. 2021, 320, 128852, https://doi.org/10.1016/j.jclepro.2021.128852.
[3] Shen, L. Bio-Based and Recycled Polymers for Cleaner Production An Assessment of Plastics and
Fibres, Utrecht University, 2011.
[4] Duhoux, T., Maes, E., Hirschnitz-Garbers, M., Peeters, K., Asscherickx, L., Christis, M., Stubbe, B.,
Colignon, P., Hinzmann, M., Sachdeva, A. Study on the Technical, Regulatory, Economic and
Environmental Effectiveness of Textile Fibres Recycling: Final Report, Commission, E., Directorate-
General for Internal Market Entrepreneurship and SMEs, I., Eds., European Commission, Brussels,
2021, https://doi.org/10.2873/828412.
[5] Koning, C., Van Duin, M., Pagnoulle, C., Jerome, R. Strategies for compatibilization of polymer
blends. Prog. Polym. Sci. 1998, 23, 707–757, https://doi.org/10.1016/S0079-6700(97)00054-3.
[6] Esteve-Turrillas, FA., de la Guardia, M. Environmental impact of recover cotton in textile industry.
Resour. Conserv. Recycl. 2017, 116, https://doi.org/10.1016/j.resconrec.2016.09.034.
[7] Aronsson, J., Persson, A. Tearing of post-consumer cotton T-shirts and jeans of varying degree of
wear. J. Eng. Fiber. Fabr. 2020, 15, https://doi.org/10.1177/1558925020901322.
[8] Klemm, D., Philipp, B., Heinze, T., Heinze, U., Wagenknecht, W. Comprehensive Cellulose Chemistry,
Volume I: Fundamentals and Analytical Methods, Wiley-VCH Verlag, Weinheim, Germany, 1998.
[9] Bodachivskyi, I., Page, CJ., Kuzhiumparambil, U., Hinkley, SFR., Sims, IM., Williams, DBG. Dissolution
of cellulose: Are ionic liquids innocent or noninnocent solvents? ACS Sustain. Chem. Eng. 2020, 8,
10142–10150, https://doi.org/10.1021/acssuschemeng.0c02204.
[10] Haslinger, S., Wang, Y., Rissanen, M., Lossa, MB., Tanttu, M., Ilen, E., Määttänen, M., Harlin, A.,
Hummel, M., Sixta, H. Recycling of vat and reactive dyed textile waste to new colored man-made
cellulose fibers. Green Chem. 2019, 21, 5598–5610, https://doi.org/10.1039/c9gc02776a.

[11]   Rosenau, T., Potthast, A., Sixta, H., Kosma, P. Cellulose solutions in N-Methylmorpholine-N-oxide (NMMO) – degradation processes and stabilizers. Prog. Polym. Sci. 2001, 26, 1763–1837, https://doi. org/10.1023/A.

[12]   Rietzler, B., Manian, AP., Rhomberg, D., Bechtold, T., Pham, T. Investigation of the decomplexation of polyamide/CaCl2 complex toward a green, nondestructive recovery of polyamide from textile waste. J. Appl. Polym. Sci. 2021, 138, 51170, https://doi.org/10.1002/app.51170.

[13]   Frisa-Rubio, A., González-Niño, C., Royo, P., García-Polanco, N., Martínez-Hernández, D., Royo-Pascual, L., Fiesser, S., Žagar, E., García-Armingol, T. Chemical recycling of plastics assisted by microwave multi-frequency heating. Clean. Eng. Technol. 2021, 5, 100297, https://doi.org/10.1016/j. clet.2021.100297.

[14]   Vonbrül, L., Cordin, M., Bechtold, T., Pham, T. Method for Separating Polyurethane from a Textile. Patent Application, 29.09.2022. EP22198684.7, 2022.

Take home messages

Circular concepts need to consider all stages of product's lifetime, raw material extraction, fibre formation, textile and garment production, consumer use, collection, sorting and recycling.

The LCI contains a database with regard to resource consumption, quantity and quality of output and impact on environment. An assessment of the data is formulated in LCA.

Future products will combine the design for recycling with appropriate and efficient recycling techniques.

Textile recycling techniques can be based on mechanical, thermal and (bio)chemical processes or combinations of these processes.

Mechanical recycling leads to fibres, fluff and dust, and the relative amounts depend on the material to be processed.

In thermal recycling, heat is applied to melt and recover thermoplastic polymers or to break down polymers into lower molecular weight building blocks for polymer synthesis.

Chemical recycling is based on the dissolution of polymers, for example cellulose fibres, or on chemical breakdown of polymers into monomers, for example hydrolysis of polyester and polyamides.

The phase of consumer use also leads to a significant contribution to overall resource (water and energy) consumption, in particular, during washing and drying operations.

Quiz

1. Give some examples for product improvement by the design for recycling.
2. Which are the four groups of recycling techniques?
3. Why mechanical recycling will not be sufficient to achieve textile circularity?
4. Why cellulose fibre recycling should go via cellulose dissolution processes?
5. What is the problem of dyes, finishes, adhesives and coatings?

Solutions

1. (Question 1): Reduction of material types used in one product; machine readable labels to support sorting and separation; and use of dyes and finishes which are compatible to the planned fibre recycling and reuse.
2. (Question 2): Mechanical recycling, thermal recycling, chemical recycling and bio-chemical recycling.
3. (Question 3): The fibre shortening during the mechanical action allows only production of rather coarse yarns. The efficiency is low, and considerable amounts of fluff and dust are generated.
4. (Question 4): Cellulose is not thermoplastic; thus, thermal processes lead to degradation. Hydrolysis to monomers is possible; however, repolymerisation to cellulose is not possible.
5. (Question 5): These chemicals must be compatible with the processing steps during recycling and must not generate problems during the production of new textiles; or these substances must be removable during polymer recycling with the use of simple and cost-efficient methods.

List of abbreviations/symbols

ΔG	Free energy
ΔH	Free enthalpy
ΔS	Entropy
LCI	Life cycle inventory
LCA	Life cycle assessment
PET	Polyethylene terephthalate
PA6	Polyamide 6
PA6.6	Polyamide 6.6
PP	Polypropylene
PBT	Polybutylene terephthalate
PPO	Poly(dimethylphenylene oxide)
PS	Polystyrene
PUR	Polyurethane
NMMO	N-Methyl-morpholine-N-oxide

Index

www.ingramcontent.com/pod-product-compliance
Lightning Source LLC
Chambersburg PA
CBHW080124220326
41598CB00032B/4950

9 783110 795691